# EDWARD I. PETERS

Department of Chemistry,
West Valley College, Saratoga, California

SAUNDERS GOLDEN SUNBURST SERIES

# Introduction to Chemical Principles

*Second Edition*

## to

## Chemical

## Principles

1978

W. B. SAUNDERS COMPANY   Philadelphia, London, Toronto

W. B. Saunders Company:   West Washington Square
Philadelphia, PA   19105

1 St. Anne's Road
Eastbourne, East Sussex BN21 3UN, England

1 Goldthorne Avenue
Toronto, Ontario M8Z 5T9, Canada

Front cover reproduction is *Color Form Synchromy (Eidos)* (1922–23), by Morgan Russell. Oil on canvas, 14½ × 10⅝″. The Museum of Modern Art. New York, Mrs. Wendell T. Bush Fund.

Introduction to Chemical Principles                              ISBN   0–7216–7209–4

Last digit is the print number:     9    8    7    6    5    4    3    2    1

To
 Marilyn Wilder—
  *a very special person, a friend to everybody;*
and to
 Catherine Brigham—
  *a student who inspires her teachers.*

# PREFACE

*"It's not what we teach that's important; it's what the student learns that counts."*

This aphorism came to my attention just before I began work on the first edition of this text. It remains the focal point of this revision, which is addressed to the student who is not adequately prepared for the full scale college level general chemistry course he will soon take. The successful student of a one semester or one quarter "prep" course using this text will be able to . . . .

(a) . . . read, write and talk about chemistry, using a basic chemical vocabulary;

(b) . . . write routine chemical formulas;

(c) . . . write and balance ordinary chemical equations;

(d) . . . set up and solve many different kinds of chemical problems, using dimensional analysis where applicable;

(e) . . . "think" chemistry in some of the simpler theoretical areas—to visualize what is happening on the atomic or molecular level.

This book retains the features of the first edition that have proved to be most helpful to beginning chemistry students, and adds some new ones as well. Most significant from both categories are the following:

1. Performance Goals appear for each important concept and skill. All objectives are in two locations, (a) immediately before the section in which a topic is introduced, where they focus the student's study of his daily assignment, and (b) in a chapter-end summary, where they may be assimilated and reviewed in preparing for a test.

2. Problem solving is presented by carefully sequenced dimensional analysis. The student is encouraged to use dimensional analysis as an analytical tool by which to *reason* through a problem logically, rather than a device by which to arrive at an answer by juggling units haphazardly.

3. Most examples within the chapters are presented in a semi-programmed format that has proved highly successful in keeping the reader *active* while studying the text. A series of questions guides the student in solving the problem for himself, building and confirming his understanding each step along the way.

4. New terms are printed in bold face in context, listed with specific page references at the end of each chapter, and, in most cases, defined fully in the Glossary at the end of the book.

5. Chapter-end Questions and Problems have been combined in this edition. More than 1100 in number, they are arranged by text section, and in matched columns. Answers for all left column questions are provided

in the back of the book. These include solution setups for most problems. Detailed answers and solutions to the right column questions are in the Teacher's Guide. A number of questions and problems, marked with an asterisk, are provided to challenge students who wish to go beyond the minimum Performance Goals.

6. An extensive review of mathematics is given in the Appendix.

7. Emphasis throughout the text is on *understanding*, rather than on memorization. For example, the student is encouraged to learn a system for writing chemical names and formulas that can be applied to ionic compounds previously unseen, not limited to those made up of ions that may have been memorized from a table. The periodic table is used repeatedly as an organizing tool and memory aid for writing formulas and electron configurations.

For this edition, many chapters have been rewritten or rearranged to provide an easier and more effective learning experience for students. Following an introduction to the study of chemistry, Chapter 2, the first chapter of substance, is about *chemistry*—not mathematics, which usually discourages the student who eagerly anticipates something new and different in his academic life. All material on atomic structure has been consolidated into a single Chapter 4, with some of the more abstract topics of the first edition shortened or deleted. Similarly all energy topics have been combined in Chapter 13, where the instructor has the option of using them or not. Stoichiometry in Chapter 8 has been expanded to include limiting reagent problems. At the recommendation of several instructors, new but optional sections have been added in different chapters on molality, colligative property problems, equivalents and normality, and elementary equilibrium calculations. This edition also includes a new Chapter 18 on nuclear chemistry.

The basic topics required for preparation for general chemistry are in Chapters 1 to 11 and 14. The sequence in the text has been selected to match a concurrent laboratory program that includes quantitative experiments early in the semester. Alternative sequences are also possible, as they are literally "written into" the book by making sure that certain broad subjects are independent of materials appearing in earlier chapters. These alternatives include most of the common orders of topics in beginning chemistry courses, and they are described in detail in the Teacher's Guide.

As with most texts, this volume contains more material than can be properly presented in a one semester or one quarter course. Everything beyond the "basic topic" chapters listed above may be considered optional. Though there are occasional references in some later chapters to material in earlier optional chapters, careful planning has made sure that study of the early topics is not necessary for understanding a later concept. This leaves the instructor full freedom of choice among the optional chapters he wishes to include toward the end of the term.

Though this book was not written with the Keller Plan in mind, its objectives and organization make it ideal for courses presented by a personalized system of instruction (PSI). If you are interested in this method of instruction and would like to see a sample of a PSI Study Guide written for this text, please let me know.

EDWARD I. PETERS

# ACKNOWLEDGMENTS

I wish to acknowledge and express my gratitude to the large number of teachers who responded to the publisher's invitation to prepare a critique of the first edition and make suggestions for the second. While it was impossible to use all of their recommendations—some of which contradicted others—the text includes numerous improvements that arose from those reviews. Thanks go also to Jeanne Rosato, who assisted by checking calculations and proofreading. My daughters, Janice Peters and Judy Serface, also read proofs at various stages. Original artwork for the second edition is by Joan Orme. Lloyd Black of the W. B. Saunders Company was outstanding in his assistance, his recommendations and his cooperation in the production process; and the entire production staff was most helpful. The greatest individual contribution came from Dr. Peter Berlow of Dawson College, Montreal, Canada, the author of the Student's Guide that accompanies this text, who read the entire book at the manuscript, galley and page proof levels, giving detailed and valuable suggestions at every step. And finally there is my most consistent supporter in all endeavors, my wife, Geb. In addition to her tangible typing and proofreading efforts, her patient support and encouragement have eased and added perspective to this writing project. I am most thankful to all of these people for joining me in preparing this book.

EDWARD I. PETERS

# CONTENTS

# 9

# 10

# 11

# 12

## 13

### ENERGY IN PHYSICAL AND CHEMICAL CHANGE ............... 296

## 14

### SOLUTIONS ............................................... 316

## 15

### CHEMICAL EQUILIBRIUM .................................. 359

## 16

### ACID-BASE (PROTON TRANSFER) REACTIONS ................... 388

## 17

### OXIDATION-REDUCTION (ELECTRON TRANSFER) REACTIONS ............................................... 412

## 18

### NUCLEAR CHEMISTRY ......................................... 433

# ABOUT BUBBLES IN FISH TANKS—A PROLOGUE

Have you ever noticed that bubbles rising from the bottom of a fish tank become larger as they approach the surface? At least I think they do. I must confess that I've never consciously observed and measured bubbles in a fish tank, but I still have full confidence that they grow larger as they rise. To explain why I have this confidence, let's consider a story—a bit of fiction that tells us something about chemistry, and, for that matter, science in general.

Once there was a man who personally *observed* the fact that bubbles increase in size as they rise from the bottom of a fish tank. Being a Curious Man, he *wondered why.* His *curiosity* drove him to find out what caused this to happen. So he *thought* about it. Eventually he developed a *hypothesis.* He guessed that the volume of the bubble was smaller at the bottom

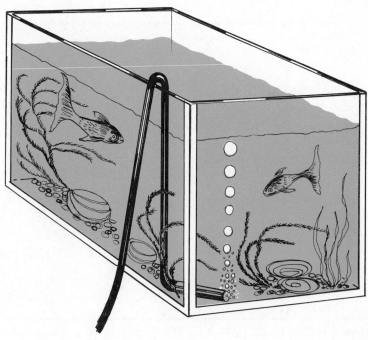

of the fish tank because the pressure was greater at the bottom, therefore "pushing" the bubble into a smaller size.

Having reasoned out a possible explanation for his observation, the Curious Man wanted to check it out. Being a practical man, as well as curious, he built a bubble. His bubble had the form of a sealed cylinder, as shown in Figure P.1. It had a snug fitting piston that was both air- and water-tight, but was free to move up and down so the pressure of the air inside would equal the pressure on the outside. The volume of the bubble was therefore governed by the pressure on top of the piston.

Gleefully our Curious Man took his bubble to the nearest lake, where he *experimented* with it. He pushed it down into the water to different depths, measuring the volume at each depth. His efforts were rewarded. He found that, like real bubbles, his artificial bubble became larger as it approached the surface. His hypothesis was correct: the volume of a bubble does depend upon the pressure on the outside. For that matter, he could say the pressure exerted by the gas inside the bubble was related to the volume, inasmuch as the inside and outside pressures were equal.

Excitedly, our Curious Man *communicated* his findings to anyone who would listen. Not many did; not many men are curious about bubbles. But one was. He was also *skeptical*. He didn't believe the reported results. Therefore he *repeated the experiments*. Curious Man No. 2 found that the experiments of Curious Man No. 1 were correct and gave *reproducible results*. Being an *intellectually honest* person, Curious Man No. 2 freely admitted his error in doubting Curious Man No. 1. The next time they gathered with their Curious Friends—curious in the same sense we have been using the term so far!—Curious Man No. 2 reported to all that he had checked the results himself.

One of the Curious Friends, Curious Woman, was not entirely satisfied. She thought there should be more. She suggested the *hypothesis* that if you

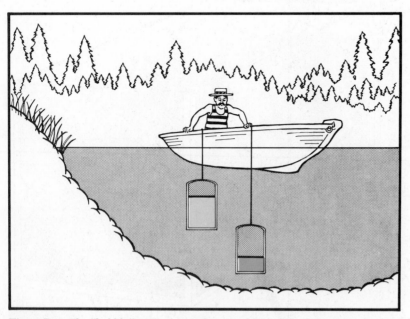

**Figure P.1** The "bubble" experiment. The size of the bubble depends on the depth of the water in which it is submerged.

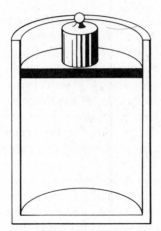 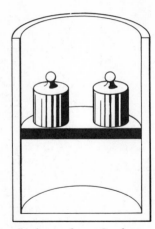

**Figure P.2**   The relationship between pressure and volume of a confined gas.

actually *measured* the pressure of the gas and its volume, you would find a quantitative relationship between them. She designed an experiment to test her hypothesis—an experiment that didn't require her to go down to the lake and get all wet, incidentally. She reasoned that you could measure the pressure more easily simply by putting weights on top of the piston as shown in Figure P.2.

One morning Curious Woman conducted her experiment and *drew a graph* of her results. Analyzing and interpreting these results, she found that if you multiply gas pressure by gas volume you always get the same answer. Wisely she *checked her results* before telling them to her Curious Friends. She checked them many times, in fact, until she was quite sure of her findings.

At the next gathering of the Curious Friends, Curious Woman reported that the product of pressure and volume of a gas is a constant. Other Curious Persons picked up the idea and tried it in their laboratories. The result was confirmed. Modifications were made. They found that temperature and quantity were variables that also had to be controlled. Experimenting with them as well, new relationships were found. Finally one Curious Person put them all together and proposed an "explanation" for these *experimental facts.* His *theory* pictured a gas as made up of many tiny particles moving about randomly and causing all the experimental results recorded by the other Curious Persons.

It's been over a hundred years since Curious Woman first proposed that pressure × volume is constant (at fixed quantity and temperature, of course). Recently the Society of Curious Persons has honored Curious Woman's relationship by elevating it to the status of a *law* of Science. It is now called Curious Woman's Law.

Now you know why I am so confident that bubbles become larger as they rise in a fish tank, even though I have not personally made this observation. It's the law. But even here, I must be cautious. Scientific laws are rarely found to be in error, but it has happened. To be absolutely certain, I'm going to conduct my own experiment. Where can I find a fish tank?

\*     \*     \*     \*     \*

In this little allegory we have tried to provide a small glimpse of the character of chemistry. The observing, hypothesizing, experimenting, testing and retesting, theorizing and finally reaching conclusions have been going on for centuries. And they continue today more actively than ever before. Collectively they are often called the *scientific method*.

There is really no rigid order to the scientific method. Looking back over the history of science, though, the features listed above always appear. They are the outcome of the day-to-day thinking of the scientifically curious person as he continually asks himself, "What do I already know that can be applied here? What is the next logical step I can take?" Out of such questions come new hypotheses, new experiments, theories and ultimately modern laws.

Our story also lists in italicized words some of the qualities and actions of the scientist. He surely is an *observer,* and he is *curious* about and *thinks* about what he sees. He develops a hypothesis as a tentative explanation of his observations. He conducts experiments to test the hypothesis. If he finds something worthy, he *communicates* with his fellow scientists, usually through scientific journals. He combines the qualities of *skepticism* with *intellectual honesty,* both of which leave him free to receive and evaluate new information that reaches him through many sources.

Our allegory furnishes one additional important insight into the nature of chemistry. Notice that the first two curious men concerned themselves primarily with *What* was taking place and *How*. Answers to these questions are considered as the *qualitative* part of chemistry. It was not until the Curious Woman entered that measurements appeared. She recognized that *What* and *How* furnished only some of the answers, but they could not be used to make reliable predictions about the extent of chemical activity. She added the vital question, *How much?* We see, then, that the study of chemistry is both qualitative and *quantitative*. In this course we shall consider both of these essential areas.

While the characters and events in this allegory are obviously fictitious, the law around which it was built is very real. It is the product of one person, not three, and was first proposed in the 17th century by Robert Boyle (1627–1691). Boyle was one of the first scientists to devote himself to orderly and carefully documented experimental investigation. You will study Boyle's Law in Chapter 11 of this text.

# INTRODUCTION TO YOUR
# STUDY OF CHEMISTRY

The next few pages may be among the most important pages that you read in this entire book. Nothing in them will be on a test question. They will teach you no chemistry. You will receive no grade for having read them. But their influence on the chemistry you learn, your performance on tests and the grades you receive will be great. Therefore, whether or not these pages are assigned, you are urged to read them and consider carefully what they suggest about the way you *study* chemistry.

The study of chemistry is not like the study of history, or sociology or English. If we were to compare the study of chemistry to the study of any other subject, the comparison would have to be made with mathematics. Chemistry, like mathematics, is cumulative in character. Nearly every concept introduced in both of these subjects is expanded and built upon in the concept that follows. Consequently, in order to comprehend what is to be presented tomorrow, it is first necessary to understand what has been presented today. If you recognize this—if you recognize that chemistry must be learned gradually in small parts, rather than in large bunches accumulated at special times (just before examinations!) scattered throughout the school term—you will have taken the first major step in making your study of chemistry fruitful, interesting and, we hope, enjoyable.

This textbook is written in a manner designed to guide you into efficient study, to maximize what you learn and to minimize the time required in learning it. You will surely agree with and share these goals! So let's take a look at some of the ways you can use your book to learn more in less time.

## STUDY TOOLS

To be used effectively, your textbook must be supported by other physical tools. You do not just *read* when you study chemistry. You do things. You think. You answer questions. You answer them, not just in your mind, but by *writing down* the answers. You simply cannot claim honestly to be studying chemistry without a pencil and plenty of paper at your fingertips. As a minimum, these two tools are absolutely essential.

Another essential tool for use with this book is an opaque shield large enough to cover the printed width of the page. The purpose of this shield will become clear as you read the next section. A piece of cardboard measuring about 2″ by 5″ or longer is ideal. In addition to its functional use, it can double as a bookmark.

You will soon learn, if you don't know it already, that the quantitative parts of chemistry require a large amount of calculation. This can become a boring and time consuming chore without mechanical assistance. At the very least you should have and be able to use a simple slide rule for multiplication and division. Today the slide rule has virtually disappeared from the classroom, replaced by the electronic calculator. These remarkable instruments come in a wide range of capability, quality and price. For use with this book a simple four-function calculator that can add, subtract, multiply and divide will be adequate. A desirable feature that is a big step forward in sophistication and price is the capability to calculate in exponential notation. This feature is recommended if you can afford it. Instruments with exponential notation usually have other desirable capabilities too, such as square root, logarithms and buttons by which a number may be raised to any power. Though not essential for an introductory course, these functions are most helpful in full-year general chemistry courses, and those beyond.

## *PERFORMANCE GOALS*

> PG  1 A   Read the performance goal or goals at the beginning of each section; study the section; re-read the performance goals; satisfy yourself that the goals have been reached; if so, go on; if not, go back—restudy the text until the goal has been met.

As you approach most sections in this text you will encounter one or more "performance goals," as you did here. In many cases the performance goal will contain language that is unfamiliar, but about to be introduced in the section. It gives you some idea of what to expect and what to look for, as you study the section. This should help you to recognize major points of the section, and enable you to direct your attention to them. This is the key to efficient study.

Notice the performance goal above is written in an "action" fashion. It describes *doing* something—performing some act. Each performance goal in this book should be thought of as the close of a sentence that begins, "After studying this section, you will be able to. . . ." The goal describes an ability or method you are to acquire in your study. Whether or not it has been acquired, you should find out immediately after completing the section. Return to the performance goals, asking yourself the question, "Am I now able to do what is expected?" Demand of yourself that the honest answer to this question be, "yes!" If it is not, go over the material again until you can answer affirmatively.

Performance goals are designated in the text by PG, followed by a number and a letter. The number identifies the chapter, and the letter identifies the performance goal within the chapter.

The performance goals appearing in the book are obviously those the author has in mind for each section. They may or may not correspond exactly with those of your instructor. If not, you should be alert to modifications he or she may introduce, and adjust your study plan to what is required.

## TEXT AND EXAMPLES

For the most part, the text section of this book is conventional. Most of the problem material, however, is presented in a self-tutorial style, a series of questions and answers designed to guide you to an understanding of the concept being presented. It will succeed only if you "play the game by the rules." The rules call for you to answer each question *yourself,* before looking at the answer that you know is a fraction of an inch down the page. This way you force yourself to *reason* through the question on the basis of the theory which underlies it, and thereby gain an understanding of the chemical concepts.

Following a discussion of a specific principle, an example problem is presented. Suggestions are made for applying the principle to the example, ending with a question that leads you to the first step in the solution of the problem. At that point a broken line extends across the page:

-------------------------------------------------------------------------------------------------

Immediately below the line is the answer to the question, followed by explanatory comments and/or the mathematical setup by which the answer is found. Further discussion and suggestions lead to a second question and another broken line, the pattern repeating itself until the example is completely solved.

It is in the working of examples that your opaque shield is used. Move the shield from the top of each page to the first broken line. Study to that point. Then, using scratch paper, or the margins of the book, if you wish, answer the question asked *without looking ahead to the answer beneath the shield.* When you have answered the question to your satisfaction, move the shield down to the next broken line. Compare your answer with the answer just exposed in the book. With the help of any explanation or problem setup in the book, be sure you understand the work to that point before proceeding to the next question.

## CHAPTER QUESTIONS AND PROBLEMS

At the end of most chapters you will find a two column set of questions and/or problems. Side by side they are matched; they involve similar reasoning and, in the case of problems, calculations. Some questions are easy, requiring only the basic information presented in the chapter. Others are more demanding, requiring you to analyze a situation, apply to it those chemical principles that are appropriate, and then explain or predict some event or calculate some result. The more difficult of these, or those clearly extending beyond stated performance goals, are identified by an asterisk *. Answers appear in the back of the book for all questions and problems in the left hand column. These include calculation setups for most problems. Questions in the right hand column are without answers; they may be used as assignment material by your instructor.

If you have difficulty working any problem from either column you will have reached a critical point. You will be tempted to return to the examples within the chapter, find one that matches your problem, and then solve the assigned problem step by step as in the completed example. Resist that temptation! If you become stuck on a problem it indicates that you *have not understood* the corresponding text example. Study the example again by itself, until you understand it thoroughly. Then return to the assigned problem with a fresh start and work it to completion without further reference to the text example.

You should approach your problem work in chemistry with a clear sense of purpose. This purpose is not simply to get the answer. It is, rather, to acquire understanding of (1) the principle that underlies each problem, and (2) the method by which it is solved. This is not achieved by duplicating, step by step, an example from the book, but rather by visualizing the whole problem conceptually so you may *reason* through the individual steps unaided by completed examples. More than anything else you may do, demanding such comprehension of yourself in each problem you attempt will assure you success in the problem solving aspects of chemistry.

## THE LANGUAGE OF CHEMISTRY

As with any area of study, chemistry has its own special language. One of the main purposes of an introductory course is to make you familiar with this language, so you will both understand it and use it properly.

A large part of this language is its vocabulary. Two features of the book aid in developing your chemical vocabulary. At the end of each chapter you will find a summary of the new terms introduced in the chapter, with a number indicating the page on which the term first appears. The presence of a word on this list does not mean its definition must be memorized. It means you should understand the expression sufficiently that it makes sense to you if you read it or hear it, and so you can use it properly in your own speaking and writing. Any definitions or concepts you must understand to the point of explaining or reproducing them are identified in the performance goals.

Quite often you encounter a word that has been introduced earlier and wish to check quickly the definition of that term. For this purpose you will find a complete glossary in the Appendix, beginning on page 515. All terms in the chapter summaries are defined in the glossary, and many other words as well.

Scientific communication should be sharp and precise in its meaning. You are encouraged to develop these qualities in your understanding as you listen and read, and in your proper use of words as you speak and write.

# 2

# MATTER AND ENERGY

In a broad sense, chemistry may be defined as the study of matter and the energy associated with chemical change. It is appropriate, then, that we begin the study of chemistry by an examination of these well known but not widely understood parts of the physical universe.

## 2.1 PHYSICAL AND CHEMICAL PROPERTIES AND CHANGES

PG   2 A   Distinguish between physical and chemical properties.

     2 B   Distinguish between physical and chemical changes.

**Matter has mass. Matter also occupies space. These two characteristics combine to define matter.** In order to examine matter, to understand what it does, or appears to do, we must be able to describe it clearly. Normally a sample of matter is described by listing its **physical properties.** Certain physical properties can be detected directly by our senses. They tell us how a material *looks* (the blackness of charcoal compared with the yellow of sulfur); *feels* (the hardness of glass compared with the softness of putty); *smells* (the odor of sour milk or the scent of a rose); or *tastes* (salt vs. sugar). Other physical properties can be measured in the laboratory. Among them are the temperatures at which materials boil or melt, called *boiling* and *melting points,* or the *density* of the material.

Changes that alter the physical form of matter *without changing its chemical identity* are called **physical changes.** The melting of ice is a physical change. The substance is water both before and after the change. Dissolving sugar in water is another example of a physical change. The form of the sugar changes, but it is still sugar. The dissolved sugar may be recovered by simply evaporating the water, another physical change.

The chemist is interested in more than the physical properties of matter. He wants to know what sort of chemical reactions it can have. Paper burns.

Iron rusts. Milk becomes sour. Eggs become rotten. Each of these is a **chemical change.** A chemical change can be recognized when one or more substances are chemically destroyed, and one or more new substances are formed. The **chemical properties** of a substance are simply a description or list of the chemical changes possible for that substance.

Chemical and physical properties lead to several ways in which matter may be classified, some of which will be considered now.

## 2.2  STATES OF MATTER: GASES, LIQUIDS, SOLIDS

PG   2 C   Identify and distinguish between three states of matter.

2 D   Explain the differences between gases, liquids and solids in terms of particle behavior.

The air we breathe, the water we drink and the food we eat are examples of the gaseous, liquid and solid **states of matter.** Water is the only substance we normally encounter in all three states, as suggested in Figure 2.1. The differences between gases, liquids and solids can be explained in terms of what is called the **kinetic molecular theory of matter** (Chapter 11). According to this theory, all matter consists of extremely tiny particles that are in constant motion. In solids at a given temperature, this movement is limited by strong attractions between the individual particles. These particles are thought to have a vibrating type of motion in which they remain in fixed positions relative to each other. This is why a solid has a definite shape and volume. (See Fig. 2.1.)

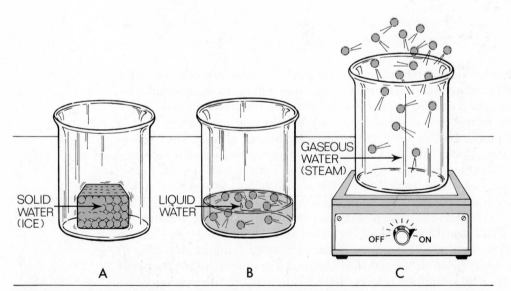

**Figure 2.1**   The three states of matter, illustrated by water. A. Solid water, ice, has a definite volume and shape. Particles vibrate in place, holding fixed positions relative to each other. B. Liquid water has a definite volume, but takes the shape of the bottom of its container to the depth necessary to hold that volume. Molecules are free to move relative to each other within that volume. C. Gaseous water, steam, has neither definite volume nor shape, but expands to fill the container if closed, and escape from it if open. Particles move in random fashion, completely independent of each other.

Interparticle attractions are weaker in liquids at the same temperature. They are strong enough to hold the particles together in a definite volume, but not strong enough to hold them in a pattern of fixed positions. The particles are free to move relative to each other, to tumble and slip about almost at will, just so they all remain together. Because the liquid particles remain together, the volume they occupy is definite, but their freedom to move allows them to adjust to the shape of the bottom of the vessel that holds them.

The attractions between gaseous particles at the same temperature are very weak—they are very close to zero. Consequently the tendency for the particles to move knows no limits. And move they do, separating from each other in a completely random fashion until they reach the walls of the container. This is why they have neither definite shape nor volume of their own, but acquire both from the container which they fill completely.

The kinetic molecular theory suggests that the "amount" of particle motion—which can be interpreted as the speed with which the particles are moving—depends on temperature. At higher temperatures the movement is more energetic, more vigorous, and at low temperatures, less energetic. This fits into our picture of solids, liquids and gases. If we consider a piece of ice, for example, we see that the limited movement at low temperatures permits the interparticle attractions to restrict the particles to the structure of a solid. As temperature is increased the more vigorous agitation overcomes those attractive forces at least to the extent of destroying the rigid structure, and we have liquid water. If temperature is raised above the boiling point the movement becomes so vigorous that the interparticle attractions are completely overcome and the particles fly apart in the gaseous state.

In summary:

(a) *A solid has definite shape and volume. Particle movement consists of vibration in place within a rigid structure.*

(b) *A liquid has definite volume, but assumes the shape of its container up to that volume. Particles are free to move among themselves so long as they remain together.*

(c) *A gas fills its container, adopting both its shape and volume. Movement of particles is completely random.*

## 2.3 HOMOGENEOUS AND HETEROGENEOUS MATTER

PG 2 E Distinguish between homogeneous and heterogeneous samples of matter.

The prefixes *homo-* and *hetero-* begin many words in the English language; they indicate *sameness* and *difference,* respectively. A **homogeneous** sample of matter has the same appearance, composition, and physical and chemical properties throughout. By contrast, **heterogeneous** materials are made up of visibly different parts, or **phases.** Composition varies from one place to another within the same sample. Each phase has its own unique properties, and, usually, appearance. Therefore a sample of matter generally may be classified as homogeneous or heterogeneous by its appearance alone.

Oil and water are each homogeneous by themselves, but when mixed

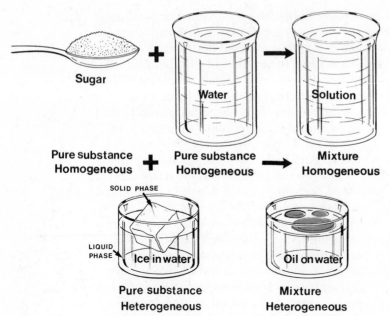

**Figure 2.2**  Examples of homogeneous and heterogeneous samples of matter.

in the same container they quickly separate into two distinct liquid phases, forming a heterogeneous mixture. Alcohol and water, on the other hand, mix completely with each other, forming a single phase, a homogeneous liquid called a solution (see Fig. 2.2). Raw milk is heterogeneous; if allowed to sit undisturbed the lower density cream rises to the top and forms a layer over the more dense milk. When the fat globules are broken down into very tiny particles and distributed uniformly, the milk is *homogenized.* This is the form in which it is usually sold today.

A sample of matter consisting of a single substance may be heterogeneous if it exists in two visibly distinct states. An ice cube floating in water is an example. In this case, we could properly refer to the two regions as different states as well as different phases.

## 2.4  PURE SUBSTANCES AND MIXTURES

PG  2 F  Distinguish between a pure substance and a mixture.

A **pure substance** is a single chemical, one kind of matter. It may be an *element* or a *compound* (see page 14). A pure substance is characterized by its own unique set of physical and chemical properties, not exactly duplicated by any other pure substance. These properties may be used to identify the substance in the laboratory.

A **mixture** is a sample of matter that contains two or more pure substances, either elements or compounds. The properties of a mixture are influenced by all components of the mixture, and they vary as the percentages of the different components change. This frequently makes it possible to

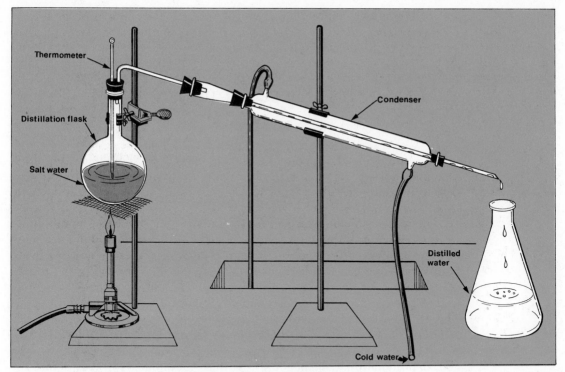

**Figure 2.3**  Laboratory distillation apparatus.

determine the relative percentages of different substances in a mixture by measuring its physical properties.

Pure water and salt water offer a good example of the differences between the properties of pure substances and mixtures. The boiling temperature of pure water at normal sea-level atmospheric pressure is 100°C. Salt water boils at a higher temperature; how much higher depends upon the concentration of salt in the solution. Moreover, in the distillation of salt water, in which pure water boils off, is removed and condensed as fresh water, the boiling temperature of the remaining solution gradually increases (see Figs. 2.3 and 2.4). This is reasonable since the removal of pure water from the mixture increases the concentration of the salt. If boiling is contin-

**Figure 2.4**  Comparison of boiling temperatures of a pure substance and an impure substance (solution).

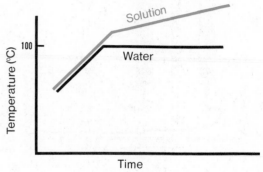

ued long enough, complete separation of the salt and water may be achieved. This illustrates another characteristic of mixtures: they may be separated into pure components by physical means alone.

## 2.5  ELEMENTS AND COMPOUNDS

PG  2 G  Distinguish between elements and compounds.

If a pure substance cannot be decomposed chemically into other pure substances, it is said to be an **element.** The fact that nobody has ever been able to decompose sulfur into two or more other substances implies that sulfur is an element.

Elements can combine chemically with one another to form other pure substances called **compounds.** Water, for example, is a compound made up of the elements hydrogen and oxygen. Water may be decomposed to its elements by passing an electric current through it, as shown in Figure 2.5.

Nature has provided us with sources of 88 elements. Nuclear research to the time of this writing has yielded 18 more, bringing the total to 106. The great bulk of our environment is made up of compounds containing a relatively small number of these elements, as shown in Table 2.1. It is natural that elements such as oxygen, silicon and aluminum, which are so abundant in nature, have become "familiar" elements. However, some elements that make up only a very small percentage of the crust of the earth

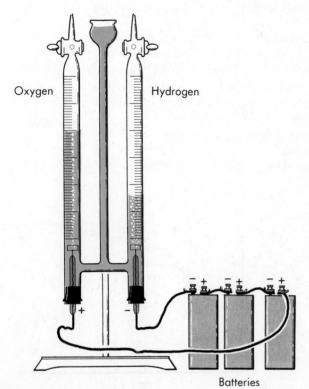

Oxygen

Hydrogen

Batteries

**Figure 2.5**  Decomposition of water into hydrogen and oxygen by means of an electric current.

are readily recognized, and have become vital to man. Absent from Table 2.1, for example, are such well known elements as chromium, copper, nickel, tin, silver and gold. Other elements that occur in lower percentages but are essential to modern technology are cobalt, vanadium, cadmium and molybdenum, to name just a few.

Only a few elements occur uncombined in nature. Most of these are gases (nitrogen, oxygen, argon) that make up over 99% of the earth's atmosphere. Gold, silver and copper are among the few solid elements that exist in nature in elemental form. To obtain pure samples of nearly all other solid elements, it is necessary to extract them from naturally occurring compounds or ores.

The vast majority of the elements are solids at normal temperatures and pressures. At 25°C and one atmosphere, only two are liquids (mercury and bromine), while 11 elements, including hydrogen, fluorine, chlorine, oxygen and nitrogen, are gases. The elements may also be divided into metals (e.g., iron, copper, magnesium, aluminum) and nonmetals (carbon, iodine). A few (boron, silicon) are often referred to as "metalloids"; these elements have properties between those of metals and nonmetals.

The properties of compounds are distinctly different from the properties of the elements from which they are formed. To illustrate, powdered iron is black in appearance, powdered sulfur, yellow. A crude mixture of these two elements, which appears gray, is readily separated into the elements. The iron in the mixture may be removed with a magnet, since it retains its magnetic properties. Sulfur, which is soluble in a liquid called carbon disulfide, can be extracted from the mixture by shaking it with that solvent; the iron does not dissolve. If the iron-sulfur mixture is heated strongly, a chemical reaction occurs to produce a compound, iron sulfide. The compound is not attracted by a magnet nor is it soluble in carbon disulfide. In other words, iron sulfide possesses neither of these properties characteristic of the elements that make it up.

Sodium chloride is another example. Neither sodium, a metallic element that reacts vigorously with both moisture and oxygen, nor chlorine, a poisonous gas with a suffocating odor, is particularly pleasant to work with. Yet the compound formed by these two elements, sodium chloride (table salt), is used daily to flavor our food.

An important experimental fact about compounds is summarized in the **Law of Definite Composition.** This law states that the percentage by weight of the elements in a compound is always the same, regardless of the source

TABLE 2.1   Composition of Earth's Crust*

| ELEMENT | PER CENT BY WEIGHT | ELEMENT | PER CENT BY WEIGHT |
|---|---|---|---|
| Oxygen | 49.2 | Sodium | 2.6 |
| Silicon | 25.7 | Potassium | 2.4 |
| Aluminum | 7.5 | Magnesium | 1.9 |
| Iron | 4.7 | Hydrogen | 0.9 |
| Calcium | 3.4 | All others | 1.7 |

*The earth's "crust" includes the atmosphere and surface waters.

or method of preparation of the compound. Water, for example, whether it comes from a pond, river or lake; in Europe, America or Asia; or is the product of a chemical reaction, always contains 11.1% hydrogen and 88.9% oxygen.

## 2.6 THE LAW OF CONSERVATION OF MASS

PG 2 H State the meaning of, or draw conclusions based upon, the Law of Conservation of Mass as it applies to a chemical change.

Early chemists, examining one of the most familiar of chemical changes, namely *burning,* concluded that since the ash remaining was so much lighter than the object burned, something was 'lost' in the reaction. Their reasoning was faulty because they did not realize that gases, which they could not see, took part in the reaction, both as reactants, or things used up, and as products, things produced. When a piece of wood is burned, gaseous oxygen is a reactant, and is consumed in the reaction. The products, in addition to ash, include two gases, carbon dioxide and water vapor. If we take careful account of all the reactants and products we find that

$$\frac{\text{Total weight of reactants}}{\text{(wood + oxygen)}} = \frac{\text{Total weight of products}}{\text{(ash + carbon dioxide + water vapor)}}$$

This type of weight balance applies to all ordinary chemical reactions. It was once called the law of conservation of matter, but now is referred to more accurately as the **Law of Conservation of Mass.** In other words, **in a chemical change, mass is conserved; it is neither created nor destroyed.** The word *mass* refers to quantity of matter, and is closely associated with the more familiar term *weight.* The distinction between them will be pointed out in Chapter 3.

## 2.7 ELECTRICAL CHARACTER OF MATTER

PG 2 I Correlate electrostatic forces of attraction and repulsion with positive and negative charges.

If you release an object held above the floor, it falls to the floor. This is caused by gravity, an invisible attractive force between all objects and the earth. Although the force is invisible, its effect is very evident. There are two other invisible forces, both capable not only of attraction but also of repulsion. You are probably familiar with one of these, magnetic force. Perhaps less well known are **electrostatic forces.**

One of the most common electrostatic phenomena occurs when you scrape your feet across a carpet on a dry day and then turn on a light switch, or perhaps "shock" another person by touching him. In the foot-scraping operation, you become "charged with electricity." In the laboratory, we can

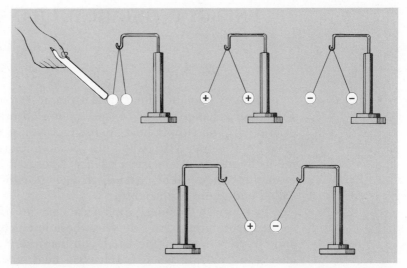

**Figure 2.6** Electrostatic attraction and repulsion. Two pith balls given identical charges repel each other (top views); pith balls with opposite charges attract each other (bottom view).

charge a rubber rod by rubbing it with fur. If a pith ball* suspended on a string is touched with the rod, the pith ball acquires the same charge as the rod (see Fig. 2.6). Two pith balls charged in the same manner repel each other, illustrating the *repulsive force between objects carrying the same kind of electrical charge.* Similarly, if two pith balls are touched with glass rods rubbed by silk, they repel each other, again illustrating repulsion between like charges. If now one of the balls touched by the rubber rod is brought near a ball touched by the glass rod, the balls are attracted toward each other. Obviously, the balls do not have like charges, which repel each other; instead they must have opposite charges. *Objects which are oppositely charged are attracted toward each other.*

These and numerous other examples lead to the conclusion that there are two, and only two, kinds of **electrical charges.** One is called **positive,** the other **negative**—adjectives assigned long before the character and source of the charges were known. The attractive and repulsive forces between charged objects are called electrostatic forces. The electrostatic force between two like charges (+ and +, or − and −) is one of repulsion; that between a positive and a negative charge is one of attraction. These relationships are illustrated in Figure 2.6.

These properties indicate that matter has electrical characteristics. We now know that all matter consists of tiny particles called atoms, and that these atoms contain within them positively charged particles called protons and negatively charged particles called electrons. In rubbing operations of the type we have described, electrons (never protons) are transferred from one object to another. The object that ends up with an excess of electrons over protons has a net negative charge, whereas the positively charged object has fewer electrons than protons. If there are an equal number of protons and electrons present, the object is said to be electrically neutral.

---

*Pith is a soft, spongy, low density substance extracted from plant fibers. Pith balls used in this experiment are typically ¼ inch in diameter.

## 2.8 ENERGY IN CHEMICAL CHANGE

PG 2 J Distinguish between exothermic and endothermic changes.

2 K Distinguish between potential energy and kinetic energy.

Changes in matter don't "just happen." Something causes the change. Frequently change occurs because matter absorbs or releases energy. For example, if you supply heat energy to a pan of water, it boils. Figure 2.5 (page 14) illustrated that water may be chemically decomposed into its elements by means of electrical energy. Energy is absorbed in each of these examples. Chemical or physical changes in which energy is *absorbed* are called **endothermic changes.**

By contrast, chemical or physical changes in which energy is *released* to the surroundings are called **exothermic changes.** If you strike a match and put your finger in the flame you learn very quickly that the chemical change in burning wood is releasing heat energy. It also releases light energy. Chemical changes in flashlight and automobile batteries release electrical energy. The physical change when steam condenses to water releases exactly the same energy that was required to boil the water in the first place. All of these are examples of exothermic changes.

Within this textbook we will be concerned with the energy released or absorbed in a chemical or physical change. Most of the time this energy will be in the form of heat. We will leave other forms of energy for more advanced courses. To form a base for understanding the source of chemical energy, however, we shall consider briefly a physicist's definition of energy and its closely associated concept, *work.*

Work, according to the physicist, is the application of a force over a distance. If you raise a book from the floor to a desk, you do work (see Fig. 2.7). You exert the force required to lift the book against the gravitational attraction of the earth. You exert this force over the distance from the floor to the

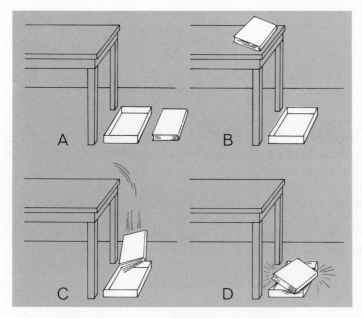

**Figure 2.7** Potential and kinetic energy. A. Book lies on floor next to box from which it has been removed. It has no potential energy relative to floor, and no kinetic energy. B. Book on table has potential energy equal to the work done to raise it from the floor. It has no kinetic energy. C. Falling book now has kinetic energy because of its motion, plus some remaining potential energy because it is still above the floor. The combined energies equal the work done raising the book to the table. D. Book now has neither kinetic nor potential energy, but has done work crushing the box and creating small amount of frictional heat when striking floor, all equal to the work done raising the book from the floor to the table.

desk. **Energy may be defined as the ability to do work.** You possessed the energy required to do the work of raising the book. In fact, you transferred this energy to the book, which has more energy on the desk than it had on the floor. The book now has **potential energy** with respect to the floor (see Fig. 2.7B). **Potential energy is energy possessed by an object because of its position in a force field.** That an object having potential energy is capable of doing work may be demonstrated by pushing the book off the desk and allowing it to fall toward the box still on the floor. In the act of falling (Fig. 2.7C) the potential energy is converted into **kinetic energy, the energy possessed by an object because of its motion.** When the book hits the box (Fig. 2.7D), it does work on it by crushing it.

Potential and kinetic energy are two forms of mechanical energy. While it is not necessary for you to know about these in detail in an introductory chemistry course, you should recognize that *potential energy is related to position in a force field,* and *kinetic energy is energy of motion.* What is called chemical energy is associated primarily with the positions of positively and negatively charged objects in an electrical force field. The energy associated with chemical changes comes from changes in these positions, just as there is a change in potential energy of a book in the gravitational force field of the earth as the book is moved between the floor and a desk.

## 2.9  THE LAW OF CONSERVATION OF ENERGY

PG  2 L   State the meaning of, or draw conclusions based upon, the Law of Conservation of Energy as it applies to a chemical change.

The various processes that take place when we drive an automobile offer a good illustration of energy conversions. The starting process uses the chemical energy of the storage battery to produce electrical energy. As we drive along, chemical energy of the fuel is being converted into both heat energy and "mechanical" energy of the engine parts. Ultimately, this mechanical energy is imparted to the car as kinetic energy, or energy of motion. Some of this energy in turn is converted to heat through friction between the tires and the road; the extent of this conversion increases dramatically if we slam on the brakes at an intersection. The potential energy, the energy associated with elevation, that the car acquires when we climb a hill is converted back into kinetic energy if we coast down to the original elevation.

Careful investigation of energy conversions such as these show that the energy "lost" in one form is always exactly equal to the energy "gained" in another form. This is the basis of another conservation law, the **Law of Conservation of Energy,** which states that **in any ordinary change energy is neither created nor destroyed.**

## 2.10  THE MODIFIED CONSERVATION LAW

Early in this century, Albert Einstein recognized the essential equivalence of mass and energy. This implies that it should be possible to convert one of these quantities into the other. Such conversions are indeed going on in the world around us. In a few cases, enough mass is converted to energy to become dramatically apparent. This happens, for example, when a hydrogen bomb explodes or a nuclear reactor is used to produce electrical energy. Einstein's famous equation relates the quantities of mass and energy:

$$\Delta E = \Delta mc^2,$$

where $\Delta E$ is the energy change, $\Delta m$ is the mass change and c is a universal constant, the speed of light.

The amount of energy potentially available from the conversion of mass is enormous. If it were possible to convert completely a given mass of coal to energy, that energy would be about 229 million times as great as the energy derived from burning that same amount of coal! This is why nuclear energy is so attractive an alternative to the traditional sources of energy, an attraction clouded by an as yet unsatisfied public demand for an assurance of safety and for an adequate disposal system for radioactive waste.

The suggestion that matter can be converted into energy, or energy into matter, might lead one to believe that the laws of conservation of mass and energy are no longer valid. In ordinary, non-nuclear, chemical reactions the matter-energy conversion is so small it cannot be measured satisfactorily, so for all practical purposes the laws are as good today as they were when first proposed. For *all* chemical changes, both nuclear and non-nuclear, the laws may be modified and combined into a single conservation law which states that the total amount of mass and energy in the universe is a constant. Expressed as an equation this becomes

$$(\text{mass} + \text{energy})_{\text{reactants}} = (\text{mass} + \text{energy})_{\text{products}}$$

# CHAPTER 2 IN REVIEW

## 2.1 PHYSICAL AND CHEMICAL PROPERTIES AND CHANGES

2 A  Distinguish between physical and chemical properties. (Page 9)

2 B  Distinguish between physical and chemical changes. (Page 9)

## 2.2 STATES OF MATTER: GASES, LIQUIDS, SOLIDS

2 C  Identify and distinguish between three states of matter. (Page 10)

2 D  Explain the differences between gases, liquids and solids in terms of particle behavior. (Page 10)

## 2.3 HOMOGENEOUS AND HETEROGENEOUS MATTER

2 E  Distinguish between homogeneous and heterogeneous samples of matter. (Page 11)

## 2.4 PURE SUBSTANCES AND MIXTURES

2 F  Distinguish between a pure substance and a mixture. (Page 12)

## 2.5  ELEMENTS AND COMPOUNDS

2 G  Distinguish between elements and compounds. (Page 14)

## 2.6  THE LAW OF CONSERVATION OF MASS

2 H  State the meaning of, or draw conclusions based upon, the Law of Conservation of Mass as it applies to a chemical change. (Page 16)

## 2.7  ELECTRICAL CHARACTER OF MATTER

2 I  Correlate electrostatic forces of attraction and repulsion with positive and negative charges. (Page 16)

## 2.8  ENERGY IN CHEMICAL CHANGE

2 J  Distinguish between exothermic and endothermic changes. (Page 18)

2 K  Distinguish between potential energy and kinetic energy. (Page 18)

## 2.9  THE LAW OF CONSERVATION OF ENERGY

2 L  State the meaning of, or draw conclusions based upon, the Law of Conservation of Energy as it applies to a chemical change. (Page 19)

## 2.10  THE MODIFIED CONSERVATION LAW (Page 19)

# TERMS AND CONCEPTS

Matter (9)
Physical property (9)
Physical change (9)
Chemical property (10)
Chemical change (10)
States of matter (10)
Kinetic Molecular Theory (10)
Homogeneous (11)
Heterogeneous (11)
Phase (11)
Pure substance (12)
Mixture (12)

Element (13)
Compound (13)
Law of Definite Composition (15)
Law of Conservation of Mass (16)
Electrostatic forces (16)
Electrical charge (17)
Endothermic change (18)
Exothermic change (18)
Energy (19)
Potential energy (19)
Kinetic energy (19)
Law of Conservation of Energy (19)

# QUESTIONS AND PROBLEMS

Section 2.1

2.1)  Classify each of the following as a physical property or a chemical property: (a) hardness of a diamond; (b) burning ability of coal; (c) tarnishability of silver; (d) flexibility of spring steel; (e) thermal conductivity of silver.

2.2)  Classify each of the following changes as physical or chemical: (a) spoiling of food; (b) vaporization of "dry ice"; (c) stretching of a rubber band; (d) dynamite explosion; (e) shattering of glass.

2.17)  Classify each of the following as a physical property or a chemical property: (a) sugar chars when heated; (b) yellow color of sulfur; (c) stability of elemental nitrogen; (d) "slipperiness" of soap; (e) wood floating on water.

2.18)  Classify each of the following changes as physical or chemical: (a) decay of dead plants; (b) extraction of iron from ore; (c) melting snow; (d) spontaneous combustion of oily rags; (e) grinding of wheat.

## Section 2.2

2.3)  List the principal differences between a gas, a liquid and a solid.

2.4)  Why is it easier to carry a plate of solid food around a corner than a shallow bowl of soup? Answer in terms of the differences between solids and liquids.

2.5)  To which one or more of the three states of matter may the word "fluid" be properly applied? Explain.

## Section 2.3

2.6)  What is the meaning of the term "homogeneous"? List several examples of homogeneous substances not mentioned in the text.

2.7)  Classify each of the following samples of matter as homogeneous or heterogeneous: (a) sand on a beach; (b) filtered air; (c) fog; (d) salt water; (e) copper wire.

## Section 2.4

2.8)  Identify the difference between a pure substance and a mixture. How do the properties of pure substances differ from the properties of mixtures? A liquid is heated until it begins to boil at temperature X. Five minutes later it is still boiling, but the temperature has changed to Y, which is higher than X. Is the liquid a pure substance or a mixture? Explain.

2.9)  The density of a liquid is determined at its boiling point. It is boiled in an open container for 15 minutes, and the density of the remaining liquid is determined again. The densities are identical. Was the beginning liquid a mixture or a pure substance? Explain.

## Section 2.5

2.10)  What is the essential difference between an element and a compound?

2.11)  Classify each of the following pure substances as elements or compounds: tin; nitrogen; carbon dioxide; iron; baking soda.

## Section 2.6

2.12)  State the Law of Conservation of Mass. Illustrate its meaning in reference to the chemical change that occurs when carbon combines with oxygen to form carbon dioxide.

2.19)  Compare the densities of gases, liquids and solids, stating any generalities that may be identified.

2.20)  Hydrogen has been used as a fuel in space. Explain why it is more convenient to carry it as a liquid at very low temperatures than as a gas at room temperature.

2.21)  We speak logically of "pouring" a liquid. Is it possible to pour a solid? Or a gas? Explain why in each case.

2.22)  Define the term "heterogeneous." List several examples of heterogeneous substances not mentioned in the text.

2.23)  Classify each of the following samples of matter as homogeneous or heterogeneous: (a) boiling water; (b) window glass; (c) ground beef; (d) pepper; (e) flawless diamond.

2.24)  The temperature at which a homogeneous liquid first begins to freeze is determined. Frozen solid is removed, and the freezing temperature of the remaining liquid determined again. The second temperature is lower than the first. Was the beginning liquid a mixture or a pure substance? Explain.

2.25)  A colored liquid evaporates in an open container. Over a period of time the color becomes darker. Was the original liquid a pure substance or mixture? Explain.

2.26)  A pure red powder darkens on heating, releases a gas, and leaves behind a silvery liquid. Is the powder a compound or an element?

2.27)  Classify each of the following pure substances as elements or compounds: sugar; helium; hydrogen chloride; lead; steam.

2.28)  Explain in terms of the Law of Conservation of Mass why, when paper combines with oxygen in burning, the remaining ash weighs less than the original paper, but when iron combines with oxygen in rusting, it gains in weight.

Section 2.7

2.13) Explain what is meant by an "electrostatic force," and under what conditions it may exist. Identify a distinguishing difference between electrostatic force and gravitational force.

Section 2.8

2.14) State the difference between exothermic changes and endothermic changes. When water boils, is the change exothermic or endothermic? Give two examples of each, neither of which is in the textbook.

2.15) Determine whether the form of mechanical energy in each of the situations described is *primarily* potential or kinetic: (a) bullet leaving the muzzle of a gun; (b) set mouse trap; (c) pendulum at the bottom of its swing; (d) water near the bottom of a waterfall; (e) the north poles of two magnets held near each other; (f) water in a lake formed by a dam; (g) compressed air.

Section 2.9

2.16) When wood burns, the evolution of heat energy is obvious. Explain the source of this energy if, according to the energy conservation law, energy may neither be created nor destroyed in an ordinary chemical change.

2.29) Certain chemical species called ions are electrically charged. Indicate whether the following pairs would attract or repel each other: (a) positively charged sodium ion and negatively charged chloride ion; (b) negatively charged chloride ion and negatively charged bromide ion. State the rule on which your answers are based.

2.30) Classify the following changes as exothermic or endothermic: (a) water freezes; (b) an explosion; (c) burning of gasoline in an automobile engine; (d) decomposing water by electrolysis (see Fig. 2.5, page 14).

2.31) Determine whether the form of mechanical energy in each of the situations described is *primarily* potential or kinetic: (a) windmill; (b) the wind that drives the windmill; (c) a wound wristwatch; (d) speedboat pulling water skier; (e) baseball bat just before hitting ball; (f) positively and negatively charged objects held close to each other; (g) roller coaster at top of first hill.

2.32) In order to decompose water by electrolysis a continuous input of electrical energy is required. In view of this, compare the amount of "chemical energy" held by a given quantity of water with the chemical energy held by an equivalent quantity of hydrogen and oxygen. Explain.

# 3

# MEASUREMENTS AND CALCULATIONS

## 3.1 INTRODUCTION TO MEASUREMENT

Chemistry is a quantitative science. In addition to investigating the qualitative character of chemistry, asking *how* and *why* chemical changes occur, the chemist also wants to know *how much*. To answer that question a scientist must make measurements. Making measurements is nothing new to you: you measure your height and weight, or the distance you drive to school or work. Your daily purchases are by measured quantities: gallons of gasoline for your car, quarts of milk, pounds of butter and feet of lumber or yards of fabric. Making and working with measurements are familiar to us all. But if you are living in the United States, the units in which you express physical and chemical measurements may very well be new to you.

Most scientific measurements are expressed in so-called **SI units**, abbreviated from the French name for the International System that has been accepted throughout the world. Metric units for fundamental measurements of mass, length and volume are parts of the SI system. In this text, metric units will be used almost exclusively. The only exceptions will be in the present chapter, which will include some of the important conversions between English and metric units. This is done to give you some "feeling" for the metric system in case you are not familiar with it. Several of the more important relationships between English and metric units are included in the conversion factor table in the Appendix (page 511).

Some of the ways in which laboratory measurements are made are suggested in Figures 3.1 and 3.2. Length is ordinarily measured with a meter stick or "ruler." Most of the common one-foot rulers sold in bookstores of American colleges today have a centimeter scale on one edge. Volumes of liquids are measured in a number of ways, depending on the precision required; a simple graduated cylinder is often satisfactory. Mass is measured on a balance; again the accuracy requirement determines the sophistication of the instrument used. A fourth fundamental measurement made in the

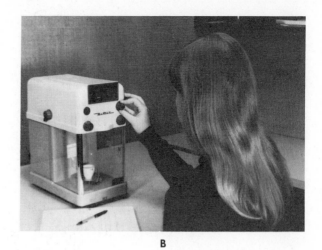

A

B

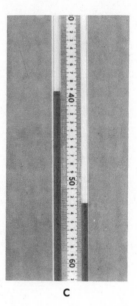

**Figure 3.1**  A. Student measures volume in graduated cylinder. B. Student measures mass on analytical balance. C. Manometer for measuring gas pressure. (Photographs by Gregory A. Peters.)

C

**Figure 3.2**  Laboratory balances. (Photographs by Gregory A. Peters.)

laboratory is temperature. Thermometers are calibrated in Celsius (formerly centigrade) degrees, rather than the Fahrenheit degrees familiar to Americans.

At the time this book goes to press, the United States and four minor and underdeveloped nations are the only countries in the world still clinging to the obsolete English system of units. In the United States, however, the long overdue change has begun. It is probably most evident on food labels, where liters and grams are printed alongside pints and pounds, and in TV weather reports where temperatures are often given in both Fahrenheit and Celsius degrees. The general conversion to metric units will probably advance significantly during the use period of this edition of your text.

## 3.2 DIMENSIONAL ANALYSIS

> PG 3 A Beginning with the "given quantity," set up and solve a problem by the dimensional analysis method.

Because our discussion of measurements will involve calculations to convert one measurement unit to another, we shall consider first the calculation method that will be used through most of this book. This method is called **dimensional analysis**.* Dimensional analysis is a logical sequence by which you reason through a problem by steps to arrive at the arithmetic setup. Once the correct setup is reached, you need only to calculate the answer. This generally is no more complicated than multiplication and division.

The method of dimensional analysis depends on three mathematical facts from your algebra courses:

1. If you multiply any quantity by 1, the product is equal to the original quantity:

$$a \times 1 = a$$

2. Any number divided by itself, or any fraction in which the numerator is equal to the denominator, is equal to 1:

$$\text{If } ab = c, \text{ then } \frac{ab}{c} = 1$$

3. In multiplication of fractions, any factor appearing in both a numerator and a denominator may be "canceled":

$$\frac{am}{b} \times \frac{1}{m} = \frac{a\cancel{m}}{b} \times \frac{1}{\cancel{m}} = \frac{a}{b}$$

This procedure is, in essence, a combination of the first two:

$$\frac{am}{b} \times \frac{1}{m} = \frac{a}{b} \times \frac{m}{m} = \frac{a}{b} \times 1 = \frac{a}{b}$$

---

*Dimensional analysis is also known as the factor-label method, unit conversion, unit cancellation and by other names.

We will illustrate the application of these facts in a problem with an "obvious" solution. The first example will be, "How many inches are in 4.5 feet?" In so apparent an example you would quickly reason, "In 1 foot there are twelve inches. Therefore in 4.5 feet there are 4.5 × 12 inches." In this reasoning you will essentially have converted 4.5 feet to inches by use of the conversion equivalence 1 foot = 12 inches, which may be expressed "12 inches per foot."

This sample problem, like most chemistry problems you will encounter, has a "given quantity." Much of your success in solving chemistry problems will depend on your ability to identify the given quantity as the starting point on which to base your subsequent calculations. In the first few examples we shall make a point of this identification. Later we shall not mention it unless there is specific reason for doing so, but to assist your thinking along this line we shall enclose the given quantity in a dotted box [____] in all examples and calculation setups throughout this book.

In our first example, "How many inches are in 4.5 feet?" there is but one quantity, and it is the given quantity for the problem. Be especially aware that it is a *quantity:* it tells how much of something is involved, and it is unique to this problem. *Conversion factors or other equivalences are not "given quantities."* They are *relationships* that connect one quantity unit to another. They frequently include the word *per;* rarely does *per* appear in a given quantity. "12 inches per foot," for example, is a relationship indicating 12 inches = 1 foot. The relationship may be applied to any number of problems; it is not unique to this example.

Many problems in chemistry may be expressed in an "equivalent interpretation," which further enhances your ability to interpret a problem correctly. The form of the equivalent interpretation is the question, "How many _____ are equivalent to _____?" The first blank is the identification of the thing sought in the problem; the second blank is the given quantity. Reworded into the equivalent interpretation form, the present example may be stated, "How many inches are equivalent to 4.5 feet?" From the equivalent interpretation you can plot a *unit path.* It goes from the units of the given quantity to the units of the answer quantity: feet → inches.

You are now ready for the problem as your first self-tutored example. We recommend that you read again the procedure you should follow, given on page 7. You should have pencil, paper and calculator ready. Your opaque shield should be placed just under the dashed line below, blocking from view everything after that line.

---

**Example 3.1**  How many inches are in 4.5 feet?

We have seen that the conversion equivalence for the unit path feet → inches is 1 foot = 12 inches. From this equation we can write two fractions with the numerator equal to the denominator, each fraction being equal to 1. Write these fractions. (At this point you should write the fractions on scratch paper, as instructed. When you have done so, move the shield to the next dotted line. The first thing you expose will be the correct answer to the question you have just answered. You should compare your answer with that in the book.)

---

**3.1a** $\dfrac{12 \text{ in}}{1 \text{ ft}}$ or $\dfrac{1 \text{ ft}}{12 \text{ in}}$

Because 12 inches = 1 foot, both fractions are equal to 1.

If the given quantity, 4.5 feet, is multiplied by 1, the product is also 4.5 feet, or its equivalent. We wish now to multiply the given quantity by 1 in the form of one of the two fractions shown above. In this process we shall use the dimensional analysis technique of *treating units just as we treat algebraic factors*. Specifically, as the "m" factor was canceled in the third of the "mathematical facts" above, so *units may be canceled in dimensional analysis*. You have two choices. You can multiply $4.5 \text{ ft} \times \dfrac{12 \text{ in}}{1 \text{ ft}}$, or $4.5 \text{ ft} \times \dfrac{1 \text{ ft}}{12 \text{ in}}$. In one multiplication the unwanted unit will cancel; in the other the surviving unit will be meaningless. Select the correct multiplication and complete the problem, including the necessary arithmetic.

---

**3.1b** 54 inches

$$\boxed{4.5 \text{ ft}} \times \frac{12 \text{ in}}{1 \text{ ft}} = 54 \text{ inches}$$

Just as the "m" factors cancel in $\dfrac{a \text{ m}}{b} \times \dfrac{1}{m}$, so the feet (or foot) factors cancel in the above example.

Had you selected the wrong factor your result would have been

$$4.5 \text{ ft} \times \frac{1 \text{ ft}}{12 \text{ in}} = 0.375 \frac{\text{ft}^2}{\text{in}}$$

Feet-squared per inch is, indeed, a rather meaningless unit!

The equivalent interpretation question for the problem is now answered: 54 inches are equivalent to 4.5 feet.

This simple example illustrates the dimensional analysis method of solving problems: starting with the given quantity (4.5 feet) we followed a unit path from feet → inches, using the conversion factor of 1 ft = 12 inches. Note the steps:

1. Begin with the *given quantity.*
2. Establish the unit path from the given quantity to the wanted quantity.
3. Establish one or more conversion factors which travel the unit path. Each conversion factor is equal to 1 because the numerator is equal to, or equivalent to, the denominator.
4. Reason your way through the problem, multiplying or dividing in a logical and meaningful sequence through each step of the unit path. Set up correctly, the units will cancel properly to yield an answer with the required units.
5. Perform the necessary *arithmetic operations*, multiplication and/or division, to get the numerical value of the answer.

6. Perform the algebraic operation of *canceling units* to obtain the units of the answer.

The results of steps 5 and 6 must match. If you come up with "strange" units because of your setup of the problem, you cannot simply return to the problem and adjust only units so they come out "right." The number work must also be adjusted in order that the numerator and denominator remain equivalent. In fact, one of the principal advantages of dimensional analysis is that faulty units in the answer warn you that your setup of the problem is wrong, and therefore your numerical answer will be incorrect.

On the presumption that a little practice brings one closer to perfection, let's try the technique on another example.

---

**Example 3.2**   In two weeks, how many times does the hour hand of a clock go around?

Can you rephrase the question into an equivalent interpretation form: How many _____ are equivalent to _____?

--------------------------------------------------------------------------------

**3.2a**   How many revolutions of the hour hand of a clock are equivalent to two weeks? You may have varied the words somewhat, but the idea should be there.

Can you identify the given quantity in this problem?

--------------------------------------------------------------------------------

**3.2b**   Two weeks. The other quantity, number of revolutions, is what you are trying to find.

Unlike the first example, you do not have a one step unit path this time, unless you "know" how many times the hour hand goes around in a week. This time we must find a unit path from weeks to revolutions of the hour hand. This requires additional information—information not given in the statement of the problem. It must be obtained from other sources. In this case the "other source" is your knowledge about clocks, hours, weeks and days. Can you suggest a unit path from the given quantity in weeks to the desired quantity in revolutions such that you know the equivalence relationship between each pair of units in the path? Write both the unit path and the equivalence relationships.

--------------------------------------------------------------------------------

**3.2c**   Weeks → days → hours → revolutions.
   1 week = 7 days; 1 day = 24 hours; 12 hours ≏ 1 revolution

The first relationship allows you to change weeks to days; days are converted to hours by the second; and the third changes hours to the required quantity, revolutions.

At this point, let's make a distinction between *equal* and *equivalent*. Two measurements may be *equal* to each other if they are measurements of the *same thing*, expressed in different units. In 1 week = 7 days, both weeks and days are units of the same thing, time. Furthermore, 1 week is *always* equal to 7 days. Similarly 1 day = 24 hours, always, and again both measurement units are units of time. But the relationship 12 hours $\simeq$ 1 revolution is true for this example only because we happen to be talking about the hour hand of a clock. One "revolution" is not a unit of time, and therefore cannot be *equal* to a measurement expressed in time units. But it can be *equivalent* to a number of time units within the limits of a specific problem, if a fixed relationship between revolutions and time can be established. Other equivalences between revolutions and time are 1 revolution $\simeq$ 1 hour for the minute hand of a clock, and 8766 hours $\simeq$ 1 revolution of the earth around the sun.

In this text we will use the symbol $\simeq$ to represent equivalent relationships between units of different measurements. The word *equivalence* will be used broadly to include both equalities and equivalent relationships.

Now let's apply these equivalences to the example. For each equivalence you have two possible fractions equal to 1. Successive multiplication by the proper three of these will yield a conversion from weeks to revolutions. The first step in the unit path converts weeks to days. Starting with the given quantity, 2 weeks, *set up* the conversion that would, if calculated, yield the number of days in two weeks, but do not solve to a numerical answer.

-------------------------------------------------------------------------------

**3.2d** $\quad \boxed{2 \text{ weeks}} \times \dfrac{7 \text{ days}}{1 \text{ week}}$

(Step 1)

Normally we don't work out intermediate answers in a dimensional analysis setup. If you did, however, and if your thought process was reasonable, the intermediate answer would be meaningful. In the setup this far, the 'weeks' cancel and the calculation yields an intermediate answer of 14 days. An incorrect setup, 2 weeks $\times \dfrac{1 \text{ week}}{7 \text{ days}}$, would result in a meaningless unit, weeks²/day, which would have signaled an error at this point.

The setup through the first multiplication indicates the number of days. Now *extend* the setup by a second multiplication by a conversion ratio that changes *days* into *hours*. Again, do not solve the setup.

-------------------------------------------------------------------------------

**3.2e** $\quad \boxed{2 \text{ weeks}} \times \dfrac{7 \text{ days}}{1 \text{ week}} \times \dfrac{24 \text{ hours}}{1 \text{ day}}$

(Step 1)  (Step 2)

Again notice that calculation of the setup this far would yield a meaningful answer in hours. This would not be true if the incorrect ratio $\dfrac{1 \text{ day}}{24 \text{ hours}}$ had been used.

The final step involves the conversion of hours to revolutions by the third equivalence relationship listed in answer 3.2c. Complete the setup, and this time solve for the numerical answer as well.

-------------------------------------------------------------------------------

**3.2f**  28 revolutions

$$\boxed{2 \text{ weeks}} \times \dfrac{7 \text{ days}}{1 \text{ week}} \times \dfrac{24 \text{ hours}}{1 \text{ day}} \times \dfrac{1 \text{ revolution}}{12 \text{ hours}} = 28 \text{ revolutions}$$

(Step 1)     (Step 2)        (Step 3)

Isolating the arithmetic from the setup, we have

$$\frac{2 \times 7 \times 24}{12} = 28$$

which may be done mentally by first seeing that 24/12 is 2. This is another advantage you should seek in a dimensional analysis setup: setting up the entire problem frequently leads to simplification in calculations, and therefore fewer errors.

A word of caution: Used unthinkingly, dimensional analysis can become a very mechanical process. Do not permit this. You will derive the most benefit from dimensional analysis if you do not juggle units randomly until they produce a wanted unit in the answer. Chart the unit path from the given quantity to the wanted quantity, and then thoughtfully consider each conversion. Reason through the steps in which you must multiply the number of weeks by the number of days in a week to convert to days, for example, and then multiply again to convert days to hours. In the final step, realize you must find out how many 12 hour "packages," each representing one revolution, are in the total number of hours, and that means divide by 12. This will make each step of every problem meaningful. It will also keep you *thinking* your way through the problem setup, and *understanding* the overall solution.

Dimensional analysis setups will be used in the unit conversions you will meet in this book. Additional practice problems may be found at the end of the chapter (page 54) if you wish to try them.

## 3.3 LENGTH AND VOLUME UNITS IN THE METRIC SYSTEM

PG 3 B Given an English-metric conversion table, convert a length measurement expressed in (a) English units, (b) meters, (c) kilometers, (d) centimeters, (e) millimeters or (f) angstroms to each of the other units.

3 C Given an English-metric conversion table, convert a volume measurement expressed in (a) English units, (b) liters, (c) milliliters or (d) cubic centimeters to each of the other units.

The standard unit of length is the **meter.**\* Initially defined as 1/10,000,000 of the length of the meridian of the earth from a pole to the equator, it is now identified more precisely as 1,650,763.73 times the wavelength of a certain line in the emission spectrum (see Chapter 4, page 65) of an elemental gas, krypton. The advantage of the modern definition over the original may not be obvious, but suffice it to say that the precision measurements of modern technology require and are greatly enhanced by such a definition. The meter is 39.37 inches long—about three inches longer than a yard (see Fig. 3.3).

As the English system uses both larger and smaller units than the yard for different purposes, so the meter must be modified for larger and smaller units. The metric system accomplishes this with multiples and submultiples

---

\*In most of the world and in many textbooks the length unit is spelled with the *r* before the *e:* metre.

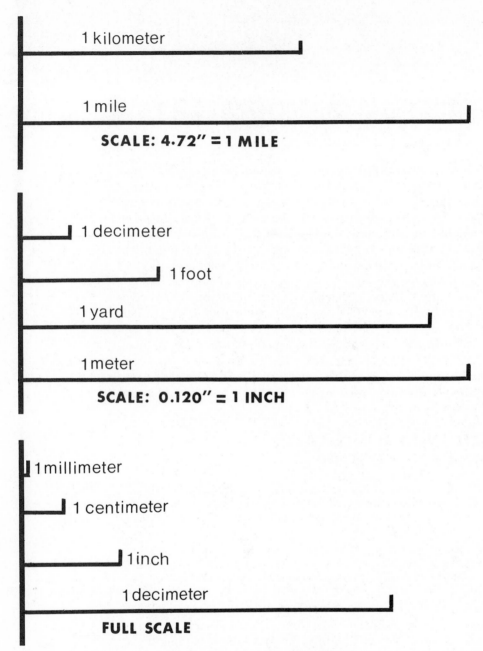

1 kilometer

1 mile

**SCALE: 4.72″ = 1 MILE**

1 decimeter

1 foot

1 yard

1 meter

**SCALE: 0.120″ = 1 INCH**

1 millimeter

1 centimeter

1 inch

1 decimeter

**FULL SCALE**

**Figure 3.3** Comparison between English and metric units of length.

of 10. One-tenth of a meter, for example, is called a decimeter; 1/100 meter is a **centimeter;** and 1/1000 meter is a **millimeter.** The prefixes deci-, **centi-** and **milli-** are applied to all measurements in the metric system, and they always identify 1/10, 1/100 and 1/1000 part, respectively, of the basic unit. Larger units are similarly identified by prefixes. The most important is the **kilo-,** which represents a quantity 1000 times as large as the basic unit. The **kilometer,** then, is 1000 meters. A partial list of metric prefixes appears in Table 3.1.

In your beginning chemistry courses you are likely to encounter only three of these metric multiples, those shown in boldface type. These pre-

TABLE 3.1  Metric Prefixes

| LARGE UNITS | | | SMALL UNITS | | |
|---|---|---|---|---|---|
| Metric Prefix | Metric Symbol | Multiple | Metric Prefix | Metric Symbol | Submultiple |
| mega- | M | $1,000,000 = 10^6$ | deci- | d | $0.1 = 10^{-1}$ |
| **kilo-** | **k** | $\mathbf{1,000} = 10^3$ | **centi-** | **c** | $\mathbf{0.01} = 10^{-2}$ |
| hecto- | h | $100 = 10^2$ | **milli-** | **m** | $\mathbf{0.001} = 10^{-3}$ |
| deka- | da | $10 = 10^1$ | micro- | $\mu$ | $0.000001 = 10^{-6}$ |
| (Unit: gram, meter, liter) | | $1 = 10^0$ | nano- | n | $0.000000001 = 10^{-9}$ |

The most important metric prefixes are printed in boldface type.

fixes are used so frequently they should become a part of your vocabulary. In the context of length measurement, these are the millimeter, centimeter and kilometer. The relationships to the meter are as follows:

$$1 \text{ kilometer} = 1000 \text{ meters} = 10^3 \text{ meters} \qquad (3.1)$$

$$1 \text{ meter} = 100 \text{ centimeters} = 10^2 \text{ centimeters} \qquad (3.2)$$

$$1 \text{ meter} = 1000 \text{ millimeters} = 10^3 \text{ millimeters} \qquad (3.3)$$

In writing these and other metric units, abbreviations are commonly used. Accordingly, the meter is designated m; the kilometer, km; the centimeter, cm; and the millimeter, mm.

The outstanding advantage of the metric system of units over the English system is the ease with which values may be converted from one unit to another. Conversions are all made by multiplying or dividing by multiples of 10—accomplished simply by moving the decimal point a certain number of places to the right or to the left. Until they become familiar with these conversions, students are sometimes uncertain which direction the decimal point should move. A dimensional analysis setup of the problem helps remove this uncertainty.

---

**Example 3.3**  How many meters are there in 28.6 cm?

First step: Identify the given quantity and unit path.

-------------------------------------------------------------------------------

**3.3a**  28.6 cm; cm → m

The given quantity is now to be multiplied by a conversion factor that will change centimeters to meters. Equation 3.2 shows that there are 100 cm in one meter. From this, two factors are possible. Select the correct one and complete the problem.

-------------------------------------------------------------------------------

**3.3b**  0.286 meter

$$\boxed{28.6\ \text{cm}} \times \frac{1\ \text{meter}}{100\ \text{cm}} = 0.286\ \text{meter}$$

If you had mistakenly multiplied by 100 instead of dividing, your units would have been meaningless ($\text{cm}^2/\text{m}$), which would have alerted you to the error.

---

**Example 3.4**  How many millimeters are in 3.04 cm?

Though the relationship between millimeters and centimeters has not been stated in an equation, you should be able to see it by comparing Equations 3.2 and 3.3. Supply the numbers to complete the following relationship:

_____ mm = _____ cm.

- - - - - - - - - - - - - - - - - - - - - - - - - - - - - - - - - - - - - - - - - - - - - - - - - - - -

**3.4a**  1000 mm = 100 cm. From Equations 3.2 and 3.3, 100 cm and 1000 mm are both equal to 1 meter, and therefore equal to each other. The relationship reduces to 10 mm = 1 cm, an equally acceptable answer.

The remainder of the solution should be apparent, with or without a dimensional analysis setup:

- - - - - - - - - - - - - - - - - - - - - - - - - - - - - - - - - - - - - - - - - - - - - - - - - - - -

**3.4b**  30.4 mm

$$\boxed{3.04\ \text{cm}} \times \frac{1000\ \text{mm}}{100\ \text{cm}} = 30.4\ \text{mm or}$$

$$\boxed{3.04\ \text{cm}} \times \frac{10\ \text{mm}}{1\ \text{cm}} = 30.4\ \text{mm}$$

---

In chemistry we are concerned with the behavior of the tiny particles (atoms, ions, molecules) of which matter is composed. Typically these particles have very small diameters, of the order of 1/20,000 to 1/10,000,000 of a centimeter. Rather than work with such awkward numbers, a different unit is used—the **angstrom,** abbreviated Å.* An angstrom is 1/10,000,000,000 meter, or $10^{-10}$ m.† The relationships

$$1\ \text{Å} = 0.0000000001\ \text{m} = 10^{-10}\ \text{m} \tag{3.4}$$

and
$$10,000,000,000\ \text{Å} = 10^{10}\ \text{Å} = 1\ \text{m} \tag{3.5}$$

allow conversion between angstrom measurements and typical metric units.

---

*There is a growing trend to discontinue the use of the angstrom and substitute in its place the SI unit *nanometer,* which is $10^{-9}$ meter. See Table 3.1, page 33.

†See Appendix ID, page 504, for information on expressing numbers in exponential notation.

**Example 3.5**  Express in meters the wavelength of a red line in the spectrum of hydrogen, 6565 Å.

From Equation 3.4 or 3.5, the setup and solution of this problem should be evident.

----------------------------------------------------------------------

**3.5a**  $6.565 \times 10^{-7}$ meter

$$\boxed{6565 \text{ Å}} \times \frac{10^{-10} \text{ m}}{1 \text{ Å}} = 6.565 \times 10^{-7} \text{ meter}$$

To convert between the English and metric systems, one of several conversion factors is required. Perhaps the most commonly used is the relationship between centimeters and inches:

$$2.54 \text{ cm} = 1 \text{ inch} \qquad (3.6)$$

**Example 3.6**  How many meters are in 7.60 feet?

A handbook may give you a direct conversion between feet and meters. But if you rely on *one* conversion factor between English and metric units of length that you know from memory, such as Equation 3.6, your unit path must be to change feet to inches, then inches to centimeters, and finally centimeters to the required meters. Start with the given quantity, set up the problem completely and solve.

----------------------------------------------------------------------

**3.6a**  2.32 meters

$$\boxed{7.60 \text{ ft}} \times \frac{12 \text{ in}}{1 \text{ ft}} \times \frac{2.54 \text{ cm}}{1 \text{ in}} \times \frac{1 \text{ m}}{100 \text{ cm}} = 2.32 \text{ m}$$

## VOLUME

The most commonly used volume unit in the metric system is the **cubic centimeter,** or **cm³.** It is the volume of a cube one centimeter on an edge. (Compare with the English unit, cubic inch, which represents the volume of a cube one inch on an edge.) Figure 3.4 illustrates the metric volume measurements, including those used for capacity, which are discussed below.

Volume conversions introduce a "different" feature that becomes apparent in the next example:

**Example 3.7**  How many cubic inches are in a box 9.00 cm long by 6.00 cm wide by 4.00 cm high?

*Solution.*  To find the volume of a rectangular box you multiply the length by the width by the height. Accordingly the volume of the box in cubic centimeters is

$$9.00 \text{ cm} \times 6.00 \text{ cm} \times 4.00 \text{ cm} = 216 \text{ cm}^3$$

To find the volume in cubic inches, each centimeter measurement could be converted to inches:

$$9.00 \cancel{\text{cm}} \times \frac{1 \text{ in}}{2.54 \cancel{\text{cm}}} \times 6.00 \cancel{\text{cm}} \times \frac{1 \text{ in}}{2.54 \cancel{\text{cm}}} \times 4.00 \cancel{\text{cm}} \times \frac{1 \text{ in}}{2.54 \cancel{\text{cm}}} = 13.2 \text{ in}^3$$

$$\text{Length} \quad \times \quad \text{Width} \quad \times \quad \text{Height} \quad = \text{Volume}$$

This expression may be simplified as

$$9.00 \cancel{\text{cm}} \times 6.00 \cancel{\text{cm}} \times 4.00 \cancel{\text{cm}} \times \left(\frac{1 \text{ in}}{2.54 \cancel{\text{cm}}}\right)^3 = 13.2 \text{ in}^3$$

or, $$9.00 \cancel{\text{cm}} \times 6.00 \cancel{\text{cm}} \times 4.00 \cancel{\text{cm}} \times \frac{1^3 \text{ in}^3}{2.54^3 \cancel{\text{cm}^3}} = 13.2 \text{ in}^3$$

*Notice that both the numbers and the units in both numerator and denominator are raised to a power when a fraction is to be raised to that power.* (The algebra is explained in greater detail on page 501 in the Appendix, if you wish to review it.)

---

Now that you've seen how volume conversions are made, try one yourself. In the following example you will not be given the precise conversion relationship that will yield a one-step unit path from the given units to the wanted. This leaves you two alternatives: first, you may put together two or more conversion relationships you do know and develop a multiple step unit path; or second, you can look up the direct conversion, if it is available. Appendix III (page 511) gives several useful equivalences for use in converting units. You just might find there the one you need. . . .

**Figure 3.4** Metric volume units in the laboratory. The large block is a cube measuring 1 decimeter (slightly less than 4 inches) on each edge— a cubic decimeter. One decimeter is 10 centimeters, so the volume of the block is also 1000 cubic centimeters (10 cm × 10 cm × 10 cm). This is equal to the capacity unit 1 liter, shown by the beaker. Separated from the block is a piece 1 cm × 1 cm × 10 cm, from which is further separated a cube 1 centimeter on each edge—a cubic centimeter, 1/1000 of a cubic decimeter. One milliliter, 1/1000 of a liter, is equal to 1 cubic centimeter. (Photograph by Judith E. Serface.)

**Example 3.8** Calculate the number of cubic meters of earth in 15.0 cubic yards of fill.

---

**3.8a** 11.6 m³

$$\boxed{15.0 \text{ yd}^3} \times \frac{1^3 \text{ m}^3}{1.09^3 \text{ yd}^3} = 11.6 \text{ m}^3$$

This setup represents the most direct unit path, yd³ → m³, using a conversion factor you looked up. If you happened to recall that there are 2.54 cm in 1 inch, and you know there are 36 inches in a yard, you could have developed a unit path yd³ → in³ → cm³ → m³:

$$15.0 \text{ yd}^3 \times \frac{36^3 \text{ in}^3}{1^3 \text{ yd}^3} \times \frac{2.54^3 \text{ cm}^3}{1^3 \text{ in}^3} \times \frac{1^3 \text{ m}^3}{100^3 \text{ cm}^3} = 11.5 \text{ m}^3$$

The small numerical difference between answers is because the conversion factors are not exact.

The English system uses special volume units for liquids, and sometimes gases, units that are not derived from length units. For example, would you be more apt to drive into a gas station and ask for 2310 cubic inches of gasoline, or for ten gallons? Gallons, quarts and pints are the customary English "capacity" units. The fundamental metric capacity unit is the **liter,** which is exactly 1000 cubic centimeters. This volume is very close to one quart—1.06 quarts, to be more exact.* The smaller unit commonly encountered in the laboratory is the **milliliter.** The prefix *milli-* again means 1/1000th part, so there are 1000 milliliters in a liter. By definition there are also 1000 cubic centimeters in a liter. This makes 1 cubic centimeter exactly equal to 1 milliliter; they are, in essence, synonymous terms. The abbreviated symbols for these capacity terms are $\ell$ for liter and ml for milliliter.

**Example 3.9** Calculate the number of liters in 0.500 cubic foot.

The necessary conversion factors are tabulated on page 511. Be careful about the final step in the setup. . . .

---

**3.9a** 14.2 $\ell$

$$\boxed{0.500 \text{ ft}^3} \times \frac{30.5^3 \text{ cm}^3}{1^3 \text{ ft}^3} \times \frac{1\ell}{1000 \text{ cm}^3} = 14.2\ell$$

Notice that the 1000 and the 1 in the final step are *not* cubed. The liter is already a volume unit, not the cube of a length unit.

---

*In countries using the British or "imperial" measure, the pint, quart and gallon are larger than their corresponding U.S. measures by the factor of 6/5; i.e., 6 U.S. units = 5 imperial units. Thus, in Canada, 1 liter = 0.88 imperial quart.

## 3.4 MASS AND WEIGHT

PG   3 D   Distinguish between mass and weight.

     3 E   Given an English-metric conversion table, convert a mass measurement, expressed in (a) English units, (b) grams, (c) kilograms, (d) centigrams or (e) milligrams, to each of the other units.

Consider a tool carried by a trio of astronauts on a trip to the moon. Suppose that tool, on earth, weighs six ounces. If it were to be weighed on the surface of the moon, it would weigh about one ounce. But halfway between the earth and moon it would be essentially weightless: released in "mid-air," it would remain there, floating until moved by one of the astronauts to some other location. Yet in all three locations it would be the same tool, having a constant quantity of matter.

**Mass is a measure of quantity of matter. Weight is a measure of the force of gravitational attraction.** The chemist is interested in mass, not weight. Weight is proportional to mass, and the proportionality between them depends upon where in the universe you happen to be. Fortunately this proportionality is essentially constant over the surface of the earth, so when you "weigh" something—measure the force of gravity for that object—you can express this weight in terms of mass. By common usage, to *weigh* an object is equivalent to measuring its mass.

The instrument used to measure mass in the laboratory is the *balance*. Most balances are constructed on the lever principle, whereby masses on opposite sides of a pivot point, or fulcrum, are compared. Figure 3.2 includes a photograph of a simple two-pan balance usually capable of weighing to the nearest decigram. In use (see Fig. 3.5) an unknown mass is placed on the left pan of the balance, and then "balanced" by placing appropriate calibrated weights on the other pan. When the pointer comes to rest at the center of the scale the masses on the two pans of the balance are equal. Most laboratory balances are more complex and more accurate than the two-pan decigram balance, but all operate on essentially the same principle.

The basic unit of mass in the metric system is the **kilogram.** A kilogram weighs somewhat more than two pounds, and therefore is too large a unit for small scale work in chemistry. The more common laboratory unit of mass is the gram, which, according to the metric prefixes established earlier, is 1/1000 kilogram. The balances commonly used in college chemistry laboratories are capable of measuring centigrams (1/100 g) or milligrams (1/1000 g).

Conversion between the metric and English systems of units may be made with either of the following factors:

$$454 \text{ grams} = 1 \text{ pound} = 16 \text{ ounces*} \tag{3.7}$$

$$1 \text{ kilogram} = 2.20 \text{ pounds} \tag{3.8}$$

---

*Ounces referred to in this text are avoirdupois ounces unless otherwise specified.

**Figure 3.5** Use of a 2-pan balance. (1) Balance is adjusted so pointer indicates zero on scale (top). (2) Object to be weighed is placed on left pan (middle). (3) Weights are placed on right pan until pointer again indicates zero. Mass of object on left pan is the same as mass of weights on right pan.

**Example 3.10**  Find the mass in kilograms of an object that weighs 13.4 ounces.

Solve the problem completely, using either metric-English conversion.

------------------------------------------------------------------------------------

**3.10a**  0.380 kg

By Equation 3.7: $\boxed{13.4 \; \cancel{oz}} \times \dfrac{1 \; \cancel{lb}}{16 \; \cancel{oz}} \times \dfrac{454 \; \cancel{g}}{1 \; \cancel{lb}} \times \dfrac{1 \; kg}{1000 \; \cancel{g}} = 0.380 \; kg$

By Equation 3.8: $\boxed{13.4 \; \cancel{oz}} \times \dfrac{1 \; \cancel{lb}}{16 \; \cancel{oz}} \times \dfrac{1 \; kg}{2.20 \; \cancel{lb}} = 0.380 \; kg$

The first setup might be shortened slightly by moving directly from ounces to grams using the extension of Equation 3.7:

$$\boxed{13.4 \; \cancel{oz}} \times \dfrac{454 \; \cancel{g}}{16 \; \cancel{oz}} \times \dfrac{1 \; kg}{1000 \; \cancel{g}} = 0.380 \; kg$$

**Example 3.11**  How many milligrams are in 34.9 grams?

This is a simple metric conversion using metric prefixes for mass measurements, just as you used them for length measurements earlier.

-------------------------------------------------------------------------

**3.11a**  34,900 mg

$$34.9 \ g \times \frac{1000 \ mg}{1 \ g} = 34,900 \ mg$$

## 3.5  DENSITY AND SPECIFIC GRAVITY

PG   3  F   Distinguish between density and specific gravity, including units of each.

3  G   Given two of the following, calculate the third: mass of a sample; volume occupied by the sample; density, or specific gravity.

The concept of *density* furnishes an example of the manner in which fundamental measurements are combined to express a physical property. The formal definition of **density is mass per unit volume.** Expressed mathematically, this becomes

$$\text{Density} = \frac{\text{Mass}}{\text{Volume}} \qquad (3.9)$$

Qualitatively we can think of density as a measure of the "heaviness" of a substance, in the sense that a block of iron is heavier than a block of aluminum of the same size.

The definition of density establishes the units in which it is measured. Mass is measured in grams; volume is measured in cubic centimeters. Therefore, according to Equation 3.9, the units of density are grams/cubic centimeter. There are, of course, other units in which density can be expressed, but they all must reflect the definition in terms of mass/volume. Examples include grams/liter, and pounds/cubic foot.

In order to determine the density of a substance it is necessary to know both the mass and volume of the same quantity of the substance. Dividing the mass by the volume yields the density.

**Example 3.12**  A 12.0 cm³ piece of magnesium weighs 20.9 grams. Find the density of magnesium.

-------------------------------------------------------------------------

**3.12a**  1.74 g/cm³

$$\frac{20.9 \ \text{grams}}{12.0 \ \text{cm}^3} = 1.74 \ \text{g/cm}^3$$

*Specific gravity* tells us how dense a substance is compared to some standard, usually water. More precisely, **specific gravity is the ratio of the density of a substance to the density of water at 4°C.** Expressed as an equation,

$$\text{Specific gravity} = \frac{\text{Density of substance}}{\text{Density of water at 4°C}} \qquad (3.10)$$

In Equation 3.10 the same units for density appear in both numerator and denominator. Therefore all units cancel, and specific gravity is a unitless or dimensionless term, i.e., a pure number.

In metric units the density of water at 4°C is 1.000 gram/cm$^3$. Dividing the density of a substance by 1.000 gram/cm$^3$ causes the specific gravity of a substance to be *numerically* equal to its density in grams per cubic centimeter. Accordingly, if the specific gravity of a substance is 6.3, we may conclude that its density is 6.3 grams/cm$^3$.

Density furnishes a useful equivalence relationship for making conversions between mass and volume. Since the density of copper, for example, is 8.9 g/cm$^3$, this means that 8.9 grams of copper represents the same quantity as 1.0 cm$^3$ of copper. Therefore we may say 8.9 g copper $\simeq$ 1.0 cm$^3$ copper. Use of density as a connecting link between mass and volume is shown in the following example.

---

**Example 3.13** The specific gravity of a certain kind of wood is 0.860. Find the mass of a block of that wood measuring 16.0 cm × 5.00 cm × 3.00 cm.

*Solution.* In this problem the volume of the wood is not given, but, as in Example 3.7, it may be found simply by multiplying the length-width-height dimensions. This product is the given quantity. Because of the *numerical* equivalence between specific gravity and density, the specific gravity of 0.860 may be interpreted as a density of 0.860 g/cm$^3$, which may be used as a conversion factor in a dimensional analysis setup:

$$16.0 \text{ cm} \times 5.00 \text{ cm} \times 3.00 \text{ cm} \times \frac{0.860 \text{ g}}{1 \text{ cm}^3} = 206 \text{ g}$$

---

Many students like to solve density problems algebraically by substituting into the defining equation. For instance, in Example 3.13,

$$\text{Density} = \frac{\text{Mass}}{\text{Volume}}$$

$$\frac{0.860 \text{ g}}{1 \text{ cm}^3} = \frac{\text{mass in grams}}{16.0 \text{ cm} \times 5.00 \text{ cm} \times 3.00 \text{ cm}}$$

$$\text{mass in grams} = 16.0 \text{ cm} \times 5.00 \text{ cm} \times 3.00 \text{ cm} \times \frac{0.860 \text{ g}}{1 \text{ cm}^3} = 206 \text{ g}$$

The resulting arithmetic is obviously identical, and the algebra approach is entirely correct. Part of the reason for introducing density at this point, however, is to give you practice in the use of dimensional analysis. We will

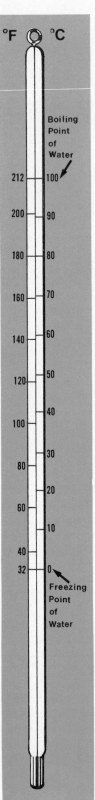

°F ◯ °C

Boiling
Point
of
Water

212 — 100

200 — 90

180 — 80

160 — 70

140 — 60

120 — 50

100 — 40

80 — 30

        20

60 — 10

40 — 0

32

Freezing
Point
of
Water

continue to use it in the coming examples, and we recommend that you do too.

---

**Example 3.14**   The density of a certain copper plating solution is 1.18 g/cm³. How many milliliters will be occupied by 150 grams of the solution?

Recall that a cubic centimeter and a milliliter are identical volumes. Starting with the given quantity and interpreting density as an equivalence between mass and volume, the setup is straightforward. Complete the problem.

---

**3.14a**   127 ml

$$150 \text{ g} \times \frac{1 \text{ ml}}{1.18 \text{ g}} = 127 \text{ ml}$$

---

## 3.6   TEMPERATURE MEASUREMENT

PG   3 H   Given a temperature in °F or °C, convert from one scale to the other.

The familiar temperature scale in the United States is the Fahrenheit scale. The temperature scale normally used for scientific purposes is the Celsius scale, formerly called the centigrade scale. The Celsius scale is based on two "fixed" points, the normal freezing and boiling points of water. On the Celsius scale, the freezing point is assigned a value of 0°C, and the boiling point 100°C. The difference between the fixed points is divided into 100 identical degrees. The corresponding temperatures on the Fahrenheit scale are 32°F for the freezing point of water and 212°F for the boiling point (see Fig. 3.6).

The Celsius and Fahrenheit temperature scales are related by the equation

$$°F - 32 = 1.8°C \qquad (3.11)$$

Temperature conversion problems are most easily solved algebraically, using Equation 3.11.

---

**Example 3.15**   What is the Celsius temperature on a comfortable 72°F day? Solve to the nearest °C.

If you are given the temperature in °F, Equation 3.11 tells you to subtract 32, equate the difference to 1.8°C, and solve for °C.

---

**Figure 3.6**  Comparison between Fahrenheit and Celsius temperature scales.

**3.15a** 22°C

The first step is to subtract 32 from the given Fahrenheit temperature: 72 − 32 = 40. This difference is then equated to 1.8°C and the equation is solved for °C:

$$1.8°C = 40$$

$$°C = \frac{40}{1.8} = 22$$

---

**Example 3.16**   It's a cold day, about −25°C. What is the equivalent Fahrenheit temperature, to the nearest degree?

If you are given the temperature in °C, Equation 3.11 says to multiply by 1.8, equate the product to °F − 32, and solve for °F.

---

**3.16a**   −13°F

The first step is to multiply °C by 1.8.     1.8(−25) = −45

Now equate the product to °F − 32 and solve:

$$°F - 32 = -45$$

$$°F = -45 + 32 = -13$$

---

SI units include a third temperature scale known as the Kelvin scale. Kelvin degrees are identical to Celsius degrees, but zero on the Kelvin scale is 273.16° below zero on the Celsius scale. The two scales are therefore related by the equation

$$°K = °C + 273.16 \qquad (3.12)$$

The Kelvin temperature scale is based on an "absolute zero," a temperature below which it is theoretically not possible to reach. All molecular movement is presumed to stop at absolute zero. We will not concern ourselves with the Kelvin scale at this time, but we will use it when studying Chapter 11 (page 234).

## 3.7   SIGNIFICANT FIGURES

### COUNTING SIGNIFICANT FIGURES

PG   3 I   Given a measured quantity, state the number of significant figures it has.

Every physical measurement has associated with it some uncertainty. This may be illustrated by the following examples in which the same board is measured by rulers calibrated in smaller and smaller subdivisions.

**Example 3.17** In the illustration below the length of a board is measured by a ruler exactly one meter long with no subdivisions marked on it. Estimate the length of the board as accurately as possible. Express your estimate as a decimal fraction.

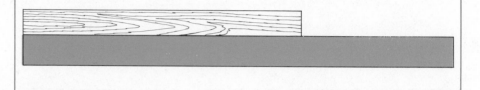

**3.17a** 0.6 or 0.7 meter. You should have *one* of these answers—not both.

Obviously the ruler shown is not a very satisfactory instrument with which to measure the board! But this is true of every measuring device, though it is not always so apparent. In this example the number of tenths of a meter are uncertain, and therefore must be estimated. Uncertainty is sometimes shown as a "plus or minus" value after the measurement. Accordingly the board length might be designated 0.6 ± 0.1 meter or 0.7 ± 0.1 meter.

A more finely divided ruler permits a finer estimate.

**Example 3.18** This time estimate the length of the board as accurately as possible with a ruler calibrated in decimeters. Include a ± uncertainty.

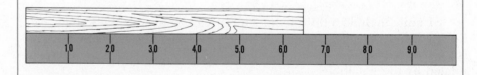

**3.18a** 0.64 ± 0.01 meter, or 0.65 ± 0.01 meter

The finer calibrations on the ruler allow the board to be measured to an additional decimal place—to the nearest hundredth of a meter. The digit value that is uncertain, the "4" in 0.64, is sometimes called the "doubtful" digit.

This answer might also be expressed in centimeters. Do so—including the ± uncertainty.

**3.18b** 64 ± 1 cm

$$0.64 \text{ m} \times \frac{100 \text{ cm}}{1 \text{ m}} = 64 \text{ cm}$$

A ruler calibrated in centimeters moves the uncertainty out one more decimal point.

**Example 3.19** Estimate the length of the board as accurately as possible. Express your answer both ways—meters and centimeters. Include the estimated uncertainty.

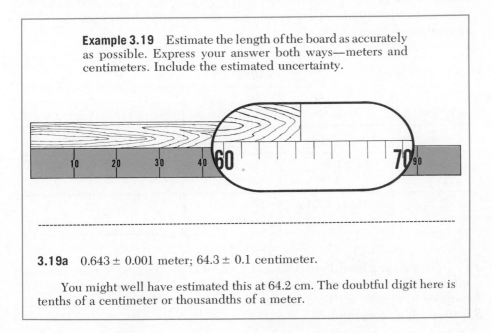

-------------------------------------------------------------------------------

**3.19a** $0.643 \pm 0.001$ meter; $64.3 \pm 0.1$ centimeter.

You might well have estimated this at 64.2 cm. The doubtful digit here is tenths of a centimeter or thousandths of a meter.

Most laboratory meter sticks are calibrated to millimeters.

**Example 3.20** Estimate the length of the board as accurately as possible. Answer in centimeters, including uncertainty.

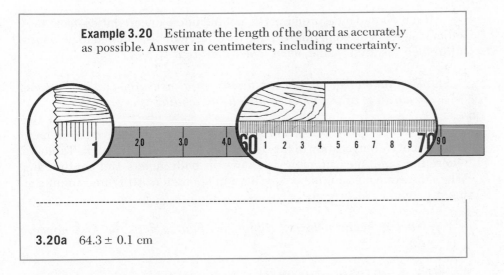

-------------------------------------------------------------------------------

**3.20a** $64.3 \pm 0.1$ cm

This answer is the same as the last one! The enlarged portion of the sketch shows that the limitation on the measurement is in such things as the roughness of the object being measured and the width of the calibration lines. Difficulty in placing the meter stick accurately on the board and unevenness at the end of the typical meter stick also make pointless any attempt to estimate tenths of a millimeter.

The foregoing examples provide a context in which to introduce the concept of **significant figures. The number of significant figures in a measurement is the number of figures that are known accurately plus one that is doubtful.** Counting significant figures always begins with the first non-zero digit and ends with the last digit shown—the doubtful digit. Thus, from the examples we have. . . .

**TABLE 3.2**

| Example | Reading | Doubtful Digit Value | Significant Figures |
|---------|---------|----------------------|---------------------|
| 3.17 | $0.6 \pm 0.1$ m | Tenths | 1 |
| 3.18 | $0.64 \pm 0.01$ m<br>$64 \pm 1$ cm | Hundredths<br>Units | 2<br>2 |
| 3.19 &<br>3.20 | $0.643 \pm 0.001$ m<br>$64.3 \pm 0.1$ cm | Thousandths<br>Tenths | 3<br>3 |

From Examples 3.17 to 3.20 and Table 3.2, the method of counting significant figures in a measurement, or recording a measurement to the proper number of significant figures, can be developed. We shall list the main features of this method:

1. *The digit value of the ± uncertainty always corresponds with the last digit value in the recorded measurement.*

Most recorded measurements do not include ± uncertainties—and they will no longer be used in this text. All readings will be assumed to be doubtful in their last digit value, *even when that digit is zero.*

2. *The number of significant figures in a measurement is the same regardless of the units in which the reading is expressed.*

The doubtful digit in a measurement is fixed by the measurement process, not by the location of the decimal point. This is evident in Table 3.2 where the same measurement is shown in both meters and centimeters. When the same measurement is written in kilometers, still three significant figures, another feature appears:

3. *Always begin counting significant figures with the first nonzero digit.*

64.3 cm = 0.643 m = 0.000643 km. In these three recordings of the same measurement, all to three significant figures, you begin counting at the "6" in each case. Specifically, you *do not begin at the decimal point and count zeros between it and the first nonzero digit.* This is one of the most common significant figure errors.

4. *If the last, or doubtful, digit happens to be zero, it must be shown if the quantity is to reflect the proper number of significant figures.*

One time in ten, on the average, the final digit in a measurement will be zero. If it is the doubtful digit, it must be shown. For example, 12 mg is a two significant figure number, with units doubtful; but 12.0 mg is a *three* significant figure number, with tenths of a unit doubtful. They have the same magnitude, but they give very different information about the preciseness of the measurement. Tail-end zeros to the right of the decimal point must be shown if they are significant.

5. *Large numbers with tail-end zeros to the left of the decimal point must be written in exponential notation.*

How many significant figures are in the quantity 12,000 grams? Which is the doubtful digit, the two thousands, the zero hundreds, the zero tens, or the zero units? If, for example, the zero hundreds is the doubtful digit and *therefore must be the last digit shown,* you cannot simply drop the last two zeros. If you did, the magnitude of the number would become one hundred twenty instead of twelve thousand. By rewriting the number in exponential notation (see Appendix IB, page 504) the tail-end zeros are to the *right* of the decimal point, and No. 4 above identifies the doubtful digit. Thus the magnitude 12,000 to
   ... two significant figures is $1.2 \times 10^4$—doubtful digit: thousands;
   ... three significant figures is $1.20 \times 10^4$—doubtful digit: hundreds;
   ... four significant figures is $1.200 \times 10^4$—doubtful digit: tens;
   ... five significant figures is $1.2000 \times 10^4$—doubtful digit: units.

If you find the rules for significant figures for large and small numbers confusing—and many beginning students do—this suggestion should help:

*Write* both *large and small numbers in exponential notation, including tail-end zeros if they are significant.*

Then all digits written are significant.
Let's see how much of all this you have mastered.

---

**Example 3.21** For each of the following quantities, write the correct number of significant figures.

| | |
|---|---|
| 45.26 g | 0.60 ft |
| 163 ml | 0.00025 kg |
| 0.109 in | $2.3569 \times 10^8$ cm |
| 62,700 cm | $5.890 \times 10^5$ liters |

----------------------------------------------------------------------------------------

**3.21a**

| | | | |
|---|---|---|---|
| 45.26 g | 4 | 0.60 ft | 2 |
| 163 ml | 3 | 0.00025 kg | 2 |
| 0.109 in | 3 | $2.3569 \times 10^8$ cm | 5 |
| 62,700 cm | ??? | $5.890 \times 10^5$ liters | 4 |

The quantity 62,700 cm is ambiguous: there is no way to tell whether the doubtful digit is in thousands, hundreds, tens or units. Exponential notation is required to remove the ambiguity. . . . The zero at the end of $5.890 \times 10^5$ liters, on the right of the decimal, shows it to be significant. The equivalent value, 589,000 liters, would be ambiguous, as is the 62,700 cm.

## *ROUNDING OFF*

**PG 3 J** Round off given numbers to a stated number of significant figures.

Sometimes when we add, subtract, multiply or divide experimentally measured quantities, our answer contains some figures which are not significant. When this happens, the result must be rounded off. Rules for rounding off are as follows:

1. If the first digit to be dropped is less than 5, leave the preceding digit unchanged.
2. If the first digit to be dropped is 5 or more, increase the preceding digit by 1.

Other rounding off rules vary in their handling of 5 if it is the first digit to be dropped. This may change the magnitude of the final digit by 1 in some cases. This is acceptable, however, because by any method only the doubtful digit is affected.

---

**Example 3.22** Round off each of the following to three significant figures:

a. 1.42752 g/cc          e. 45,853 cm³
b. 643.349 cm²         f. 0.0394498 m
c. 0.0074562 kg       g. $3.605 \times 10^{-7}$ cm
d. $2.103 \times 10^4$ cm     h. 3.5000 sec

- - - - - - - - - - - - - - - - - - - - - - - - - - - - - - - - - - - - - - - - - - - - - - - - - - - - - - - - - - - - - - -

**3.22a**

a. 1.43 g/cc              e. $4.59 \times 10^4$ cm³
b. 643 cm²              f. 0.0394 or $3.94 \times 10^{-2}$ m
c. 0.00746 or $7.46 \times 10^{-3}$ kg    g. $3.61 \times 10^{-7}$ cm
d. $2.10 \times 10^4$ cm       h. 3.50 sec

## *ADDITION AND SUBTRACTION*

**PG 3 K** Add or subtract given measurements and express the result to the proper number of significant figures.

The **significant figure rule for addition and subtraction** can be stated as follows: **Round off the answer to the first column that has a doubtful digit.** Example 3.23 shows how this rule is applied:

**Example 3.23**  A student weighs four different chemicals into a pre-weighed beaker. The individual weights and their sum are as follows:

| | | |
|---|---|---|
| Beaker | 319.542 | grams |
| Chemical A | 20.460 | grams |
| Chemical B | 0.0639 | gram |
| Chemical C | 38.2 | grams |
| Chemical D | 4.073 | grams |
| Total | 382.3389 | grams |

Express the sum to the proper number of significant figures.

This sum is to be rounded off to the first column that has a doubtful digit. What column is this: hundreds, tens, units, tenths, hundredths, thousandths or ten thousandths?

----------------------------------------------------------------------------

**3.23a**  Tenths. The doubtful digit in 38.2 is in the tenths column. In all other numbers the doubtful digit is in the hundredths column or smaller.

According to the rule, the answer must now be rounded off to the nearest number of tenths. What answer do you report?

----------------------------------------------------------------------------

**3.23b**  382.3 grams

An alternate procedure is to round off before adding. In the example given, first rounding off each number to tenths would give 382.4 as the answer:

$$319.5$$
$$20.5$$
$$0.1$$
$$38.2$$
$$4.1$$
$$\overline{382.4}$$

Between these two methods, the majority opinion seems to favor performing calculations with all the data available and then rounding off the final answer, as in the example. This also seems to be the "natural" procedure students tend to follow. *But you must remember to round off that final answer.* Either approach to the above example yields uncertainty in the tenths digit, so both answers are acceptable.

Example 3.23 may be used to justify the rule for addition and subtraction. A sum or difference number must be doubtful if any number entering into that sum or difference is doubtful or unknown. In the left addition below all doubtful digits are shown in color, and the *hundredths* in 38.2 are unknown.

|  |  |
|---|---|
| 319.542 | 319.5\|42 |
| 20.4609 | 20.4\|609 |
| 0.063 | 0.0\|63 |
| 38.2 | 38.2\| |
| 4.073 | 4.0\|73 |
| 382.3389 | 382.3\|389 |

Recalling that the last digit shown must be the first doubtful digit, the sum must be rounded off to tenths: 382.3. The addition at the right above shows another way of locating the first doubtful digit in a summation. If you draw a dotted line just to the right of the last column that has every space occupied—or just to the left of the first column with an open space—the summation digit just to the left of that line will be the first doubtful digit to which the answer must be rounded off.

The same rule, procedure and rationalization hold for subtraction.

---

**Example 3.24** In an experiment in which oxygen is produced by heating potassium chlorate in the presence of a catalyst, a student assembled and weighed a test tube, test tube holder and catalyst. He then added potassium chlorate and weighed the assembly again. The data were as follows:

Test tube, test tube holder, catalyst
and potassium chlorate      26.255 grams

Test tube, test tube holder and catalyst      24.05 grams

The weight of potassium chlorate is the difference between these numbers. Express this weight in the proper number of significant figures.

---

**3.24a** 2.21 grams

$$
\begin{array}{l}
26.255 \text{ grams} \\
\underline{24.05\ \ \text{ grams}} \\
2.205 \text{ grams} = 2.21 \text{ grams}
\end{array}
$$

The 24.05 is doubtful in the hundredths column, so the difference is rounded off to hundredths.

---

## MULTIPLICATION AND DIVISION

PG 3 L Multiply or divide given measurements and express the result in the proper number of significant figures.

The **significant figure rule for multiplication and division** is as follows: **Round off the answer to the *same number* of significant figures as the *smallest number* of significant figures in any factor.**

Again application will be illustrated by example:

---

**Example 3.25**  The density of a certain gas is 1.436 grams per liter. Find the mass of 0.0573 liter of the gas. Express the answer in the proper number of significant figures.

The mass is found by multiplying the volume by the density. The setup of the problem is

$$0.0573 \text{ liter} \times \frac{1.436 \text{ grams}}{1 \text{ liter}}$$

According to the rule, the product may have no more significant figures than the least number of significant figures in any factor. What is the least number of significant figures in any factor?

------------------------------------------------------------

**3.25a**  Three. 0.0573 is a three significant figure number: $5.73 \times 10^{-2}$.

Now calculate the answer and round off to three significant figures.

------------------------------------------------------------

**3.25b**  0.0823 gram. If you showed 0.082, you forgot that counting significant figures begins at the first nonzero digit.

---

Example 3.25 may be used to justify the rule for multiplication and division. If both quantities have final digits that are one number higher than their true values, the true answer would be $0.0572 \times 1.435 = 0.0821$. If both are too low by 1, the problem is $0.0574 \times 1.437 = 0.0825$. Uncertainty appears in the third significant figure, just as predicted. Alternatively, each product number into which a doubtful multiplier enters must itself be doubtful. Colored numbers indicated the doubtful digits in the detailed multiplication:

$$
\begin{array}{r}
1.436 \\
\times\ 0.0573 \\
\hline
4308 \\
10052 \\
7180 \\
\hline
0.0822828
\end{array}
$$

---

**Example 3.26**  Assuming the numbers are derived from experimental measurements, solve

$$\frac{(2.86 \times 10^4)\,(3.163 \times 10^{-2})}{1.8}$$

and express the answer in the correct number of significant figures.

------------------------------------------------------------

**3.26a** $5.0 \times 10^2$. The answer should not be shown as 500, as the number of significant figures could be read as one, two or three. Two significant figures are set by the 1.8.

---

**Example 3.27** How many millimeters are in 0.6294 centimeter? Express the answer in the proper number of significant figures.

With ten millimeters per centimeter, the arithmetic is simple enough. But how many significant figures? How would you state the answer?

-----------------------------------------------------------------------------------------------

**3.27a** 6.294 millimeters. See explanation below.

---

In Example 3.27 you are multiplying by ten—exactly 10. By definition there are *exactly* 10 mm in 1 cm. Similarly there are exactly 12 inches in 1 foot. Whenever a multiplier is an exact number, that factor is *infinitely* significant, and never establishes the limit on the number of significant figures in an answer. Significant figures and doubtful digits arise in *measured quantities only,* not in definitions. Realize, though, that some conversion factors between the English and metric systems are by measurement, not definition. The relationship 454 grams = 1 pound gives to three significant figures the closest number of grams that are equal to exactly 1 pound.

### *SIGNIFICANT FIGURES AND THIS BOOK*

Most modern calculators report answers in all the digits they are able to display, usually eight or more. Rarely does a chemistry problem justify that many significant digits in a calculated result; such an answer is simply unrealistic and should never be used. Therefore you should assume that all the data given in a problem are based on measurements expressed to the proper number of significant figures, and write your calculated answers to the number of significant figures justified by the data. As an arbitrary standard, all calculations in this book have been made using all the figures given in all measurements. Final answers have then been rounded off to the number of significant figures allowed by the data and the calculation rules stated earlier in this section. This generally yields answers to three significant digits.

## CHAPTER 3 IN REVIEW

## 3.3 LENGTH AND VOLUME UNITS IN THE METRIC SYSTEM

3 B Given an English-metric conversion table, convert a length measurement expressed in (a) English units, (b) meters, (c) kilometers, (d) centimeters, (e) millimeters or (f) angstroms to each of the other units. (Page 31)

3 C Given an English-metric conversion table, convert a volume measurement expressed in (a) English units, (b) liters, (c) milliliters or (d) cubic centimeters to each of the other units. (Page 31)

## 3.4 MASS AND WEIGHT

3 D Distinguish between mass and weight. (Page 38)

3 E Given an English-metric conversion table, convert a mass measurement expressed in (a) English units, (b) grams, (c) kilograms, (d) centigrams or (e) milligrams to each of the other units. (Page 38)

## 3.5 DENSITY AND SPECIFIC GRAVITY

3 F Distinguish between density and specific gravity, including units of each. (Page 40)

3 G Given two of the following, calculate the third: mass of a sample; volume occupied by the sample; density, or specific gravity. (Page 40)

## 3.6 TEMPERATURE MEASUREMENT

3 H Given a temperature in °F or °C, convert from one scale to the other. (Page 42)

## 3.7 SIGNIFICANT FIGURES

3 I Given a measured quantity, state the number of significant figures it has. (Page 43)

3 J Round off given numbers to a stated number of significant figures. (Page 48)

3 K Add or subtract given measurements and express the result to the proper number of significant figures. (Page 48)

3 L Multiply or divide given measurements and express the result to the proper number of significant figures. (Page 50)

# TERMS AND CONCEPTS

SI units (24)
Dimensional analysis (26)
"Given quantity" (27)
"Unit path" (27)
Meter (31)
Metric prefixes: centi-, milli-, kilo- (32)
Ångstrom (34)
Liter (37)
Mass (38)

Weight (38)
Kilogram (38)
Density (40)
Specific gravity (41)
Fahrenheit (42)
Celsius (42)
Significant figures (46)
Doubtful digit (46)

# QUESTIONS AND PROBLEMS

*For each of the problems that follow, write the calculation setup and solve for the numerical answer, even if either seems obvious. Remember that, at this point, the interpretation and setup of the problem are more important than the solution. Use dimensional analysis wherever possible.*

Section 3.2

3.1)  Express 122 quarts in gallons: (4 qts = 1 gal).

3.2)  How many ounces are in 0.35 ton? (1 ton = 2000 lbs; 1 lb = 16 oz)

3.3)  A bicycle manufacturer plans to produce 1500 bicycles per month over a 5 month period. How many wheels must the production department schedule for the entire run?

3.4)  How many minutes does it take a car traveling 35 miles per hour to cover a distance of 2500 feet?

3.5)  Certain bolts are zinc plated in batches averaging 75 pounds each. If the plating cycle is 12 minutes for each batch, and the cost of operating the plating system is $18/hour, what is the cost of plating 2.0 pounds of bolts, expressed in cents?

3.6)  How many cubic feet of earth are in 3.6 cubic yards?

3.7)  What is the area of a football field in square feet? (The field is 100 yards long × 53.3 yards wide.)

Section 3.3

3.8)  A kitchen table is 28.0″ high. Express this measurement in centimeters.

3.9)  A highway sign indicates you are 148 kilometers from your destination. How many miles must you yet travel?

3.10)  A football kicker averages 40.8 yards per punt. Express this distance in meters.

3.11)  What are the centimeter dimensions of a two-by-four (2″ × 4″) piece of lumber?

3.43)  If a British penny = 1.6¢, express 54 pence in American cents.

3.44)  How many rulers are in 4 gross of rulers? (1 gross = 12 dozen)

3.45)  A laboratory hot plate has three identical heating elements. How long will an inventory of 558 elements last at a production rate of 40 hot plates per week?

3.46)  A large reception is planned. Five cakes are to be baked as part of the refreshments. If the recipe for each cake calls for three eggs, and eggs are selling at 84¢ per dozen, calculate the cost of eggs in dollars.

3.47)  A certain type of copper rivet is sold at 85¢/lb, and the average number of rivets in 1 pound is 34. What is the maximum number of rivets that can be purchased for fifty dollars?

3.48)  A tank measures 72″ × 30″ × 24″. What is its volume in cubic feet?

3.49)  How many square feet are in a field 12.5 rods long × 8.9 rods wide? (1 rod = 16.5 ft)

3.50)  What will be the dimensions in meters of a 12.0′ × 14.0′ bedroom (12.0 feet wide, 14.0 feet long)?

3.51)  A man's size 15 shirt has a neckband 15.0 inches in circumference. What size will he wear by metric measurement (centimeters)?

3.52)  The speed limit on a highway is 55 miles per hour. To what number of kilometers per hour will the speedometer of a 1988 car point when moving at the same speed?

3.53)  A sheet of 4′ × 8′ plywood will have what dimensions to the nearest centimeter?

3.12) Fractional measurements that make up a large portion of the engineering design data of the United States will require a major adjustment in thinking and specifications for replacement parts. How many millimeters, for example, are equal to ⅛ inch?

3.13) A machine part is shown to be 3⅝″ long on an engineering drawing. What is its millimeter measurement? Careful . . .

3.14) How many millimeters are in 0.786 meter?

3.15) Calculate the number of centimeters in the wavelength of the $5.9 \times 10^3$ Å line in the yellow portion of the light spectrum.

3.16) In planting a new lawn, the American homeowner today might order 2.50 cubic yards of topsoil. What equivalent amount in cubic meters will his son order in 1994?

3.17) In continental Europe today, gasoline is sold by the liter. How many liters are in 5.00 gallons?

## Section 3.4

3.18) The terms *weight* and *mass* are commonly used interchangeably. We also say an astronaut is "weightless" when traveling in space. Would it be proper to say instead that he is "massless"? Explain why or why not.

3.19) Calculate the number of grams in 4.80 ounces.

3.20) Find the mass in grams of a 0.85 carat diamond if 1 carat = 200 milligrams by definition.

3.21) Today the price of rice is 79¢ per pound. Assuming no inflation between now and 1988, what will you then pay in dollars per kilogram?

## Section 3.5

3.22) The specific gravity of a certain substance is 0.89. What is its density? Compare the density of the substance with the density of water.

3.23) Find the mass of 50.0 ml carbon tetrachloride if its specific gravity is 1.60.

3.24) Among natural minerals, gold is one of the most dense, at 19.3 g/cm³. Find the volume occupied by 68.3 grams of gold.

3.25) Mercury is a dense material by any standard, and particularly as a liquid. 150.0 milliliters of mercury weighs $2.04 \times 10^3$ grams. Calculate the specific gravity of mercury.

3.54) Beauty contest winners tend to be tall girls. How would you classify a girl whose height is 176 centimeters: short, tall, or in between? Express her height in inches.

3.55) It's easier to convert English measurements to metric than metric to English. To see the difference, express 175 millimeters to the nearest 16th inch.

3.56) Find the number of kilometers in $4.85 \times 10^4$ centimeters.

3.57) An atom of copper has a diameter of $1.28 \times 10^{-8}$ centimeter. Express this diameter in angstroms.

3.58) The area of the new lawn in Problem 3.16 is 219 square meters, by 1994 measurements. In 1978 the same lawn would be calculated to be what number of square feet?

3.59) If gasoline today costs 61.9¢ per gallon, calculate its price in cents per liter, using 3.785 liters = 1 gallon.

3.60) When you "weigh" yourself on a typical bathroom scale, are you actually measuring mass or force? Explain your answer.

3.61) Calculate the mass in kg of a $3.0 \times 10^3$ pound automobile.

3.62) If our beauty of Problem 3.54 has a mass of 55.5 kilograms, find her weight in pounds.

3.63) The former luxury ocean liner "Queen Mary," now a tourist attraction in Long Beach, California, weighs $8.2 \times 10^4$ tons. Express her mass in megagrams (1 megagram = $10^6$ grams).

3.64) The density of bismuth is 608 pounds per cubic foot in English units. If the density of water is 62.4 pounds per cubic foot, calculate the specific gravity of bismuth.

3.65) What is the volume of 255 grams of ethyl alcohol if its specific gravity is 0.791?

3.66) Balsa wood, the soft, light wood used in making model airplanes, has a density of 0.125 g/cm³. Find the mass of 135 cm³ of balsa wood.

3.67) A typical ice cube from the refrigerator measures 4.0 cm × 3.5 cm × 3.0 cm and weighs 38.5 grams. Calculate the density of ice.

3.26) To fully appreciate the density of mercury, calculate the weight in pounds of one quart of mercury.

3.68) Milk is among the heavier items carried home from the grocery store. Find the mass of ½ gallon of milk in pounds. (Specific gravity = 1.03.)

## Section 3.6

3.27) Table salt melts at 805°C. What is the equivalent Fahrenheit temperature?

3.69) What Celsius temperature is equivalent to 0°F?

3.28) Most people find it "uncomfortable" to touch a piece of metal if its temperature is above 120°F. What is the corresponding Celsius temperature?

3.70) In fractional distillation of petroleum, the components are separated by their boiling temperatures. One component of gasoline boils at 116°C. Express this in °F.

3.29) Convert the following temperatures to degrees Celsius: −320°F; 205°F; 2460°F.

3.71) Express the following temperatures in Celsius degrees: 3190°F; −179°F; −87°F.

3.30) Find the Fahrenheit equivalent of each of the following temperatures: −244°C; 46°C; 1520°C.

3.72) Change each of the following to degrees Fahrenheit: 2190°C; −61°C; −192°C.

## Section 3.7

*3.31–3.33 and 3.73–3.75: Indicate the number of significant figures in which each of the following quantities is stated:*

3.31)   4.060 liters
3.32)   $1.6 \times 10^{-4}$ gram
3.33)   312°C

3.73)   10.40 kilograms
3.74)   0.0026 liter
3.75)   19,623.0 milligrams

*3.34–3.36 and 3.76–3.78: Round off each of the following quantities to three significant figures:*

3.34)   8.3562 g/cm³
3.35)   124.563 g
3.36)   $5.0635 \times 10^{-4}$ liter

3.76)   21.639 ml
3.77)   0.000439841 kg
3.78)   4200 mg

3.37) The following quantities of a precious metal compound are drawn from a stock bottle for four different purposes: 0.475 gram; 3.40 grams; 1.8 grams; 12.92 grams. Find the total usage, expressed in the proper number of significant figures.

3.79) Over a given period of time a farmer's five cows produce the following quantities of milk: 34.3 gallons; 41.07 gallons; 38 gallons; 36.294 gallons; and 35.0 gallons. State the total milk volume in the correct number of significant figures.

*3.38, 3.39, 3.80 and 3.81: From the data given, express the results required in the proper number of significant figures:*

3.38) What volume will be occupied by 85.0 grams of magnesium if its density is 1.74 grams/cubic centimeter?

3.80) What is the weight in pounds of a mass of 1862 grams if 454 grams = 1 pound?

3.39) A certain liquid dispensing unit delivers 10.36 ml of a solution each time it is used. If it is used twice each hour for 12 hours, how many milliliters of solution will it deliver?

3.81) A pump in a chemical plant delivers 14.6 liters of hydrochloric acid each hour to a continuous industrial process. If the pump operates 518.6 hours in one working month, what will be the total volume of acid consumed?

*The remaining problems extend beyond the performance goals stated in the chapter, and combine concepts from two or more sections. Throughout the text such problems are marked with an asterisk (\*). For this group of problems, assume all values are expressed in the proper number of significant figures, and then write your final answer in the number of significant figures justified by the data.*

3.40)\* If the density of a certain alloy is 519 pounds per cubic foot, find its density in grams per cubic centimeter by dimensional analysis.

3.82)\* The average density of a certain soil is 3.29 g/cm³. Express that density in tons per cubic yard.

3.41)*  If the specific gravity of an industrial acid is 1.29, calculate the tons of acid in a $3.0 \times 10^3$ gallon storage tank.

3.83)*  A 42 inch diameter cylindrical storage tank is to be built for the storage of fuel oil in a home heating system. How long must the tank be if it is to hold 18.0 tons of oil having a density of 0.880 g/cm³? (The area of a circle is $\pi r^2$, or $\dfrac{\pi d^2}{4}$.)

3.42)*  An empty graduated cylinder has a mass of 62.9 grams. When 47.2 milliliters of a liquid are placed in the cylinder, the total mass is 113 grams. Find the density of the liquid.

3.84)*  An empty 55 gallon drum weighs 19 pounds. When filled with a machine oil the total mass of the drum and its contents is 193 kilograms. Find the specific gravity of the oil.

# 4

# ATOMIC STRUCTURE

## 4.1 INTRODUCTION TO ATOMIC STRUCTURE

Atoms. Are they real? Is there such a thing as an ultimate particle that cannot be divided into two smaller particles? The early Greeks talked about such topics as early as 400 B.C. In fact, it is from their word *atomos*, meaning indivisible, or uncuttable, that our word, atom, is derived. But the atomic theory did not grow among the Greeks—a rather influential gentleman named Aristotle didn't believe in it. Consequently the idea lay quietly in the minds of philosophers for centuries. And that's exactly where "science" was during those nonproductive and nonprogressive centuries—in the minds of philosophers.

It was not until the 17th century that *observation* and *experimentation* were added to pure thinking. Then things started to move, slowly at first, but quickening to a pace that, at times, has been difficult to understand. Space and the scope of an introductory text do not permit a thorough presentation of this topic. This is unfortunate, for it is an exciting story about people and society, their strengths and their weaknesses, their wars and politics and religions and prejudices, all of which at one time or another encouraged or held back the advance of research related to the atom. Nor has the story ended. Today the nucleus of the atom is at the same time our greatest threat for destruction and our greatest hope for providing energy to the world's growing population. We must hope that future research will reveal how to reap the benefits of the atom in a way that is completely safe, and that mankind will never again release the power of the atom for the purpose of destruction, as it was released in 1945.

In this chapter we will outline the major concepts of the atom that have developed over the past 170 years, concentrating more on the conclusions reached than the way they were arrived at. If you ever have some spare time in a library, you might find that a book on atomic history and the people behind it makes interesting reading.*

---

*One such book is Fine, *Chemistry Decoded*, Oxford University Press, 1976.

## 4.2 THE ATOMIC THEORY

PG   4 A   Identify the postulates of Dalton's atomic theory.

Early in the 19th century John Dalton, an English chemist and school-teacher, did a lot of thinking about the Law of Definite Composition (page 15) and the Law of Conservation of Mass (page 16). In an attempt to explain these laws he proposed his atomic theory in 1808. The primary postulates of that theory are:

1. Each element is made up of tiny, individual, unit particles called atoms.
2. Atoms are indivisible; they can neither be created nor destroyed.
3. All atoms of each element are identical in every respect.
4. Atoms of each element are different from the atoms of any other element.
5. Atoms of one element may combine with atoms of another element as individual units, usually in the ratio of small, whole numbers, to form chemical compounds.

The agreement between Dalton's theory and the Law of Conservation of Mass is readily apparent. If reacting chemicals consist of atoms of different elements, each with its unique mass, and these atoms can neither be created nor destroyed in a chemical reaction, but merely rearranged, it follows that the total mass after the reaction must equal the total mass before the reaction. The Law of Definite Composition is also compatible with the theory. If there is a fixed atom ratio by which two elements combine, and if all atoms of a given element have the same mass, then the mass ratio must also be fixed.

As with many newly proposed explanations for natural phenomena, Dalton's suggestions were not readily accepted. From them, however, came a third law which *must* be true if the atomic theory is correct. The **Law of Multiple Proportions** states that when two elements combine to form more than one compound, the different weights of one that combine with a fixed weight of the other are in a simple ratio of whole numbers. This law is described in Figure 4.1, on the next page. Experimental efforts to test this prediction confirmed it, and the atomic theory became the most promising of the ideas then entertained about the structure of matter.

## 4.3 SUBATOMIC PARTICLES

PG   4 B   Identify the postulates of the original Dalton atomic theory that are no longer considered valid and explain why.

    4 C   Identify the three major subatomic particles by charge and mass.

    4 D   Explain what isotopes of an element are and how they differ from each other.

Despite the success of the Law of Multiple Proportions in promoting acceptance of the atomic theory in general, the theory was soon to be challenged in some of its details. As early as the 1830s laboratory experiments suggested that the atom, while still the smallest particle of an element that

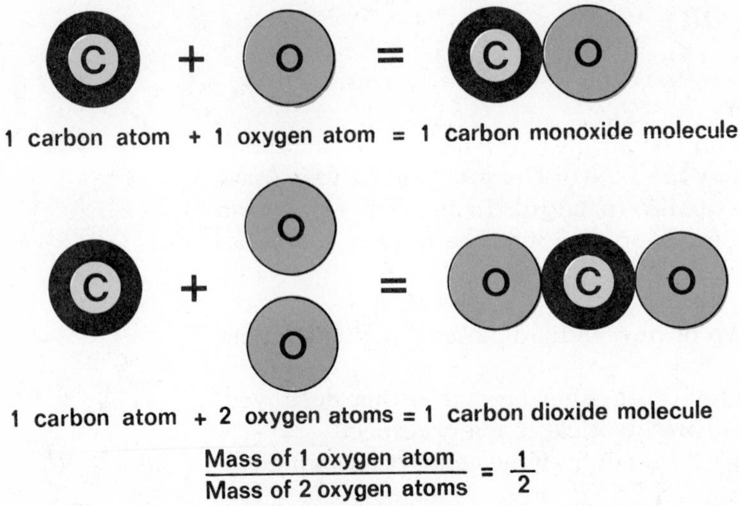

1 carbon atom + 1 oxygen atom = 1 carbon monoxide molecule

1 carbon atom + 2 oxygen atoms = 1 carbon dioxide molecule

$$\frac{\textbf{Mass of 1 oxygen atom}}{\textbf{Mass of 2 oxygen atoms}} = \frac{1}{2}$$

**Figure 4.1** Example of the Law of Multiple Proportions. Carbon and oxygen combine to form two compounds, carbon monoxide and carbon dioxide. A unit particle of a chemical compound is called a molecule (see page 95). A carbon monoxide molecule consists of one carbon atom and one oxygen atom; a carbon dioxide molecule has one carbon atom and two oxygen atoms. Considering both molecules, for a fixed number of carbon atoms—one in each molecule—the ratio of oxygen atoms is one to 2, or $\frac{1}{2}$. If all oxygen atoms have the same mass, which will be represented by M, the mass ratio is also $\frac{1}{2}$:

$$\frac{\text{M grams (one atom)}}{2\text{ M grams (two atoms)}} = \frac{1}{2}.$$

The same ratio results from any number, N, or carbon atoms. N carbon atoms in carbon monoxide will combine with N oxygen atoms having a total mass of $1 \times N \times M$; N carbon atoms (an equal mass of carbon) in carbon dioxide will combine with $2 \times N$ oxygen atoms having a total mass equal to $2 \times N \times M$. In the two compounds the ratio of oxygen masses for any fixed number (mass) of carbon atoms is $\frac{1 \times N \times M}{2 \times N \times M} = \frac{1}{2}$.

retains the identity of the element, is made up of even smaller particles, or subatomic particles. Today there is overwhelming evidence that atoms of all elements are made up of different combinations of many kinds of subatomic particles. Only three are of interest within the scope of an introductory chemistry course. These are the **electron, proton and neutron.**

The electron was first suspected to be a part of an atom because of the electrochemical experiments of Michael Faraday through the 1830s. For want of a better term, he called them "charged corpuscles." The electron was first identified experimentally by J. J. Thomson in 1897. It is a negatively charged particle, and its charge is apparently the smallest unit of electric charge possible. That charge is assigned the value −1 electron unit.

The proton was isolated in 1919 by Ernest Rutherford. It is about 1837 times as massive as the electron, and carries a positive charge, +1, equal in magnitude to the negative charge of the electron. The neutron was officially discovered by James Chadwick in 1932. As its name suggests, it is electrically neutral, and it has a mass just slightly greater than that of the proton. Actual masses, symbols by which the subatomic particles are represented, and other information are summarized in Table 4.1.

It took more than a hundred years for another of Dalton's major postulates to fall. All atoms of a given element are not identical; some have

TABLE 4.1   Subatomic Particles

| SUBATOMIC PARTICLE | SYMBOL | FUNDA- MENTAL CHARGE | MASS | | LOCATION | DISCOVERED |
|---|---|---|---|---|---|---|
| | | | *Grams* | *amu($C^{12}$ = 12.0000)* | | |
| Electron | $e^-$ | $-1$ | $9.107 \times 10^{-28}$ | $0.000549 \approx 0$ | Outside nucleus | 1897 Thomson |
| Proton | p or $p^+$ | $+1$ | $1.672 \times 10^{-24}$ | $1.00728 \approx 1$ | Inside nucleus | 1919 Rutherford |
| Neutron | n or $n^0$ | $0$ | $1.675 \times 10^{-24}$ | $1.00867 \approx 1$ | Inside nucleus | 1932 Chadwick |

more mass than others. This is because the number of neutrons in one atom of an element may be different from the number of neutrons in another atom of the same element. Neutrons being one of the more massive particles in an atom, differences in their number account for the differences in atomic mass. **Atoms of the same element that have different masses are called isotopes of that element.** We shall consider isotopes more closely when we examine atomic weight in Chapter 6.

All atoms of an element have the same number of protons, that number being called the **atomic number** of the element. In a neutral atom the number of electrons is the same as the number of protons—the atom would not be neutral if they were different.

## 4.4   THE PERIODIC TABLE

PG 4 E   Referring to a periodic table, distinguish between periods and groups, and identify them by number.

At this point we interrupt briefly the discussion of atomic structure to introduce the **periodic table.** At least what is written here will seem to be an interruption because there will be no mention of the atom. When you reach the end of the chapter, though, you will see the rather remarkable connection between this table and the structure of the atom.

As knowledge of atoms and elements increased during the 19th century it was logical for chemists to seek a pattern in which this information could be ordered and organized. Their efforts led, in 1869, to one of those remarkable coincidences that are found throughout the history of science. Two men, Dmitri Ivanovich Mendeleev and Lothar Meyer, working independently, announced almost simultaneously that when elements are arranged according to increasing atomic weight (see page 118), certain properties recur in a periodic fashion. Mendeleev reached this conclusion by studying chemical properties, while Meyer examined physical properties. This generalization has since become known as the Periodic Law.

Based on their observations, Mendeleev and Meyer arranged the elements in tables such that elements having similar properties appeared in the same column or row—the first Periodic Tables of the elements. The arrangements were not perfect: it was noted that, in order for all the elements to fall into the proper groups it was necessary to interchange a few elements, interrupting the orderly increase in atomic weights. This was partly because of errors in atomic weights as they were known in 1869. More importantly, nearly 50 years later it was found that the correct ordering property was the atomic number, the number of protons in the atoms of each element, rather than atomic weight.*

As Dalton used atomic theory to predict successfully the Law of Multiple Proportions, so Mendeleev used his periodic table to predict the existence and properties of elements unknown in 1869. Noting certain "blanks" in the table, he reasoned that the blank space was there simply because nobody had yet discovered the corresponding element. He then proceeded to forecast the properties the element would have when discovered, doing this by averaging the properties of the elements above and below or on each side of the unknown element. One of the elements about which he made these predictions is germanium. The predictions and the presently accepted values of the properties of germanium are summarized in Table 4.2.

The modern periodic table may be found inside the front cover of this textbook. Each element is assigned a box in this table. Each box contains the atomic number of the element on top, the symbol that represents the element in the middle, and the atomic weight on the bottom (see Fig. 4.2). Elements of similar chemical properties are arranged vertically in the periodic table in **chemical families** or **groups**. The groups are numbered across the top of the table. Group IA elements, at the far left, are known as the alkali metal family, and the group of elements at the far right are the noble gases. Horizontal rows in the table are called **periods**. The periods vary in length, as may be seen by examining the table.

Later in this chapter you will see that the rather unusual shape of the periodic table is dictated by the arrangement of electrons in the atom. In the next chapter we'll look more closely into elemental symbols and atomic weights. You will also learn to use the table as a source of information extending well beyond the three items printed in each box.

---

*The atomic weight and atomic number sequences are perfectly matched except for three pairs of elements. These are argon (at. no. 18–at. wt. 39.948) and potassium (at. no. 19–at. wt. 39.102); cobalt (at. no. 27–at. wt. 58.9332) and nickel (at. no. 28–at. wt. 58.71); and tellurium (at. no. 52–at. wt. 127.60) and iodine (at. no. 53–at. wt. 126.9044).

TABLE 4.2   The Predicted and Observed Properties of Germanium

| PROPERTY | PREDICTED BY MENDELEEV | OBSERVED |
|---|---|---|
| Atomic weight | 72 | 72.60 |
| Density of metal | $5.5 \text{ g/cm}^3$ | $5.36 \text{ g/cm}^3$ |
| Color of metal | dark gray | gray |
| Formula of oxide | $GeO_2$ | $GeO_2$ |
| Density of oxide | $4.7 \text{ g/cm}^3$ | $4.703 \text{ g/cm}^3$ |
| Formula of chloride | $GeCl_4$ | $GeCl_4$ |
| Density of chloride | $1.9 \text{ g/cm}^3$ | $1.887 \text{ g/cm}^3$ |
| Boiling point of chloride | below 100°C | 86°C |
| Formula of ethyl compound | $Ge(C_2H_5)_4$ | $Ge(C_2H_5)_4$ |
| Boiling point of ethyl compound | 160°C | 160°C |
| Density of ethyl compound | $0.96 \text{ g/cm}^3$ | Slightly less than $1.0 \text{ g/cm}^3$ |

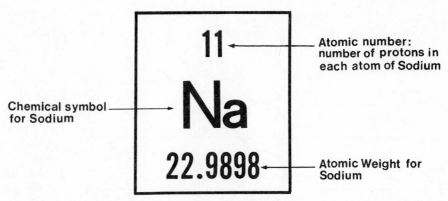

**Figure 4.2**  Sample box from the periodic table, showing the atomic number, symbol and atomic weight of the element sodium.

## 4.5   THE NUCLEAR ATOM

PG   4  F   Describe the nuclear model of the atom, based on the Rutherford scattering experiment.

In 1911, Ernest Rutherford and his students performed a series of experiments that are described by Figures 4.3 and 4.4. The results of these experiments led to the following conclusions:

1.  Every atom contains an extremely small, extremely dense nucleus.

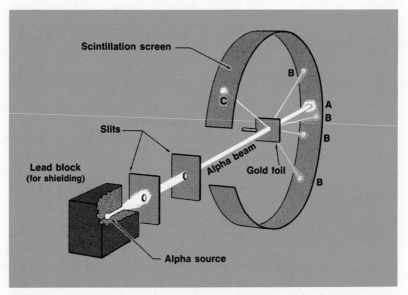

**Figure 4.3**  Rutherford scattering experiment. Using a natural radioactive source, a narrow beam of alpha particles (helium atoms stripped of their electrons) was directed at a very thin gold foil. Most of the particles passed right through the foil, striking a fluorescent screen at A, causing it to glow. Some of the particles were deflected, striking the screen at points such as those labeled B. The larger deflections were surprises, but the 0.001% of the total that were reflected at acute angles, C, were totally unexpected. Similar results were observed using foils of other metals.

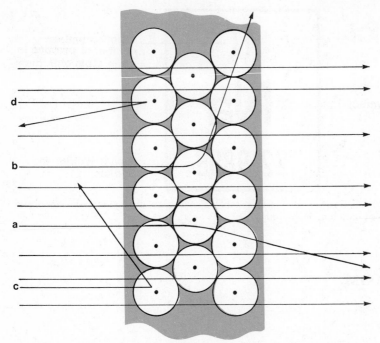

**Figure 4.4** Interpretation of the Rutherford scattering technique. The atom is pictured as consisting mostly of open space. At the center is a tiny and extremely dense nucleus that contains all of the positive charge of the atom and nearly all of the mass. The electrons are thinly distributed throughout the open space. Most of the positively charged alpha particles pass through the open space undeflected, not coming near any gold nuclei. Those that would pass close to a nucleus (a and b) are repelled by electrostatic force and thereby deflected. The few particles that are on a "collision course" with gold nuclei are repelled backward at acute angles (c and d). Calculations based on the results of the experiment indicated that the diameter of the open-space atom is from 10,000 to 100,000 times greater than the diameter of the nucleus.

2. The nucleus accounts for all of the positive charge and nearly all of the mass of the atom.
3. The nucleus is surrounded by a much larger volume of nearly empty space.
4. The space outside the nucleus is very thinly populated by electrons, the total charge of which exactly balances the positive charge of the nucleus.

This description of an atom is called the **nuclear model of the atom.**

No one was more startled at the unexpected results of the Rutherford experiment than Rutherford himself. According to the then popular concept of an atom, some slight deflection of the alpha particle beam was anticipated. But bouncing almost directly back off the gold foil was totally unexpected. Describing the experiment, Rutherford is quoted as saying, "It was almost as incredible as if you fired a 15-inch shell at a piece of tissue paper and it came back and hit you." The conclusions demanded by the experiment— the relative size of the whole atom compared with its nucleus—were no less inconceivable. If the nucleus of an atom were the size of a pea, the distance between it and its nearest nucleus neighbor would be as much as 6/10 of a mile, or nearly a kilometer. Furthermore, if it were possible to eliminate all the space around the nucleus and fill a sphere the size of a period on this page with gold nuclei, that sphere would weigh more than one million tons!

When protons and neutrons were later discovered it was concluded that these relatively massive particles make up the nucleus of the atom. But the electrons were already known in 1911, and it was natural to wonder what they did in that vast open space they occupied. The most widely held opinion was that they traveled in circular orbits around the nucleus, much as planets move in orbits around the sun. The atom then would have the character of a miniature solar system. This is called the **planetary model of the atom.**

## 4.6 THE BOHR MODEL OF THE HYDROGEN ATOM

PG  4 G   Describe the Bohr model of the hydrogen atom.

   4 H   Explain the meaning of quantized energy levels in an atom and show how these levels are related to the discrete lines in the spectrum of that atom.

   4 I   Distinguish between ground state and excited state.

In 1913 the Danish physicist Niels Bohr proposed a model of the hydrogen atom which proved to be a major breakthrough in thinking about atomic structure. His suggestion drew upon many things, some of them seeming to be quite removed from the arrangement of particles in the atom. Among them were the following:

1.  When white light is passed through a prism, it produces a **continuous spectrum** (spreading out of a beam of light in an orderly arrangement of the component colors of the beam according to wavelength). When light originating in a particular element passes through a prism it yields a spectrum consisting of separate, individually distinct or *discrete* lines. See Figures 4.5 and 4.6.

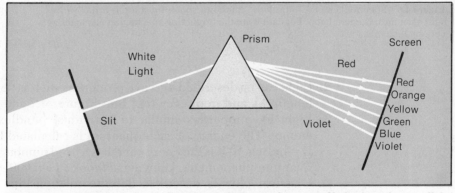

**Figure 4.5**  Spectrum of white light. White light is a combination of light waves having all the colors of the rainbow. When the light passes through a prism, the colors are separated into a spectrum, which may be projected onto a screen.

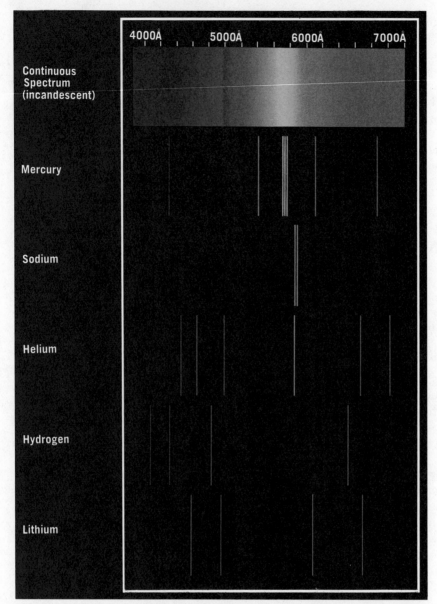

**Figure 4.6** Spectrum chart. Continuous "rainbow" spectrum at top is from "white" light of an incandescent lamp. Beneath it are the bright line spectra that characterize certain elements.

2. Light can be described by wave properties such as velocity (c), wavelength ($\lambda$), and frequency ($\nu$).* See Figures 4.7 and 4.8.
3. Light has properties similar to individual bundles, or quanta, of energy. The energy of each quantum is calculated by the equation $E = h\nu$, where E is the energy and h is a fixed number associated with quantum phenomena, known as *Planck's constant.*
4. Classical physics describes definite relationships between radii, speed, energy and forces when one object moves in an orbit around

---

*$\lambda$ is the Greek letter lambda, and $\nu$ is the Greek letter nu. Do not confuse $\nu$ with v.

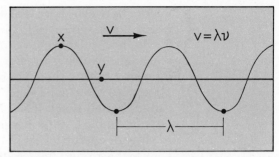

**Figure 4.7** Wave properties. Waves may be represented mathematically by the curve shown. Water waves approximate the mathematical representation physically, while other wave phenomena have quite different appearances. All waves have measurements such as wavelength, $\lambda$ (the Greek letter lambda), the distance between corresponding points on consecutive waves; velocity, v, or, in the case of light, c, the linear speed of a point x on a wave; and frequency, $\nu$ (the Greek letter nu), the number of wave cycles that pass a point, y, each second. Velocity, wavelength and frequency are related by the equation v = $\lambda \nu$, or, in the case of light, c = $\lambda \nu$.

another, as in the planetary model of the atom. Physics also furnishes information about electrostatic attraction between plus and minus charges.

From these beginnings Bohr made a bold assumption. He assumed that the energy possessed by the electron and the radius of its orbit are **quantized. Quantized energy levels mean that the electron in the atom may have, at any instant, any one of certain discrete energies, but at no time may it have an**

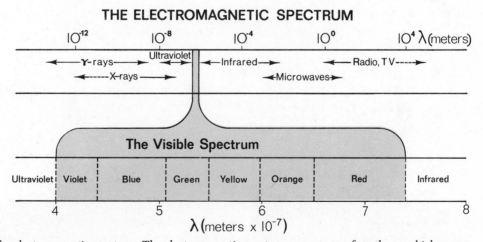

**Figure 4.8**   The electromagnetic spectrum. The electromagnetic specture covers a span from the very high energy shortwave gamma rays ($\lambda = 10^{-10}$ meters or less) to the low energy longwave radio waves at $10^4$ meters or more. The visible spectrum is but a small portion of the electromagnetic spectrum, covering wavelengths of $4 \times 10^{-4}$ to $7 \times 10^{-4}$ meters, approximately.

**energy between those discrete energies.** An example of something that is quantized is shown in Figure 4.9.

Bohr calculated the values of his quantized radii and energy levels using equations involving an integer—1, 2, 3, . . . and so forth. The results for the integers 1 to 4 are shown in Figure 4.10. The lowest energy level, when n = 1, is called the **ground state.** Higher energy levels, when n = 2 or more, are called **excited states.** The electron is normally found in the ground state, but if the atom absorbs energy the electron is "excited," or raised to one of the higher energy levels. It cannot stay at that level, and immediately falls back to the ground state, releasing in the process the energy hν that is equal to the difference in energy between the two energy levels. Sometimes this energy falls in the visible range of the spectrum, and this accounts for some of the discrete lines observed in the spectra of many elements. Using different values of n, Bohr was able to calculate and predict some of these energies that had not yet been observed. (This is the third time in this chapter that predictions of the unobserved have been mentioned—the Law of Multiple Proportions and undiscovered elements being the others.) When these predictions were confirmed experimentally, his calculations were found to be correct to within one part per thousand. It certainly seemed that Bohr had found the answer to the structure of the atom.

There were problems, however. First, hydrogen is the *only* neutral element that fits the Bohr model of the atom. Thus the model fails for an atom with more than one electron. Second, it was a well known fact that a charged body moving in a circle must radiate energy. This meant the electron itself should lose energy, and promptly—in about 0.0000000001 second!—crash into the nucleus. To do otherwise would be to violate the Law of Conservation of Energy (page 19). So for a period of 13 years science did what it does so often when faced with contradictions. It accepted a theory known to be not completely correct. It worked on the faulty parts, modified them, improved them and ultimately replaced them with better concepts.

Niels Bohr made two huge contributions to the development of modern atomic theory, contributions without which one can only wonder how our

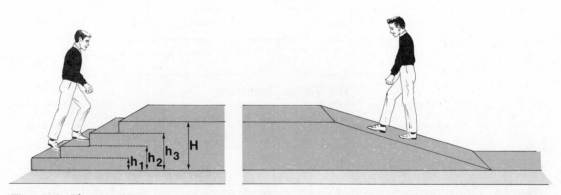

**Figure 4.9** The quantization, or quantum, concept. Man on ramp can stop at any level above ground; his elevation above ground is not quantized. Man on stairs can stop only on a step; his elevation is quantized at $h_1$, $h_2$, $h_3$ or H.

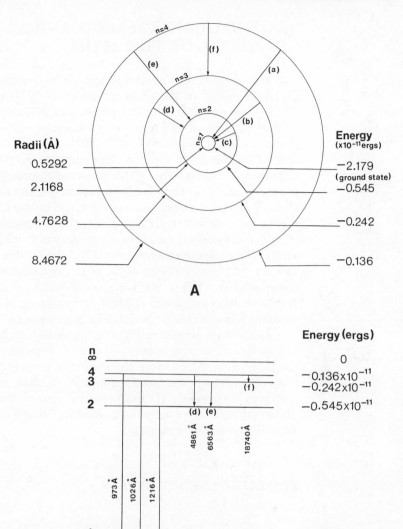

**Figure 4.10** The hydrogen atom based on the Bohr model. According to this model the electron was "allowed" to circle the nucleus only in certain radii, the first four of which are shown in A. At these radii the electron possesses energies indicated in A and plotted in B. An electron in the ground state—n =1 level— can absorb the exact amount of energy to elevate it to any other level, e.g., n = 2, 3 or 4. An electron at an excited state is unstable and drops back to the n = 1 level by any combination of energy jumps possible, radiating energy with each jump. Energy jumps a–c are in the ultraviolet portion of the spectrum, d and e are in the visible range, and f is in the infrared region.

present concepts would have evolved. First, he proposed a reasonable explanation for the atomic spectra of the elements and showed that those spectra can be interpreted in terms of electron energies. Second, he introduced the idea of quantized electron energy levels in the atom. These levels are retained in modern theory as **principal energy levels,** and identified by the **principal quantum number,** n.

## 4.7 THE QUANTUM MECHANICAL MODEL OF THE ATOM

The first major breakthrough in overcoming the shortcomings of the Bohr model of the atom occurred in 1924 when a French physicist, De Broglie, suggested that matter in motion has properties normally associated with waves, and that these properties were significant in subatomic-size particles. The period 1925–1928 saw the development of the **quantum mechanical model** of the atom, the heart of which is the Schroedinger wave equation. As of this writing, the quantum mechanical model has been tested for nearly half a century. It explains more satisfactorily than any other theory all experimental observations to date, and no experimental exceptions to the theory have appeared. It represents the accepted theory of 1978.

Unfortunately the quantum mechanical model does not give a very good idea of the appearance of an atom. One physicist notes, "Indeed, the modern theoretical physicist goes so far as to say that the question, 'What does an atom look like?' has no meaning, much less an answer."* The quantum concept is abstract and mathematical, rather than physical. You might keep this in mind when reading the following paragraphs.

As its name would imply, the quantum mechanical model of the atom retains the concept of quantized energy levels introduced by Bohr. It proposes, however, that more than just the principal quantum number is required to identify the electron energy. In fact, it requires three quantum numbers that specify (1) the principal energy level, (2) the sublevel, and (3) the orbital. These are summarized in Table 4.3, which you may follow as each is examined separately.

### PRINCIPAL ENERGY LEVELS

> PG 4 J Identify the principal energy levels within an atom and state the energy trend among them.

Following the Bohr model, principal energy levels are designated by the principal quantum number, n. Thus we have the first (n = 1), second (n = 2), third (n = 3) and up to seventh (n = 7) principal energy levels, the seventh level being the highest occupied by any ground state electrons among the elements now known. (An older system used the letters K, L, M . . . to designate principal energy levels.)

Within the atom the energy possessed by a specific electron depends upon the principal energy level it is in. An electron in the n = 2 level has a higher energy than one in the n = 1 level; one in n = 3 is at higher energy than one in n = 2. This relationship holds generally, but as we shall see, the principal energy levels for all elements other than hydrogen actually represent a *range* of energies. Beginning with n = 3 and n = 4 the ranges overlap. It is therefore possible for a high n = 3 electron to have more energy than a low n = 4 electron.

---

*White, *Modern College Physics*, D. VanNostrand Co., 1962.

TABLE 4.3   Principal Energy Levels, Sublevels and Orbitals in the
Quantum Mechanical Model of the Atom

| PRINCIPAL ENERGY LEVELS | NUMBER OF SUBLEVELS | IDENTIFICATION OF SUBLEVELS | | | | | | |
|---|---|---|---|---|---|---|---|---|
| n = 1 | 1 | 1s | | | | | | |
| | Orbitals per sublevel | 1 | | | | | | |
| n = 2 | 2 | 2s | 2p | | | | | |
| | Orbitals per sublevel | 1 | 3 | | | | | |
| n = 3 | 3 | 3s | 3p | 3d | | | | |
| | Orbitals per sublevel | 1 | 3 | 5 | | | | |
| n = 4 | 4 | 4s | 4p | 4d | 4f | | | |
| | Orbitals per sublevel | 1 | 3 | 5 | 7 | | | |
| n = 5 | 5 | 5s | 5p | 5d | 5f | 5g | | |
| | Orbitals per sublevel | 1 | 3 | 5 | 7 | 9 | | |
| n = 6 | 6 | 6s | 6p | 6d | 6f | 6g | 6h | |
| | Orbitals per sublevel | 1 | 3 | 5 | 7 | 9 | 11 | |
| n = 7 | 7 | 7s | 7p | 7d | 7f | 7g | 7h | 7i |
| | Orbitals per sublevel | 1 | 3 | 5 | 7 | 9 | 11 | 13 |

Principal energy levels are identified by numbers, 1, 2, 3, . . . 7. They provide for all elements known today. Each principal energy level has a number of sublevels equal to the identifying number of the principal energy level. Only four sublevels, s, p, d and f, are required for the elements now known, but if more are discovered the electrons will occupy the g, h and i sublevels predicted. Sublevels are divided into orbitals. Each s sublevel has 1 orbital, each p sublevel has 3 orbitals, a d sublevel has 5 orbitals, and an f sublevel has 7 orbitals. The sublevels shown in color are not required to accommodate ground state electrons of elements now known.

## SUBLEVELS

PG  4 K   For each principal energy level, state the number of sublevels, identify them by letter, and state the energy trend among them.

According to the quantum mechanical model of the atom, principal energy levels consist of one or more **sublevels** (see Table 4.3). These are identified by the letters **s, p, d** and **f**, letters having their origin in spectroscopy, but now obsolete in that sense. **The total number of sublevels within a given principal energy level is equal to n, the principal quantum**

**number.** Thus for the first principal energy level (n = 1) there is only one sublevel. It is designated the 1s sublevel. The second principal level (n = 2) is subdivided into two sublevels, an s sublevel and a p sublevel, identified as 2s and 2p sublevels, respectively. The d sublevel appears for the first time in the third principal level (n = 3) where there are three sublevels, 3s, 3p and 3d. Finally, the fourth principal level (n = 4) is the first to contain all four sublevels, 4s, 4p, 4d and 4f.

Notice that, *at any given principal energy level,* the *first* sublevel is always an *s* sublevel; the *second* sublevel is a *p* sublevel; the *third* is a *d* sublevel; and the *fourth* is an *f* sublevel, up to the number of sublevels required by the principal quantum number. Observe also the manner in which sublevels are identified. First the principal energy level is designated by number, followed immediately by the letter identifying the sublevel within that principal energy level. When we speak of the "2p" or "4p" sublevels, we refer to the p sublevels within the 2nd and 4th principal levels, respectively. The energy of an electron is specified in the same manner. An electron located in the 2s sublevel would be called a "2s electron"; a "3p electron" would be found in the p sublevel of the 3rd principal level.

Just as electron energies increase through the principal energy levels, so they increase through the sublevels. Within a given principal energy level, electrons in the s sublevel are at lower energy than those in the p sublevel, and the p electron is lower in energy than the d, and the d is lower than the f. This is how it is possible for the principal energy levels to overlap. The highest n = 3 electron is a 3d electron, while the lowest n = 4 electron is a 4s electron. A 3d electron is at higher energy than a 4s electron. But among s electrons, energy always increases from 1s to 2s to 3s. . . . And the same is true among p electrons, as well as d and f electrons, of the increasing principal energy levels.

### ELECTRON ORBITALS

PG  4 L   Sketch the shapes of s and p orbitals.

     4 M   State the number of orbitals in each sublevel.

Being a mathematical rather than a physical model, modern atomic theory tells us that we cannot specify with any reasonable degree of certainty either the position of an electron at a given instant or the "path" that an electron travels. We can, however, develop mathematical equations which predict the *probability* of finding an electron in a particular region of space. These equations picture the electron as forming a negatively charged "cloud" concentrated in a certain location near the nucleus. The shape of the charge cloud depends on the sublevel occupied by the electron. Three dimensional representations of these orbitals, as they are called, are shown in Figure 4.11. An orbital may be considered as a region in which there is a "high probability" of finding an electron at any given instant. Note the uncertainty of the orbital description, stated in terms of probability, compared with the preciseness of the Bohr orbit, which implies exactly where the electron is, where it has been, and where it is going.

Table 4.3 summarizes the sublevels and orbitals in the first seven principal energy levels. Each principal energy level has one s orbital.

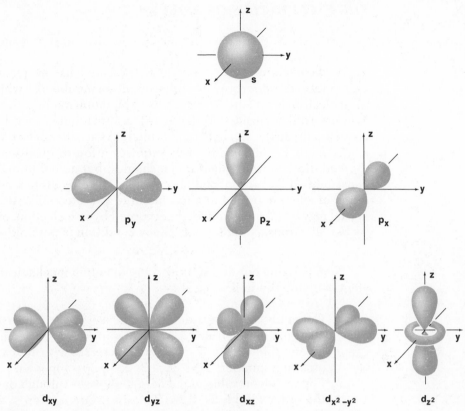

**Figure 4.11** Shapes of electron orbitals according to the quantum mechanical model of the atom.

Figure 4.11 shows all s orbitals to be *spherical* in shape. Within any atom, the 1s orbital is the smallest of the s orbitals. Both the size of an s orbital and the energy of an electron within the orbital increase as the principal quantum number increases from 1s to 2s to 3s . . . up to 7s, the largest s orbital required by any known element.

Table 4.3 and Figure 4.11 indicate three p orbitals when n is equal to or greater than 2. Each p orbital consists of a pair of lobes oriented at right angles to the other two pairs at their common center, the nucleus of the atom. These orientations can be described as lying along the x, y and z axes of the usual coordinate system. As with the s orbitals, both size and energy increase with increasing principal quantum number, n.

The d orbitals shown in Figure 4.11 are more complex, and their shapes will not be of concern within the scope of this text. It is important to note, however, that the first d orbitals appear at n = 3, and they are five in number (see Table 4.3). Furthermore, the energies of the d orbitals increase in the order 3d, 4d, 5d. . . .

Figure 4.11 does not show the f orbitals. Their shapes are quite complex, and very difficult to illustrate clearly on a page. Table 4.3 shows that there are seven f orbitals for each principal quantum number, beginning at n = 4. As with the s, p and d orbitals the energy of the f orbitals increases with increasing values of n.

### THE PAULI EXCLUSION PRINCIPLE

In the beginning of this section we indicated that the quantum mechanical model of the atom provides three quantum numbers by which the energy of an electron is specified. The principal quantum number, n, was identified. While we did not name the other two quantum numbers, we described them conceptually in the sublevels and orbitals. As a matter of fact, what is known as the Pauli Exclusion Principle requires a fourth quantum number, not derived directly from quantum mechanics, but added to make the theory consistent with experimental observations. Its effect is to restrict the population of any one orbital to a maximum of two electrons. At any time, any orbital may be (a) unoccupied, (b) occupied by one electron, or (c) occupied by two electrons. No other occupancy condition is possible.

\* \* \* \* \* \* \* \* \* \*

Summarizing our description of the quantum mechanical model of the atom, we note the following:

1. Generally speaking, energy increases with increasing principal quantum number: $n = 1 < n = 2 < n = 3$. . . . Specifically, for any given sublevel, s, p, d or f, energy increases with increasing principal quantum number: $1s < 2s < 3s$ . . . ; $2p < 3p < 4p$ . . . ; etc.
2. For any given value of n, energy increases through the sublevels in the order s p d f: $2s < 2p$; $3s < 3p < 3d$; $4s < 4p < 4d < 4f$; etc.
3. For any value of n there are n sublevels:

   | n | *Sublevels* |
   |---|---|
   | 1 | 1:  s |
   | 2 | 2:  s, p |
   | 3 | 3:  s, p, d |
   | 4 | 4:  s, p, d, f |

4. The number of orbitals for each sublevel is

   | *Sublevel* | *Orbitals* |
   |---|---|
   | s | 1 |
   | p | 3 |
   | d | 5 |
   | f | 7 |

5. An orbital may be occupied by 0, 1 or 2 electrons, but never more than 2.


## 4.8  ELECTRON CONFIGURATION

### ELECTRON ENERGY LEVEL DIAGRAM

From the summary of the foregoing section we may now construct an electron energy level diagram in which electron orbitals are arranged in order of increasing energy. This is essentially a plot of relative orbital energies versus quantum number, as shown in Figure 4.12. The lowest energy orbital of all is the 1s orbital, and it is the only orbital at the first principal

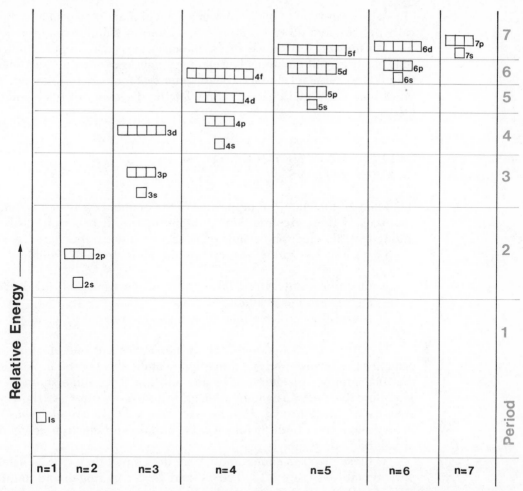

**Figure 4.12** Electron energy level diagram. Each column contains one box for each orbital in the n = 1, n = 2, n = 3 . . . n = 7 principal energy levels. The boxes are grouped by sublevels s, p, d and f, as far as required within each principal energy level. All sublevels are positioned vertically on the page according to a general energy level scale shown at the left. Any orbital that is higher on the page than a second orbital is therefore higher in energy than the second orbital. In filling orbitals from the lowest energy level, each orbital will generally hold 2 electrons before any orbital higher in energy accepts an electron. The colored lines and the scale at the right correlate energies of the sublevels with the periods of the periodic table.

energy level. This orbital is represented by a box at the bottom of the n = 1 column of the diagram.

The next lowest principal energy level is n = 2, where we must account for two sublevels, 2s and 2p. Between them, the 2s sublevel has the lower energy, so it is placed below the 2p sublevel in the n = 2 column. Again a single box is shown for the single 2s orbital, but three boxes are provided for the 2p sublevel because there are three 2p orbitals.

When n = 3 we provide one orbital box for the single 3s orbital, three for the 3p orbitals and five for the 3d orbitals, each positioned at increasing energies. For n = 4 we must extend the sequence to seven boxes for the seven 4f orbitals. It is here that we encounter the principal energy level overlap mentioned previously. Specifically, the highest n = 3 sublevel, 3d, has a higher energy than the lowest n = 4 sublevel, 4s. This, as well as several other overlaps in the higher values of n, appears in Figure 4.12.

As we continue to build our energy level diagram we should theoretically provide five sublevels for n = 5, six for n = 6 and seven for n = 7. We find this unnecessary, however, inasmuch as we can account for the ground state electron configuration of all elements now known at energies no higher than the 6d sublevel. In fact, the diagram that appears in Figure 4.12 provides enough orbital boxes to allow for the discovery of twelve new elements.

## DEVELOPMENT OF ELECTRON CONFIGURATIONS

Many of the chemical properties of the elements may be explained on the basis of their **electron configuration,** or the distribution of electrons throughout the electron orbitals of a gaseous atom in its ground state.

Two principles guide our assignment of electrons to orbitals:

1. At ground state, the total number of electrons will fill the *lowest* energy orbitals available.
2. No orbital can have more than two electrons.

**1s.** An atom of hydrogen (H, at. no. 1) has but one electron, and it occupies the lowest energy orbital in any atom, the 1s orbital. We indicate the total number of electrons in any sublevel by a superscript number. Therefore the electron configuration of hydrogen is written $1s^1$, showing one electron in the 1s orbital. Helium (He, at. no. 2) has two electrons in the neutral atom, and both fit into the 1s orbital. The electron configuration of helium is therefore $1s^2$.

As these electron configurations are developed they will be listed in a periodic table, Figure 4.13. The $1s^1$ and $1s^2$ for hydrogen and helium are shown in their spaces in the table.

**2s.** Lithium (Li, at. no. 3) has three electrons. The first two will fill the 1s orbital as before. The third electron goes to the next orbital up the energy scale with a vacancy, which Figure 4.12 shows to be the 2s orbital. The electron configuration for lithium is therefore $1s^2 2s^1$. In a similar manner beryllium (Be, at. no. 4) divides its four electrons between the two lowest orbitals, filling them both: $1s^2 2s^2$. These two configurations are entered into Figure 4.13 also.

**2p.** The first four electrons of boron (B, at. no. 5) will fill 1s and 2s orbitals. The fifth electron will go to the next highest level, the 2p, according to Figure 4.12. The configuration for boron is $1s^2 2s^2 2p^1$. Similarly carbon (C, at. no. 6) has a configuration $1s^2 2s^2 2p^2$. Though we shall not emphasize the point, it is recognized that the two 2p electrons will occupy different p orbitals. Electrons will half-fill all the orbitals in a sublevel before completely filling any of them. The next four elements increase the number of electrons in the three 2p orbitals until they are filled with six electrons for neon (Ne, at. no. 10): $1s^2 2s^2 2p^6$. All these configurations appear in Figure 4.13.

**3s AND 3p.** The first ten electrons of sodium (Na, at. no. 11) will distribute themselves among the 1s, 2s and 2p orbitals, as did those of neon. The eleventh will be in the 3s orbital, giving the total configuration $1s^2 2s^2 2p^6 3s^1$.

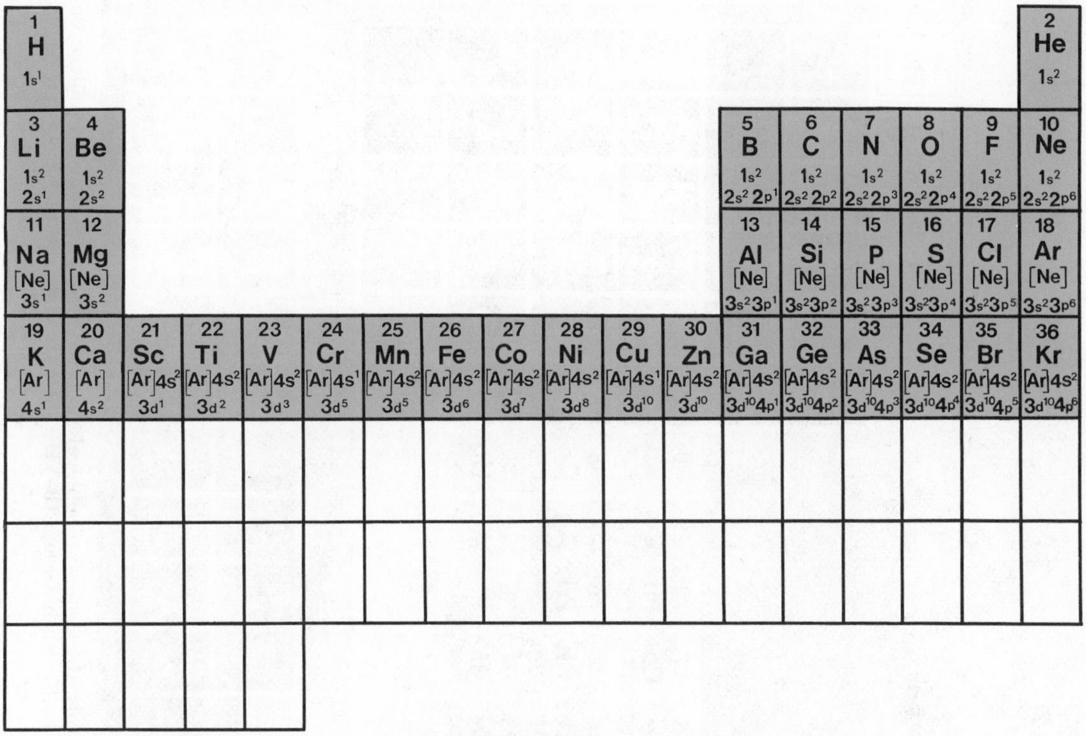

| 1 | 2 | 3 | 4 | 5 | 6 | 7 | 8 | 9 | 10 | 11 | 12 | 13 | 14 | 15 | 16 | 17 | 18 |
|---|---|---|---|---|---|---|---|---|---|---|---|---|---|---|---|---|---|
| 1 **H** $1s^1$ | | | | | | | | | | | | | | | | | 2 **He** $1s^2$ |
| 3 **Li** $1s^2$ $2s^1$ | 4 **Be** $1s^2$ $2s^2$ | | | | | | | | | | | 5 **B** $1s^2$ $2s^2 2p^1$ | 6 **C** $1s^2$ $2s^2 2p^2$ | 7 **N** $1s^2$ $2s^2 2p^3$ | 8 **O** $1s^2$ $2s^2 2p^4$ | 9 **F** $1s^2$ $2s^2 2p^5$ | 10 **Ne** $1s^2$ $2s^2 2p^6$ |
| 11 **Na** [Ne] $3s^1$ | 12 **Mg** [Ne] $3s^2$ | | | | | | | | | | | 13 **Al** [Ne] $3s^2 3p^1$ | 14 **Si** [Ne] $3s^2 3p^2$ | 15 **P** [Ne] $3s^2 3p^3$ | 16 **S** [Ne] $3s^2 3p^4$ | 17 **Cl** [Ne] $3s^2 3p^5$ | 18 **Ar** [Ne] $3s^2 3p^6$ |
| 19 **K** [Ar] $4s^1$ | 20 **Ca** [Ar] $4s^2$ | 21 **Sc** [Ar] $4s^2$ $3d^1$ | 22 **Ti** [Ar] $4s^2$ $3d^2$ | 23 **V** [Ar] $4s^2$ $3d^3$ | 24 **Cr** [Ar] $4s^1$ $3d^5$ | 25 **Mn** [Ar] $4s^2$ $3d^5$ | 26 **Fe** [Ar] $4s^2$ $3d^6$ | 27 **Co** [Ar] $4s^2$ $3d^7$ | 28 **Ni** [Ar] $4s^2$ $3d^8$ | 29 **Cu** [Ar] $4s^1$ $3d^{10}$ | 30 **Zn** [Ar] $4s^2$ $3d^{10}$ | 31 **Ga** [Ar] $4s^2$ $3d^{10}4p^1$ | 32 **Ge** [Ar] $4s^2$ $3d^{10}4p^2$ | 33 **As** [Ar] $4s^2$ $3d^{10}4p^3$ | 34 **Se** [Ar] $4s^2$ $3d^{10}4p^4$ | 35 **Br** [Ar] $4s^2$ $3d^{10}4p^5$ | 36 **Kr** [Ar] $4s^2$ $3d^{10}4p^6$ |

**Figure 4.13** Ground state electron configurations of neutral gaseous atoms.

Because all electron configurations for elements having atomic numbers greater than 10 begin with the configuration for neon, $1s^2 2s^2 2p^6$, this part of such configurations is frequently shortened to the **neon core**, [Ne], and is so recorded in Figure 4.13.

The development of the configuration for sodium at the 3s level is a repetition of the development for lithium at the 2s level. In fact, all the elements of the third period repeat the elements of the second period just above them, yielding the configurations shown in Figure 4.13.

**4s.** Potassium (K, at no. 19) repeats at the 4s level the development of sodium at the 3s level. Its complete configuration is $1s^2 2s^2 2p^6 3s^2 3p^6 4s^1$. All configurations for atomic numbers greater than 18 distribute their first 18 electrons in the configuration of argon (Ar, at. no. 18), $1s^2 2s^2 2p^6 3s^2 3p^6$, which may be shortened to the **argon core**, [Ar]. Accordingly the configuration for potassium may be written [Ar]$4s^1$, and calcium (Ca, at. no. 20) is [Ar]$4s^2$.

**3d.** Figure 4.12 predicts that five 3d orbitals are next available for electron occupancy, and at two electrons per orbital they should accommodate the next ten elements. The first three of these, scandium, Sc, titanium, Ti, and vanadium, V, atomic numbers 21–23, fill in order as predicted. The configuration for vanadium is $1s^2 2s^2 2p^6 3s^2 3p^6 4s^2 3d^3$, or [Ar]$4s^2 3d^3$.* Chromium (Cr, at. no. 24) breaks the orderly sequence in which lowest energy orbitals

*Some chemists prefer to write this configuration $1s^2 2s^2 2p^6 3s^2 3p^6 3d^3 4s^2$, or [Ar]$3d^3 4s^2$, putting the 3d before the 4s. This is an equally acceptable alternative. There is perhaps some advantage at this point in listing the sublevels in the order in which they fill, so we shall continue to use this order in this text.

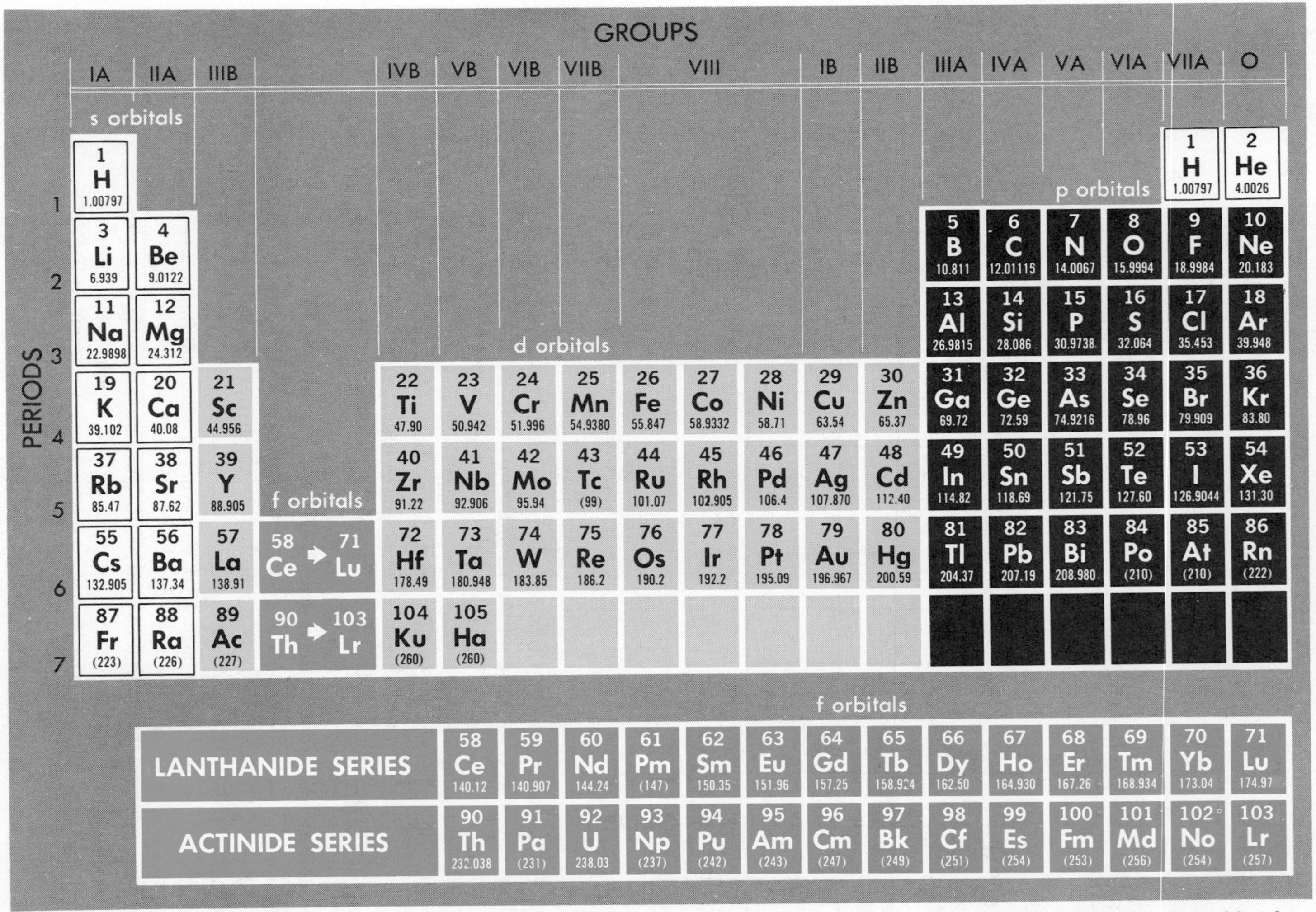

**Figure 4.14** Arrangement of periodic table according to atomic sublevels. Highest energy sublevels occupied at ground state are s sublevels in Groups IA and IIA; p sublevels in Groups IIIA–O, except for helium; d sublevels in Groups IB–VIIB and VIII; and f sublevels in the lanthanide and actinide series.

are filled first. Its configuration is [Ar]$4s^1 3d^5$, rather than [Ar]$4s^2 3d^4$, as would be expected. This is generally attributed to a unique stability associated with a sublevel being half-filled or completely filled. Manganese (Mn, at. no. 25) puts us back on the track, only to be derailed again at copper (Cu, at. no. 29): [Ar]$4s^1 3d^{10}$. Zinc (Zn, at. no. 30) has the expected configuration of [Ar]$4s^2 3d^{10}$. Examine the sequence for atomic numbers 21–30 in Figure 4.13 and note the two exceptions.

**4p.** By now the pattern should be clear; you should expect atomic numbers 31–36 to fill in sequence the next orbitals available, which are the 4p orbitals. They do, as shown in Figure 4.13.

We shall end our consideration of electron configurations with atomic number 36, krypton. Were we to continue, we would find the higher s and p orbitals would fill in perfect accord with the procedures used thus far. The 4d, 4f, 5d and 5f orbitals have several variations such as we encountered with chromium and copper, so the only reliable way to determine configurations of those elements is to look them up. But you should be able to reproduce easily the configurations for the first 36 elements—not from memory, but from their obvious correlation with the periodic table.

Notice first that Figure 4.13 shows that specific sublevels are being filled in different regions of the periodic table. This is indicated by colors in Figure 4.14. In Groups IA and IIA the s sublevels are being filled. Disregard the Group VIIA appearance of hydrogen, H, which has some properties similar to elements in that column as well as those in Group IA. The p orbitals are filled in sequence across Groups IIIA-O. (Helium, He, a noble gas without p electrons, is an exception.) In the B groups and Group VIII of the table the d electrons appear. Finally, though we shall not be concerned with them, the f electrons appear in those elements isolated as the lanthanide and actinide series of the table.

Notice also that the periodic table, when read from left to right across the periods, gives the order of increasing sublevel energy. The first period gives only the 1s sublevel. Period 2 takes you through 2s (atomic numbers 3 and 4), and 2p (atomic numbers 5–10). Similarly the third period covers the 3s and 3p sublevels. Period 4 starts with 4s, and then follows with 3d (atomic numbers 21–30), before finishing with 4p. These relationships are summarized in Table 4.4. With minor variations the correspondence between the sublevels taken from the periodic table and from the electron energy diagram (Fig. 4.12, page 75) is perfect. *The periodic table therefore replaces the electron energy diagram as a reference point in determining the order of increasing sublevel energy.*

TABLE 4.4  Sublevel Energies

| PERIOD | 1 | 2 | | 3 | | 4 | | |
|---|---|---|---|---|---|---|---|---|
| SUBLEVELS IN INCREASING ENERGY | 1s | 2s | 2p | 3s | 3p | 4s | 3d | 4p |
| ATOMIC NUMBERS | 1, 2 | 3, 4 | 5–10 | 11, 12 | 13–18 | 19, 20 | 21–30 | 31–36 |

If you are ever required to list the sublevels in order of increasing energy without reference to a periodic table, the diagram produced by the sublevels in Table 4.3 is helpful and is easily recalled. Beginning at the upper left, the diagonal lines pass through the sublevels in the sequence required.

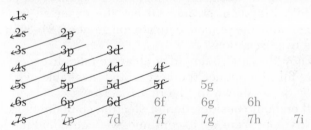

The sublevels shown in color are not required for the elements known today, but the mathematics of the quantum mechanical model predict the order of increasing energy indefinitely.

Finally, observe that, with the exception of chromium and copper, the number of electrons in each s, p and d sublevel increases in numerical sequence from the left to right in any period of the periodic table. As a consequence every element in Group IA has an electron configuration ending in $ns^1$, where n is the principal quantum number of the highest occupied energy level of the atom in its ground state. For atomic number 3 this would be $2s^1$; for atomic number 11, $3s^1$; for atomic number 19, $4s^1$, and so on. Similarly the characteristic highest occupied energy level configuration for Group IIA elements is $ns^2$; for Group IIIA, $ns^2np^1$, and so forth to Group O at $ns^2np^6$.

## WRITING ELECTRON CONFIGURATIONS

PG  4 O   Referring only to a periodic table, write the ground state electron configuration of a gaseous atom of any element up to atomic number 36.

If you can list the sublevels in order of increasing energy, and if you can count, you can write electron configurations. To illustrate, use only the periodic table inside the front cover of your text while considering these examples:

**Example 4.1**  Write the complete electron configuration for chlorine. Cl, atomic number 17. (Do not use the neon core.)

*Solution.*   A stepwise sequence is helpful in writing electron configurations:

1. *Locate the element (chlorine) in the periodic table.*

2. *List the sublevels in order of increasing energy until you reach the orbital that contains the element. Leave space for superscripts.* Do this by reading across the periods in the table from left to right. Atomic numbers 1 and 2 constitute the 1s orbital. In the second period atomic numbers 3 and 4 make up the 2s orbital, and numbers 5–10 constitute the 2p orbital. In row 3 we find atomic numbers 11 and 12 for 3s, and 13–18 for 3p. Chlorine is atomic number 17. Therefore the list is 1s 2s 2p 3s 3p.

3. *Fill in as superscripts the maximum electron populations of all sublevels*

*except the last one* (last two, if the element happens to be chromium or copper). You probably know the maximum electron populations of the s, p and d sublevels by now (2, 6 and 10), but if you forgot you can find them simply by counting across the periodic table. For chlorine this step yields $1s^22s^22p^63s^23p$.

4. *Determine the population of the final sublevel and insert that number as the last superscript.* Start at the beginning of that "sublevel group" in the table and count until reaching the element in question. In this instance chlorine is in the number 13–18 sublevel group, so atomic number 13 would be counted as "one," 14 as "two," 15 as "three," 16 as "four," and 17 as "five." There are five electrons in the 3p sublevel, so the complete configuration is $1s^22s^22p^63s^23p^5$.

5. *Add the superscripts and check the total against the atomic number.* They should be equal, since the number of protons is the same as the number of electrons in a neutral atom. $2 + 2 + 6 + 2 + 5 = 17$, the atomic number of chlorine.

In using a neon or argon core you would locate your element in the table and write the core symbol for that Group O element that lies *before* it in the atomic number sequence. For chlorine, atomic number 17, you would use neon, atomic number 10. Then list the sublevels as in the sequence above. For chlorine: $[Ne]3s^23p^5$.

If your element happens to be chromium or copper you must remember that the 4s orbital has 1 electron and the 3d orbital has 5 electrons for chromium and 10 for copper, both being one more than the counting sequence would suggest.

Now you try one:

---

**Example 4.2** Derive the complete (no Group O core) electron configuration for the element potassium (K, at. no. 19).

First locate potassium in the periodic table. Then list the sublevels in sequence until reaching the highest occupied sublevel for potassium.

---

**4.2a** 1s 2s 2p 3s 3p 4s

Now fill in the maximum electron populations of all sublevels except the last.

---

**4.2b** $1s^22s^22p^63s^23p^64s$

Finally, starting with the beginning of the 4s sublevel in the periodic table, count off until reaching potassium. Fill in the last superscript.

---

**4.2c** $1s^22s^22p^63s^23p^64s^1$

**Example 4.3** Write the complete electron configuration for oxygen, O, atomic number 8.

You may be able to complete this example without intermediate steps.

-----------------------------------------------------------------------------------

**4.3a** $1s^2 2s^2 2p^4$

Listing sublevels until reaching oxygen gives 1s 2s 2p. The 1s and 2s orbitals are filled with two electrons each, and counting from boron (at. no. 5) to oxygen gives 4 electrons in the 2p sublevel.

**Example 4.4** Develop the electron configuration for cobalt, Co, atomic number 27.

This is the first example in which you encounter d electrons. The procedure is the same; go all the way in one step, using a Group O core this time.

-----------------------------------------------------------------------------------

**4.4a** $[Ar]4s^2 3d^7$

The noble gas prior to cobalt in the periodic table is argon. Reading from left to right through the sublevels of the fourth period until reaching cobalt we have 4s and 3d. The 4s orbital is filled with 2 electrons. Counting through the 3d sublevel we have seven, beginning with scandium (at. no. 21) and ending with cobalt.

## 4.9 TRENDS IN THE PERIODIC TABLE

When the periodic table was introduced in Section 4.4 it was noted that Mendeleev and Meyer conceived it in an attempt to "organize" some of the recurring chemical and physical properties of the elements known in 1869. They suggested no reason for these regularities. We now recognize modern atomic theory as the *explanation* that neither Mendeleev nor Meyer could foresee. That they, knowing nothing of electrons, protons, nuclei, quantized energy levels or wave equations, were able to suggest a table that would correspond so perfectly with all of these things nearly 60 years later stands as one more tribute to the genius expressed by man.

In this section we will identify some of the recurring properties that appear in the periodic table. This will lead to the recognition of chemical families, elements that have similar chemical properties. Other regions in the table will be identified, and some trends across the rows and down the columns of the table will be pointed out.

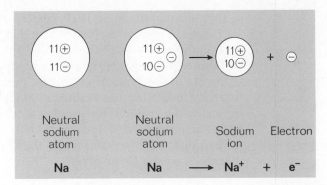

**Figure 4.15** The formation of a sodium ion from a sodium atom.

## IONIZATION ENERGY

A neutral atom of sodium, Na, atomic number 11, has eleven protons in its nucleus and eleven electrons outside the nucleus. (We'll not be concerned with the 12 neutrons also present in most sodium atoms.) Mentally isolate one of the electrons from the rest of the atom. Figure 4.15 will aid you in this exercise. The middle sketch shows the electron still as part of the atom, but separated from the rest of it. The electron has a charge of −1. The rest of the atom, consisting of eleven protons and ten electrons (11 plus charges and 10 minus charges) has a net charge of +1. Being oppositely charged, there is an attractive force between the electron and the rest of the atom (see page 17). Work must be done against this attractive force (see page 18) to separate the charges and remove the electron completely from the atom. The "thing" that is left, with its net charge caused by an unequal number of protons and electrons, is called an **ion.** More generally an ion may be defined as **an atom or group of atoms that has a positive or negative charge because of an excess or deficiency of electrons compared to protons.** The ion that remains after removing an electron from a sodium atom is called a sodium ion. Its symbol is Na⁺.

**The amount of energy required to remove one electron from a neutral gaseous atom of an element is the ionization energy, or ionization potential, of that element.** Ionization potential is one of the more striking examples of the periodic recurrence of elemental properties, particularly when graphed, as in Figure 4.16. Notice the similarity of the shape of the graph between

**Figure 4.16** Ionization potential, plotted as a function of atomic number, to show periodic properties of elements.

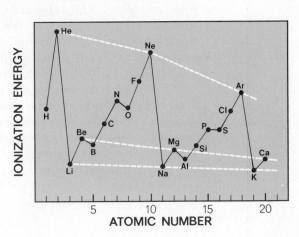

atomic numbers 3 and 10 (Period 2 in the periodic table) and atomic numbers 11 and 18 (Period 3). Notice also that the three peaks are elements in Group O at the right end of the table, and the three low points are from Group IA. Observe the trends: without exception, if you draw lines connecting elements in the same group in the periodic table the lines slant down to the right, indicating lower ionization energies with increasing atomic number within the group. If the graph is extended to include higher atomic numbers the same general shapes and trends are found through the s and p sublevel portions of all periods, while the d and f sublevels introduce less distinct trends among themselves.

### CHEMICAL FAMILIES

PG 4 P Explain, from the standpoint of electron configuration, why certain groups of elements constitute chemical families.

4 Q Identify within the periodic table the following chemical families: alkali metals; alkaline earths; halogens; noble gases.

We will see in Chapter 9 that the formation of chemical bonds involves the "high energy" electrons found in the outer energy levels of the bonded atoms. While d electrons do participate in bond formation, our consideration will be limited to s and p electrons, which are frequently called **valence electrons.** We find that atoms having similar s and p electron configurations appear among the elements in the vertical groups of the periodic table. These groups are therefore called **chemical families.**

ALKALI METALS. The elements in Group IA, with the exception of hydrogen, are known as **alkali metals** (see Fig. 4.17). (Francium, Fr, at. no. 87, is a radioactive element about which we know little, but which we presume behaves as an alkali metal.) Their $ns^1$ electron in the highest occupied energy level is rather loosely held, and therefore easily lost, yielding an ion having a +1 charge. This accounts for most of the chemical properties of the family. As you move down the column in the table the highest energy s electron is found farther from the nucleus. Therefore it is lost more easily and the element is more reactive. This accounts for the lower ionization energy as atomic number increases, indicated above and in Figure 4.16.

As noted, Group IA elements are metals, though they don't normally look like metals. This is because exposure to air leads to the formation of an oxide coating that conceals the bright metallic luster that is visible in a freshly cut sample. The group IA elements possess other common metallic properties too, such as being good conductors of heat and electricity and having good malleability and ductility.

Table 4.5 lists some of the physical properties of the alkali metals, as well as those of other chemical families we will consider. Notice the steady increase in density as the atomic number increases, and the decline in melting and boiling points, including one exception in the regular pattern. (Exceptions are always challenging, and frequently instructive when one seeks out an explanation for them.)

ALKALINE EARTHS. Next to the alkali metals in the periodic table are the **alkaline earths,** a group of elements that are quite similar to the alkali

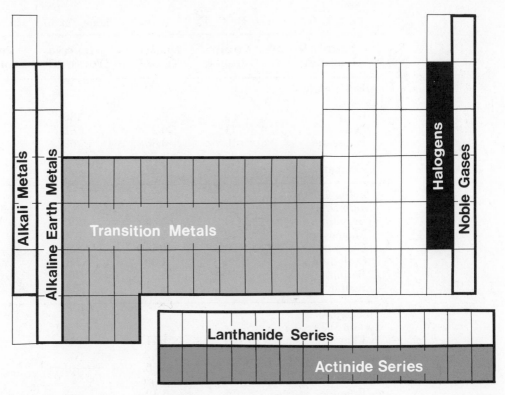

**Figure 4.17** Chemical families and regions in the periodic table.

metals, only "in moderation." All members of Group IIA have an ns² configuration at their highest energy level: 3s² for magnesium, 4s² for calcium and 6s² for barium, to list the best known of the family. Though not as easily lost as the ns¹ electron of the alkali metals, the electrons are given up readily, forming ions with a +2 charge.

Trends comparable to those noted with the alkali metals are apparent also with the alkaline earth metals. Reactivity is again greater for the elements of higher atomic number. Indeed, the reaction between magnesium and water is virtually nonexistent. Physical property trends may also be noted (see Table 4.5).

**HALOGENS.** Four elements of Group VIIA, fluorine, chlorine, bromine and iodine, constitute the family known as the **halogens,** or *salt formers.* (Astatine, a radioactive element about which we know little, probably would fit into the group if we had sufficient information to classify it.) The highest energy electron configuration is of the ns²np⁵ variety: 2s²2p⁵ for fluorine, 3s²3p⁵ for chlorine, etc. These elements tend to form ions by *gaining* one electron, thereby acquiring a −1 charge.

Reactivity trends comparable to but in reverse of those of the alkali and alkaline earth metals may be observed. In general with the nonmetals, reactivity decreases with increasing atomic number. Fluorine is thus the most reactive of the group, and iodine the least. Physical property trends in density and melting and boiling points also appear as shown in Table 4.5.

TABLE 4-5   Selected Physical Properties of Some Families of Elements

| FAMILY & ELEMENTS | ATOMIC NUMBER | DENSITY (g/cm³) | MELTING POINT (°C) | BOILING POINT (°C) |
|---|---|---|---|---|
| *Alkali Metals— Group IA* | | | | |
| Lithium | 3 | 0.53 | 180 | 1326 |
| Sodium | 11 | 0.97 | 98 | 889 |
| Potassium | 19 | 0.86 | 63 | 757 |
| Rubidium | 37 | 1.53 | 39 | 679 |
| Cesium | 55 | 1.87 | 29 | 690 |
| *Alkaline Earths— Group IIA* | | | | |
| Beryllium | 4 | 1.86 | 1283 | 2970 |
| Magnesium | 12 | 1.74 | 650 | 1120 |
| Calcium | 20 | 1.54 | 850 | 1490 |
| Strontium | 38 | 2.60 | 770 | 1384 |
| Barium | 56 | 3.5 | 704 | 1140 |
| Radium | 88 | 5 | 700 | 1500 |
| *Halogens— Group VIIA* | | | | |
| Fluorine | 9 | 1.11 (liq.) | −218 | −188 |
| Chlorine | 17 | 1.56 (liq.) | −101 | −34 |
| Bromine | 35 | 3.12 (liq.) | −7 | 59 |
| Iodine | 53 | 4.93 (sol.) | 114 | 184 |
| *Noble Gases— Group O* | | (g/l, STP) | | |
| Helium | 2 | 0.18 | −270* | −269 |
| Neon | 10 | 0.90 | −249 | −246 |
| Argon | 18 | 1.78 | −189 | −186 |
| Krypton | 36 | 3.74 | −157 | −153 |
| Xenon | 54 | 5.86 | −112 | −108 |
| Radon | 86 | 9.9 | −71 | −62 |

*Measured at ten atmospheres of pressure.

**NOBLE GASES.** Group O in the periodic table constitutes the noble gas family, so called because its elements are generally unreactive. Indeed, until quite recently, it was believed that these elements were totally inert chemically. Since the early 1960s, however, a few compounds of the heavier noble gases have been prepared (e.g., $XeF_6$, $XeF_4$, $KrF_4$); nevertheless, as a group, they are much less reactive than any other elements in the table, a fact which stands out as their singular unifying chemical property. Family trends are clearly evident in the physical properties in Table 4.5.

The chemical inertness of the noble gases is attributed to their $ns^2np^6$ electron configuration at the highest occupied energy level. (Helium, with only two electrons, is obviously an exception. Its electron configuration is $1s^2$.) There appears to be some unique stability—a minimization of energy, perhaps—associated with filled s and p orbitals at the highest energy level. As we shall see, this is the basis of much of our understanding of chemical bonding.

## THE TRANSITION METALS

PG 4 R  Identify the transition elements in the periodic table.

Though not a chemical family in the sense we have been using the word, the elements in the B groups and Group VIII of the periodic table are often referred to as the **transition metals.** These are the elements whose 3d, 4d and 5d sublevels are being filled as one progresses through the electron configuration scheme, thereby "expanding" the periodic table. Included among the transition elements are some of our better known metals: for example, iron, gold, copper, silver, chromium, zinc, nickel and mercury are some. Although the elements within a given transition series differ considerably from each other in chemical properties, they are, as a group, distinguished from the preceding IA and IIA metals by being somewhat less reactive. For example, none of the elements in the first transition series (atomic number 21–30) react with water to produce hydrogen as do potassium and calcium.

Another "expansion" of the periodic table occurs in the 6th and 7th periods. It is fourteen elements long, reflecting the seven f orbitals that are filled in from atomic numbers 58–71, called the *lanthanide series,* and 90–103, the *actinide series.* To save space, these elements are listed separately at the bottom of the table.

## ATOMIC SIZE

PG 4 S  Predict how and explain why atomic size varies as a function of position in the periodic table.

Figure 4.18 shows the sizes of neutral atoms as a function of position in the periodic table. It is clear from the figure that as we move across the

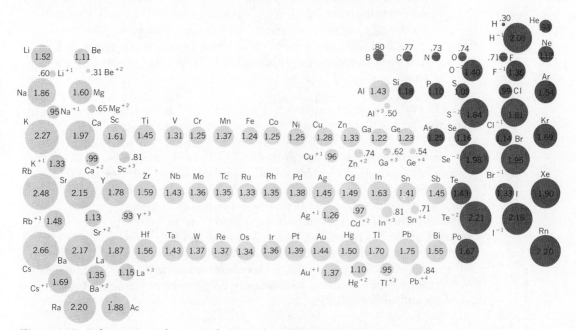

**Figure 4.18**  Relative sizes of atoms and ions in the periodic table. (From Campbell, J. A., *Chemical Systems: Energetics, Dynamics, Structure.* San Francisco: W. H. Freeman & Co., 1970, with permission of the author.)

table from left to right the atoms become smaller. This trend is perhaps most obvious in the 2nd period, where the element at the far left, lithium, has a radius more than twice that of fluorine at the far right. As we move down within a given group in the table, atoms ordinarily increase in size, at least for the Group A elements. We see, for example, that cesium, the last element listed in Group IA, has a radius nearly twice that of lithium.

These observations about atomic size are believed to be the result of three influences:

1. *Number of occupied principal energy levels.* As the number of occupied principal energy levels increases, the atom becomes larger. Electrons in higher energy levels are generally farther from the nucleus. This influence alone accounts for the increase in size as you move down a column in the periodic table.
2. *Nuclear charge.* Moving from left to right across a row of the periodic table—and the second and third rows are the best examples—the highest energy electrons are all in the same principal energy level. This removes from consideration the first influence described above. As the number of protons in the nucleus increases across the row, the magnitude of positive nuclear charge increases, which causes the nucleus to exert a stronger attraction for the outermost electrons, pulling them closer to the nucleus. This accounts for smaller atomic size from left to right across a period of the periodic table.
3. *Shielding effect.* The attraction of the nucleus for the outermost electrons is partially counteracted, or "shielded," by the repulsion of electrons in lower energy levels. This no doubt contributes to the fact that the number of occupied energy levels is more important than nuclear charge in determining atomic size in any group in the periodic table. Sodium, for example, has nearly three times as much nuclear charge as lithium; but the attraction for the outermost sodium electron is partially countered by 10 lower energy 1s, 2s and 2p electrons that are closer to the nucleus, whereas there are only two 1s electrons between the lithium nucleus and its highest energy electron.

One must exercise caution in interpreting tables or drawings that compare the radii or diameters of atomic-size particles. An atom is not something that can be placed on a table and measured directly with a ruler. After all, how can you measure to the end of a cloud—and, relatively speaking, the "edge" of a so-called electron cloud is probably less well defined than that of a sky-type cloud? The measurements are, rather, calculations of internuclear distances from experimental evidence obtained primarily from *bonded* atoms. The sizes of noble gas atoms are particularly questionable; *bonded* noble gas atoms are virtually nonexistent, to say nothing of atomic size information derived therefrom. Other data lead us to the prediction of the size of noble gas atoms under different conditions. Comparison of sizes from the two sources—atoms in two different environments—is misleading and of doubtful value.

## METALS AND NONMETALS

PG   4  T   Identify metals and nonmetals in the periodic table.

Both physically and chemically the alkali metals, alkaline earths and transition elements are metals. The chemical criterion for a metal is its ability to lose one or more electrons and become a positively charged ion; and that the larger the atom, the more easily the outermost electron is removed. Therefore the metallic character of elements in a group increases as you go down a column in the periodic table. It was also noted that, as atomic number increases across a period in the table, the number of protons in the nucleus increases, causing the size of the atom to become smaller. Both reduced size and an increasing number of protons lead to a stronger hold on the outermost electron, making it more difficult for that electron to be lost. Consequently the metallic character of elements *decreases* as you go from left to right in the same period of the table.

The distinction between metals, elements that lose electrons in chemical reactions, and nonmetals, those that do not, is not a sharp one, but it can be drawn as a stair-step line beginning between atomic numbers 4 and 5 in Period 2, and ending between 84 and 85 in Period 6 (see inside front cover and Fig. 4.19). Most elements on either side of this line enter into some compounds as metals, and others as nonmetals. They are therefore sometimes loosely classified as **metalloids**.

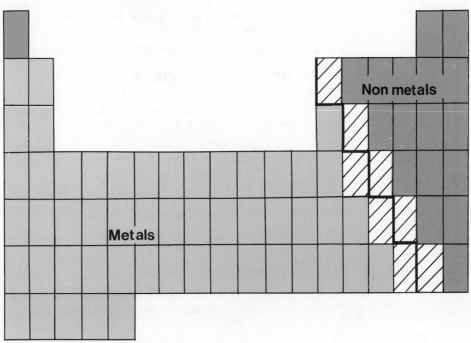

**Figure 4.19** Metals and nonmetals. Shaded elements are metalloids, having properties intermediate between those of metals and nonmetals.

# CHAPTER 4 IN REVIEW

## 4.1 INTRODUCTION TO ATOMIC STRUCTURE

## 4.2 THE ATOMIC THEORY

4 A   Identify the postulates of Dalton's Atomic Theory. (Page 59)

## 4.3 SUBATOMIC PARTICLES

4 B   Identify the postulates of the original Dalton Atomic Theory that are no longer considered valid and explain why. (Page 59)

4 C   Identify the three major subatomic particles by charge and mass. (Page 59)

4 D   Explain what isotopes of an element are, and how they differ from each other. (Page 59)

## 4.4 THE PERIODIC TABLE

4 E   Referring to a periodic table, distinguish between periods and groups, and identify them by number. (Page 61)

## 4.5 THE NUCLEAR ATOM

4 F   Describe the nuclear model of the atom, based on the Rutherford scattering experiment. (Page 63)

## 4.6 THE BOHR MODEL OF THE HYDROGEN ATOM

4 G   Describe the Bohr model of the hydrogen atom. (Page 65)

4 H   Explain the meaning of quantized energy levels in an atom and show how these levels are related to the discrete lines in the spectrum of that atom. (Page 65)

4 I   Distinguish between ground state and excited state. (Page 65)

## 4.7 THE QUANTUM MECHANICAL MODEL OF THE ATOM

4 J   Identify the principal energy levels within an atom and state the energy trend among them. (Page 70)

4 K   For each principal energy level, state the number of sublevels, identify them by letter and state the energy trend among them. (Page 71)

4 L   Sketch the shapes of the s and p orbitals. (Page 72)

4 M   State the number of orbitals in each sublevel. (Page 72)

4 N   State the restrictions on the electron population of an orbital. (Page 74)

## 4.8 ELECTRON CONFIGURATION

4 O   Referring only to a periodic table, write the ground state electron configuration of a gaseous atom of any element up to atomic number 36. (Page 80)

## 4.9 TRENDS IN THE PERIODIC TABLE

4 P   Explain, from the standpoint of electron configuration, why certain groups of elements constitute chemical families. (Page 84)

4 Q    Identify within the periodic table the following chemical families: alkali metals; alkaline earths; halogens; noble gases. (Page 84)

4 R    Identify the transition elements in the periodic table. (Page 87)

4 S    Predict how and explain why atomic size varies as a function of position in the periodic table. (Page 87)

4 T    Identify metals and nonmetals in the periodic table. (Page 89)

## TERMS AND CONCEPTS

Law of Multiple Proportions (59)
Electron (60)
Proton (60)
Neutron (60)
Isotopes (61)
Atomic number (61)
Periodic table (61)
Chemical family (62)
Group (in periodic table) (62)
Period (in periodic table) (62)
Nucleus (63)
Nuclear model of the atom (63)
Orbit (65)
Planetary model of the atom (65)
Bohr model of the atom (65)
Spectrum (65)
Discrete (65)
Quantized energy level (67)
Ground state (68)
Excited state (68)

Principal energy level (69)
Principal quantum number (69)
Quantum mechanical model (70)
Sublevel (71)
Orbital (72)
Pauli exclusion principle (74)
Electron configuration (74)
Neon core (77)
Argon core (77)
Ion (83)
Ionization energy (potential) (83)
Valence electrons (84)
Alkali metals (84)
Alkaline earth metals (84)
Halogens (85)
Noble gases (86)
Transition elements (metals) (87)
Metal (chemical criterion) (87)
Metalloids (89)

## QUESTIONS AND PROBLEMS

An asterisk (*) identifies a question that is relatively difficult, or that extends beyond the performance goals in the chapter.

### Section 4.2

4.1)  List the major points in Dalton's atomic theory.

4.2)*  Sodium oxide and sodium peroxide are two compounds made up of the elements sodium and oxygen. Sixty-two grams of sodium oxide consist of 46 grams of sodium and 16 grams of oxygen; 78 grams of sodium peroxide are made up of 46 grams of sodium and 32 grams of oxygen. Show how these figures confirm the Law of Multiple Proportions.

4.27)  Show how the atomic theory "explains" the laws of definite composition and conservation of mass.

4.28)*  Two compounds of mercury and chlorine are mercury(I) chloride and mercury(II) chloride. The quantity of mercury(I) chloride that contains 71 grams of chlorine holds 402 grams of mercury; the amount of mercury(II) chloride having 71 grams of chlorine has 201 grams of mercury. Show how the Law of Multiple Proportions is illustrated by these figures.

### Section 4.3

4.3)  Identify the parts of Dalton's original atomic theory that are now known to be in error.

4.29)  Explain why Dalton's theory is incorrect in the parts identified in Question 4.3.

4.4) Compare the relative masses of the three major components of an atom.

4.5) In a neutral atom, compare the number of protons with the number of electrons. Do the same for protons and neutrons; electrons and neutrons.

## Section 4.4

4.6) Write the symbols of the elements in Group IIA of the periodic table. Write the atomic numbers of the elements in Period 3.

4.7) Locate on a periodic table each pair of elements whose atomic numbers are given and classify them as belonging to the same period or the same chemical family: (a) 12 and 16; (b) 7 and 33; (c) 2 and 10; (d) 42 and 51.

## Section 4.5

4.8) How can we account for the fact that in the Rutherford scattering experiment, most of the alpha particles passed directly through a solid sheet of gold?

4.9) What major conclusions were drawn from the Rutherford scattering experiment?

4.10) The Rutherford experiment was performed and its conclusions were reached before protons and neutrons were discovered. Why, when they were found, was it believed they were in the nucleus of the atom?

## Section 4.6

4.11) Distinguish between a continuous spectrum and a discrete line spectrum.

4.12) "Electron energies are quantized within the atom." What is the meaning of this statement?

4.13) Compare an atom in the ground state with an atom in an excited state.

4.14) Draw a sketch of an atom according to the Bohr model. Describe the atom with reference to the sketch.

4.15) Identify the shortcomings of the Bohr theory of the atom.

## Section 4.7

4.16) What is the meaning of the "principal energy levels" of an atom? Compare the relative energies of the principal energy levels within the same atom.

4.30) Compare the electrical charges of the three major components of an atom.

4.31) A neutral atom of oxygen has 8 protons, 8 electrons and 8 neutrons. State the number of protons, electrons and neutrons that might exist in an isotope of the atom described.

4.32) How many elements are in Period 5 of the periodic table? Write the atomic numbers of the elements in Group IB.

4.33) Locate on a periodic table each element whose atomic number is given and identify first the number of the period it is in, and then the number of the group: (a) 20; (b) 14; (c) 43.

4.34) How can we account for the fact that in the Rutherford scattering experiment, some of the alpha particles were deflected from their paths through the sheet of gold, and some even bounced back at various angles?

4.35) What name is given to the central part of an atom?

4.36) Describe the activity of electrons according to the planetary model of the atom that arose from the Rutherford scattering experiment.

4.37)* What is meant by the statement that "something behaves like a wave"?

4.38) What experimental evidence leads us to believe in quantization of electron energy levels? Explain.

4.39) Which atom has the potential of emitting spectral light, one in the ground state or one in an excited state? Why?

4.40) Using a sketch of the Bohr model of an atom, explain the source of the observed lines in the spectrum of the hydrogen atom.

4.41) Identify the major advances arising from the Bohr model.

4.42) What are sublevels and how are they identified? State the relationship between each principal energy level and the number of its sublevels.

4.17)   Distinguish between electron orbits and electron orbitals. Draw sketches of the s and p orbitals.

4.18)   Identify the three possible populations of an electron orbital.

4.43)   How many orbitals are in the s sublevel? the p sublevel? the d sublevel? the f sublevel?

4.44)   What is the maximum total electron population of the s orbital within a given principal energy level? of the p orbitals? of the d orbitals? of the f orbitals?

Section 4.8

*4.19, 4.20, 4.45 and 4.46: Write electron configurations for neutral atoms of the elements identified.*

4.19)   Nitrogen (N, at. no. 7) and titanium (Ti, at. no. 22).

4.20)   Potassium (K, at. no. 19) and manganese (Mn, at. no. 25).

4.45)   Magnesium (Mg, at. no. 12) and nickel (Ni, at. no. 28).

4.46)   Vanadium (V, at. no. 23) and selenium (Se, at. no. 34).

Section 4.9

4.21)   Explain the meaning of *ionization potential*, or *ionization energy*.

4.22)   Compare values of the ionization potentials of calcium (Ca, at. no. 20) and strontium (Sr, at. no. 38). Why are these values as they are?

4.23)   Explain how the electron configurations of magnesium and calcium are responsible for the many chemical similarities between these elements.

4.47)*   Why is the definition of atomic number based on the number of protons in an atom rather than the number of electrons?

4.48)*   Compare values of the ionization potentials of aluminum (Al, at. no. 13) and chlorine (Cl, at. no. 17). Why are these values as they are?

4.49)   Account for the chemical similarities between chlorine and iodine in terms of their electron configurations.

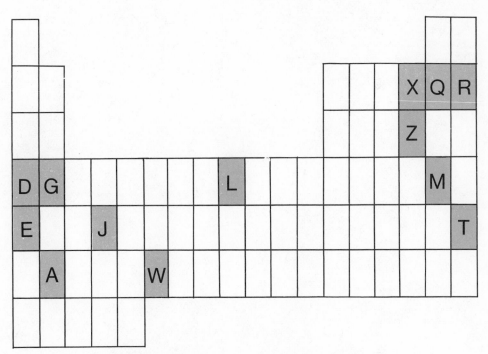

**Figure 4.20**   Chart for questions 4.24–4.26 and 4.50–4.52.

*4.24–4.26 and 4.50–4.52: Fictitious symbols have been entered into the skeleton periodic table in Figure 4.20. Assume the elements represented by these symbols have electron configurations and other characteristics corresponding to their locations in the table. Answer the following questions with the fictitious symbols only.*

4.24) Give the symbols for those elements that are (a) halogens and (b) alkali metals.

4.25) List the symbols of the transition elements.

4.26) List the elements D, E and G in order of decreasing atomic size (largest element first).

4.50) Give the symbols for those elements that are (a) alkaline earth metals and (b) noble gases.

4.51) List the symbols of all nonmetals.

4.52) List the elements Q, X and Z in order of increasing atomic size (smallest element first).

# 5

# INTRODUCTION TO
# CHEMICAL FORMULAS

## 5.1 THE STRUCTURE OF PURE SUBSTANCES

PG 5 A Distinguish between atoms and molecules.

5 B Distinguish between molecular compounds and ionic compounds.

A pure substance falls under one of two classifications: it may be an element, or it may be a compound. Elementary substances are composed of individual atoms as the ultimate structural unit. Some elements, however, are not stable as independent atoms, but form stable particles consisting of two or more atoms that are chemically bonded to each other. Such particles are called **molecules.** Oxygen, hydrogen and chlorine are examples, as shown in Figure 5.1.

The character of molecules is most clearly seen in gases, which consist of individual and distinct particles that fly about in a random fashion filling the container that holds them. These particles are molecules. Most gaseous molecules consist of two or more atoms bonded together, as described above. Certain gaseous elements, however,—the *noble gases,* so-called because they are almost completely unreactive—exist as individual atoms. Accordingly, noble gas atoms are sometimes referred to as **monatomic molecules,** *mono-* being a prefix meaning *one.*

Dalton's atomic theory pictures atoms of different elements combining to form compounds. This often leads to the formation of individual stable particles that comprise the fundamental building blocks of the compound. These particles are also molecules, consisting this time of two or more atoms of *different* elements. Such compounds are called **molecular compounds.** Water and carbon dioxide are well known molecular compounds. Water molecules consist of two atoms of hydrogen and one of oxygen; the carbon dioxide molecule has one carbon atom and two oxygen atoms.

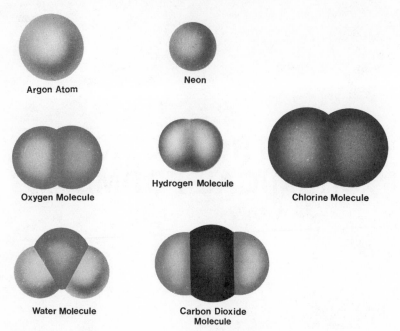

**Figure 5.1**   Illustration of (a) monatomic molecules of argon and neon, (b) diatomic elemental molecules of chlorine, hydrogen and oxygen, and (c) compound molecules of water and carbon dioxide. Drawings are all to the same scale.

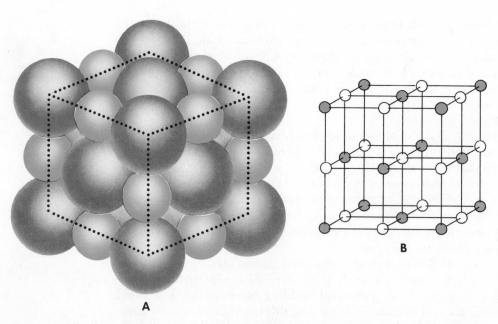

**Figure 5.2**   Two representations of the crystal structure of sodium chloride. *A* is a "space-filling" model showing the arrangement of the sodium (smaller) ions and chloride (larger) ions. *B* is a ball-and-stick model showing the cubic geometry of the crystal. The open circles represent sodium ions, and the colored circles represent chlorine ions.

Other compounds do not have molecules as structural units, but consist instead of a vast assembly of positively and negatively charged particles called *ions* (see Fig. 5.2). These compounds are called **ionic compounds.** An ion may be derived from a single atom, called a **monatomic ion,** or from two or more atoms, known as a **polyatomic ion.** Polyatomic ions are usually from two elements. Sodium chloride (table salt) and calcium carbonate (limestone) are common ionic compounds. Ions and ionic compounds will be described later in this chapter.

## 5.2   THE SYMBOLS OF CHEMICAL ELEMENTS

PG   5 C   Referring to a periodic table, given the name, symbol or atomic number of any element shown in Figure 5.3, write the other two. (Elements shown may be varied by your instructor.)

As communication between chemists began to expand in the early 19th century, there was a need for some sort of universally understood chemical shorthand for describing chemical changes. It was John Dalton, of atomic theory fame, who responded to this need by developing a series of simple symbols to identify the elements. Some of these reflected the character of the element: a solid black circle represented carbon, to suggest charcoal or coal itself. Compounds were represented by drawing the symbols of the elements touching each other.

Dalton's symbols were perhaps reasonably satisfactory when the number of elements was small. As knowledge grew, however, something more than little sketches became necessary. The symbols took the form of initial letters of the names of the elements. When more than one element began with the same letter, two letters were used. The association of an element's symbol with one or two letters of its name eases the task of learning the symbols in most instances, but not all. A significant number of the symbols are derived from Latin names, rather than the more familiar English names. The symbol for iron, for example, comes from the Latin *ferrum:* Fe. The symbol for gold is Au, derived from *aurum;* for sodium, Na, derived from *natrium;* and for silver, Ag, derived from *argentum.*

An alphabetical list of the elements and their symbols, as well as other information, appears inside the back cover of this book. A far more valuable reference, however, is the periodic table inside the front cover. You are strongly urged to make the periodic table your *only* reference for the atomic number, atomic weight and symbol of the common elements you encounter in this course.

Only about one third of the elements now known are normally mentioned in an introductory chemistry course. The effort and time you invest in memorizing their names and symbols will be more than repaid in time and effort saved as you begin writing chemical formulas and equations. The task is not difficult, considering that most of the symbols correspond with the initial letter of the name of the element.

Figure 5.3 consists of a partial periodic table containing only the symbols and atomic numbers of 35 elements you will encounter in the next few chapters, followed by an alphabetical list of those elements. Use it as an

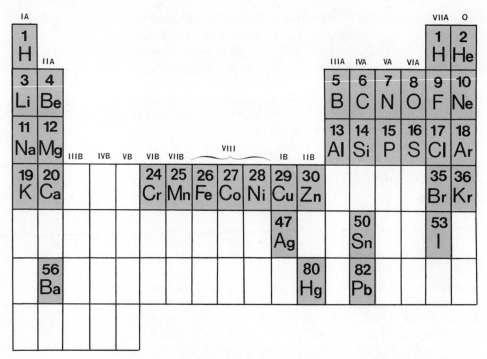

**Figure 5.3** Partial periodic table showing the symbols and locations of the more common elements. The symbols above and the list that follows identify the elements you should be able to recognize or write, referring only to a complete periodic table. Associating the names and symbols with the table makes learning them much easier. The elemental names are:

| | | | | | | |
|---|---|---|---|---|---|---|
| aluminum | bromine | chromium | iodine | magnesium | nitrogen | silver |
| argon | calcium | copper | iron | manganese | oxygen | sodium |
| barium | carbon | fluorine | krypton | mercury | phosphorus | sulfur |
| beryllium | chlorine | helium | lead | neon | potassium | tin |
| boron | cobalt | hydrogen | lithium | nickel | silicon | zinc |

"association aid" in memorizing elemental names and symbols. By associating the symbol of an element with its position in the periodic table you will develop more easily the ability to state the symbol when given the name, or vice versa. Furthermore, as you become familiar with the locations of the different elements on the periodic table, you prepare yourself to use the table as a source for atomic weights.

## 5.3 CHEMICAL FORMULAS

PG 5 D Identify the elements that form diatomic molecules and write their formulas.

5 E Given the formula of a molecular compound, state the number of atoms of each element in a molecule; given the number of atoms of each element in a molecule, write the formula of the compound.

5 F Given the name (or formula) of water, carbon monoxide, carbon dioxide or ammonia, write its formula (or name).

In the written language of chemistry, particularly in chemical equations, elements and compounds are usually shown by their chemical symbols or formulas. The formula of a chemical species identifies the elements it contains and the number of atoms of each element in the "formula unit." This may be illustrated for each type of species described in the previous sections.

**ELEMENTS.** The formula unit of most elements is the atom. The chemical formula of such elements is simply the symbol of the element. For carbon, the formula is C; for sodium, Na. The formula unit of elements that form molecules containing more than one atom is the molecule. These *molecular formulas* reflect the number of atoms in the molecule; this number is written as a subscript after the elemental symbol. The molecular formula for oxygen is $O_2$; for phosphorus, $P_4$.

In writing chemical equations it is essential that formulas of elemental gases that form diatomic molecules—molecules having two atoms—be written as molecular formulas. These elements are nitrogen, oxygen, hydrogen, fluorine, chlorine, bromine and iodine. (Bromine and iodine are not actually gases at room temperature and pressure, but they are readily converted to gases by heating.) The molecular formulas for these elements are $N_2$, $O_2$, $H_2$, $F_2$, $Cl_2$, $Br_2$, and $I_2$. To assist you in remembering which elements form diatomic molecules, visualize them in their positions in the periodic table and think of them in the following groups: the halogens (fluorine, chlorine, bromine and iodine), the two elements in water (hydrogen and oxygen), and the two most abundant gases in air (nitrogen and oxygen—the latter an obvious duplication).

**MOLECULAR COMPOUNDS.** The formula unit for a molecular compound is the molecule. The chemical formula therefore shows the number of atoms of each element in the molecule. The formula for water, for example, is $H_2O$. It tells us that the water molecule contains two atoms of hydrogen and one of oxygen. Note that the subscript 1 is omitted if there is only one atom of an element in a molecule. The formula for carbon monoxide is CO, the molecule containing one atom each of carbon and oxygen; for carbon dioxide it is $CO_2$, showing one atom of carbon and two of oxygen in the molecule. An ammonia molecule is made up of one nitrogen atom and three hydrogen atoms. Its formula is $NH_3$. These four compounds are so common their names and formulas should hereafter be a part of your chemical vocabulary.

---

**Example 5.1**  Give the atomic composition of the following molecules: (a) HCl; (b) $CH_3Cl$.

---

**5.1a**  (a) One atom of hydrogen, one atom of chlorine.
(b) One atom of carbon, three atoms of hydrogen and one atom of chlorine. This compound happens to have three elements, but the interpretation is the same.

**Example 5.2** For each of the following molecular compounds, write the formula:
    (a) Sulfur trioxide, one atom of sulfur and three of oxygen per molecule.
    (b) Dinitrogen pentoxide, where the molecule consists of two atoms of nitrogen and five of oxygen.

---

**5.2a** (a) $SO_3$   (b) $N_2O_5$

**IONIC COMPOUNDS.** The formulas of ionic compounds will be described in Section 5.7.

## 5.4 MONATOMIC IONS

PG 5 G   Distinguish between an atom and a monatomic ion.

5 H   Distinguish between a cation and an anion.

5 I   Given a periodic table and the name and atomic number of a Group A element, or a source from which that name and number may be found, write the name and symbol of the monatomic ion formed by that element.

An individual atom is electrically neutral. In order for this to be true the number of protons (positive charges) must equal the number of electrons (negative charges). The number of protons in an atom of an element cannot change without changing the identity of the element, inasmuch as the number of protons establishes that identity. But the number of electrons may change under certain circumstances. If the atom loses electrons, there are fewer electrons than protons, so it becomes a positively charged ion (see page 83). If the atom gains electrons, there are more negative than positive charges, so the ion is negatively charged. An ion with a positive charge is called a **cation;** an ion with a negative charge is an **anion.**

The manner by which a Group A element forms monatomic ions is a property of the chemical family to which it belongs. The alkali metals—the Group IA elements—form monatomic ions with a +1 charge by losing one electron per atom. See Figure 4.15 on page 83 for an illustration of the process.

Ionization may also be shown by a **chemical equation.** Chemical equations will be considered in detail in Chapter 7, but because they can be useful to us here they will be introduced at this time. An equation is a written description of a chemical change in which the formulas of the reacting substances, called the **reactants,** are placed on the left side of an arrow, and the formulas of the new substances formed, called the **products,** are written on the right. For example, the element zinc reacts with the element sulfur to form the compound zinc sulfide. Zinc and sulfur are the reactants, and zinc sulfide is the product. The equation for the reaction is

$$Zn + S \rightarrow ZnS$$

When there is more than one reactant or product, their formulas are separated by plus signs, as on the left in this equation. To read the above equation you would say, "zinc plus sulfur yields zinc sulfide." The arrow is usually read "yields," but "produces" or "forms" may also be used.

The formation of a sodium ion by the loss of the highest energy electron in a sodium atom may be represented by the equation

$$Na \rightarrow Na^+ + e^-$$

Notice the difference between the symbol of the neutral atom and the symbol of the ion. The atom symbol is simply the symbol of the element; the ion symbol is the elemental symbol with a superscript indicating the charge on the ion. The symbol $e^-$ represents an electron with its negative charge. This equation is typical of the ionization equations for all Group IA elements. For lithium, $Li \rightarrow Li^+ + e^-$; for potassium, $K \rightarrow K^+ + e^-$; and so forth. We therefore observe that *all alkali metals (Group IA elements) form monatomic ions with a +1 charge by losing one electron per atom.* **The name of a positively charged monatomic ion is simply the name of the element with the word "ion" attached:** sodium ion, potassium ion, etc.

The formation of monatomic cations by Group IIA elements is similar to that for the Group IA elements, except that two electrons are lost by the neutral atom. The formation of the magnesium ion, for example, is

$$Mg \rightarrow Mg^{2+} + 2\ e^-$$

Notice that the double positive charge of the magnesium ion is indicated as a 2+. The coefficient 2 in front of $e^-$ indicates that two electrons are released. Other members of the family form ions in a similar fashion, leading to the generalization that *alkaline earth metals (Group IIA elements) form monatomic ions with a +2 charge by losing two electrons per atom.*

To the extent that they form monatomic ions, Group IIIA elements behave in a similar manner, but they lose three electrons and form cations with a +3 charge. The formation of the aluminum ion is the best example:

$$Al \rightarrow Al^{3+} + 3\ e^-$$

The elements at the top of Group IVA are nonmetals. They do not form monatomic cations by losing electrons, but enter into molecular compounds instead. The metals at the bottom of the group form monatomic ions, but they will not be considered until Chapter 10.

In Group VA, VIA and VIIA we encounter primarily nonmetals that form monatomic ions by *gaining* electrons rather than by losing them. As a consequence the ions are negatively charged. There are only a few compounds involving anions of nitrogen and phosphorus (Group VA), and we will not be concerned with them. Elements in Group VIA produce anions with $-2$ charges by gaining two electrons per atom. For sulfur this may be represented by the equation

$$S + 2\ e^- \rightarrow S^{2-}$$

The $S^{2-}$ ion is called the *sulfide* ion. **In naming monatomic anions the name of the element is modified by the addition of an -*ide* suffix.**

The elements in Group VIIA form anions with a −1 charge. For an *atom* of chlorine this would be represented by the equation

$$Cl + e^- \rightarrow Cl^-$$

and the ion would be called the chloride ion. (While this is a correct equation and it satisfies the immediate purpose of identifying the chloride ion, one doesn't usually find atomic chlorine in abundance. The stable form of chlorine is in diatomic molecules, from which the formation of the chloride ion is shown by $Cl_2 + 2\,e^- \rightarrow 2\,Cl^-$.) The fluoride, bromide and iodide ions, $F^-$, $Br^-$ and $I^-$ respectively, are similarly formed by gaining one electron per atom. Even hydrogen, not like a halogen in other respects, forms a hydride ion in a small number of compounds.

The immediate purpose of this section is to introduce you to the nomenclature of ions and to show you how to use the periodic table in determining the name and formula, including charge, of monatomic ions formed by Group A elements. To that end, let's summarize:

1. Positively charged monatomic ions are called cations; negatively charged monatomic ions are called anions.
2. Monatomic cations are called by the elemental name, followed by "ion," as in "potassium ion."
3. Monatomic anions are named by modifying the elemental name with an -*ide* suffix, followed by "ion," as in "oxide ion," or "fluoride ion."
4. The symbols for monatomic ions are the symbols of the elements from which they are formed followed by the ionic charge written as a superscript. (IMPORTANT: The charge *must* be shown when

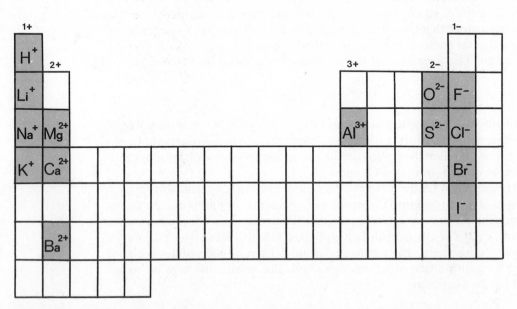

**Figure 5.4** Partial periodic table showing the symbols and locations of some of the more important monatomic ions. This is the minimum list of monatomic ions you should be able to use in writing formulas of ionic compounds, referring only to a complete periodic table. Associating the charges on the ions with their location in the periodic table removes the need for memorizing ionic charges.

representing an ion. Na, elemental sodium, and $Na^+$, sodium ion, are two very different substances with very different properties.)

5. Group IA elements form cations with a +1 charge.
6. Group IIA elements form cations with a +2 charge.
7. Group IIIA elements form cations with a +3 charge.
8. Group VIA elements form anions with a −2 charge.
9. Group VIIA elements form anions with a −1 charge.

Figure 5.4 illustrates items 5–9 of the above summary.

---

**Example 5.3**  Referring only to the periodic table inside the front cover of this text, write the chemical symbols of (a) calcium atoms and calcium ions; (b) fluorine atoms and fluoride ions.

---

**5.3a**  (a) Ca and $Ca^{2+}$; (b) F and $F^-$. All ions in Group IIA, such as magnesium, have a +2 charge, and their symbols therefore carry a 2+ superscript. Similarly, all ions from Group VIIA have a −1 charge, and therefore have a minus sign as a superscript.

---

**Example 5.4**  Referring only to the periodic table inside the front cover of this text, write the symbol of the monatomic ion formed by the following elements, and classify the ion as a cation or an anion: barium; bromine; oxygen; potassium.

---

**5.4a**

| Element | Ion Symbol | Anion or Cation | Element | Ion Symbol | Anion or Cation |
|---------|-----------|-----------------|---------|-----------|-----------------|
| Barium | $Ba^{2+}$ | Cation | Oxygen | $O^{2-}$ | Anion |
| Bromine | $Br^-$ | Anion | Potassium | $K^+$ | Cation |

## 5.5  THE HYDROGEN ION

PG  5 J  Explain how the hydrogen and chloride ions are derived from the ionization of hydrogen chloride.

5 K  Given the name (or formula) of hydrochloric acid, write its formula (or name).

The term "hydrogen ion" refers to $H^+$, as would be expected by the position of hydrogen in Group IA of the periodic table. This species does not exist as such in pure compounds. It is said to be present in acids which, for the moment, can be considered simply as solutions formed by the reaction of water with certain hydrogen-bearing compounds. Hydrogen chloride, a gaseous compound having the formula HCl, is typical. When bubbled through water its reaction may be represented by

$$HCl + H_2O \rightarrow H \cdot H_2O^+ + Cl^-$$

The species $H \cdot H_2O^+$ is more commonly written $H_3O^+$, and is called the **hydronium ion.** Chemists frequently simplify the equation by omitting the water from each side,

$$HCl \rightarrow H^+ + Cl^-$$

In this form the hydrogen ion appears. When referring to acids it is understood that the hydrogen ion present is "hydrated," or attached to one or more water molecules, even though it may be written simply as $H^+$.

An acid solution is electrically neutral. Therefore it must contain a negatively charged ion to balance the positive charge of the hydrogen ion. In the solution of hydrogen chloride, the anion is the chloride ion. The water solution of hydrogen and chloride ions is hydrochloric acid, one of our most important laboratory and industrial chemicals. Its formula is usually represented as HCl, the same as the formula of the compound from which it is derived. Chemists frequently differentiate between gaseous hydrogen chloride and its water solution by adding a "g" in parentheses after the formula, HCl(g), for the gas, and an "aq" (aqueous, from the Latin *aqua* for water), HCl(aq), for the solution. We will do this when the distinction is required.

## 5.6 POLYATOMIC IONS

PG   5 L   Given the name (or formula) of nitric, sulfuric, carbonic or phosphoric acid, write the formula (or name) of the acid, and the name and formula of the anion derived from its total ionization.

5 M   Given the name (or formula) of the ammonium or hydroxide ions, write the formula (or name).

Just as the chloride ion results from the ionization of hydrogen chloride in hydrochloric acid, several of the more common polyatomic ions can be related to the ionization of **oxyacids,** acids that contain oxygen. Equations representing these, other oxyacids, and other ionizations to be considered in this chapter are shown in Table 5.1.

Compare the ionization equations for hydrochloric acid, HCl, and nitric acid, $HNO_3$, in Table 5.1. They are identical except for the substitution of the nitrate ion, $NO_3^-$ for the chloride ion, $Cl^-$. The nitrate ion serves to illustrate the naming of polyatomic anions. *If the name of the "source acid" is simply the name of the element written between hydrogen and oxygen in the formula—nitrogen in this case—modified with an -ic suffix, the name of the anion is the name of the element modified with an -ate suffix.* The -ate suffix replaces the -ic suffix of the acid. Thus nitric acid yields nitrate ion. Caution: this generalization does not include *hydro–ic* acids, such as *hydro*chloric acid. Hydro–ic acids contain no oxygen, and yield monatomic anions on total ionization.

TABLE 5.1   Polyatomic Ions

| SOURCE COMPOUND | TOTAL IONIZATION EQUATION | ION NAME |
|---|---|---|
| Hydrochloric acid* | $HCl \rightarrow H^+ + Cl^-$ | Chloride |
| Nitric acid | $HNO_3 \rightarrow H^+ + NO_3^-$ | Nitrate |
| Sulfuric acid | $H_2SO_4 \rightarrow 2\ H^+ + SO_4^{2-}$ | Sulfate |
| Carbonic acid† | $H_2CO_3 \rightarrow 2\ H^+ + CO_3^{2-}$ | Carbonate |
| Phosphoric acid† | $H_3PO_4 \rightarrow 3\ H^+ + PO_4^{3-}$ | Phosphate |
| Water§ | $HOH \rightarrow H^+ + OH^-$ | Hydroxide |
| Ammonia | $NH_3 + H^+ \rightarrow NH_4^+$ | Ammonium |

*Hydrochloric acid is included for comparison. The chloride ion is not polyatomic.

†The carbonic and phosphoric acid ionizations occur only slightly in water solutions. They are used here to illustrate the derivation of the formulas and names of the carbonate and phosphate ions, both of which are quite abundant from sources other than their parent acids.

§The water ionization is very slight but extremely important in many chemical phenomena. The hydroxide ion occurs widely in nature from other sources.

In the case of sulfuric acid there are two hydrogens that may be released by ionization. Each of them carries a +1 charge, meaning that the neutral $H_2SO_4$ molecule must leave behind a polyatomic ion having a charge of −2 if both hydrogens are removed. This is what is meant by the *total* ionization of the acid: all the ionizable hydrogens are, in fact, ionized. Only total ionization will be considered in this chapter; stepwise ionization will be discussed in Chapter 10. The name *sulfuric acid* is the name of an element, *sulfur,* modified by an *-ic* suffix. The name of $SO_4^{2-}$ is *sulfate,* the *-ate* suffix of the ion replacing the *-ic* suffix of the acid.

"Carbonic acid," $H_2CO_3$, central element carbon, is a compound that exists only in solutions, if at all. Nevertheless its total ionization, to the extent that it occurs, is identical in principle to the ionization of sulfuric acid, and it yields the name and formula of the carbonate ion, $CO_3^{2-}$. This ion occurs in large quantities in nature, limestone and coral reefs being two examples. Phosphoric acid, $H_3PO_4$, with phosphorus as the central element, undergoes total ionization in water to only a small extent; but to that extent it yields the phosphate ion, $PO_4^{3-}$, which is also abundant in nature. In each case the negative charge on the anion matches the number of hydrogens removed from the neutral molecule: −2 for the carbonate ion after removing two hydrogens from $H_2CO_3$, and −3 for the phosphate ion after removing three hydrogens from $H_3PO_4$.

Another polyatomic anion that is so important that it must be included in this introduction is the hydroxide ion, $OH^-$. While available from many sources, it can be thought of as arising from the ionization of *one* hydrogen from a molecule of water, written as HOH to emphasize the similarity to the other ionizations shown in Table 5.1.

There is only one positively charged polyatomic ion that is of importance at this point. It comes from the compound ammonia, $NH_3$. If a neutral ammonia molecule *gains* a hydrogen ion, $H^+$, the resulting species must have a +1 charge. The product is the $NH_4^+$ ion, known as the ammonium ion. Note the small but important differences between the names and formulas of these closely related species: the molecule is ammonia, $NH_3$; the ion is ammonium, $NH_4^+$.

For purposes of further building your chemical vocabulary, it is recommended that you become thoroughly familiar with the formulas and names of the "source compounds" of the six polyatomic ions introduced in this section. From them you should be able to derive the formula and name of the polyatomic ion associated with each compound.

As you study this section, and others later that involve descriptive material, we urge you to *learn* a few *systems*, rather than *memorize* many specific *examples*. Some memorization, such as the names and formulas of the substances listed in Table 5.1 is unavoidable. You must also remember the two rules that have been illustrated in the discussion above:

1.  To name a polyatomic anion derived from the total ionization of an oxyacid whose name ends in the suffix -*ic*, replace the -*ic* with the anion suffix, -*ate*.
2.  The negative charge on the anion derived from the total ionization of an acid is equal to the number of hydrogens that ionize.

If you have command of these basic facts it is not necessary to memorize the names or charges of the polyatomic anions themselves. Furthermore, you can apply the rules to unfamiliar ions. For example, if $HBrO_3$ is the formula of bromic acid, what are the name and formula of the polyatomic negatively charged ion derived from its total ionization?*

## 5.7 NAMES AND FORMULAS OF IONIC COMPOUNDS

PG 5 N  Given a periodic table and the name (or formula) of any ionic compound whose ions are from the following, write its formula (or name):

Monatomic cation from Group IA, IIA or IIIA of the periodic table, or the ammonium cation;

Monatomic anion from Group VI or VII of the periodic table, or the nitrate, sulfate, carbonate, phosphate or hydroxide anion.

In both written and oral communication, the chemist must be able to identify chemical substances by name and by formula. There must be no ambiguity in either. Accordingly he employs a set of rules of nomenclature which relate name and formula to each other. At this time only the rules for ionic compounds will be presented.

Ionic compounds are identified by two names. The first is the name of

---

*Bromate ion, $BrO_3^-$.

the cation (+); the second is the name of the anion (−). In writing formulas of ionic compounds, the formula of the cation is written first, followed by the formula of the anion, complete with subscripts required, if any, to make the formula electrically neutral. This combination of ions is the formula unit of an ionic compound. Naming ionic compounds from their formulas is simply a matter of recognizing the ions that make up the compound and naming them in the order in which they appear.

---

**Example 5.5** For each of the following ionic compounds, identify the cation and the anion, and write the name of the compound: KCl; AlPO₄; BaSO₄; NaF.

- - - - - - - - - - - - - - - - - - - - - - - - - - - - - - - - - - - - - - - - - - - - - - - - - - - - - - - - -

**5.5a**

| FORMULA | CATION | ANION | NAME |
|---------|--------|-------|------|
| KCl | Potassium, $K^+$ | Chloride, $Cl^-$ | Potassium chloride |
| AlPO₄ | Aluminum, $Al^{3+}$ | Phosphate, $PO_4^{3-}$ | Aluminum phosphate |
| BaSO₄ | Barium, $Ba^{2+}$ | Sulfate, $SO_4^{2-}$ | Barium sulfate |
| NaF | Sodium, $Na^+$ | Fluoride, $F^-$ | Sodium fluoride |

Chemical compounds are electrically neutral. If these compounds are made up of charged particles, such as ions, it follows that the total charge contributed by the negative ions must equal the total charge of the positive ions. Notice that this is true in the above example. For KCl and NaF a +1 charge is balanced by a −1 charge; for BaSO₄ a +2 charge is balanced by a −2 charge; and for AlPO₄ a +3 charge is balanced by a −3 charge. But what happens if the charges of the positive and negative ions have different magnitudes? How about barium chloride, made up of $Ba^{2+}$ and $Cl^-$ ions, for example?

If two ions of unequal charges make up an ionic compound, they occur in such numbers as to achieve electrical neutrality. In the case of barium chloride, electrical balance can be achieved by having twice as many chloride ions ($Cl^-$) as barium ions ($Ba^{2+}$). The formula for barium chloride is therefore BaCl₂. Notice that the formula is *not* $Ba^{2+}Cl_2^-$. Ordinarily we do not show ionic charge in writing formulas of ionic compounds. If you find them helpful in learning to write formulas, use them for that purpose—but they should be erased from the final formula.

Balancing ionic charges is the key to writing correct formulas for ionic compounds. The principle may be illustrated for all combinations of ions up to +3 and −3 by completing the table in the following example:

---

**Example 5.6** Assume $A^+$, $B^{2+}$ and $C^{3+}$ are three positively charged ions, and $X^-$, $Y^{2-}$ and $Z^{3-}$ are three negatively charged ions. Write the formulas for the compounds that may be formed by these ions in all possible combinations, placing the formula in the appropriate box in the table. The compound that corresponds with barium chloride, $B^{2+}$ for $Ba^{2+}$ and $X^-$ for $Cl^-$, is entered in its appropriate place as an example.

|  | $X^-$ | $Y^{2-}$ | $Z^{3-}$ |
|---|---|---|---|
| $A^+$ |  |  |  |
| $B^{2+}$ | $BX_2$ |  |  |
| $C^{3+}$ |  |  |  |

---

**5.6a**

|  | $X^-$ | $Y^{2-}$ | $Z^{3-}$ |
|---|---|---|---|
| $A^+$ | $AX$ | $A_2Y$ | $A_3Z$ |
| $B^{2+}$ | $BX_2$ | $BY$ | $B_3Z_2$ |
| $C^{3+}$ | $CX_3$ | $C_2Y_3$ | $CZ$ |

The only entries in this table that are likely to be troublesome are $C_2Y_3$ and $B_3Z_2$. In these examples the ionic charges are 2 and 3. Balance is achieved by taking three of the ions with a charge of 2, giving $3 \times 2 = 6$ of one charge; and two of the ions with a charge of 3, giving $2 \times 3 = 6$ of the other charge.

Now we will apply this procedure to writing the formulas of real compounds.

**Example 5.7** Write the formulas of sodium chloride and sodium hydroxide.

---

**5.7a** NaCl and NaOH. The ions are $Na^+$, $Cl^-$ and $OH^-$. Each compound is a +1 and −1 combination, corresponding to AX in Example 5.6.

**Example 5.8** Write the formulas of calcium chloride and calcium hydroxide.

This question has a small catch to it—something you haven't encountered before in this text. Nevertheless, try to write the formulas without further instructions.

---

**5.8a** $CaCl_2$ and $Ca(OH)_2$

Did you by any chance write $CaOH_2$ for the second compound? That formula indicates an atomic ratio of one atom of calcium, one atom of oxygen and two atoms of hydrogen. The compound is actually made up of one calcium ion for every two *hydroxide ions:* $Ca^{2+}$ and $OH^-$ and $OH^-$. *To show more than one polyatomic ion in a formula we enclose that ion in parentheses, and indicate the number of these ions with a subscript after the parentheses.* The subscript, then, applies to everything in the parentheses immediately preceding it. [Caution: parentheses are used only to enclose polyatomic ions, never monatomic ions. The formula for calcium chloride, for example, is $CaCl_2$, not $Ca(Cl)_2$.]

---

**Example 5.9** Write the formulas for sodium nitrate and calcium nitrate.

Think carefully about the formulas you have written already, and what has just been said about parentheses . . .

------------------------------------------------------------------------

**5.9a** $NaNO_3$ and $Ca(NO_3)_2$. The ions are $Na^+$, $Ca^{2+}$ and $NO_3^-$. Sodium nitrate is a one-and-one combination, just like NaOH in Example 5.7. Calcium nitrate is a one-and-two combination, just like calcium hydroxide in Example 5.8. The subscript 3 at the end of the $NO_3^-$ symbol does not make the handling of the nitrate different from the hydroxide. The entire symbol is enclosed in parentheses.

---

**Example 5.10** Write the formulas of ammonium nitrate and ammonium carbonate.

Again think carefully in applying the rules you have learned.

------------------------------------------------------------------------

**5.10a** $NH_4NO_3$ and $(NH_4)_2CO_3$. The formula for ammonium nitrate has, as with all other ionic compounds, the formula of the positive ion followed by that of the negative ion. This is not altered just because both ions happen to contain the same element (nitrogen). In $(NH_4)_2CO_3$ a polyatomic ion is taken twice, and therefore must be enclosed in parentheses as in $Ca(NO_3)_2$ above. That it is a cation rather than an anion does not change the rule.

---

**Example 5.11** Write the formula of barium phosphate.

------------------------------------------------------------------------

**5.11a** $Ba_3(PO_4)_2$. This is a +2 and −3 combination, comparable to $B_3Z_2$ in Example 5.6.

---

**Example 5.12**  Referring only to the periodic table inside the front cover, for each compound listed below, write the formula if the name is given, and the name if the formula is given.

| | |
|---|---|
| Calcium oxide | $Ba(OH)_2$ |
| Potassium sulfide | $KNO_3$ |
| Magnesium bromide | $(NH_4)_3PO_4$ |
| Aluminum sulfate | NaBr |

-------------------------------------------------------------------

**5.12a**

| | | | |
|---|---|---|---|
| Calcium oxide | CaO | $Ba(OH)_2$ | Barium hydroxide |
| Potassium sulfide | $K_2S$ | $KNO_3$ | Potassium nitrate |
| Magnesium bromide | $MgBr_2$ | $(NH_4)_3PO_4$ | Ammonium phosphate |
| Aluminum sulfate | $Al_2(SO_4)_3$ | NaBr | Sodium bromide |

A final but important comment before leaving this section: We have been practicing with the system by which names and formulas of *ionic* compounds are written. Molecular compounds are governed by a different system, which you will study in Chapter 10. The source compounds in Table 5.1 (page 105), HCl, $HNO_3$, $H_2SO_4$, $H_2CO_3$, $H_3PO_4$ and $NH_3$, are all molecular compounds, and are named accordingly. In particular, the name of HOH is *water, NEVER hydrogen hydroxide*, as if it were a combination of hydrogen and hydroxide ions. Sometimes the systems yield identical names, as *hydrogen chloride* for HCl; but that conclusion is properly reached only by the molecular binary system, not the ionic system. While water, $H_2O$, and ammonia, $NH_3$, could be called by their formal names, dihydrogen oxide and nitrogen trihydride, the compounds are so well known that their common names are always used. Traditionally the oxyacids are called by their acid names, not as *hydrogen nitrate, hydrogen sulfate* or *hydrogen phosphate*, all of which would suggest incorrectly that they are ionic compounds.

## 5.8  HYDRATES

PG  5 O  Given the formula of a hydrate, state the number of water molecules associated with each formula unit of the anhydrous compound.

5 P  Given the formula (or name) of a hydrate, write its name (or formula). (This performance goal is limited to hydrates of ionic compounds discussed elsewhere in this chapter.)

Some compounds, when crystallized from water solutions, form solids that include water molecules as a part of the crystal structure. Such water is referred to as **water of crystallization,** or **water of hydration.** The compound is said to be **hydrated,** and it is called a **hydrate.** Hydration water can usually be driven from a compound by heating, leaving the **anhydrous** compound.

Let us illustrate how formulas of hydrates are written by considering the compound copper sulfate. Anhydrous copper sulfate, a nearly white powder, has the formula $CuSO_4$. The hydrate forms large, dark blue crystals of distinct shape. Its formula is $CuSO_4 \cdot 5H_2O$. The number of water molecules that crystallize with each formula unit of the anhydrous compound is shown after the anhydrous formula, separated by a dot. This number may be indicated in the name of the compound by using number prefixes shown in the Appendix (page 513). $CuSO_4 \cdot 5H_2O$ is, by this system, called copper sulfate pentahydrate, *penta-* being the prefix signifying 5. Hydrates are also named by stating the number of water molecules, as copper sulfate 5-water or copper sulfate 5-hydrate.

The number of water molecules associated with one formula unit of the anhydrous compound can vary depending on external conditions of temperature and pressure. Some compounds form only one stable hydrate; for other compounds several different hydrates are known. Sodium carbonate, for example, crystallizes from water solution at room temperature as a decahydrate, $Na_2CO_3 \cdot 10H_2O$. At higher temperatures the heptahydrate, $Na_2CO_3 \cdot 7H_2O$, and the monohydrate, $Na_2CO_3 \cdot H_2O$ are stable.

---

**Example 5.13**   (a) How many molecules of water are associated with each formula unit of anhydrous barium nitrate in $Ba(NO_3)_2 \cdot H_2O$? Name the hydrate. (b) Write the formula for the hexahydrate of nickel chloride if the formula of the anhydrous compound is $NiCl_2$.

--------------------------------------------------------------------------------

**5.13a**   (a) One. Barium nitrate monohydrate.
   (b) $NiCl_2 \cdot 6H_2O$. "Hexa-" is the prefix for six.

---

# CHAPTER 5 IN REVIEW

**5.1   THE STRUCTURE OF PURE SUBSTANCES**

   5 A   Distinguish between atoms and molecules. (Page 95)

   5 B   Distinguish between molecular compounds and ionic compounds. (Page 95)

**5.2   THE SYMBOLS OF THE CHEMICAL ELEMENTS**

   5 C   Referring to a periodic table, given the name, symbol or atomic number of any element shown in Figure 5.3, write the other two. (Page 97)

### 5.3 CHEMICAL FORMULAS

5 D   Identify the elements that form diatomic molecules and write their formulas. (Page 98)

5 E   Given the formula of a molecular compound, state the number of atoms of each element in a molecule; given the number of atoms of each element in a molecule, write the formula of the compound. (Page 98)

5 F   Given the name (or formula) of water, carbon monoxide, carbon dioxide or ammonia, write its formula (or name). (Page 98)

### 5.4 MONATOMIC IONS

5 G   Distinguish between an atom and a monatomic ion. (Page 100)

5 H   Distinguish between a cation and an anion. (Page 100)

5 I   Given a periodic table and the name and atomic number of a Group A element, or a source from which that name and number may be found, write the name and symbol of the monatomic ion formed by that element. (Page 100)

### 5.5 THE HYDROGEN ION

5 J   Explain how the hydrogen and chloride ions are derived from the ionization of hydrogen chloride. (Page 103)

5 K   Given the name (or formula) of hydrochloric acid, write its formula (or name). (Page 103)

### 5.6 POLYATOMIC IONS

5 L   Given the name (or formula) of nitric, sulfuric, carbonic or phosphoric acid, write the formula (or name) of the acid, and the name and formula of the anion derived from its total ionization. (Page 104)

5 M   Given the name (or formula) of the ammonium or hydroxide ion, write the formula (or name). (Page 104)

### 5.7 NAMES AND FORMULAS OF IONIC COMPOUNDS

5 N   Given a periodic table and the name (or formula) of any ionic compound whose ions are from the following, write its formula (or name):

Monatomic cation from Group IA, IIA or IIIA of the periodic table, or the ammonium cation;

Monatomic anion from Group VI or VII of the periodic table, or the nitrate, sulfate, carbonate, phosphate or hydroxide anion. (Page 106)

### 5.8 HYDRATES

5 O   Given the formula of a hydrate, state the number of water molecules associated with each formula unit of the anhydrous compound. (Page 110)

5 P   Given the formula (or name) of a hydrate, write its name (or formula). (This performance goal is limited to hydrates of ionic compounds discussed elsewhere in this chapter.) (Page 110)

# TERMS AND CONCEPTS

Molecule (95)
Molecular compound (95)
Ionic compound (97)
Monatomic ion (97)
Polyatomic ion (97)
Cation (100)
Anion (100)
Chemical equation (100)

Reactant (100)
Product (100)
Hydronium ion (104)
Oxyacid (104)
Total ionization (105)
Water of hydration (crystallization) (111)
Hydrate-hydrated (111)
Anhydrous (111)

# QUESTIONS AND PROBLEMS

### Section 5.1

5.1)  Give an example of a substance whose fundamental unit particle is an atom.

5.2)  What is (are) the distinguishing feature(s) between atoms and molecules?

5.3)  What feature(s) distinguish molecular compounds from ionic compounds?

5.20)  Give an example of a substance whose fundamental unit particle is a molecule.

5.21)  Predict whether each substance in the following list is most likely to be made up of molecules or atoms: (a) oxygen; (b) water; (c) sulfur dioxide; (d) iron; (e) carbon; (f) carbon tetrachloride.

5.22)  We never speak of charged particles in molecular compounds, but we do speak of them in ionic compounds. How can this be, when all compounds are electrically neutral?

### Section 5.2

5.4)  The names, atomic numbers or symbols of some of the most common elements—those whose names and symbols you should recognize at sight—are shown in Table 5.2 (next page). Fill in the open spaces, referring only to the periodic table for any information you require.

### Section 5.3

5.5)  Name the elements that normally exist as diatomic molecules.

5.6)  For each of the following molecular compounds, state the identity of each element in the compound and the number of atoms of that element in the molecule: (a) $N_2O$; (b) $PCl_3$; (c) $CH_3CHO$.

5.7)  Write the names of $H_2O$ and $CO$.

5.23)  Write the formulas of the following elemental gases: neon, oxygen, nitrogen, helium and fluorine.

5.24)  The number of atoms of each element in a molecule of several molecular compounds is given below. Write the molecular formula for the compound. (a) 1 nitrogen, 3 hydrogen; (b) 2 hydrogen, 1 sulfur, 3 oxygen; (c) 3 carbon, 8 hydrogen, 1 oxygen.

5.25)  Write the formulas of carbon dioxide and ammonia.

### Section 5.4

5.8)  Explain how monatomic ions are formed from atoms.

5.26)  How does a monatomic anion differ from a monatomic cation?

TABLE 5.2  Table of Elements

| NAME OF ELEMENT | ATOMIC NUMBER | SYMBOL OF ELEMENT | NAME OF ELEMENT | ATOMIC NUMBER | SYMBOL OF ELEMENT |
|---|---|---|---|---|---|
| Sodium | | | | | Mg |
| | | Pb | | 8 | |
| Aluminum | | | Phosphorus | | |
| | 26 | | | | Ca |
| | | F | Zinc | | |
| Boron | | | | | Li |
| | 18 | | Nitrogen | | |
| Silver | | | | 16 | |
| | 6 | | | 53 | |
| Copper | | | Barium | | |
| | | Be | | | K |
| Krypton | | | | 10 | |
| Chlorine | | | Helium | | |
| | 1 | | | | Br |
| | | Mn | | | Ni |
| | 24 | | Tin | | |
| Cobalt | | | | 14 | |
| | 80 | | | | |

5.9)  Referring only to the periodic table, fill in the blanks in the following table with the name, ion symbol and classification as anion or cation, of the monatomic ions derived from the elements whose atomic numbers are given.

| ATOMIC NUMBER | ION NAME | ION SYMBOL | ANION OR CATION |
|---|---|---|---|
| 11 | | | |
| 53 | | | |
| 3 | | | |
| 20 | | | |
| 16 | | | |

*Questions continued on page 116.*

*Questions continued on page 116.*

TABLE 5.3   Formula Writing and Nomenclature Exercise Number 1

*Instructions:* For each box write the chemical formula and name of the compound formed by the cation at the head of the column and the anion at the left of the row. Correct formulas and names are listed on page 527.

|  | $Li^+$ | $Mg^{2+}$ | $NH_4^+$ | $Al^{3+}$ | $Na^+$ | $Ba^{2+}$ | $K^+$ | $Ca^{2+}$ |
|---|---|---|---|---|---|---|---|---|
| $Br^-$ | 1 | 2 | 3 | 4 | 5 | 6 | 7 | 8 |
| $SO_4^{2-}$ | 9 | 10 | 11 | 12 | 13 | 14 | 15 | 16 |
| $OH^-$ | 17 | 18 | 19 | 20 | 21 | 22 | 23 | 24 |
| $F^-$ | 25 | 26 | 27 | 28 | 29 | 30 | 31 | 32 |
| $O^{2-}$ | 33 | 34 | 35 | 36 | 37 | 38 | 39 | 40 |
| $NO_3^-$ | 41 | 42 | 43 | 44 | 45 | 46 | 47 | 48 |
| $PO_4^{3-}$ | 49 | 50 | 51 | 52 | 53 | 54 | 55 | 56 |
| $Cl^-$ | 57 | 58 | 59 | 60 | 61 | 62 | 63 | 64 |
| $S^{2-}$ | 65 | 66 | 67 | 68 | 69 | 70 | 71 | 72 |
| $I^-$ | 73 | 74 | 75 | 76 | 77 | 78 | 79 | 80 |
| $CO_3^{2-}$ | 81 | 82 | 83 | 84 | 85 | 86 | 87 | 88 |

5.27)  Refer only to the periodic table in answering this question. The element whose atomic number is 34 is selenium; the element whose atomic number is 37 is rubidium. For each of these elements, write (1) the symbol of the monatomic ion formed, (2) the name of the monatomic ion, and (3) its classification as an anion or cation.

| ELEMENT | ION SYMBOL | ION NAME | ANION OR CATION |
|---------|------------|----------|-----------------|
| Selenium | | | |
| Rubidium | | | |

## Section 5.5

5.10)  Write the formula of hydrochloric acid.

5.28)  Show how the hydrogen and chloride ions are derived from the total ionization of hydrochloric acid.

## Section 5.6

5.11)  Write the formula of nitric acid, the formula of the anion derived from the total ionization of nitric acid, and the name of the anion.

5.12)  Write the name of the acid $H_2CO_3$, the formula of the anion that would be derived from its total ionization, and the name of the anion.

5.13)  What is the name of the $OH^-$ ion?

5.29)  Write the name of the acid $H_2SO_4$, the formula of the anion derived from the total ionization of $H_2SO_4$, and the name of the anion.

5.30)  Write the formula of phosphoric acid, the formula of the anion that would be derived from the total ionization of phosphoric acid, and the name of the anion.

5.31)  What is the formula of the ammonium ion?

## Section 5.7

*The formula writing exercises on pages 115 and 117 are the principal questions for Section 5.7. For questions 5.14–5.17 and 5.32–5.35: Given the name of an ionic compound, write its formula; or given the formula, write the name. Refer only to the periodic table in answering these questions.*

5.14)  Lithium chloride; CaS

5.15)  Ammonium nitrate; $BaCO_3$

5.16)  Magnesium bromide; $K_3PO_4$

5.17)  Barium phosphate; $(NH_4)_2SO_4$

5.32)  KF; magnesium oxide

5.33)  NaOH; aluminum phosphate

5.34)  $CaI_2$; sodium sulfate

5.35)  $Al_2(SO_4)_3$; lithium sulfide

## Section 5.8

5.18)  How many water molecules are associated with one anhydrous formula unit of calcium chloride in $CaCl_2 \cdot 2H_2O$? Write the name of the compound.

5.19)  One hydrate of barium hydroxide contains eight molecules of water per formula unit of the anhydrous compound. Write the formula of and name the hydrate.

5.36)  How many water molecules are associated with one anhydrous formula unit of magnesium sulfate in $MgSO_4 \cdot 7H_2O$? Write the name of the compound.

5.37)  Write the formulas of ammonium phosphate trihydrate and potassium sulfide 5-hydrate.

TABLE 5.4  Formula Writing Exercise Number 2

*Instructions:* For each box write the chemical formula of the compound formed by the cation at the head of the column and the anion at the left of the row. Refer only to the periodic table in completing this exercise. Correct formulas are listed on page 528.

| | CAL-CIUM | POTAS-SIUM | MAGNE-SIUM | AMMO-NIUM | LITH-IUM | ALUMI-NUM | BAR-IUM | SODIUM |
|---|---|---|---|---|---|---|---|---|
| Hydroxide | 1 | 2 | 3 | 4 | 5 | 6 | 7 | 8 |
| Bromide | 9 | 10 | 11 | 12 | 13 | 14 | 15 | 16 |
| Sulfate | 17 | 18 | 19 | 20 | 21 | 22 | 23 | 24 |
| Fluoride | 25 | 26 | 27 | 28 | 29 | 30 | 31 | 32 |
| Carbonate | 33 | 34 | 35 | 36 | 37 | 38 | 39 | 40 |
| Oxide | 41 | 42 | 43 | 44 | 45 | 46 | 47 | 48 |
| Nitrate | 49 | 50 | 51 | 52 | 53 | 54 | 55 | 56 |
| Phosphate | 57 | 58 | 59 | 60 | 61 | 62 | 63 | 64 |
| Iodide | 65 | 66 | 67 | 68 | 69 | 70 | 71 | 72 |
| Sulfide | 73 | 74 | 75 | 76 | 77 | 78 | 79 | 80 |
| Chloride | 81 | 82 | 83 | 84 | 85 | 86 | 87 | 88 |

# 6

# CHEMICAL FORMULA CALCULATIONS

Chapter 5 included an introduction to chemical nomenclature, the system of naming compounds and writing their formulas. This *mini*nomenclature system consisted of the names and formulas of five acids (hydrochloric, HCl; nitric, $HNO_3$; sulfuric, $H_2SO_4$; carbonic, $H_2CO_3$; and phosphoric, $H_3PO_4$), four molecular compounds (water, $H_2O$; carbon monoxide, CO; carbon dioxide, $CO_2$; and ammonia, $NH_3$) and 88 ionic compounds. You will require the formulas of many of these compounds as you study the next few chapters. To help you strengthen your formula writing skills within this narrow range of compounds, we will identify them by name only, leaving it to you to furnish the formulas. We encourage you to try to write all formulas as they are required, using a periodic table as your only reference. If you need assistance you can refer to this page, the first page in Chapter 6, for the acids and molecular compounds. The 88 ionic compound formulas are tabulated as Formula Writing Exercise Number 2 in the answer section at the back of the book, on page 528.

## 6.1 ATOMIC WEIGHT

Atoms have mass, which gives them "weight." For the most part this mass is simply the combined mass of the protons and neutrons, which, as indicated in Chapter 4 (page 61) are the "heavier" parts of the atom. Even these parts are so tiny that they cannot be "weighed" in the usual sense of the word, but their masses can be determined by their behavior while moving through a magnetic field in an instrument known as a mass spectrometer. With this instrument it is possible also to determine the masses of individual atoms (actually the masses of ions derived from atoms) and the relative abundance of isotopes in the sample of an element. Let's think about isotopes for a moment to establish a base from which to approach the subject of atomic weight.

## *ISOTOPES*

PG 6 A Referring to a periodic table and given one of the following, state the other two: (a) nuclear symbol; (b) number of protons and neutrons in the nucleus; or (c) atomic number and mass number.

Isotopes are atoms of the same element that have different masses (see page 61). The masses of the three principal parts of an atom are listed in Table 4.1. All atoms of an element have the same number of protons; this is the atomic number, sometimes designated Z, which identifies the element. The number of electrons in a neutral atom is the same as the number of protons, but their masses are so small that their total contribution to atomic mass is negligible. We therefore conclude that the differences in mass between individual atoms of an element must be caused by a variation in the number of neutrons in the nucleus. A neutron has about the same mass as a proton.

An isotope of an element is identified in writing by its **nuclear symbol.** In general the nuclear symbol of an isotope is

$$_Z^A Sy \qquad \text{or} \qquad ^A Sy$$

where Sy is the symbol of the element and A is the **mass number, the total number of protons plus neutrons in the nucleus.** As the symbol alone is sufficient to identify the element, it is not necessary to include the atomic number, Z; but it is useful in considering nuclear reactions (Chapter 18). Expressed as an equation, the definition of mass number becomes

$$\text{Mass number} = \text{number of protons} + \text{number of neutrons} \qquad (6.1)$$

$$A = Z + \text{number of neutrons}$$

The name of an isotope is the elemental name followed by the mass number. $_8^{16}O$, for example, is called "oxygen sixteen," and written oxygen-16.

The two major naturally-occurring isotopes of carbon are $_6^{12}C$ and $_6^{13}C$, or carbon-12 and carbon-13. From the name and symbol of the isotope and Equation 6.1, it is possible to find the number of neutrons in the nucleus of an atom. In carbon-12, for example, if you subtract the atomic number (protons) from the mass number (protons + neutrons) you get the number of neutrons: $12 - 6 = 6$. In carbon-13 there are 7 neutrons: $13 - 6 = 7$.

If you know the proton and neutron count in an atom you can readily determine its mass number and nuclear symbol. For example, a nucleus having 76 protons and 116 neutrons has a mass number, according to Equation 6.1, of 192: $76 + 116 = 192$. From the periodic table, the element with atomic number 76 has the symbol Os (osmium). The nuclear symbol is therefore $_{76}^{192}Os$.

---

**Example 6.1** List in order the number of protons, neutrons and electrons, and the name of a neutral atom of $_{20}^{42}Ca$.

---

**6.1a** 20 protons, 22 neutrons and 20 electrons, calcium-42

The number of protons is the atomic (subscript) number. The number of neutrons is the difference between the superscript (mass) number and the atomic number: $42 - 20 = 22$. The number of electrons is equal to the number of protons in a neutral atom. The name of an isotope is its elemental name followed by its mass number.

**Example 6.2** Write the symbol of the isotope of sulfur that has 18 neutrons in the nucleus.

Think first, what is the atomic number? (How will you find it?) Then think, how can I find the mass number from the atomic number and the number of neutrons? Write the symbol.

---

**6.2a** $^{34}_{16}S$

From the periodic table the atomic number of sulfur is 16, meaning 16 protons in the nucleus. The mass number is the sum of the protons plus neutrons, or $16 + 18 = 34$.

**Example 6.3** Write the nuclear symbol and name of the isotope having in its nucleus 82 protons and 122 neutrons.

---

**6.3a** $^{204}_{82}Pb$; lead-204

## DETERMINATION OF ATOMIC WEIGHT

PG 6 B Define the atomic mass unit, amu.

6 C Given the relative abundance of the naturally occurring isotopes of an element, calculate the atomic weight.

6 D Given the name or symbol of an element you know, or the atomic number of any element, and a periodic table, state the atomic weight of the element.

Long before it was possible to measure the masses of atomic and subatomic particles, chemists had developed highly useful information about "atomic weights." Their methods did not yield actual masses, but rather a table of *relative* masses of the atoms of the different elements, all compared to some standard having an arbitrarily assigned value. Even though we now have the means to determine atomic masses in grams, the numbers are so small we find it more convenient to continue using, in effect, the relative scale. This is done by comparing the mass of all atoms to the mass of one

TABLE 6.1   Percent Abundance of Some Natural Isotopes

| SYMBOL | PERCENT | SYMBOL | PERCENT | SYMBOL | PERCENT |
|---|---|---|---|---|---|
| $^{1}_{1}H$ | 99.985 | $^{24}_{12}Mg$ | 78.70 | $^{40}_{20}Ca$ | 96.97 |
| $^{2}_{1}H$ | 0.015 | $^{25}_{12}Mg$ | 10.13 | $^{42}_{20}Ca$ | 0.64 |
| $^{3}_{2}He$ | 0.00013 | $^{26}_{12}Mg$ | 11.17 | $^{43}_{20}Ca$ | 0.145 |
| $^{4}_{2}He$ | 99.99987 | $^{27}_{13}Al$ | 100 | $^{44}_{20}Ca$ | 2.06 |
| $^{6}_{3}Li$ | 7.42 | $^{28}_{14}Si$ | 92.21 | $^{46}_{20}Ca$ | 0.0033 |
| $^{7}_{3}Li$ | 92.58 | $^{29}_{14}Si$ | 4.70 | $^{48}_{20}Ca$ | 0.18 |
| $^{12}_{6}C$ | 98.89 | $^{30}_{14}Si$ | 3.09 | $^{54}_{26}Fe$ | 5.82 |
| $^{13}_{6}C$ | 1.11 | $^{31}_{15}P$ | 100 | $^{56}_{26}Fe$ | 91.66 |
| $^{14}_{7}N$ | 99.63 | $^{32}_{16}S$ | 95.0 | $^{57}_{26}Fe$ | 2.19 |
| $^{15}_{7}N$ | 0.37 | $^{33}_{16}S$ | 0.76 | $^{58}_{26}Fe$ | 0.33 |
| $^{16}_{8}O$ | 99.759 | $^{34}_{16}S$ | 4.22 | $^{63}_{29}Cu$ | 69.09 |
| $^{17}_{8}O$ | 0.037 | $^{36}_{16}S$ | 0.014 | $^{65}_{29}Cu$ | 30.91 |
| $^{18}_{8}O$ | 0.204 | $^{35}_{17}Cl$ | 75.53 | $^{64}_{30}Zn$ | 48.89 |
| $^{19}_{9}F$ | 100. | $^{37}_{17}Cl$ | 24.47 | $^{66}_{30}Zn$ | 27.81 |
| $^{20}_{10}Ne$ | 90.92 | $^{39}_{19}K$ | 93.10 | $^{67}_{30}Zn$ | 4.11 |
| $^{21}_{10}Ne$ | 0.257 | $^{40}_{19}K$ | 0.0118 | $^{68}_{30}Zn$ | 18.57 |
| $^{22}_{10}Ne$ | 8.82 | $^{41}_{19}K$ | 6.88 | $^{70}_{30}Zn$ | 0.62 |
| $^{23}_{11}Na$ | 100. | | | $^{79}_{35}Br$ | 50.54 |
| | | | | $^{81}_{35}Br$ | 49.46 |

carbon-12 atom, to which is assigned the value of exactly 12 atomic units. Consequently, **one atomic mass unit is defined as exactly 1/12 of the mass of one atom of carbon-12.**

As a practical matter, almost every sample of an element used by a chemist, either in its elemental state or in a compound, consists of all the naturally occurring isotopes of that element. Fortunately the percentage, or fraction, of each isotope is generally the same, regardless of the source of the sample. Table 6.1 lists the isotopes of some of the more common elements and their relative abundance in nature. From these it is possible to calculate the **atomic weight of an element,** which is defined as **the average mass of the atoms of an element compared to an atom of carbon-12 at exactly 12 atomic mass units.** The method of calculation is as follows:

---

**Example 6.4** The natural distribution of isotopes of magnesium is $^{24}_{12}$Mg: 78.70%; $^{25}_{12}$Mg: 10.13%; $^{26}_{12}$Mg: 11.17%. Calculate the atomic weight of magnesium.

*Solution:* Percentage values will be expressed as decimal fractions of an "average" atom; that atom is made up of 0.7870 of an atom having a mass of 24 amu, 0.1013 of an atom having a mass of 25 amu, and 0.1117 of the isotope with mass 26 amu. The total mass of the "average" atom is therefore

$$
\begin{aligned}
0.7870 \times 24 \text{ amu} &= 18.89 \text{ amu} \\
0.1013 \times 25 \text{ amu} &= 2.533 \text{ amu} \\
0.1117 \times 26 \text{ amu} &= \underline{2.904 \text{ amu}} \\
\overline{1.000} \text{ "av" atom} &= 24.327 \text{ amu}
\end{aligned}
$$

---

The periodic table lists the atomic weight of magnesium as 24.305, a difference of 0.02 in the fourth significant figure. This is because of round-offs in the masses of the isotopes and in the calculations. As a rule, atomic masses may be regarded as integers to at least two significant figures. *For our purposes we shall use atomic weights from the periodic table to three significant figures; or, if the atomic weight is less than 10 (hydrogen, helium, lithium or beryllium), to the first digit beyond the decimal.*

Now you try an atomic weight calculation:

---

**Example 6.5** Calculate the atomic weight of potassium if the natural distribution of its isotopes is $^{39}_{19}$K: 93.10%; $^{40}_{19}$K: 0.0118%; $^{41}_{19}$K: 6.88%.

------------------------------------------------------------------------------------------

**6.5a** 39.13 amu

$$
\begin{aligned}
0.9310 \times 39 \text{ amu} &= 36.31 \text{ amu} \\
0.000118 \times 40 \text{ amu} &= 0.00472 \text{ amu} \\
0.0688 \times 41 \text{ amu} &= \underline{2.82 \text{ amu}} \\
& \quad\; 39.13 \text{ amu}
\end{aligned}
$$

---

It has been indicated previously that you will use the periodic table as your source for atomic weights. To be sure you are familiar with this application of the table, let's try an example:

---

**Example 6.6** Referring only to a periodic table, list the atomic weights of the following elements to the first digit past the decimal: calcium; bromine; at. no. 48; at. no. 55.

-------------------------------------------------------------------------------------

**6.6a** Calcium: 40.1 amu    At. no. 48: 112.4 amu
Bromine: 79.9 amu    At. no. 55: 132.9 amu

---

A final comment: the term *atomic weight* is really an incorrect name for the thing we have been considering. It should be *atomic mass,* consistent with the distinction between *weight* and *mass* made on page 38. You perhaps noticed that atomic *weight* was defined as "the average *mass* . . ." Some textbooks do refer to atomic mass, and to "molecular mass" or "molar mass," extending the idea to concepts we are about to discuss. But the term *weight* is used so much more generally we shall continue to use it in this text.

## 6.2   MOLECULAR WEIGHT; FORMULA WEIGHT

PG   6 E   Distinguish between the following terms: atomic weight; molecular weight; formula weight.

The atomic weight scale furnishes a relative scale of the masses of atoms of all elements. But what about compounds? To extend the atomic weight concept to *molecular* compounds, chemists use a term that is precisely parallel to atomic weight: **Molecular weight is the average mass of the molecules of a molecular substance compared to the mass of an atom of carbon-12 at exactly 12 atomic mass units.**

As pointed out in Chapter 5, not all compounds form individual molecules. *Ionic* compounds are vast assemblies of ions arranged in a precise numerical ratio which is expressed in their chemical formulas. It would not be appropriate to refer to the *molecular* weight of such a compound. Instead the term *formula weight* is used: **Formula weight is the average mass of one formula unit of a compound compared to the mass of an atom of carbon-12 at exactly 12 atomic mass units.** The "formula unit" is the imaginary unit consisting of the number of atoms of each element that appears in the chemical formula of the compound.

## 6.3 THE MOLE CONCEPT

PG 6 F Define the mole.

Chemistry is essentially a counting game. Dalton's atomic theory states that compounds consist of whole atoms of different elements combined in the ratio of small whole numbers. Therefore we must "count out" the proper number of atoms of the different elements to form a compound. But if we wished to count out equal numbers of oxygen and carbon atoms so they could be reacted to produce carbon monoxide, it would be a time-consuming process because of the tremendous number of atoms of each element in the smallest of samples. "There must be a better way"—and chemists have found it.

It is a fact that, as atomic weights compare the masses of individual atoms of different elements, they compare also the masses of *equal numbers* of atoms of those elements. If one atom of carbon has a mass of 12.0 amu, and one atom of oxygen has a mass of 16.0 amu, their mass ratio is

$$\frac{1 \text{ atom of O}}{1 \text{ atom of C}} = \frac{16.0 \text{ amu}}{12.0 \text{ amu}} = \frac{16.0}{12.0}$$

Similarly Y atoms of carbon have a mass of $12.0 \times Y$, and Y atoms of oxygen have a mass of $16.0 \times Y$, where Y can be any number. Their mass ratio is

$$\frac{Y \text{ atoms of O}}{Y \text{ atoms of C}} = \frac{16.0 \times Y \text{ amu}}{12.0 \times Y \text{ amu}} = \frac{16.0}{12.0}$$

The ratios are identical. It follows that the grams of oxygen that numerically correspond with its atomic weight, 16.0 grams, contain the same number of atoms as the grams of carbon that match its atomic weight, 12.0 grams. We can therefore "count out" an equal number of atoms of the two elements by weighing out 16.0 grams of oxygen and 12.0 grams of carbon. If we weighed out 8.0 grams of oxygen and 6.0 grams of carbon, we would have half as many atoms of the two elements, and the numbers would again be equal. These weights, 8.0 grams of oxygen and 6.0 grams of carbon, are in the same ratio as the atomic weights, 16.0 and 12.0, respectively. They therefore must contain the same numbers of atoms.

Chemists have organized the "counting by weighing" idea into what is called the mole concept. A **mole** is formally defined as **that quantity of any species that contains the same number of units as the number of atoms in exactly 12 grams of carbon-12.** This number is given a name. It is called **Avogadro's Number, N,** in honor of the man whose interpretation of experiments with gases led to an early method of estimating atomic weights.

Notice that the definition of the mole does not identify the number of atoms in exactly 12 grams of carbon-12. That number must be found in the laboratory. Since a carbon atom is extremely small we would expect the number of atoms in exactly 12 grams of carbon-12 to be huge. This is the case: from experiment the number has been estimated to be 602,000,000,000,000,000,000,000. In exponential notation this is $6.02 \times 10^{23}$. The following equivalence results from the definition of a mole:

$$\text{One mole of any species} = 6.02 \times 10^{23} \text{ units of that species} \quad (6.2)$$

To get a better idea of what is meant by a mole, let us consider another counting unit with which we are more familiar, the dozen. Suppose we define a dozen as the number of eggs in the egg carton in which they are sold at the local supermarket. By experiment (walking down to the store, opening a box of eggs and counting them) you can determine that this number is 12. In buying eggs, you would rarely use the number 12 directly: rather, you would think in terms of dozens. When is the last time, for example, you saw a shopping list with "24 eggs" written on it? Would it not be written simply as two dozen eggs? Just as two dozen eggs expresses a quantity of eggs, two moles of carbon expresses a quantity of carbon, a quantity that can be thought of as containing a certain number of carbon atoms. If you ever have difficulty visualizing the significance of a *mole* in a sentence, substitute the word *dozen* and the meaning will be the same.

Let's think for a moment about this number—Avogadro's Number. Its magnitude is tremendous: it staggers the imagination. Many attempts have been made to create a word picture of it, but they can only hint at its vast size by leaving the reader with another huge number that likewise defies the imagination. For example, if you were to embark on the task of counting the atoms in exactly 12 grams of carbon-12, and did so at the rate of 100 atoms per minute without interruption 24 hours per day, 365 days per year, the task would require 11,500,000,000,000,000 years. This is about two million times as long as the earth has been in existence! If we enlisted the aid of the entire earth's population, about 3 billion people, all counting at the same rate, the time required could be brought down to a somewhat more reasonable 3,800,000 years.

Is it any wonder the chemist prefers to think in terms of moles rather than dealing with actual numbers of molecules in his quantitative work?

## 6.4 MOLAR WEIGHT

PG 6 G Define molar weight.

**Molar weight is the weight in grams of one mole of any species.** As defined, the units of molar weight are grams per mole. The concept is particularly useful in finding the number of grams in a specified number of moles, or vice versa, as we shall see in Section 6.7.

*Molar weight* is a general term that may be applied to any species, atoms, molecules or formula units. Many chemists and many textbooks use separate terms to identify the molar weights of these three species. For example, **the gram atomic weight of an element is the weight in grams that contains the same number of atoms as the number of atoms in exactly 12 grams of carbon-12.** If you substitute the word "units" for the first "atoms," the back part of that definition becomes "... that contains the same number of [units] as the number of atoms in exactly 12 grams of carbon-12," which is the definition of a mole given on page 124. Gram atomic weight, then, is the weight in grams of one mole of atoms of an element, which is what we have called *molar weight.* Identical comments may be made about the terms **gram molecular weight** and **gram formula weight,** so the term *molar weight* does indeed include all three.

There is one situation in which the distinction between *gram atomic weight* and *gram molecular weight* has an advantage over *molar weight*. It is when applied to the gaseous elements that form diatomic molecules. For example, the "molar weight of oxygen" is ambiguous. Does it mean the gram atomic weight, 16.0 grams, or the gram molecular weight, 32.0 grams for $O_2$? The difficulty is removed if you state clearly by word or formula the species whose molar weight is intended, as molar weight of O, oxygen atoms, or molar weight of $O_2$, oxygen molecules.

Another term widely used by chemists and in textbooks is *gram atom*. It refers to the quantity of an element whose weight is equal to the gram atomic weight. This quantity is the same as one mole, which has been identified as the official unit of *amount of substance* in the SI system of units. *Gram molecule* and *gram formula unit*, the latter term not widely used, are comparable. In this text the word *mole* will be used exclusively for the amount of a substance.

There is an important relationship you should catch at this point. Because the definitions of atomic weight, mole and molar weight are all tied to "exactly 12 grams of carbon-12," atomic weight and molar weight of atoms are *numerically* equal. The atomic weight of sodium is 23.0 amu; the molar weight of sodium atoms is 23.0 grams per mole. Consequently the periodic table becomes a source of molar weights of atoms in grams per mole, as well as atomic weights in atomic mass units. Similarly, the molecular weight of molecular compounds and the formula weight of ionic compounds are numerically equal to the molar weights of those compounds.

## CALCULATION OF MOLAR WEIGHT

> PG   6 H   Given the chemical formula of any species, or the name from which the formula may be determined, calculate the molar weight of that species.

What is the molar weight of CO? Based on the time-honored truth that the whole is equal to the sum of its parts, the weight of one mole of CO is equal to the sum of the weights of all the atoms it contains. Each molecule of CO contains one carbon atom and one oxygen atom. Therefore one mole of CO molecules contains one mole of carbon atoms (12.0 grams) and one mole of oxygen atoms (16.0 grams). The molar weight of CO is therefore 12.0 g C + 16.0 g O, or 28.0 grams per mole. Similarly . . . .

---

**Example 6.7**   Calculate the molar weight of sodium fluoride, NaF.

The procedure is straightforward. Solve the problem, remembering that the periodic table is your best source for atomic weights (molar weights of atoms).

--------------------------------------------------------------------------------

**6.7a**   42.0 g NaF/mole

$$23.0 \text{ g Na} + 19.0 \text{ g F} = 42.0 \text{ g NaF/mole}$$

Did you come up with 41.8 g/mole? If so, it's because you rounded off the atomic weight of sodium (22.98977) to 22.9, and of fluorine (18.99840) to 18.9. Always watch those round-offs! The answer given is both to three significant figures and to the first decimal, our arbitrarily stated standard for calculations based on atomic weight.

Now that the procedure is established, try another that is a bit more complicated. . . .

---

**Example 6.8**   Calculate the molar weight of calcium fluoride.

No formula this time. This you must furnish. What is it? Refer to periodic table for help, if necessary.

------------------------------------------------------------

**6.8a**   $CaF_2$ (Recall Example 5.8, page 108.)

Now you should be asking yourself how many moles of calcium and how many moles of fluorine are in one mole of $CaF_2$ units. Answer, please.

------------------------------------------------------------

**6.8b**   One mole of calcium and two moles of fluorine.

Now find the grams of calcium fluoride per mole.

------------------------------------------------------------

**6.8c**   78.1 g $CaF_2$/mole.

$$Ca: \quad 1 \times 40.1 = 40.1 \text{ g Ca}$$
$$F: \quad 2 \times 19.0 = \underline{38.0} \text{ g F}$$
$$78.1 \text{ g } CaF_2$$

---

**Example 6.9**   Find the molar weight of magnesium hydroxide.

You know the procedure. Although this problem does introduce a variation not met previously, you should be able to think it through. All the way to the answer, please. Refer to Example 5.8, page 108, if you have trouble with the formula.

------------------------------------------------------------

**6.9a**   58.3 g $Mg(OH)_2$/mole

$$Mg: \quad 1 \times 24.3 = 24.3 \text{ g Mg}$$
$$O: \quad 2 \times 16.0 = 32.0 \text{ g O}$$
$$H: \quad 2 \times 1.0 = \underline{2.0} \text{ g H}$$
$$58.3 \text{ g } Mg(OH)_2$$

This problem requires you to recognize that one mole of $Mg(OH)_2$ contains one mole of magnesium ions and two moles of hydroxide ions. Each hydroxide ion contains one mole of oxygen atoms and one mole of hydrogen atoms. Hence there are two moles of oxygen and hydrogen in the calculations.

---

**Example 6.10** Calculate the molar weight of aluminum sulfate.

------------------------------------------------------------

**6.10a** 342 g $Al_2(SO_4)_3$/mole

$$
\begin{array}{lll}
\text{Al:} & 2 \times 27.0 = & 54.0 \text{ g Al} \\
\text{S:} & 3 \times 32.1 = & 96.3 \text{ g S} \\
\text{O:} & 12 \times 16.0 = & \underline{192 \quad} \text{ g O} \\
& & 342 \quad \text{g } Al_2(SO_4)_3
\end{array}
$$

Each sulfate ion contains four atoms of oxygen, and there are three such ions in the formula, yielding $3 \times 4 = 12$ moles of oxygen atoms. The answer has been rounded off to three significant figures.

Let's try a hydrate:

---

**Example 6.11** Find the molar weight of $CuSO_4 \cdot 5H_2O$. (Cu is copper, at. no. 29.)

Recall the meaning of the hydrate formula: the indicated number of water molecules is locked into the crystal with one formula unit of the compound. Suggested procedure: First find the molar weight of $CuSO_4$.

------------------------------------------------------------

**6.11a** 160 g $CuSO_4$/mole

$$
\begin{array}{lll}
\text{Cu:} & 1 \times 63.5 = & 63.5 \text{ g Cu} \\
\text{S:} & 1 \times 32.1 = & 32.1 \text{ g S} \\
\text{O:} & 4 \times 16.0 = & \underline{64.0} \text{ g O} \\
& & 160 \quad \text{g } CuSO_4
\end{array}
$$

Now the molar weight of $H_2O$

------------------------------------------------------------

**6.11b**   18.0 g $H_2O$/mole

$$
\begin{array}{lll}
\text{H:} & 2 \times\ \ 1.0 = & 2.0 \text{ g H} \\
\text{O:} & 1 \times 16.0 = & \underline{16.0} \text{ g O} \\
& & 18.0 \text{ g } H_2O
\end{array}
$$

The formula unit consists of one $CuSO_4$ and 5 $H_2O$. The total molar weight is . . . .

-------------------------------------------------------------------------------------

**6.11c**   250 g $CuSO_4 \cdot 5H_2O$/mole

$$
\begin{array}{lll}
CuSO_4\text{:} & 1 \times 160\ \ = 160 & \text{g } CuSO_4 \\
H_2O\text{:} & 5 \times\ \ 18.0 = & \underline{90.0} \text{ g } H_2O \\
& & 250\ \ \ \text{g } CuSO_4 \cdot 5H_2O
\end{array}
$$

One more simple question—with a catch to it!

**Example 6.12**   Find the molar weight of bromine vapor.

-------------------------------------------------------------------------------------

**6.12a**   160 g $Br_2$/mole

If you reported 79.9 you forgot that bromine forms diatomic molecules (see Sec. 5.3, Fig. 5.4, page 99). Failure to recognize diatomic molecules is a frequent source of error in later problems. Be alert to it. If you recorded 159.8 grams/mole you were correct, but more correct than necessary to meet the three significant figure standard. An answer of $1.60 \times 10^2$ grams/mole would be preferable to the one given, since it would indicate clearly the number of significant figures.

# 6.5   PERCENTAGE COMPOSITION OF A COMPOUND

PG   6 I   Given the chemical formula of any compound, or the name from which the formula may be determined, calculate the percentage of each element in the compound.

The percentage of a component can always be computed by dividing the quantity of that component by the total quantity and multiplying by 100:

$$
\frac{\text{Quantity of component}}{\text{Total quantity}} \times 100 = \text{Percent} \qquad (6.3)
$$

Unless otherwise specified, in this book "percent" is understood to mean percent by weight. The percentage by weight of an element follows directly from the concept of molar weight. In Example 6.10 above, it was found that one mole of $Al_2(SO_4)_3$ is made up of 54.0 grams of aluminum, 96.3 grams of sulfur, and 192 grams of oxygen, a total of 342 grams per mole. The percentage composition of $Al_2(SO_4)_3$ is therefore as follows:

$$Al: \quad \frac{54.0 \text{ grams Al}}{342 \text{ grams Al}_2(SO_4)_3} \times 100 = \quad 15.8\% \text{ Al}$$

$$S: \quad \frac{96.3 \text{ grams S}}{342 \text{ grams Al}_2(SO_4)_3} \times 100 = \quad 28.2\% \text{ S}$$

$$O: \quad \frac{192 \text{ grams O}}{342 \text{ grams Al}_2(SO_4)_3} \times 100 = \quad 56.1\% \text{ O}$$

$$\text{Total percent} \qquad\qquad = 100.1$$

The sum of the percentages must total 100.0. Deviations of 0.1 or 0.2 may be attributed to rounding off.

The procedure illustrated is applicable to any compound.

---

**Example 6.13** Calculate the percentage composition of calcium nitrate.

The formula is needed, then the figures leading to the molar weight. Finally you must calculate the percent of each element from those figures. Carry it all the way, and check to see if the percentages total 100.0.

-------------------------------------------------------------------------------------

**6.13a** 24.4% calcium; 17.1% nitrogen; 58.5% oxygen

Formula: $Ca(NO_3)_2$

| Element | Grams | Percent |
|---------|-------|---------|
| Ca | $1 \times 40.1 = \quad 40.1$ g Ca | $\frac{40.1}{164} \times 100 = \quad 24.4\%$ Ca |
| N | $2 \times 14.0 = \quad 28.0$ g N | $\frac{28.0}{164} \times 100 = \quad 17.1\%$ N |
| O | $6 \times 16.0 = \quad 96.0$ g O | $\frac{96.0}{164} \times 100 = \quad 58.5\%$ O |
|  | $\overline{164.1 \text{ g Ca(NO}_3)_2}$ | $\overline{100.0\%}$ |

## 6.6 CONVERSION BETWEEN MASS AND NUMBER OF MOLES

PG 6 J  Given the number of grams (or moles) of a chemical species of known or calculable molar weight, find the number of moles (or grams).

The "counting by weighing" value of the mole concept is based on the ready conversion of the mass of a chemical species into a number of moles, and vice versa. This simple operation is a part of many chemical problems, and is one of the most important skills to be acquired in the introductory course. The operation is based upon the equivalence between grams and moles of a chemical that emerges from the mole concept. If MW is the molar weight of any species, then

$$\text{MW grams} \doteq 1 \text{ mole} \tag{6.4}$$

The following examples illustrate the two possible conversions.

---

**Example 6.14**  How many moles of aluminum sulfate, $Al_2(SO_4)_3$ are in 150 grams? The molar weight, from Example 6.10, is 342 grams/mole.

Applying Equation 6.4 to the problem at hand yields the equivalence 342 grams $Al_2(SO_4)_3 \doteq 1$ mole $Al_2(SO_4)_3$. Your unit path is simply grams to moles. If one mole is 342 grams, then the given quantity of 150 grams is . . . .

---

**6.14a**  0.439 mole $Al_2(SO_4)_3$

$$150 \text{ grams } Al_2(SO_4)_3 \times \frac{1 \text{ mole } Al_2(SO_4)_3}{342 \text{ grams } Al_2(SO_4)_3} = 0.439 \text{ mole } Al_2(SO_4)_3$$

---

**Example 6.15**  You are carrying out a laboratory reaction that requires 0.0250 mole of $NiCl_2 \cdot 6 H_2O$. How many grams of the compound do you weigh out?

First, find the molar weight of $NiCl_2 \cdot 6 H_2O$. (Ni is nickel, at. no. 28).

---

**6.15a**  238 g $NiCl_2 \cdot 6 H_2O$/mole

Use molar weight to convert from the given quantity of moles to grams. If 1 mole is 238 grams, 0.0250 mole is . . . .

---

**6.15b**  5.95 grams $NiCl_2 \cdot 6 H_2O$

$$0.0250 \text{ mole } NiCl_2 \cdot 6 H_2O \times \frac{238 \text{ grams } NiCl_2 \cdot 6 H_2O}{1 \text{ mole } NiCl_2 \cdot 6 H_2O}$$

$$= 5.95 \text{ grams } NiCl_2 \cdot 6 H_2O$$

## 6.7 NUMBER OF ATOMS, MOLECULES OR FORMULA UNITS IN A SAMPLE

PG 6 K   Given the mass of a pure substance of known or calculable molar weight, or the number of moles, find the number of atoms, molecules or formula units; or, given the number of atoms, molecules or formula units, find the mass or the number of moles.

Equation 6.2, 1 mole $\simeq 6.02 \times 10^{23}$ units, furnishes a simple one-step conversion between moles and units. In principle the following example is no more complicated then telling how many eggs are in four dozen.

---

**Example 6.16**   How many sodium atoms are in 4.00 moles of sodium atoms?

------------------------------------------------------------------------

**6.16a**   $24.1 \times 10^{23}$ sodium atoms

$$4.00 \text{ moles Na} \times \frac{6.02 \times 10^{23} \text{ Na atoms}}{1 \text{ mole Na}} = 24.1 \times 10^{23} \text{ Na atoms}$$

---

To find the number of units of a substance in a given mass, you must begin the setup one step earlier with a grams-to-moles conversion.

---

**Example 6.17**   How many molecules are in 500 grams of water (about one pint)?

*Solution.*   Using Equation 6.4 (page 131) first, and then 6.2, the unit path begins with the given quantity and goes grams → moles → molecules:

$$500 \text{ g } H_2O \times \frac{1 \text{ mole } H_2O}{18.0 \text{ g } H_2O} \times \frac{6.02 \times 10^{23} H_2O \text{ molecules}}{1 \text{ mole } H_2O}$$

$$= 1.67 \times 10^{25} H_2O \text{ molecules}$$

---

In the reverse problem, finding the mass of a given number of units, the procedure is logically reversed. . . .

---

**Example 6.18**   What is the mass of one billion billion ($1.00 \times 10^{18}$) molecules of ammonia, $NH_3$?

At one point the molar weight of ammonia will be required. Find it first.

------------------------------------------------------------------------

**6.18a**  17.0 g $NH_3$/mole

Now, starting with the given quantity, convert molecules to moles, and then change moles to mass in grams.

-----------------------------------------------------------------------------------------

**6.18b**  $2.82 \times 10^{-5}$ gram $NH_3$

$$1.00 \times 10^{18} \text{ molecules NH}_3 \times \frac{1 \text{ mole NH}_3}{6.02 \times 10^{23} \text{ molecules NH}_3} \times \frac{17.0 \text{ g NH}_3}{\text{mole NH}_3}$$

$$= 2.82 \times 10^{-5} \text{ gram NH}_3.$$

This very small mass, about $\frac{6}{100,000,000}$ of a pound, suggests again the enormity of the number of molecules in a mole.

# 6.8  EMPIRICAL FORMULA OF A COMPOUND

## EMPIRICAL FORMULAS AND MOLECULAR FORMULAS

> PG  6 L  Distinguish between an empirical formula and a molecular formula.

The percentage composition of the compound ethene (also called ethylene) is 85.7% carbon and 14.3% hydrogen. Its chemical formula is $C_2H_4$. The percentage composition of the compound propene (also called propylene) is likewise 85.7% carbon and 14.3% hydrogen. Its formula is $C_3H_6$. These are, in fact, but two of a whole series of compounds having the general formula $C_nH_{2n}$, where n is an integer. In ethene and propene, n = 2 and 3, respectively. All compounds in this series have the same percentage composition.

$C_2H_4$ and $C_3H_6$ are typical **molecular formulas.** They show the number of carbon and hydrogen atoms actually present in a molecule of ethene and propene. They are formulas of real chemical substances. If, in the general formula $C_nH_{2n}$, we let n be 1, the resulting formula is $CH_2$. This formula is the **empirical formula** for all compounds having the general formula $C_nH_{2n}$. The **empirical formula shows the simplest ratio of atoms of the elements in the compound.** All subscripts are reduced to their lowest terms: they have no common divisors. Empirical formulas can be determined from percentage composition data—as we shall shortly see—which are found by chemical analysis. Hence the name *empirical,* which means that which is based on experience and/or experiment.

Empirical formulas may or may not be molecular formulas of real chemical compounds. There happens to be no known stable compound that has the formula $CH_2$—and there is good reason to believe that no such compound can exist. On the other hand, the molecular formula of dinitrogen tetroxide is $N_2O_4$. The subscript numbers have a common divisor, 2. Dividing by 2 we reach the empirical formula, $NO_2$. This empirical formula is also the molecular formula of a real chemical compound, nitrogen dioxide. In other words, $NO_2$ is *both* the empirical formula and the molecular formula for nitrogen dioxide, as well as the empirical formula for dinitrogen tetroxide.

---

**Example 6.19** For each formula below that could be an empirical formula, write EF; for each formula that could not be an empirical formula, write the empirical formula corresponding to the formula shown: $C_4H_{10}$; $C_2H_6O$; $Hg_2Cl_2$; $(CH)_6$.

---

**6.19a** $C_4H_{10}$ ___$C_2H_5$___          $Hg_2Cl_2$ ___HgCl___

$C_2H_6O$ ___EF___          $(CH)_6$ ___CH___

---

## DETERMINATION OF AN EMPIRICAL FORMULA

PG 6 M   Given data from which the ratio of relative weights of the elements in a compound may be determined, calculate the empirical formula of the compound.

To determine the empirical formula of a compound from its percent composition we must find the whole number ratio of atoms of the elements in the compound. The numbers in this ratio become the subscripts in the empirical formula of the compound. The procedure by which this is done is as follows:

1. Determine the relative weights of different elements in the compound.
2. Convert these relative weights to the relative numbers of moles of atoms of the elements.
3. Express the number of moles of atoms as the smallest possible ratio of integers.

The integers found in the last step are the subscripts in the empirical formula.

It is sometimes helpful to organize the calculations in an empirical formula problem in a table. We shall follow this course. The headings in the table will be

ELEMENT   GRAMS   MOLES   MOLE RATIO   FORMULA RATIO

We will illustrate the procedure for finding empirical formulas with the compound ethene. As noted, ethene is 85.7% carbon and 14.3% hydrogen. The first step in the above procedure calls for converting the percentage figures into relative weights of the elements. Thinking of percent as the number of parts of one element per 100 parts of the compound—or number of *grams* of one element per 100 *grams* of the compound—it follows that a 100 gram sample of the unknown contains 85.7 grams of carbon and 14.3 grams of hydrogen. From this we see that *percentage composition figures may always be interpreted directly as the relative weights of different elements* in satisfying the first step of the procedure. The elemental symbol and the relative grams of each element are entered into the table:

| Element | Grams | Moles | Mole Ratio | Formula Ratio |
|---|---|---|---|---|
| C | 85.7 | | | |
| H | 14.3 | | | |

We are now ready to find the number of moles of atoms of each element, step 2 in the procedure. This is a direct one-step conversion from grams to moles, as in Section 6.6.

| Element | Grams | Moles | Mole Ratio | Formula Ratio |
|---|---|---|---|---|
| C | 85.7 | $\dfrac{85.7 \text{ g C}}{12.0 \text{ g/mole}} = 7.14$ | | |
| H | 14.3 | $\dfrac{14.3 \text{ g H}}{1.01 \text{ g/mole}} = 14.3$ | | |

Students sometimes question the conversion of grams of hydrogen, or any elemental gas that forms diatomic molecules, to moles of atoms. They tend to divide by the molar weight of *molecules*—the *gram molecular weight*—rather than the molar weight of *atoms*—the *gram atomic* weight. But a chemical formula expresses the ratio of moles of *atoms* of the different elements, so the molar weight of atoms, or atomic weight, must be used.

It is the ratio of these moles of atoms that must now be expressed in the smallest possible ratio of integers, step 3 in the procedure. This is most easily done by *dividing each number of moles by the smallest number of moles*. In this problem the smallest number of moles is 7.14. Thus,

| Element | Grams | Moles | Mole Ratio | Formula Ratio |
|---|---|---|---|---|
| C | 85.7 | 7.14 | $\dfrac{7.14}{7.14} = 1.00$ | 1 |
| H | 14.3 | 14.2 | $\dfrac{14.2}{7.14} = 1.99$ | 2 |

*The ratio of moles of atoms of the different elements is equal to the ratio of atoms of the elements*—the formula ratio, as we have called it here. To see this in a more familiar setting, the numerical ratio of handlebars to wheels in bicycles is $\frac{1}{2}$. In four dozen bicycles there are four dozen handlebars and eight dozen wheels. The ratio $\dfrac{4 \text{ dozen handlebars}}{8 \text{ dozen wheels}}$ is also equal to $\frac{1}{2}$. Thus the numbers in the Mole Ratio column are in the same ratio as the subscripts in the empirical formula.

The Formula Ratio column, which is filled in above, expresses the mole ratio numbers as integers. Adjustments of *no more than* a few hundredths of a unit are often necessary to correct for experimental error or roundoff and significant figure variations. Interpretation of 1.99 as 2 falls into either of these categories. We therefore conclude that the empirical formula of the compound is $CH_2$.

If either quotient in the Mole Ratio column had been other than an integer, the formula ratio would be found by multiplying both quotients by a suitable number. This is shown in the example that follows.

---

**Example 6.20** A piece of iron (at. no. 26) weighs 1.34 grams. Exposed to oxygen under conditions in which oxygen combines with all of the iron to form a pure oxide of iron, the final weight increases to 1.92 grams. Find the empirical formula of the compound.

As before, relative weights of the elements contained in the compound are required—but this time they are not obtained from percentage composition values. The number of grams of iron in the final compound is given in the problem. How many grams of oxygen combine with 1.34 grams of iron if the iron oxide produced weighs 1.92 grams? Set up the tabulation, determine the grams of oxygen, and enter both elemental masses into the table.

---

**6.20a**  0.58 gram of oxygen

Grams oxygen = grams iron oxide − grams iron

$$= 1.92 \text{ grams} \quad - 1.34 \text{ grams} = 0.58 \text{ gram}$$

| ELEMENT | GRAMS | MOLES | MOLE RATIO | FORMULA RATIO |
|---------|-------|-------|------------|---------------|
| Fe      | 1.34  |       |            |               |
| O       | 0.58  |       |            |               |

Step 2 is to compute the number of moles of atoms of each element. Do so, and place the results into the table.

---

**6.20b**

| ELEMENT | GRAMS | MOLES | MOLE RATIO | FORMULA RATIO |
|---------|-------|-------|------------|---------------|
| Fe      | 1.34  | $\dfrac{1.34 \text{ g Fe}}{55.8 \text{ g/mole}} = 0.0240$ | | |
| O       | 0.58  | $\dfrac{0.58 \text{ g O}}{16.0 \text{ g/mole}} = 0.036$ | | |

Recalling that the mole ratio figures are obtained by dividing each number of moles by the smallest, complete step 3 and place the result in the table.

---

**6.20c**

| ELEMENT | GRAMS | MOLES | MOLE RATIO | FORMULA RATIO |
|---------|-------|-------|------------|---------------|
| Fe | 1.34 | 0.0240 | $\dfrac{0.0240}{0.0240} = 1.00$ | |
| O | 0.58 | 0.036 | $\dfrac{0.036}{0.0240} = 1.5$ | |

This time the numbers in the mole ratio column are not both integers or very close to integers. But they can be changed to integers and kept in the same ratio by multiplying both of them by the same small integer. Find the smallest whole number that will yield integers when used as a multiplier for 1.00 and 1.5; use it to obtain the formula ratio figures; complete the table; and write the empirical formula of the compound.

------------------------------------------------------------------------------

**6.20d** $Fe_2O_3$

| ELEMENT | GRAMS | MOLES | MOLE RATIO | FORMULA RATIO |
|---------|-------|-------|------------|---------------|
| Fe | 1.34 | 0.0240 | 1.00 | 2 |
| O | 0.58 | 0.036 | 1.5 | 3 |

Multiplication of both 1.00 and 1.5 by 2 yields $1.00 \times 2 = 2$ and $1.5 \times 2 = 3$, both whole numbers. The decimal part of the mole ratio determines the multiplier. Had the mole ratio been 1.33 or 1.67, multiplication by 3 would have yielded 4 or 5 for the formula ratio column: $1.33 \times 3 = 4$, and $1.67 \times 3 = 5$. A mole ratio of 1.25 or 1.75 could be multiplied by 4 to produce 5 or 7 as integer products: $1.25 \times 4 = 5$, and $1.75 \times 4 = 7$. You are unlikely to encounter more complicated ratios in the beginning course.

The procedure is the same for compounds containing more than two elements.

**Example 6.21** An organic compound analyzes 20.0% carbon, 2.2% hydrogen and 77.8% chlorine. Determine the empirical formula of the compound.

------------------------------------------------------------------------------

**6.21a** $C_3H_4Cl_4$

| ELEMENT | GRAMS | MOLES | MOLE RATIO | FORMULA RATIO |
|---------|-------|-------|------------|---------------|
| C | 20.0 | $\dfrac{20.0 \text{ g C}}{12.0 \text{ g/mole}} = 1.67$ | $\dfrac{1.67}{1.67} = 1.00$ | 3 |
| H | 2.2 | $\dfrac{2.2 \text{ g H}}{1.01 \text{ g/mole}} = 2.2$ | $\dfrac{2.2}{1.67} = 1.3$ | 4 |
| Cl | 77.8 | $\dfrac{77.8 \text{ g Cl}}{35.5 \text{ g/mole}} = 2.19$ | $\dfrac{2.19}{1.67} = 1.31$ | 4 |

Notice that the experimental results in this problem yield mole ratio figures 1.3 and 1.31, both close to .33 in the decimal part of the number. Multiplication of the mole ratio figures by 3 yields integers for the formula ratio column: $1.00 \times 3 = 3$; $1.3 \times 3 = 3.9$ or 4; $1.31 \times 3 = 3.93$ or 4.

The determination of an empirical formula is an important step in finding the molecular formula of an unknown compound if the molar weight of the unknown can be established. For example, suppose we know the empirical formula of a compound is $CH_2$, as in ethene. Suppose also we can determine that the molar weight of the compound is 70.0 grams per mole. What value of n in $C_nH_{2n}$—or $(CH_2)_n$, which is a somewhat more convenient form for this purpose—would yield a formula with a molar weight of 70.0? The molar weight of the empirical formula, $CH_2$, is 14 grams per mole. Dividing 70 by 14 gives five empirical formula units in one mole of unknown. Therefore $(CH_2)_5$, or $C_5H_{10}$ is the molecular formula with molar weight 70.0 grams per mole and empirical formula $CH_2$. The compound is called pentene. Complete examples of this procedure appear in Chapter 11, where you will learn how to find the molecular weight of a compound from experimental data.

# CHAPTER 6 IN REVIEW

### 6.1  ATOMIC WEIGHT

6 A    Referring to a periodic table and given one of the following, state the other two: (a) nuclear symbol; (b) number of protons and neutrons in the nucleus; (c) atomic number and mass number. (Page 119)

6 B    Define the atomic mass unit, amu. (Page 120)

6 C    Given the relative abundance of the naturally occurring isotopes of an element, calculate the atomic weight. (Page 120)

6 D    Given the name or symbol of an element you know, or the atomic number of any element, and a periodic table, state the atomic weight of the element. (Page 120)

### 6.2  MOLECULAR WEIGHT; FORMULA WEIGHT

6 E    Distinguish between the following terms: atomic weight; molecular weight; formula weight. (Page 123)

### 6.3  THE MOLE CONCEPT

6 F    Define the mole. (Page 124)

### 6.4  MOLAR WEIGHT

6 G    Define molar weight. (Page 125)

6 H    Given the chemical formula of any species, or the name from which the formula may be determined, calculate the molar weight of that species. (Page 126)

### 6.5  PERCENTAGE COMPOSITION OF A COMPOUND

6 I    Given the chemical formula of any compound, or the name from which the formula may be determined, calculate the percentage of each element in the compound. (Page 129)

## 6.6 CONVERSION BETWEEN MASS AND NUMBER OF MOLES

**6 J**    Given the number of grams (or moles) of a chemical species of known or calculable molar weight, find the number of moles (or grams). (Page 130)

## 6.7 NUMBER OF ATOMS OR MOLECULES IN A SAMPLE

**6 K**    Given the mass of a pure substance of known or calculable molar weight, or the number of moles, find the number of atoms, molecules or formula units; or, given the number of atoms, molecules or formula units, find the mass or the number of moles. (Page 132)

## 6.8 EMPIRICAL FORMULA OF A COMPOUND

**6 L**    Distinguish between an empirical formula and a molecular formula. (Page 134)

**6 M**    Given data from which the ratio of relative weights of the elements in a compound may be determined, calculate the empirical formula of the compound. (Page 134)

# TERMS AND CONCEPTS

Nuclear symbol (119)
Mass number (119)
Atomic mass unit, amu (122)
Atomic weight (122)
Molecular weight (123)
Formula weight (123)

Mole (124)
Avogadro's number, $6.02 \times 10^{23}$ (124)
Molar weight (125)
Percent composition of a compound (129)
Molecular formula (133)
Empirical formula (133)

# QUESTIONS AND PROBLEMS

*Recall that formulas included in the mininomenclature system introduced in Chapter 5 are summarized for reference on pages 118 and 528. An asterisk (\*) identifies a question that is relatively difficult, or that extends beyond the performance goals in the chapter.*

Section 6.1

6.1) A natural isotope of lithium has the symbol $^{6}_{3}$Li. State the following about this isotope: atomic number; mass number; number of protons; number of neutrons.

6.2) An element has an atomic number of 24, and its nucleus contains 28 neutrons. State the name and mass number and write the symbol of the isotope. (Refer to the periodic table.)

6.3) The mass number of a certain isotope of zinc is 68. For an atom of this isotope list in order the number of protons, neutrons and electrons, the nuclear symbol, and the name.

6.50) The most common isotope of lithium has four neutrons in its nucleus. State the mass number and write the nuclear symbol of the isotope.

6.51) 75 is the mass number of the only natural isotope of the element having atomic number 33. State the number of neutrons in the nucleus, write the nuclear symbol of the isotope, and state its name. (Use both periodic table and Table of Elements inside back cover of your book if necessary.)

6.52) One of the most hazardous species in nuclear fallout is strontium-90. Given the hint that strontium is an alkaline earth metal, list the nuclear symbol for this isotope and the number of neutrons, protons and electrons in the nucleus. Refer only to the periodic table.

6.4) What is an atomic mass unit? What advantage does it have over grams as a unit of mass when speaking of atomic and subatomic particles?

6.5) Calculate to two decimals the atomic weight of chlorine from the data in Table 6.1, page 121.

6.6) 51.82% of the atoms of a certain element in nature have a mass number of 107, and the balance are 109. Calculate the average atomic weight from these data and, by comparing with a periodic table, identify the element.

6.7) Referring only to a periodic table, list the atomic weights of the following elements, which are identified by name or atomic number: potassium; sulfur; at. no. 24; at. no. 50.

Section 6.2

6.8) Why is it incorrect to refer to the atomic weight of sodium chloride?

6.9) It may be said that because atomic, molecular and formula weights are all comparative weights, they are conceptually alike. What, then, is their difference?

Section 6.3

6.10) Define the term *mole*.

6.11) Give a definition of Avogadro's Number N.

Section 6.4

6.12) Define *molar weight*.

6.13–6.18 and 6.62–6.67: For each substance identified by name or formula, calculate the molar weight:

6.13) Lithium bromide

6.14) Chlorine gas

6.15) Calcium sulfate

6.16) Ammonium nitrate

6.17) Benzene, $C_6H_6$

6.18) Sodium carbonate decahydrate

6.53) (a) What is the average mass of boron atoms in atomic mass units? (b)* The mass of an "average atom" of a certain element is 3.34 times as great as the mass of an atom of carbon-12. Identify the element.

6.54) Using data from Table 6.1, page 121, find the atomic weight of sulfur to two decimal places.

6.55) A certain element has five isotopes, distributed as follows in nature:

| Mass no. | 58 | 60 | 61 | 62 | 64 |
|---|---|---|---|---|---|
| Percent | 67.88 | 26.23 | 1.19 | 3.66 | 1.08 |

Calculate the average atomic weight and identify the element from these data.

6.56) Using only a periodic table for reference, list the atomic weights of the elements below, which are identified by name or atomic number: helium; aluminum; at. no. 29; at. no. 82.

6.57) Why is it proper to speak of the molecular weight of water, but not the molecular weight of potassium nitrate?

6.58) Other than sodium chloride, potassium nitrate and water, list two species to which *atomic weight* applies, two that have a *molecular weight*, and two for *formula weight*.

6.59) What do quantities representing one mole of sodium, one mole of water and one mole of sodium chloride have in common?

6.60) Give the name and value of the number associated with the mole.

6.61) Why is the expression "molar weight of hydrogen" ambiguous, and how may the ambiguity be removed?

6.62) Sodium chloride

6.63) Fluorine gas

6.64) Potassium nitrate

6.65) Ammonium phosphate

6.66) Ethanol, $C_2H_5OH$

6.67) Cobalt sulfate hexahydrate, $CoSO_4 \cdot 6\ H_2O$

Section 6.5

6.19–6.22 and 6.68–6.71: Calculate the percentage compositions of the following compounds, selected from Problems 6.13–6.18 and 6.62–6.67:

6.19)  Lithium bromide

6.20)  Calcium sulfate

6.21)  Benzene, $C_6H_6$

6.22)  Calculate the percentage water in sodium carbonate decahydrate.

6.68)  Sodium chloride

6.69)  Potassium nitrate

6.70)  Ethanol, $C_2H_5OH$

6.71)  Calculate the percentage anhydrous cobalt sulfate in $CoSO_4 \cdot 6\ H_2O$.

## Section 6.6

*6.23–6.25 and 6.72–6.74: For each given number of grams of substances selected from Problems 6.13–6.18 and 6.62–6.67, calculate the number of moles:*

6.23)  68.4 g LiBr

6.24)  17.2 g $Cl_2$

6.25)  34.1 g $CaSO_4$

6.72)  19.7 g $(NH_4)_3PO_4$

6.73)  28.9 g $C_2H_5OH$

6.74)  108 g $CoSO_4 \cdot 6\ H_2O$

*6.26–6.28 and 6.75–6.77: For each given number of moles of substances selected from Problems 6.13–6.18 and 6.62–6.67, calculate the number of grams:*

6.26)  0.345 mole $NH_4NO_3$

6.27)  1.82 moles $C_6H_6$

6.28)  0.791 mole $Na_2CO_3 \cdot 10\ H_2O$

6.75)  1.40 moles NaCl

6.76)  2.19 moles $F_2$

6.77)  0.108 mole $(NH_4)_3PO_4$

## Section 6.7

*6.29–6.33 and 6.78–6.82: How many atoms are in each of the following:*

6.29)  1.24 moles Mg

6.30)  0.713 mole $Br_2$ (caution!)

6.31)  29.6 g Na

6.32)  3.40 g Ca

6.33)  38.1 g $N_2$

6.78)  0.436 mole P

6.79)  $1.41 \times 10^{-5}$ mole Rb (at. no. 37)

6.80)  41.2 g Al

6.81)  0.000162 g K

6.82)  0.00461 g $Cl_2$

*6.34–6.38 and 6.83–6.87: How many molecules or formula units are in each of the following:*

6.34)  0.521 mole $H_2S$

6.35)  0.0626 mole $Br_2$

6.36)  12.4 g $N_2$

6.37)  6.45 g CO

6.38)  13.6 g LiBr

6.83)  0.00331 mole Ne

6.84)  6.53 moles $N_2O_5$

6.85)  0.106 g $Cl_2$

6.86)  40.2 g $CO_2$

6.87)  30.9 g NaCl

*6.39, 6.40, 6.88 and 6.89: Calculate the number of moles in each of the following:*

6.39)  $2.35 \times 10^{21}$ Ba atoms

6.40)  $1.09 \times 10^{23}$ $Br_2$ molecules

6.88)  $1.84 \times 10^{24}$ Pb atoms

6.89)  $6.91 \times 10^{20}$ formula units of $CuSO_4$

*6.41–6.43 and 6.90–6.92: Calculate the mass of:*

6.41)  $7.06 \times 10^{23}$ He atoms

6.42)  $4.06 \times 10^{22}$ molecules of $O_2$

6.43)  $1.19 \times 10^{23}$ formula units of KI

6.90)  $1.07 \times 10^{23}$ Mg atoms

6.91)  $8.19 \times 10^{24}$ CO molecules

6.92)  $7.23 \times 10^{23}$ formula units of NaF

Section 6.8

6.44)   Explain why $C_6H_{10}$ must be a molecular formula, while $C_7H_{10}$ could be either a molecular formula or an empirical formula.

6.45)   Analysis of an organic compound shows it to be 40.0% carbon, 6.7% hydrogen and 53.3% oxygen. Calculate the empirical formula of the compound.

6.46)   28.8 grams of a certain hydrocarbon is found to contain 25.3 grams of carbon and the balance hydrogen. Find the empirical formula of the compound.

6.47)   A compound analyzes as 32.4% sodium, 22.6% sulfur and 45.0% oxygen. Find the empirical formula of the compound.

6.48)   The white rock dolomite, used in gardens, is 13.2% magnesium, 21.8% calcium, 13.0% carbon and 52.1% oxygen. Find the empirical formula of dolomite.

6.49)*   A compound is, by analysis, 5.00% hydrogen and 95.00% fluorine. Its molar weight is 40.0 grams per mole. Find both the empirical and molecular formulas of the compound.

6.93)   From the following list, identify each formula that could be an empirical formula. Write the empirical formulas of the other compounds. $C_2H_5OH$; $Na_2O_2$; $CH_3COOH$; $N_2O_5$.

6.94)   The percentage composition of an organic compound is 48.6% carbon, 8.1% hydrogen and 43.2% oxygen. Determine the empirical formula of the compound.

6.95)   Analysis of a sample of a hydrocarbon shows it consists of 17.9 grams of carbon and 4.5 grams of hydrogen. Determine the empirical formula of the compound.

6.96)   The percentage composition of a compound is 26.1% nitrogen, 7.5% hydrogen and 66.4% chlorine. Write the empirical formula of the compound.

6.97)   When 7.30 grams of iron powder react and combine with 6.30 grams of powdered sulfur, a sulfide of iron is formed. Determine the empirical formula of the compound.

6.98)*   A 3.80 gram sample of a pure compound of sulfur and fluorine is found by analysis to contain 1.37 grams of sulfur. Another experiment establishes that the molar weight of the compound is 178 grams/mole. Find the empirical and molecular formulas of this fluoride of sulfur.

# CHEMICAL REACTIONS AND EQUATIONS

In Chapter 7 we continue to encourage the development of your formula writing skills by furnishing only the names of compounds whose formulas were in the mininomenclature system introduced in Chapter 5. If you need assistance, you may find the acid and molecular compound formulas summarized on page 118, and the formulas of ionic compounds on page 528.

## 7.1 EVOLUTION OF A CHEMICAL EQUATION

If sodium is placed in water, the substances react vigorously, producing hydrogen gas and a solution of sodium hydroxide, and releasing heat (see Fig. 7.1). In the foregoing sentence 22 words have been used to describe a chemical reaction. With many reactions to consider, the chemist will seek a shorter description. He writes instead a *chemical equation*, as introduced in Section 5.4, page 100. You will study chemical equations and learn how to write them in this chapter.

The minimum "word equation" for the reaction referred to above is "sodium plus water yields hydrogen plus sodium hydroxide solution plus heat." The corresponding chemical equation reads exactly the same way:

$$Na + H_2O \rightarrow H_2 + NaOH \ (aq) + heat \tag{7.1}$$

Nearly all chemical reactions involve some transfer of energy, usually in the form of heat, as shown here. Energy terms are generally omitted from equations unless there is a specific reason for including them. Equations that include the gain or loss of heat will be considered in Chapter 13.

The word-equation states that a product of the reaction is "sodium hydroxide solution." The (aq) following NaOH indicates the sodium hydroxide

**Figure 7.1** Sodium reacting with water. A, Small piece of sodium dropped into beaker of water. B, Sodium forms "ball" that dashes erratically over water surface, releasing hydrogen as it reacts. C, Solution of sodium hydroxide, NaOH (aq), which is hot because of heat released in the reaction. WARNING: Do not "try" this experiment, as it is dangerous, potentially splattering hot alkali into eyes and onto skin and clothing.

is in *aqueous* (water) solution. There are many times that the state or form of a species in a reaction is important, and it may be shown by an appropriate "state designation," in parentheses, following the symbol of the species. We shall, as a matter of policy, include state designations in all equations in this book, directing attention to those reactions in which the state is particularly significant. In addition to the aqueous solution form of a reaction species, the three normal states of matter, solid (s), liquid (l) and gas (g), are identified. In the reaction just considered all four states appear. The chemical equation, including states, is:

$$Na\ (s)\ +\ H_2O\ (l) \rightarrow H_2\ (g)\ +\ NaOH\ (aq) \qquad (7.2)$$

Sodium at room conditions is a solid; water at room temperature and atmospheric pressure is a liquid (water vapor does exist at room temperature, and sometimes the distinction between liquid and gaseous water is important); and hydrogen is a gas.

Two quantitative interpretations of a chemical equation will be described in the next section, but one must be introduced here. From a "molecular" point of view, the above reaction might be considered as one in which atoms of sodium react with molecules of water, producing molecules of hydrogen and "formula units" of sodium hydroxide—an ionic compound—in aqueous solution. But the equation does not say *how many* of each of these species are involved. It cannot be one of each. This would violate the law of conservation of mass, as well as its counterpart in Dalton's atomic theory that says atoms are neither created nor destroyed in chemical reactions. Count up the atoms of each element and you will see that this is so.

There is a single sodium atom on each side of the equation. There is also one oxygen atom. But the left side of the equation has two hydrogen atoms, whereas the right side has three.

How do we account for this seeming discrepancy? The answer is that the chemical species do not take part in the reaction on a simple 1:1:1:1 ratio. The problem can be resolved by taking one or more of the chemicals in multiple—done by adjusting coefficients of the four species—until the atoms of each element are "balanced" on each side of the equation. It is the hydrogen that is unbalanced; the left side is short by one. Therefore let's try two water molecules:

$$\text{Na (s)} + 2\ \text{H}_2\text{O (l)} \rightarrow \text{H}_2 \text{(g)} + \text{NaOH (aq)} \qquad (7.3)$$

At first glance, this hasn't helped; indeed it seems to have made matters worse. The hydrogen is still out of balance (four on the left, three on the right); and furthermore, oxygen is now unbalanced (two on the left, one on the right). We are short one oxygen and one hydrogen atom on the right side. But look closely. Oxygen and hydrogen are part of the same compound on the right, and there is one atom of each in that compound. If we take two NaOH units,

$$\text{Na (s)} + 2\ \text{H}_2\text{O (l)} \rightarrow \text{H}_2 \text{(g)} + 2\ \text{NaOH (aq)} \qquad (7.4)$$

there are four hydrogens and two oxygens on both sides of the equation. These elements are now in balance. But, alas, the sodium has been *un*balanced. Correction of this condition, however, should be obvious:

$$2\ \text{Na (s)} + 2\ \text{H}_2\text{O (l)} \rightarrow \text{H}_2 \text{(g)} + 2\ \text{NaOH (aq)} \qquad (7.5)$$

The equation is now balanced; it satisfies the law of conservation of mass, reflected in Dalton's statement that atoms are neither created nor destroyed in chemical reactions. Note that, in the absence of a numerical coefficient, as with $\text{H}_2$, the coefficient is assumed to be 1.

Balancing an equation involves some important do's and don'ts that are apparent in this example:

DO:    *Balance the equation* entirely *by use of coefficients placed before the different chemical formulas.*

DON'T: *Change a correct chemical formula in order to make an element balance; and*

DON'T: *Add some real or imaginary chemical species to either side of the equation just to make an element balance.*

There is a great temptation toward either of the "don'ts," and a moment's reflection should show why they are improper. The original equation expresses the *correct* formula for each species present. If you change or add a formula, even if the new formula is for a real chemical, it is *not a species in the given reaction* and does not belong in the equation. A chemical equation must correspond to reality; that is, it must describe as accurately as possible what actually happens when we carry out a reaction. Coefficients alone must be used in balancing equations.

## 7.2 MEANING OF A CHEMICAL EQUATION

PG 7 A Given a chemical equation, interpret it in terms of (a) atoms, molecules and/or formula units and (b) moles.

The "molecular" meaning of a chemical equation guided the balancing procedure used in Section 7.1. A complete interpretation of the balanced equation would be as follows: two atoms of sodium react with two molecules of water to produce one molecule of hydrogen and two formula units of sodium hydroxide (i.e., 2 $Na^+$ ions and 2 $OH^-$ ions). But it would be equally true that *four* atoms of sodium react with *four* molecules of water to produce *two* molecules of hydrogen and *four* formula units of sodium hydroxide:

$$4 \text{ Na (s)} + 4 \text{ H}_2\text{O (l)} \rightarrow 2 \text{ H}_2 \text{ (g)} + 4 \text{ NaOH (aq)} \qquad (7.6)$$

Equation 7.6, which is simply Equation 7.5 multiplied by 2, conforms to the law of conservation of mass and to the atomic theory. As in algebra, both sides of a chemical equation may be multiplied by any number without destroying its validity.

Using other multipliers on Equation 7.5 we find that

$$2 \text{ (12) Na (s)} + 2 \text{ (12) H}_2\text{O (l)} \rightarrow 1 \text{ (12) H}_2 \text{ (g)} + 2 \text{ (12) NaOH (aq)}$$

This could be read or written correctly as

$$2 \text{ dozen Na (s)} + 2 \text{ dozen H}_2\text{O (l)} \rightarrow 1 \text{ dozen H}_2 \text{ (g)} + 2 \text{ dozen NaOH (aq)}$$

Similarly,

$$2 \text{ (6.02} \times 10^{23}\text{) Na (s)} + 2 \text{ (6.02} \times 10^{23}\text{) H}_2\text{O (l)} \rightarrow$$
$$1 \text{ (6.02} \times 10^{23}\text{) H}_2 \text{ (g)} + 2 \text{ (6.02} \times 10^{23}\text{) NaOH (aq)}$$

may be thought of as

$$2 \text{ moles Na (s)} + 2 \text{ moles H}_2\text{O (l)} \rightarrow 1 \text{ mole H}_2 \text{ (g)} + 2 \text{ moles NaOH (aq)}$$

We may refer to this last equation as a "molar" interpretation of a chemical equation. With the understanding that the coefficients refer to *moles* rather than atoms, molecules or formula units, the interpretation of the equation may be applied directly to Equation 7.5.

The molar interpretation provides another option in the balancing of chemical equations. Returning to the unbalanced equation,

$$\text{Na (s)} + \text{H}_2\text{O (l)} \rightarrow \text{H}_2 \text{ (g)} + \text{NaOH (aq)} \qquad (7.2)$$

we find that the left side of the equation has two moles of hydrogen atoms, whereas the right side has three moles of hydrogen atoms. A single coefficient, ½, applied to hydrogen balances the equation:

$$\text{Na (s)} + \text{H}_2\text{O (l)} \rightarrow \tfrac{1}{2} \text{ H}_2 \text{ (g)} + \text{NaOH (aq)} \qquad (7.7)$$

This says that one mole of sodium reacts with one mole of water to form half a mole of hydrogen plus 1 mole of sodium hydroxide. There are now two moles of hydrogen atoms on each side of the equation, which is balanced in other atoms as well. From a molecular standpoint you cannot properly think of half of a hydrogen molecule, or any fractional part of any atom or molecule, any more than you can think of half an egg: but half a *mole* of molecules is no less acceptable than half a dozen eggs.

In chemistry, as in algebra, equations are *usually* written with the lowest set of whole number (integer) coefficients possible, and you should conform to this practice. There are occasions, however, when fractional coefficients or coefficients all divisible by the same integer are required.

## 7.3 WRITING CHEMICAL EQUATIONS

The procedure for writing chemical equations may be expressed in two steps:

1. Write the correct chemical formula for each reactant on the left and each product on the right.
2. *Using coefficients only* (do NOT change a formula, or add another), balance the number of atoms of each element on each side of the equation.

There are numerous "techniques" that may be used in writing equations, and certain types of reactions are readily recognized. Both techniques and types will be introduced through a series of examples.

### COMBINATION REACTIONS

> PG 7 B Given the identity of a compound that is formed from two or more simpler substances, write the equation for the reaction.

Reactions in which two or more substances, often elements, combine to form a compound are called combination, or synthesis, reactions.

---

**Example 7.1** When charcoal (carbon) is burned completely in air (see Fig. 7.2), carbon dioxide is formed. Write the equation for the reaction.

Step 1 requires the correct formulas for each reactant and product. The formula of carbon, as an element, is simply C. Identifying the second reactant requires knowledge of what happens chemically in the process of "burning in air." Oxygen in air is capable of combining chemically with many things, sometimes slowly and unnoticed, sometimes rapidly, with evolution of heat and light. When a substance combines with oxygen we say **oxidation** has occurred; the substance has been **oxidized**.* If oxidation is accompanied by heat and light we call the process burning.

---

*In Chapter 17 you will learn a broader and more useful meaning for oxidation and oxidized.

**Figure 7.2** Charcoal (carbon) burning in air. "Burning in air" is interpreted chemically as "reacting or combining with oxygen." Carbon combines with oxygen to produce carbon dioxide, $CO_2$.

Carbon and oxygen, then, are the two reactants in this burning equation. The single product of the reaction is carbon dioxide, $CO_2$. With this information you should be able to complete the first step by writing the unbalanced equation.

-----------------------------------------------------------------------------------------------------------

**7.1a** $$C\ (s) + O_2\ (g) \rightarrow CO_2\ (g)$$

The second step is to balance the number of atoms of each element *by use of coefficients only*. Complete the equation.

-----------------------------------------------------------------------------------------------------------

**7.1b** $C\ (s) + O_2\ (g) \rightarrow CO_2\ (g)$. Sometimes balancing is easy, as when all the coefficients are 1!

Limiting the quantity of air (oxygen) in the burning of carbon brings about a different result. . . .

**Example 7.2** When carbon is burned in a limited quantity of air, poisonous carbon monoxide, $CO\ (g)$, is produced. Write the equation.

First the unbalanced equation. . . .

-----------------------------------------------------------------------------------------------------------

**7.2a**         $C (s) + O_2 (g) \rightarrow CO (g)$      (Unbalanced)

Now count up the atoms of each element on each side of the equation and balance by inspection, using coefficients only.

-------------------------------------------------------------------

**7.2b**                   $2 C (s) + O_2 (g) \rightarrow 2 CO (g)$ or

$$C (s) + \tfrac{1}{2} O_2 (g) \rightarrow CO (g)$$

As noted previously, unless there is some specific reason for doing otherwise, the smallest integer coefficients are usually used. The first equation is preferred.

If, as in this example, you have an equation involving fractional coefficients and wish to convert them to integers, multiplication by the lowest common denominator will accomplish this conversion. Thus, for $C (s) + \tfrac{1}{2} O_2 (g) \rightarrow$ $CO (g)$, multiplication by two yields $2 \times \tfrac{1}{2} = 1$ for the coefficient of oxygen, and $2 \times 1 = 2$ for the coefficients of carbon and carbon monoxide.

---

**Example 7.3**   Write the equation for the formation of sodium chloride by direct combination of its elements.

-------------------------------------------------------------------

**7.3a**   $2 Na (s) + Cl_2 (g) \rightarrow 2 NaCl (s)$. You did recall, did you not, that chlorine is one of those gaseous elements that form diatomic molecules?

---

## DECOMPOSITION REACTIONS

PG  7 C  Given the identity of a compound that is decomposed into simpler substances, either compounds or elements, write the equation for the reaction.

Decomposition reactions are the opposite of combination reactions, in that chemical compounds break down into simpler substances, usually elements. For example . . . .

---

**Example 7.4**   When water is electrolyzed it decomposes into its elements (see Fig. 2.5, page 14). Write the equation.

"Electrolyzing" a substance involves running an electric current through it, a process used in many chemical reactions. You know the formulas for water and the elements in it, so you should be able to produce the unbalanced equation readily. Balancing is straightforward. Go all the way . . . .

-------------------------------------------------------------------

**7.4a**                   $2 H_2O (l) \rightarrow 2 H_2 (g) + O_2 (g)$

Decomposition reactions do not always go all the way to the elements, as the following example illustrates:

---

**Example 7.5**  Heating potassium chlorate, $KClO_3$ (s), releases oxygen, leaving solid potassium chloride (see Fig. 7.3). Write the equation.

From your knowledge of the formulas of ionic compounds (Chapter 5), you should be able to produce the formula of potassium chloride. From the description of the reaction you should be able to identify reactants and products. Therefore you should be able to write the unbalanced equation . . . .

---

**7.5a**  $$KClO_3 \text{ (s)} \rightarrow KCl \text{ (s)} + O_2 \text{ (g)} \qquad \text{(Unbalanced)}$$

Inspection balancing is a bit more complicated this time. A quick glance shows the potassium and chlorine already balanced; only the oxygen remains. There are three oxygen atoms on the left, and two on the right. You have encountered a 3-and-2 combination before: remember the $C^{3+}$ and $Y^{2-}$ hypothetical formula for an ionic compound (see Example 5.6, page 107)? You found the fact that $3 \times 2 = 6$ handy at that time. Could be here, too. With that hint, see if you can balance the oxygen, disregarding potassium and chlorine.

---

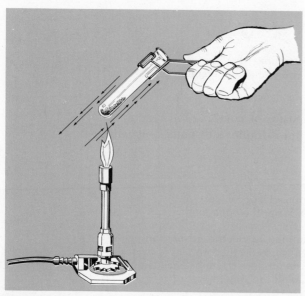

**Figure 7.3**  Decomposition of potassium chlorate, $KClO_3$, into potassium chloride, KCl, and oxygen in the presence of a catalyst.

**7.5b**        $2 KClO_3 (s) \rightarrow KCl (s) + 3 O_2 (g)$      (Unbalanced)

Two potassium chlorates, each with 3 oxygens, give six atoms of oxygen; three oxygen molecules, with atoms two at a time, also give six atoms of oxygen. The oxygen is in balance.
   Completing the equation follows . . . .

------------------------------------------------------------

**7.5c**              $2 KClO_3 (s) \rightarrow 2 KCl (s) + 3 O_2 (g)$

   Now a slightly different approach: Going back to the unbalanced equation,

$$KClO_3 (s) \rightarrow KCl (s) + O_2 (g)      \text{(Unbalanced)}$$

and noting again that oxygen is the only unbalanced element, observe also that oxygen is in elemental form on the right side of the equation. Any time all elements are in balance except one that is in elemental form, the only necessary step is to apply the proper coefficient to that element to complete the balancing. With three oxygen atoms on the left, how many oxygen *molecules* must be used to obtain three oxygen *atoms* on the right?

------------------------------------------------------------

**7.5d**  3/2. 3/2 oxygen molecules $\times$ 2 atoms/molecule gives three oxygen atoms. The improper fraction form, 3/2, is more useful than the mixed number form, 1½, in seeing the implied multiplication, $^3/_2 \times 2 = 3$.

   Using the fractional coefficient so found, write the balanced equation.

------------------------------------------------------------

**7.5e**              $KClO_3 (s) \rightarrow KCl (s) + {}^3/_2 O_2 (g)$

   Now multiplication of the entire equation by 2 will reproduce the balanced equation with integral coefficients:

$$2 KClO_3 (s) \rightarrow 2 KCl (s) + 3 O_2 (g).$$

------------------------------------------------------------

**Example 7.6**  Lime, CaO (s), and carbon dioxide gas, $CO_2$, are the products of the thermal decomposition of limestone, $CaCO_3$ (s) (see Fig. 7.4). Write the equation.

------------------------------------------------------------

**7.6a**              $CaCO_3 (s) \rightarrow CaO (s) + CO_2 (g)$

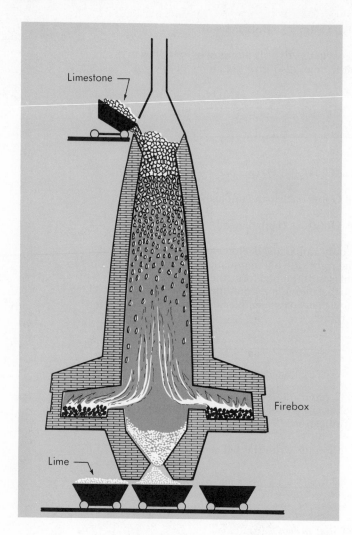

**Figure 7.4** Lime kiln in which limestone, calcium carbonate, $CaCO_3$, is decomposed into lime, CaO, and carbon dioxide, $CO_2$, by heating.

Limestone

Firebox

Lime

## COMPLETE OXIDATION OR BURNING OF ORGANIC COMPOUNDS

PG 7 D Write the equation for the complete burning of any compound consisting only of carbon, hydrogen and possibly oxygen.

A large number of organic compounds consist of two or three elements: carbon and hydrogen; or carbon, hydrogen and oxygen. When such compounds are burned in an excess of air, or otherwise oxidized completely, the end products are always the same: carbon dioxide, $CO_2$ (g), and steam, $H_2O$ (g).* When the steam condenses it becomes the "white smoke" so commonly seen rising from chimneys. In writing the following equations, recall the chemical interpretation of "burning in air"—reacting with oxygen.

---

*In some cases you may wish to consider *liquid* water, $H_2O$ (l), as the product. We will use either in the coming pages, while recognizing that the distinction is important when writing thermochemical equations (see page 308).

**Example 7.7**  Write the equation for the complete burning of the liquid hydrocarbon pentane, $C_5H_{12}$, in air.

Reactants and products are both known, so the unbalanced equation should be written readily.

----

**7.7a**   $C_5H_{12}$ (l) $+ O_2$ (g) $\rightarrow CO_2$ (g) $+ H_2O$ (g)   (Unbalanced)

Quite often in balancing equations it is advisable to balance first those elements other than hydrogen and oxygen; then balance hydrogen, and finally oxygen. This is particularly true when elemental oxygen is in the equation, as it is here. Working in the order, carbon, hydrogen, oxygen, you should be able to balance this equation.

----

**7.7b**         $C_5H_{12}$ (l) $+ 8\ O_2$ (g) $\rightarrow 5\ CO_2$ (g) $+ 6\ H_2O$ (g)

Five carbons in $C_5H_{12}$ require 5 carbon dioxides. Twelve hydrogens in $C_5H_{12}$ call for 6 waters. This adds up to a total of 10 oxygens in $CO_2 + 6$ in $H_2O$, or 16 atoms of oxygen, all to come from the oxygen of the air. Thus $8\ O_2$ molecules are required.

----

**Example 7.8**  Write the equation for the complete burning of ethane, $C_2H_6$ (g).

----

**7.8a**         $2\ C_2H_6$ (g) $+ 7\ O_2$ (g) $\rightarrow 4\ CO_2$ (g) $+ 6\ H_2O$ (g)

Working with one molecule of ethane produces

$C_2H_6$ (g) $+ O_2$ (g) $\rightarrow 2\ CO_2$ (g) $+ 3\ H_2O$ (g)   (Unbalanced)

with the oxygen yet to balance. There are $4 + 3$ or 7 oxygen atoms on the right, which takes $^7/_2\ O_2$ molecules on the left ($^7/_2 \times 2 = 7$). Therefore

$C_2H_6$ (g) $+ {}^7/_2\ O_2$ (g) $\rightarrow 2\ CO_2$ (g) $+ 3\ H_2O$ (g)

Multiplying by 2 to clear the fractional coefficient yields the answer given.

----

Try this one, but be careful.

----

**Example 7.9**  Write the equation for the complete burning of propanol, $C_3H_5OH$ (l), in air.

----

> **7.9a** $\qquad$ $C_3H_5OH$ (l) + 4 $O_2$ (g) $\rightarrow$ 3 $CO_2$ (g) + 3 $H_2O$ (g)
>
> If you did not get this answer you probably counted only five hydrogen atoms in $C_3H_5OH$ (there are six), or overlooked the fact that the one oxygen in $C_3H_5OH$ takes care of one of the nine oxygens on the right, leaving only eight to come from $O_2$.

## *REACTIONS OCCURRING IN WATER SOLUTIONS*

> PG 7 E  Given the identity of reactants and products of a reaction that occurs in water solution, write the equation.
>
> 7 F  Given the reactants in a neutralization reaction, write the equation.

The reactions we have considered in Examples 7.1 through 7.9 have all dealt with pure substances in the gas, liquid or solid state. Most of the reactions you will study in the laboratory, and many that occur in the world around us, take place in aqueous solutions. Such reactions may be described by equations similar to those we have been writing for reactions between pure substances. They may also be described in the form of *net ionic equations*, which we will consider in Chapter 14. Each type of equation has its unique value; later you will choose between them, depending on your particular need at the time. In this chapter we will confine ourselves to the conventional equation for describing reactions in aqueous solutions. In so doing, the symbol (aq) will be used to identify dissolved substances.

SINGLE REPLACEMENT (REDOX) REACTIONS. Many elements are capable of replacing other elements from aqueous solutions. These are sometimes called *single replacement reactions*. More formally, they are one kind of an oxidation-reduction, or "redox" reaction, which, in Chapter 17, we will consider from the standpoint of predicting whether or not the reaction will occur. The next three examples illustrate this kind of redox reaction.

> **Example 7.10** Elemental calcium reacts with hydrochloric acid to produce hydrogen gas and a solution of calcium chloride (see Fig. 7.5).
>
> To begin, write the unbalanced equation, showing the formula of each reactant and each product, including its state. Remember what you learned about hydrochloric acid back in Section 5.5 (page 103).
>
> -----------------------------------------------------------------------------------

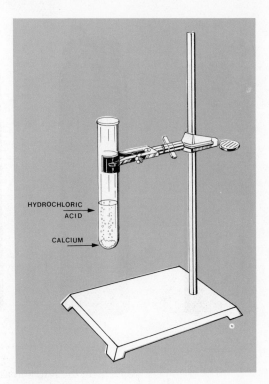

**Figure 7.5** Calcium reacts with hydrochloric acid to release hydrogen gas. WARNING: Do not "try" this experiment, as it is potentially dangerous.

HYDROCHLORIC
ACID

CALCIUM

**7.10a**    $Ca$ (s) $+$ HCl (aq) $\rightarrow$ $H_2$ (g) $+$ $CaCl_2$ (aq)    (Unbalanced)

Hydrochloric acid, you will recall, is a solution of hydrogen chloride. They have the same formula, HCl.

Balancing the equation is easy . . . .

---------------------------------------------------------------------------------------------------------

**7.10b**            $Ca$ (s) $+$ 2 HCl (aq) $\rightarrow$ $H_2$ (g) $+$ $CaCl_2$ (aq)

**Example 7.11**    Copper reacts with a solution of silver nitrate, $AgNO_3$, to produce silver and a solution of copper(II) nitrate,* $Cu(NO_3)_2$, as shown in Figure 7.6.

The unbalanced equation, please . . . .

---------------------------------------------------------------------------------------------------------

**7.11a**    $Cu$ (s) $+$ $AgNO_3$ (aq) $\rightarrow$ $Ag$ (s) $+$ $Cu(NO_3)_2$ (aq)    (Unbalanced)

An important technique appears in this equation. Notice that the nitrate ion, $NO_3^-$, appears on both sides of the equation: as a part of silver nitrate,

---

*Copper and some other elements to be identified in Chapter 10 vary in the manner in which they form compounds. These variations are indicated by Roman numerals in the name.

**Figure 7.6** Reaction between copper wire and solution of silver nitrate, $AgNO_3$, in which silver metal grows on wire in thin needles.

$AgNO_3$, on the left, and as a part of $Cu(NO_3)_2$, on the right. When a polyatomic ion is unchanged in a reaction it may be balanced as a distinct unit, just as atoms of an element are balanced. The equation is balanced for that matter, except for the nitrate ion; there is one nitrate on the left, and there are two on the right. Balance them as the next step . . . .

-------------------------------------------------------------------------------------

**7.11b**   $Cu\ (s) + 2\ AgNO_3\ (aq) \rightarrow Ag\ (s) + Cu(NO_3)_2\ (aq)$       (Unbalanced)

The equation now has the nitrates balanced, but it is unbalanced in another respect. Completion of the balancing, however, is obvious.

-------------------------------------------------------------------------------------

**7.11c**       $Cu\ (s) + 2\ AgNO_3\ (aq) \rightarrow 2\ Ag\ (s) + Cu(NO_3)_2\ (aq)$

**Example 7.12**   If chlorine gas is bubbled through a solution of sodium bromide, an aqueous solution of bromine and sodium chloride results.

This equation is straightforward. Take it all the way.

-------------------------------------------------------------------------------------

**7.12a**       $Cl_2\ (g) + 2\ NaBr\ (aq) \rightarrow Br_2\ (aq) + 2\ NaCl\ (aq)$

**PRECIPITATION REACTIONS.** When solutions of two soluble ionic compounds are mixed, the combination of the positive ion from one solution and the negative ion of the other solution may be insoluble in water. They therefore unite to form a solid compound, which is called a **precipitate.** A precipitation reaction is one of several kinds of ion exchange reactions in which ions of two compounds appear to change partners. Such reactions are sometimes called *double replacement* or metathesis reactions. The next two examples illustrate chemical precipitations.

---

**Example 7.13** If solutions of sodium chloride and silver nitrate, $AgNO_3$, are combined, silver chloride precipitates (see Fig. 7.7). Write the equation.

The word description of the reaction may appear incomplete—and, in fact, it is incomplete. What happens to the sodium and nitrate ions on the product side of the equation? These ions remain in solution, effectively as a solution of sodium nitrate. In solution reactions, any ions not otherwise accounted for may be considered as a compound in solution.

With this explanation you should be able to write the equation.

---

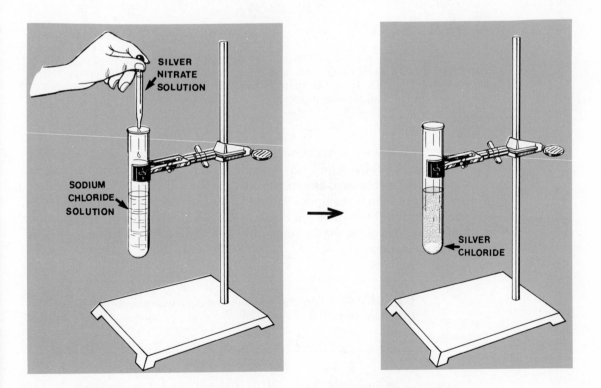

**Figure 7.7** Silver chloride, AgCl, precipitates when solutions of silver nitrate, $AgNO_3$, and sodium chloride, NaCl, are combined.

**7.13a**     $NaCl \, (aq) + AgNO_3 \, (aq) \rightarrow AgCl \, (s) + NaNO_3 \, (aq)$

The equation is balanced with all coefficients equal to 1. In checking this balance the nitrate ion may again be regarded as a distinct unit. There is one nitrate on each side of the equation.

---

**Example 7.14** Aluminum hydroxide precipitates on combining solutions of potassium hydroxide and aluminum nitrate. Write the equation.

Recalling what was said above about ions unaccounted for in precipitation reactions, write the unbalanced equation.

-------------------------------------------------------------------------------

**7.14a**
$KOH \, (aq) + Al(NO_3)_3 \, (aq) \rightarrow Al(OH)_3 \, (s) + KNO_3 \, (aq)$     (Unbalanced)

In this equation there are two polyatomic ions, $OH^-$ and $NO_3^-$, which appear unchanged in the reaction. They may be treated as units in balancing the equation. Doing so makes the remainder of the balancing routine.

-------------------------------------------------------------------------------

**7.14b**     $3 \, KOH \, (aq) + Al(NO_3)_3 \, (aq) \rightarrow Al(OH)_3 \, (s) + 3 \, KNO_3 \, (aq)$

The three hydroxides in $Al(OH)_3$ on the right side of the unbalanced equation require three KOH on the left; and the three nitrates in $Al(NO_3)_3$ on the left require three $KNO_3$ on the right. Balancing the polyatomic ions leaves the potassium and aluminum in balance.

NEUTRALIZATION.    A third kind of aqueous solution reaction called neutralization occurs when a hydroxide, either as a solid or in solution, is treated with an acid such as hydrochloric acid or sulfuric acid. The products are water and an ionic compound, called a **salt,** usually in solution. The water comes from the hydroxide ion combining with the hydrogen ion from the acid; the salt accounts for the remaining ions. Two examples illustrate neutralization reactions.

---

**Example 7.15** Write the equation for the reaction between hydrochloric acid and a solution of sodium hydroxide.

-------------------------------------------------------------------------------

**7.15a** $NaOH \, (aq) + HCl \, (aq) \rightarrow HOH \, (l) + NaCl \, (aq)$. The equation is balanced with 1 for all coefficients.

Water is a molecular compound, and its formula is usually written $H_2O$. In the above equation it has been shown as HOH, as if it were made up of a combination of $H^+$ and $OH^-$ ions. Indeed, the ions do combine to form water, but the compound is molecular. The HOH formula will be used freely in this text because it simplifies equation balancing, permitting treatment of $OH^-$ as a polyatomic unit. Remember, though, that HOH and $H_2O$ are both formulas for *water*, which is a *molecular* compound, and the name "hydrogen hydroxide" is *not* a correct alternative.

---

**Example 7.16** Write the equation for the reaction between sulfuric acid and solid aluminum hydroxide.

The character of the products here is the same as before. Write the unbalanced equation.

---

**7.16a** $H_2SO_4$ (aq) + $Al(OH)_3$ (s) $\rightarrow$ $Al_2(SO_4)_3$ (aq) + HOH (l)     (Unbalanced)

Balancing this time is a bit trickier than before, but everything falls into place if you treat the polyatomic ions, $SO_4^{2-}$ and $OH^-$, as distinct units in balancing.

---

**7.16b**     3 $H_2SO_4$ (aq) + 2 $Al(OH)_3$ (s) $\rightarrow$ $Al_2(SO_4)_3$ (aq) + 6 HOH (l)

Starting from the unbalanced equation and balancing first the sulfate ions, three on the right in $Al_2(SO_4)_3$ require three $H_2SO_4$:

③ $H_2$ ⟨$SO_4$⟩ (aq) + $Al(OH)_3$ (s) $\rightarrow$ $Al_2$ ⟨$SO_4$⟩ $_3$ (aq) + HOH (l)  (Unbalanced)

Next, two aluminums in $Al_2(SO_4)_3$ on the right require two $Al(OH)_3$ on the left:

3 $H_2SO_4$ (aq) + ②$Al$ $(OH)_3$ (s) $\rightarrow$ ⟨$Al_2$⟩$(SO_4)_3$ (aq) + HOH (l)  (Unbalanced)

Two aluminum hydroxides make six hydroxide ions available for the formation of six water molecules:

3 $H_2SO_4$ (aq) + ② $Al$⟨$(OH)_3$⟩ (s) $\rightarrow$ $Al_2(SO_4)_3$ (aq) + ⑥ H⟨OH⟩ (l)

The hydroxide step also balances the hydrogens other than those in the hydroxide ions, six from 3 molecules of $H_2SO_4$ on the left, and six from six water molecules on the right. All oxygen is in polyatomic ions, so it too should be in balance. It is, at 18 atoms on each side.

Your sequence of steps in balancing this equation may have been different, but the final result should be the same.

## OTHER REACTIONS

PG  7 G   Given a word description of a chemical reaction in which all reactants and products are identified, write the chemical equation for the reaction.

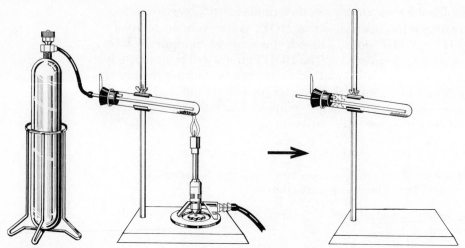

**Figure 7.8**   Reaction of copper(II) oxide, CuO, with hydrogen, $H_2$. Oxygen from the copper oxide (black substance in left test tube) combines with hydrogen to form water (note drops in right test tube), leaving copper (colored substance) behind.

In addition to those already considered, there are a vast number of reactions between different combinations of elements and compounds that fall into other classifications, or no classification. We shall use a few of these to provide further practice in balancing equations.

**Example 7.17**   If solid copper(II) oxide (CuO) is heated in a stream of hydrogen gas, metallic copper and water vapor are produced (see Fig. 7.8). Write the equation.

-------------------------------------------------------------------------------------

**7.17a**          $CuO \text{ (s)} + H_2 \text{ (g)} \rightarrow Cu \text{ (s)} + H_2O \text{ (g)}$

Note $H_2O$ (g) is used for water vapor.

In this simple reaction the equation is balanced with all coefficients having the value of 1. Not so with the following similar reaction:

**Example 7.18**   When iron(III) oxide, $Fe_2O_3$ (s), is treated with hydrogen gas, it is reduced to metallic iron, with water vapor as a second product. Write and balance the equation.

-------------------------------------------------------------------------------------

**7.18a**          $Fe_2O_3 \text{ (s)} + 3 H_2 \text{ (g)} \rightarrow 2 Fe \text{ (s)} + 3 H_2O \text{ (g)}$

From the unbalanced equation, $Fe_2O_3 \text{ (s)} + H_2 \text{ (g)} \rightarrow Fe \text{ (s)} + H_2O \text{ (g)}$, the iron is readily balanced by a coefficient of 2 before Fe on the right side. This leaves only oxygen unbalanced. Three waters take care of the oxygen, but unbalance the hydrogen. Hydrogen, being in elemental form on the left, is readily restored to balance by a coefficient of 3. Whenever an equation can be balanced completely except for some elemental species, balancing that element last is readily accomplished by whatever coefficient is necessary.

---

**Example 7.19**  Carbon disulfide, $CS_2$ (l), reacts with oxygen gas to produce two gaseous products, carbon dioxide, $CO_2$, and sulfur dioxide, $SO_2$. Write the equation.

---

**7.19a**  $$CS_2 \text{ (l)} + 3\ O_2 \text{ (g)} \rightarrow CO_2 \text{ (g)} + 2\ SO_2 \text{ (g)}$$

There is one carbon on each side of the unbalanced equation. Sulfur is readily balanced by a coefficient of 2 for the sulfur dioxide. Elemental oxygen is balanced last.

---

**Example 7.20**  Balance the following equation:

$$PCl_5 \text{ (s)} + H_2O \text{ (l)} \rightarrow H_3PO_4 \text{ (aq)} + HCl \text{ (aq)}$$

---

**7.20a**  $$PCl_5 \text{ (s)} + 4\ H_2O \text{ (l)} \rightarrow H_3PO_4 \text{ (aq)} + 5\ HCl \text{ (aq)}$$

Taking chlorine first, one $PCl_5$ on the left requires 5 HCl on the right. Next, phosphorus is already balanced. Then hydrogen: three from $H_3PO_4$ and five from HCl is a total of eight on the right, requiring 4 $H_2O$ on the left. Oxygen, as frequently happens when not present in elemental form, is balanced by the time we get to it.

## 7.4  MISCELLANEOUS SYMBOLS USED IN CHEMICAL EQUATIONS

When it is necessary to convey more information about a chemical reaction than the basic equation provides, other symbols are sometimes employed. State designations, (s), (l) and (g), and for solutions, (aq), which are used throughout this book, are examples. An alternate way of indicating a formation of a precipitate is to place next to the formula of the precipitate an arrow pointing down, as in the equation of Example 7.13:

$$NaCl + AgNO_3 \rightarrow AgCl \downarrow + NaNO_3$$

Formation of a gas may be shown by an arrow pointing up, as in the equation of Example 7.10:

$$Ca + 2\ HCl \rightarrow H_2\uparrow + CaCl_2$$

If a reaction is brought about by the application of heat, as in Example 7.6, it may be indicated by placing a Δ above the arrow:

$$CaCO_3 \xrightarrow{\Delta} CaO + CO_2$$

The use of other reaction conditions is also shown above, and sometimes above and below, the arrow in the equation. The electrolysis of water (Example 7.4) may be designated

$$2\ H_2O \xrightarrow{\text{electrolysis}} 2\ H_2 + O_2$$

The use of manganese dioxide, $MnO_2$, as a catalyst* in the thermal decomposition of potassium chlorate (Example 7.5) may be shown:

$$2\ KClO_3 \xrightarrow{MnO_2} 2\ KCl + 3\ O_2$$

Both heat and the catalyst may be indicated:

$$2\ KClO_3 \xrightarrow[MnO_2]{\Delta} 2\ KCl + 3\ O_2$$

*A *catalyst* is a substance that speeds up the rate of a chemical reaction without being consumed itself.

# CHAPTER 7 IN REVIEW

**7.1 EVOLUTION OF A CHEMICAL EQUATION (Page 143)**

**7.2 MEANING OF A CHEMICAL EQUATION**

**7 A** Given a chemical equation, interpret it in terms of (a) atoms, molecules and/or formula units and (b) moles. (Page 146)

**7.3 WRITING CHEMICAL EQUATIONS**

**7 B** Given the identity of a compound that is formed from two or more simpler substances, write the equation for the reaction. (Page 147)

**7 C** Given the identity of a compound that is decomposed into simpler substances, either compounds or elements, write the equation for the reaction. (Page 149)

**7 D** Write the equation for the complete burning of any compound consisting only of carbon, hydrogen and possibly oxygen. (Page 152)

**7 E** Given the identity of reactants and products of a reaction that occurs in water solution, write the equation. (Page 154)

7  F  Given the reactants in a neutralization reaction, write the equation. (Page 154)

7  G  Given a word description of a chemical reaction in which all reactants and products are identified, write the chemical equation for the reaction. (Page 159)

7.4  MISCELLANEOUS SYMBOLS USED IN CHEMICAL EQUATIONS (Page 161)

## TERMS AND CONCEPTS

Aqueous (144)
State designation (144)
"Balanced" equation (145)
Combination reaction (147)
Burning (in air) (147)
Oxidation–oxidize (147)

Decomposition reaction (149)
Single replacement (redox) reaction (154)
Precipitation reaction (157)
Precipitate (157)
Neutralization (158)
Salt (158)

## EQUATION BALANCING EXERCISE

*Instructions:* Balance the following equations, for which correct chemical formulas are already written. Balanced equations are on pages 530 and 531.

1) $Na + O_2 \rightarrow Na_2O$
2) $H_2 + Cl_2 \rightarrow HCl$
3) $P + O_2 \rightarrow P_2O_3$
4) $KClO_4 \rightarrow KCl + O_2$
5) $Sb_2S_3 + HCl \rightarrow SbCl_3 + H_2S$
6) $NH_3 + H_2SO_4 \rightarrow (NH_4)_2SO_4$
7) $CuO + HCl \rightarrow CuCl_2 + H_2O$
8) $Zn + Pb(NO_3)_2 \rightarrow Zn(NO_3)_2 + Pb$
9) $AgNO_3 + H_2S \rightarrow Ag_2S + HNO_3$
10) $Cu + S \rightarrow Cu_2S$
11) $Al + H_3PO_4 \rightarrow H_2 + AlPO_4$
12) $NaNO_3 \rightarrow NaNO_2 + O_2$
13) $Mg(ClO_3)_2 \rightarrow MgCl_2 + O_2$
14) $H_2O_2 \rightarrow H_2O + O_2$
15) $BaO_2 \rightarrow BaO + O_2$
16) $H_2CO_3 \rightarrow H_2O + CO_2$
17) $Pb(NO_3)_2 + KCl \rightarrow PbCl_2 + KNO_3$
18) $Al + Cl_2 \rightarrow AlCl_3$
19) $P + O_2 \rightarrow P_2O_5$
20) $NH_4NO_2 \rightarrow N_2 + H_2O$
21) $H_2 + N_2 \rightarrow NH_3$
22) $Cl_2 + KBr \rightarrow Br_2 + KCl$
23) $BaCl_2 + (NH_4)_2CO_3 \rightarrow BaCO_3 + NH_4Cl$
24) $MgCO_3 + HCl \rightarrow MgCl_2 + CO_2 + H_2O$
25) $P + I_2 \rightarrow PI_3$
26) $PbO_2 \rightarrow PbO + O_2$
27) $Al + HCl \rightarrow AlCl_3 + H_2$
28) $Fe_2(SO_4)_3 + Ba(OH)_2 \rightarrow BaSO_4 + Fe(OH)_3$
29) $Al + CuSO_4 \rightarrow Al_2(SO_4)_3 + Cu$

30) $KClO_3 \rightarrow KCl + O_2$

31) $Mg + N_2 \rightarrow Mg_3N_2$

32) $C_6H_{14} + O_2 \rightarrow CO_2 + H_2O$

33) $FeCl_2 + Na_3PO_4 \rightarrow Fe_3(PO_4)_2 + NaCl$

34) $Li_2O + HOH \rightarrow LiOH$

35) $HgO \rightarrow Hg + O_2$

36) $CaSO_4 \cdot 2\,H_2O \rightarrow CaSO_4 + H_2O$

37) $C_3H_7CHO + O_2 \rightarrow CO_2 + H_2O$

38) $NaHCO_3 + HCl \rightarrow NaCl + H_2O + CO_2$

39) $Bi(NO_3)_3 + NaOH \rightarrow Bi(OH)_3 + NaNO_3$

40) $FeS + HBr \rightarrow FeBr_2 + H_2S$

41) $Zn(OH)_2 + H_2SO_4 \rightarrow ZnSO_4 + HOH$

42) $P_4O_{10} + H_2O \rightarrow H_3PO_4$

43) $C_4H_9OH + O_2 \rightarrow CO_2 + H_2O$

44) $CaC_2 + H_2O \rightarrow C_2H_2 + Ca(OH)_2$

45) $CaCO_3 + H_3PO_4 \rightarrow Ca_3(PO_4)_2 + CO_2 + H_2O$

46) $PCl_5 + H_2O \rightarrow H_3PO_4 + HCl$

47) $CaI_2 + H_2SO_4 \rightarrow HI + CaSO_4$

48) $C_3H_7COOH + O_2 \rightarrow CO_2 + H_2O$

49) $Mg(CN)_2 + HCl \rightarrow HCN + MgCl_2$

50) $(NH_4)_2S + HgBr_2 \rightarrow NH_4Br + HgS$

---

# QUESTIONS AND PROBLEMS

*As in Chapter 6, only names are given for those compounds whose formulas are included in the nomenclature introduction developed in Chapter 5. For reference, you may find the formulas of the acids and molecular compounds on page 118, and the formulas of ionic compounds on page 118. As before, you are encouraged to write formulas with no assistance other than a periodic table, using the reference pages only if necessary. This is the way you will build and improve your formula writing skills.*

Section 7.2

7.1) Write a sentence giving the "molecular" meaning of the equation for the complete oxidation of benzene:

$$2\,C_6H_6 + 15\,O_2 \rightarrow 12\,CO_2 + 6\,H_2O$$

7.31) State the "molar" interpretation of the equation in 7.1. Explain the difference in the two interpretations.

Section 7.3

*7.2–7.6 and 7.32–7.36: Write balanced chemical equations for the combination reactions described below.*

7.2) Calcium forms calcium oxide by reaction with the oxygen in the air.

7.3) $P_4O_{10}$ is the product of the direct combination of phosphorous and oxygen.

7.4) When potassium contacts fluorine gas—two highly reactive elements—potassium fluoride is produced.

7.32) Lithium combines with oxygen to form lithium oxide.

7.33) Boron (at. no. 5) combines with oxygen to form $B_2O_3$.

7.34) Calcium combines with bromine to make calcium bromide.

7.5)   Silicon reacts with chlorine to form silicon tetrachloride, $SiCl_4$.

7.6)   Magnesium nitride is one of the few metallic nitrides. Its formula is what you would expect from the positions of the elements in the periodic table, and it is formed by direct combination of the elements.

7.35)   Phosphorus tribromide, $PBr_3$, is produced by direct combination of phosphorus and bromine.

7.36)   There are only a few carbides—compounds made up of a metal and carbon. Though aluminum carbide is not ionic, its formula is as would be expected from the periodic table positions of the elements, counting carbon at $-4$ to balance aluminum's $+3$. Write the equation for its formation by direct combination of the elements.

*7.7–7.10 and 7.37–7.40: Write the equations for the following decomposition reactions:*

7.7)   Melted crystals of table salt, sodium chloride, may be decomposed to its elements by electrolysis.

7.8)   Mercury (at. no. 80) and oxygen gas are produced by heating mercury(II) oxide, $HgO$.

7.9)   Carbonic acid is an unstable compound, decomposing to carbon dioxide and water.

7.10)   Hydrogen peroxide, $H_2O_2$, the familiar bleaching compound, decomposes slowly to water and oxygen.

7.37)   Pure hydrogen iodide, HI, decomposes spontaneously to its elements.

7.38)   Silver oxide, $Ag_2O$, may be decomposed to its elements by heating.

7.39)   Sulfurous acid, $H_2SO_3$, decomposes spontaneously to water and sulfur dioxide.

7.40)   When calcium hydroxide—sometimes called slaked lime—is heated, it decomposes to calcium oxide—or lime—and water vapor.

*7.11–7.14 and 7.41–7.44: Write equations for the complete oxidation of the following organic compounds:*

7.11)   Propane, $C_3H_8$

7.12)   Acetylene, $C_2H_2$

7.13)   Acetaldehyde, $CH_3CHO$

7.14)   Sugar, $C_{12}H_{22}O_{11}$

7.41)   Butane, $C_4H_{10}$

7.42)   Ethyl alcohol, $C_2H_5OH$

7.43)   Acetic acid, $CH_3COOH$

7.44)   Glycerine, $C_3H_8O_3$

*7.15–7.26 and 7.45–7.56: Write chemical equations for the following reactions that occur in water solution:*

7.15)   Hydrogen gas and a solution of magnesium sulfate are the result of the reaction that occurs when metallic magnesium is placed into sulfuric acid.

7.16)   When barium metal is placed into water, hydrogen gas bubbles off and a solution of barium hydroxide remains.

7.17)   A strip of zinc (at. no. 30) replaces the silver from a solution of silver nitrate, $AgNO_3$, leaving needle-like crystals of silver (at. no. 47) and a solution of zinc nitrate, $Zn(NO_3)_2$.

7.18)   Magnesium metal will replace the nickel from a solution of nickel(II) chloride, $NiCl_2$, yielding metallic nickel (at. no. 28) and a solution of magnesium chloride.

7.19)   Bromine liberates iodine from a solution of sodium iodide and leaves a solution of sodium bromide.

7.20)   If silver nitrate, $AgNO_3$, and potassium bromide solutions are combined, solid silver bromide, $AgBr$, precipitates.

7.45)   Calcium metal reacts with hydrobromic acid, HBr, to form hydrogen gas and a solution of calcium bromide.

7.46)   Potassium and water react to produce a solution of potassium hydroxide and hydrogen gas.

7.47)   If zinc (at. no. 30) is placed into a solution of copper(II) nitrate, $Cu(NO_3)_2$, metallic copper (at. no. 29) and a solution of zinc nitrate, $Zn(NO_3)_2$, result.

7.48)   If metallic manganese (at. no. 25) is placed in a solution of chromium(III) chloride, $CrCl_3$, solid chromium (at. no. 24) and manganese(II) chloride solution, $MnCl_2$ (aq), result.

7.49)   If chlorine gas is bubbled through a solution of potassium iodide, a solution of potassium chloride and elemental iodine are the products.

7.50)   Lead nitrate solution, $Pb(NO_3)_2$ (aq), reacts with sodium iodide solution to precipitate lead iodide, $PbI_2$.

7.21) Combination of solutions of lead nitrate, $Pb(NO_3)_2$, and copper(II) sulfate, $CuSO_4$, yields lead sulfate, $PbSO_4$, as a precipitate, and a solution of copper(II) nitrate, $Cu(NO_3)_2$.

7.22) Magnesium chloride solution combined with a solution of sodium fluoride yields a precipitate of magnesium fluoride.

7.23) If sodium sulfide solution is poured into silver nitrate solution, $AgNO_3$ (aq), silver sulfide, $Ag_2S$, precipitates.

7.24) Sodium hydroxide solution added to magnesium bromide solution yields a precipitate of magnesium hydroxide.

7.25) Potassium hydroxide solution neutralizes nitric acid.

7.26) Solid magnesium hydroxide reacts with hydrochloric acid.

7.51) Combining solutions of barium chloride and sodium sulfate produces a precipitate of barium sulfate.

7.52) Calcium fluoride precipitates from combining solutions of calcium nitrate and potassium fluoride.

7.53) Ammonium sulfide solution added to a solution of copper(II) nitrate, $Cu(NO_3)_2$, yields a precipitate of copper(II) sulfide, CuS.

7.54) Zinc chloride solution, $ZnCl_2$ (aq), produces a precipitate of zinc hydroxide, $Zn(OH)_2$, when added to a solution of potassium hydroxide.

7.55) Sulfuric acid neutralizes barium hydroxide solution.

7.56) Solid nickel(II) hydroxide, $Ni(OH)_2$, reacts with sulfuric acid. Nickel(II) sulfate solution, $NiSO_4$ (aq), is one product.

*7.27–7.30 and 7.57–7.60: For each general reaction described below, write the chemical equation:*

7.27) Metallic zinc (at. no. 30) reacts with steam at high temperatures to produce solid zinc oxide, ZnO, and hydrogen gas.

7.28) When solid barium oxide is placed into water, a solution of barium hydroxide is produced.

7.29) Igniting a mixture of powdered iron(II) oxide, FeO, and aluminum produces a vigorous reaction in which aluminum oxide and iron are the products.

7.30) Solid iron(III) oxide, $Fe_2O_3$, reacts with gaseous carbon monoxide to produce iron and carbon dioxide.

7.57) A solid oxide of iron, $Fe_3O_4$ (s), and hydrogen are the products of the reaction between iron (at. no. 26) and steam.

7.58) Sulfuric acid is produced when gaseous sulfur trioxide, $SO_3$ (g), reacts with water.

7.59) At high temperatures carbon monoxide and steam react to produce carbon dioxide and hydrogen.

7.60) Magnesium nitride, $Mg_3N_2$, and hydrogen are the products of the reaction between magnesium and ammonia.

# QUANTITY RELATIONSHIPS
# IN CHEMICAL REACTIONS

In this chapter, as in the previous two, we continue to assume that you are able to write the formulas of those compounds appearing in the nomenclature introduction in Chapter 5. If you require assistance for any of these compounds, recall that the formulas of the acids and molecular compounds may be found on page 118, and of the ionic compounds on page 528.

## 8.1  QUANTITATIVE INTERPRETATION
## OF A CHEMICAL EQUATION

> PG  8 A   Given a chemical equation, or a reaction for which the equation can be written, and the number of moles of one species in the reaction, calculate the number of moles of any other species.

A balanced chemical equation tells how many moles of individual reactants are required for the reaction, as well as the number of moles of different species that will be produced. Using the equation

$$PCl_5 \text{ (s)} + 4 \text{ } H_2O \text{ (l)} \rightarrow H_3PO_4 \text{ (aq)} + 5 \text{ } HCl \text{ (aq)},$$

we see that four moles of water are required to react with each mole of $PCl_5$. *For this reaction* there is, in effect, an equivalence between $PCl_5$ and $H_2O$: 1 mole $PCl_5 \simeq 4$ moles $H_2O$. Furthermore, for this reaction, each mole of $PCl_5$ that reacts will produce one mole of $H_3PO_4$ and *five* moles of HCl. The equivalence may thus be extended to the products:

$$1 \text{ mole } PCl_5 \simeq 4 \text{ moles } H_2O \simeq 1 \text{ mole } H_3PO_4 \simeq 5 \text{ moles } HCl \quad (8.1)$$

Notice that *the numbers in the series of equivalent quantities derived from an equation correspond with the coefficients in the equation.*

If the number of moles of one species in a reaction, either reactant or product, is known, the equivalences establish a unit path from the given quantity to the moles of any other species. A single step conversion is involved. If, for example, 3.20 moles of $PCl_5$ react according to the above equation, how many moles of HCl will be produced? If 5 moles of HCl are produced by one mole of $PCl_5$ (1 mole $PCl_5 \backsimeq 5$ moles HCl), then the given quantity of $PCl_5$ (3.20 moles) will yield

$$3.20 \text{ moles } PCl_5 \times \frac{5 \text{ moles HCl}}{1 \text{ mole } PCl_5} = 16.0 \text{ moles HCl}$$

The procedure may be applied to any equation:

---

**Example 8.1**   How many moles of oxygen are required to burn 2.40 moles of ethane, $C_2H_6$, according to the equation

$$2 \ C_2H_6 \ (g) + 7 \ O_2 \ (g) \rightarrow 4 \ CO_2 \ (g) + 6 \ H_2O \ (l)$$

From the equivalences available in the equation, what unit path and what specific equivalent relationship are required for this problem?

------------------------------------------------------------------------------------------------

**8.1a**      Moles $C_2H_6 \rightarrow$ moles $O_2$; 2 moles $C_2H_6 \backsimeq 7$ moles $O_2$.

The problem requires conversion from moles of ethane to moles of oxygen, so only those two species are needed. For every 2 moles of $C_2H_6$, 7 moles of $O_2$ are required.

Complete the problem.

------------------------------------------------------------------------------------------------

**8.1b**   8.40 moles $O_2$

$$2.40 \text{ moles } C_2H_6 \times \frac{7 \text{ moles } O_2}{2 \text{ moles } C_2H_6} = 8.40 \text{ moles } O_2$$

---

**Example 8.2**   In the reaction $N_2$ (g) + 3 $H_2$ (g) $\rightarrow$ 2 $NH_3$ (g), how many moles of hydrogen are required to produce 4.20 moles of ammonia ($NH_3$)?

------------------------------------------------------------------------------------------------

**8.2a**         $$4.20 \text{ moles } NH_3 \times \frac{3 \text{ moles } H_2}{2 \text{ moles } NH_3} = 6.3 \text{ moles } H_2$$

## 8.2 MASS CALCULATIONS: STOICHIOMETRY

PG  8 B    Given a chemical equation, or a reaction for which the equation can be written, and the number of grams or moles of one species in the reaction, find the number of grams or moles of any other species.

Section 6.6 (page 130) presented the method by which the number of grams of a chemical of known formula can be converted to moles, or the number of moles converted to grams. Unit paths were *grams → moles* or *moles → grams*, and the conversion ratio was molar weight, grams/mole, in both directions. In Section 8.1 we have just seen how to follow a unit path of *moles (given species) → moles (wanted species)*, using coefficients from the equation for the conversion ratio. By combining these techniques we have a unit path from the given quantity to the wanted quantity: *grams (given species) → moles (given species) → moles (wanted species) → grams (wanted species)*. If you know the grams of any one substance in a reaction, you can, in three steps, calculate the grams of any other.

The determination of a quantity of one substance in a reaction from a known quantity of another is called **stoichiometry**. "Quantity" may be expressed in a number of ways besides grams, as will be seen in the coming chapters. Regardless of the units in which quantities are expressed, there is a single "stoichiometry pattern" for the solution of *all* stoichiometry problems. It consists of three steps, as follows:

1. Convert the quantity of the given species to moles (as in grams → moles, Section 6.6, page 131).
2. Convert the moles of given species to moles of wanted species (moles given → moles wanted, Section 8.1).
3. Convert the moles of wanted species to the quantity units required (as in moles → grams, Section 6.6, page 131).

If you learn how to use the stoichiometry pattern now, with grams as the only quantity unit, you will find that you already know how to solve problems in which other units appear later.

Occasionally you may be given the quantity of one substance in grams and asked to find the number of moles of a second species. In this case the first two steps of the stoichiometry pattern complete the problem. Or you may be given the moles of one substance and asked to find the grams of another. Steps two and three accomplish this objective.

The burning of ethane may be used to illustrate the method of stoichiometry:

**Example 8.3**  Calculate the number of grams of oxygen that are required to burn 150 grams of ethane, $C_2H_6$, in the reaction $2 C_2H_6 (g) + 7 O_2 (g) \rightarrow 4 CO_2 (g) + 6 H_2O (l)$.

*Solution.* Step 1 calls for the conversion of the given quantity to moles—in this case, the conversion of 150 grams of ethane to moles of ethane as in Section 6.6, page 131. The setup begins

$$\boxed{150 \text{ grams } C_2H_6} \times \frac{1 \text{ mole } C_2H_6}{30.0 \text{ grams } C_2H_6}$$

In step 2 the setup is extended to convert moles of $C_2H_6$ to moles of $O_2$, as in Section 8.1. The conversion equivalence is 7 moles $O_2 \simeq 2$ moles $C_2H_6$, from the equation.

$$\boxed{150 \text{ grams } C_2H_6} \times \frac{1 \text{ mole } C_2H_6}{30.0 \text{ grams } C_2H_6} \times \frac{7 \text{ moles } O_2}{2 \text{ moles } C_2H_6}$$

Finally, in step 3, the moles of oxygen are converted to grams, as in Section 6.6:

$$\boxed{150 \text{ g } C_2H_6} \times \frac{1 \text{ mole } C_2H_6}{30.0 \text{ g } C_2H_6} \times \frac{7 \text{ moles } O_2}{2 \text{ moles } C_2H_6} \times \frac{32.0 \text{ g } O_2}{1 \text{ mole } O_2} = 560 \text{ grams } O_2$$

Now you try one . . . .

**Example 8.4** The equation for burning heptane, $C_7H_{16}$, is

$$C_7H_{16} (l) + 11 O_2 (g) \rightarrow 7 CO_2 (g) + 8 H_2O (l)$$

How many grams of oxygen are required to burn 3.50 moles of $C_7H_{16}$?

The given quantity is *moles* of heptane. In other words, we already have moles of the given substance, so the first step in the stoichiometry pattern is effectively completed. The equation gives us the molar equivalence between heptane and oxygen. Start with the second step, and set up, but do not calculate, the number of moles of the wanted substance, oxygen.

-------------------------------------------------------------------------------

**8.4a**

$$\boxed{3.50 \text{ moles } C_7H_{16}} \times \frac{11 \text{ moles } O_2}{1 \text{ mole } C_7H_{16}}$$

From the equation, 11 moles $O_2 \simeq 1$ mole $C_7H_{16}$. This establishes the conversion factor.

Changing from moles of $O_2$ to grams of $O_2$ via molar weight is step 3 in the stoichiometric sequence, and finishes the problem. Complete the setup and solve for the answer.

-------------------------------------------------------------------------------

**8.4b** $1.23 \times 10^3$ grams $O_2$

$$\boxed{3.50 \text{ moles } C_7H_{16}} \times \frac{11 \text{ moles } O_2}{1 \text{ mole } C_7H_{16}} \times \frac{32.0 \text{ g } O_2}{1 \text{ mole } O_2} = 1.23 \times 10^3 \text{ grams } O_2$$

**Example 8.5** How many moles of $H_2O$ will be produced in the heptane burning reaction that also yields 115 grams of $CO_2$? (See equation in Example 8.4.)

This time we are given grams of one species, and must find moles of another. The first two steps in the stoichiometric pattern are all that is necessary. Set up, but do not solve, the first step, beginning with the given quantity.

------------------------------------------------------------------------

**8.5a**
$$\boxed{115 \text{ g } CO_2} \times \frac{1 \text{ mole } CO_2}{44.0 \text{ g } CO_2}$$

By step 2, and by drawing the proper molar equivalence from the equation, extend the setup and solve the problem.

------------------------------------------------------------------------

**8.5b** 2.99 moles $H_2O$

$$\boxed{115 \text{ g } CO_2} \times \frac{1 \text{ mole } CO_2}{44.0 \text{ g } CO_2} \times \frac{8 \text{ moles } H_2O}{7 \text{ moles } CO_2} = 2.99 \text{ moles } H_2O$$

Now we'll go all the way from grams to grams.

**Example 8.6** How many grams of $CO_2$ will be produced by burning 66.0 grams of heptane? (See equation in Example 8.4.)

Beginning with the given quantity, set up, but do not solve, the first step, the conversion of 66.0 grams of heptane to moles.

------------------------------------------------------------------------

**8.6a**
$$\boxed{66.0 \text{ g } C_7H_{16}} \times \frac{1 \text{ mole } C_7H_{16}}{100 \text{ g } C_7H_{16}}$$

Step 2 converts the moles of given species (heptane) to moles of wanted species (carbon dioxide). Extend the setup that far.

------------------------------------------------------------------------

**8.6b**
$$\boxed{66.0 \text{ g } C_7H_{16}} \times \frac{1 \text{ mole } C_7H_{16}}{100 \text{ g } C_7H_{16}} \times \frac{7 \text{ moles } CO_2}{1 \text{ mole } C_7H_{16}}$$

Finally, convert moles of $CO_2$ to grams, step 3. Solve to the numerical answer.

------------------------------------------------------------------------

**8.6c**   203 grams $CO_2$

$$\boxed{66.0 \text{ g } C_7H_{16}} \times \frac{1 \text{ mole } C_7H_{16}}{100 \text{ g } C_7H_{16}} \times \frac{7 \text{ moles } CO_2}{1 \text{ mole } C_7H_{16}} \times \frac{44.0 \text{ g } CO_2}{1 \text{ mole } CO_2} = 203 \text{ g } CO_2$$

## 8.3  PERCENTAGE YIELD

PG  8 C   Given two of the following, or information from which two of the following may be determined, calculate the third: theoretical yield, actual yield and percentage yield.

Example 8.6 indicates that the burning of 66.0 grams of heptane theoretically will produce 203 grams of $CO_2$. This is called a *theoretical yield,* where *yield* is the amount produced. In actual practice, factors such as impure reactants, incomplete reactions and side reactions cause the *actual yield* to be less than the calculated yield. Knowing the actual yield found in the laboratory and the theoretical yield calculated from the equation for the reaction, we can compute the **percentage yield:**

$$\frac{\text{actual yield}}{\text{theoretical yield}} \times 100 = \text{percentage yield} \tag{8.2}$$

If only 180 grams of carbon dioxide resulted in Example 8.6 instead of the theoretical 203 grams, the percentage yield would be

$$\frac{\text{actual yield}}{\text{theoretical yield}} \times 100 = \frac{180 \text{ g}}{203 \text{ g}} \times 100 = 88.7\%$$

**Example 8.7**   A solution containing excess* sodium sulfate is added to a second solution containing 3.18 grams of barium nitrate. Barium sulfate precipitates. (a) Calculate the theoretical yield of barium sulfate. (b) If the actual yield is 2.69 grams, calculate the percentage yield.

Your first requirement for any stoichiometry problem is the relationship between given and wanted species, expressed in the reaction equation. Write that equation, please.

-------------------------------------------------------------------------------

**8.7a**    $Ba(NO_3)_2 \text{ (aq)} + Na_2SO_4 \text{ (aq)} \rightarrow BaSO_4 \text{ (aq)} + 2 \text{ NaNO}_3 \text{ (aq)}$

Calculation of the theoretical yield is a typical stoichiometry problem. Solve part (a).

-------------------------------------------------------------------------------

*The word "excess" as used here means "more than enough" ... more than enough sodium sulfate than is required to precipitate all the barium in 3.18 grams of barium nitrate.

**8.7b**   2.84 grams $BaSO_4$

$$\boxed{3.18 \text{ g } Ba(NO_3)_2} \times \frac{1 \text{ mole } Ba(NO_3)_2}{261 \text{ g } Ba(NO_3)_2} \times \frac{1 \text{ mole } BaSO_4}{1 \text{ mole } Ba(NO_3)_2} \times \frac{233 \text{ g } BaSO_4}{1 \text{ mole } BaSO_4}$$

$$= 2.84 \text{ g } BaSO_4$$

Now, if the actual yield is 2.69 grams, find the percentage yield by substituting into Equation 8.2.

------------------------------------------------------------------------

**8.7c**   94.7%

$$\frac{2.69 \text{ g } BaSO_4 \text{ actual}}{2.84 \text{ g } BaSO_4 \text{ theoretical}} \times 100 = 94.7\%$$

**Example 8.8**   In the precipitation of calcium carbonate by the reaction between solutions of calcium chloride and sodium carbonate, the percentage yield is known to be 96.8%, based on the starting mass of dissolved sodium carbonate. How many grams of $Na_2CO_3$ were initially present in solution if the actual yield of $CaCO_3$ is 1.56 grams? The equation is

$$CaCl_2 \text{ (aq)} + Na_2CO_3 \text{ (aq)} \rightarrow CaCO_3 \text{ (s)} + 2 \text{ NaCl (aq)}$$

In this problem you are given the *actual* yield and the known *percentage* yield, presumably based on prior experience with the same reaction. To find the amount of reactant, you must first know the *theoretical* amount of product—the theoretical yield. In other words, 1.56 grams of calcium carbonate is 96.8% of what quantity? You have three approaches to this problem: (1) you may solve Equation 8.2 for theoretical yield, substitute the given values, and solve; (2) you may express percentage yield as a decimal fraction and solve; or (3) you may use the percentage equivalence between theoretical yield and actual yield, and solve by dimensional analysis. Details on the second and third methods may be found in Appendix I, page 507, if you need them. Choose a method and find the theoretical yield of $CaCO_3$.

------------------------------------------------------------------------

**8.8a**   1.61 g $Na_2CO_2$

From Equation 8.2,

$$\text{theoretical yield} = \frac{\text{actual yield}}{\text{percentage yield}} \times 100 = \frac{1.56 \text{ g } CaCO_3}{96.8} \times 100$$

$$= 1.61 \text{ g } CaCO_3$$

or, by converting percentage yield to a decimal fraction,

$$\text{theoretical yield} = \frac{1.56 \text{ g } CaCO_3}{0.968} = 1.61 \text{ g } CaCO_3$$

or, by the dimensional analysis equivalence 96.8 g CaCO$_3$ actual $\simeq$ 100 g CaCO$_3$ theoretical,

$$\boxed{1.56 \text{ g CaCO}_3 \text{ actual}} \times \frac{100 \text{ g CaCO}_3 \text{ theoretical}}{96.8 \text{ g CaCO}_3 \text{ actual}} = 1.61 \text{ g CaCO}_3 \text{ theoretical}$$

Now that the theoretical quantity of calcium carbonate product is known, the number of grams of sodium carbonate reactant may be found by the usual stoichiometric procedure. Complete the problem.

---

**8.8b**   1.71 g Na$_2$CO$_3$

$$\boxed{1.61 \text{ g CaCO}_3} \times \frac{1 \text{ mole CaCO}_3}{100 \text{ g CaCO}_3} \times \frac{1 \text{ mole Na}_2\text{CO}_3}{1 \text{ mole CaCO}_3} \times \frac{106 \text{ g Na}_2\text{CO}_3}{1 \text{ mole Na}_2\text{CO}_3}$$
$$= 1.71 \text{ g Na}_2\text{CO}_3$$

Note that, by dimensional analysis, the problem may be solved in a single setup:

$$1.56 \text{ g CaCO}_3 \text{ actual} \times \frac{100 \text{ g CaCO}_3 \text{ theoretical}}{96.8 \text{ g CaCO}_3 \text{ actual}} \times \frac{1 \text{ mole CaCO}_3}{100 \text{ g CaCO}_3} \times$$

$$\frac{1 \text{ mole Na}_2\text{CO}_3}{1 \text{ mole CaCO}_3} \times \frac{106 \text{ g Na}_2\text{CO}_3}{1 \text{ mole Na}_2\text{CO}_3} = 1.71 \text{ g Na}_2\text{CO}_3$$

## 8.4  LIMITING REAGENT PROBLEMS

PG 8 D   Given a chemical equation or information from which it may be determined, and initial quantities of two or more reactants, (a) identify the limiting reagent; (b) calculate the theoretical yield of a specified product, assuming complete consumption of the limiting reagent; and (c) calculate the quantity of the reactant initially in excess that remains unreacted.

Zinc reacts with sulfur to form zinc sulfide: Zn (s) + S (s) → ZnS (s). Suppose you put three moles of zinc and two moles of sulfur into a reaction vessel and caused them to react until one is totally used up. How many moles of zinc sulfide will result? Also, how many moles of which element will remain unreacted?

This question is something like asking: how many pairs of gloves can you assemble out of 20 left gloves and 30 right gloves, and how many unmatched gloves, and for which hand, will be left over? The answer, of course, is 20 pairs of gloves. After you have assembled 20 pairs you run out of left gloves, even though you have 10 right gloves remaining. Similarly, in the zinc sulfide question where the elements combine on a 1-to-1 mole ratio, you will run out of sulfur after two moles of zinc sulfide form. One mole of zinc, the excess reactant, will remain unreacted. Sulfur, **the reactant that is completely consumed by the reaction, is called the limiting reagent.**

The reasoning is a bit more complicated if we consider the reaction of hydrogen and oxygen to form water: 2 H$_2$ (g) + O$_2$ (g) → 2 H$_2$O (l). If you have two moles of oxygen and three of hydrogen, how many moles of water can be formed? Also, how many moles of which element will remain?

Don't jump too quickly at identifying oxygen as the element in short supply just because there are fewer moles of it than there are of hydrogen. You must consider the ratio in which they are used. It's like being in the bicycle business: if you have three dozen wheels and two dozen handlebars, and enough of everything else, how many dozen bicycles can you assemble? The answer is one and a half dozen because you are going to run out of wheels, taken two at a time, before handlebars, taken only one at a time. Similarly, taking hydrogen molecules two at a time, you will use them all before you exhaust the initially smaller supply of oxygen molecules taken one at a time. Wheels and hydrogen are the limiting species in these examples.

Notice that the entire discussion leading to the identification of the limiting reagent has been in terms of *moles*, not grams. Returning to the Zn (s) + S (s) → ZnS (s) reaction, suppose you had 65 grams of zinc and 48 grams of sulfur. How many grams of zinc sulfide will form, and how many grams of the element in excess will remain after all of the other is gone? First, which element is the limiting reagent? You cannot look at the smaller number of grams of sulfur and conclude that it is the limiting species. Observe: Using the atomic weight—the molar weight of atoms—of each element,

$$65 \text{ g Zn} \times \frac{1 \text{ mole Zn}}{65 \text{ g Zn}} = 1.0 \text{ mole Zn}$$

$$48 \text{ g S} \times \frac{1 \text{ mole S}}{32 \text{ g S}} = 1.5 \text{ moles S}$$

The one mole of zinc will require 1.0 mole of sulfur to react with it, and 1.5 moles are available. Sulfur is in excess, and zinc is the limiting reactant. The same conclusion is reached from the sulfur standpoint: 1.5 moles of sulfur require 1.5 moles of zinc to react with—but only 1.0 mole of zinc is available, so zinc is the limiting reagent. The grams of zinc sulfide produced may be calculated from the limiting reagent:

$$1.0 \text{ mole Zn} \times \frac{1 \text{ mole ZnS}}{1 \text{ mole Zn}} \times \frac{97 \text{ g ZnS}}{1 \text{ mole ZnS}} = 97 \text{ g ZnS}$$

The grams of sulfur in excess may be found by determining the moles in excess and converting to grams:

$$\begin{array}{r} 1.5 \text{ moles S available} \\ -1.0 \text{ mole \ \ S react} \\ \hline 0.5 \text{ mole \ \ S unreacted} \end{array}$$

$$0.5 \text{ mole S excess} \times \frac{32 \text{ g S}}{1 \text{ mole S}} = 16 \text{ g S excess}$$

This entire example may be summarized in "equation" form:

$$1.0 \text{ mole Zn} + 1.5 \text{ moles S} \rightarrow 1.0 \text{ mole ZnS} + 0.5 \text{ mole S remain}$$

$$65 \text{ g Zn} + 48 \text{ g S} \rightarrow 97 \text{ g ZnS} + 16 \text{ g S remain}$$

In the foregoing paragraph we have illustrated the four-step procedure by which limiting reagent problems may be solved. These steps are,

1. Express each of the given quantities in *moles*. (This is the first step in the stoichiometry pattern.)
2. Using either molar quantity, identify the limiting reagent.
3. From the moles of the limiting reagent calculate the quantity of wanted species in the units required. (This requires the second and third steps in the stoichiometry pattern.)
4. If asked, determine the moles of excess reactant that remain unreacted and convert to the units required.

You can use this procedure in solving the following example:

**Example 8.9**   Calculate the number of grams of calcium fluoride that will precipitate after combining a solution containing 39.0 grams of calcium nitrate with a solution containing 33.0 grams of sodium fluoride. Also determine which of the reactants is in excess, and by how many grams. The equation is

$$Ca(NO_3)_2 \text{ (aq)} + 2 \, NaF \text{ (aq)} \rightarrow CaF_2 \text{ (s)} + 2 \, NaNO_3 \text{ (aq)}$$

You begin by converting the grams of each substance to moles.

-------------------------------------------------------------------------------

**8.9a**   0.238 mole $Ca(NO_3)_2$ and 0.786 mole NaF

$$39.0 \text{ g } Ca(NO_3)_2 \times \frac{1 \text{ mole } Ca(NO_3)_2}{164 \text{ g } Ca(NO_3)_2} = 0.238 \text{ mole } Ca(NO_3)_2$$

$$33.0 \text{ g } NaF \times \frac{1 \text{ mole } NaF}{42.0 \text{ g } NaF} = 0.786 \text{ mole } NaF$$

Now determine which species is the limiting reagent. Recall you may use either starting point for your thought process.

-------------------------------------------------------------------------------

**8.9b**   $Ca(NO_3)_2$ is the limiting reagent.

$$0.238 \text{ mole } Ca(NO_3)_2 \times \frac{2 \text{ moles } NaF}{1 \text{ mole } Ca(NO_3)_2} = 0.476 \text{ mole } NaF$$

0.476 mole NaF is required to react with 0.238 mole of $Ca(NO_3)_2$, and 0.786 mole NaF, an excess, is available. $Ca(NO_3)_2$ limits. Alternatively,

$$0.786 \text{ mole } NaF \times \frac{1 \text{ mole } Ca(NO_3)_2}{2 \text{ moles } NaF} = 0.393 \text{ mole } Ca(NO_3)_2$$

0.393 mole $Ca(NO_3)_2$ is required to react with 0.786 mole NaF, but only 0.238 mole $Ca(NO_3)_2$ is available. $Ca(NO_3)_2$ limits.

Having identified the limiting reactant and calculated the moles present, you can calculate the grams of calcium fluoride formed by using the last two steps of the stoichiometry pattern.

-------------------------------------------------------------------------------

**8.9c**   18.6 $CaF_2$

$$0.238 \text{ mole } Ca(NO_3)_2 \times \frac{1 \text{ mole } CaF_2}{1 \text{ mole } Ca(NO_3)_2} \times \frac{78.1 \text{ g } CaF_2}{1 \text{ mole } CaF_2} = 18.6 \text{ g } CaF_2$$

Now return to the sodium fluoride. Find the number of moles that are in excess, and convert that quantity to grams.

------------------------------------------------------------------------------------------------

**8.9d**   0.310 mole or 13.0 grams NaF in excess

0.786 mole NaF available − 0.476 mole NaF used = 0.310 mole NaF excess

$$0.310 \text{ mole NaF} \times \frac{42.0 \text{ g NaF}}{1 \text{ mole NaF}} = 13.0 \text{ g NaF in excess}$$

This time take it all the way without assistance . . . .

**Example 8.10**   Aluminum chloride may be made by direct combination of its elements. If 16.9 grams of aluminum and 85.1 grams of chlorine are brought together in a manner that yields complete conversion of the limiting reagent, how many grams of aluminum chloride will form? Also identify the excess reactant and calculate the grams that remain unreacted.

------------------------------------------------------------------------------------------------

**8.10a**   83.3 g $AlCl_3$; 18.5 g $Cl_2$ remain unreacted

Reaction equation: $2 \text{ Al (s)} + 3 \text{ Cl}_2 \text{ (g)} \rightarrow 2 \text{ AlCl}_3 \text{ (s)}$

$$16.9 \text{ g Al} \times \frac{1 \text{ mole Al}}{27.0 \text{ g Al}} = 0.626 \text{ mole Al available}$$

$$85.1 \text{ g Cl}_2 \times \frac{1 \text{ mole Cl}_2}{71.0 \text{ g Cl}_2} = 1.20 \text{ moles Cl}_2 \text{ available}$$

$0.626 \text{ mole Al} \times \frac{3 \text{ moles Cl}_2}{2 \text{ moles Al}} = 0.939 \text{ mole Cl}_2$ required to react with 0.626 mole of Al. 1.20 moles of $Cl_2$ are available. Therefore Al is the limiting reagent.

Alternatively, $1.20 \text{ moles Cl}_2 \times \frac{2 \text{ moles Al}}{3 \text{ moles Cl}_2} = 0.800 \text{ mole Al}$ required to react with 1.20 moles $Cl_2$. Only 0.626 mole Al is available. Al is the limiting reagent.

$$0.626 \text{ mole Al} \times \frac{2 \text{ moles AlCl}_3}{2 \text{ moles Al}} \times \frac{133 \text{ g AlCl}_3}{1 \text{ mole AlCl}_3} = 83.3 \text{ g AlCl}_3$$

1.20 moles $Cl_2$ available − 0.939 mole reacts = 0.261 mole $Cl_2$ remains

$$0.261 \text{ mole Cl}_2 \times \frac{71.0 \text{ g Cl}_2}{1 \text{ mole Cl}_2} = 18.5 \text{ g Cl}_2 \text{ remain unreacted}$$

# CHAPTER 8 IN REVIEW

---

### 8.1 QUANTITATIVE INTERPRETATION OF A CHEMICAL EQUATION

**8 A**  Given a chemical equation, or a reaction for which the equation can be written, and the number of moles of one species in the reaction, calculate the number of moles of any other species. (Page 167)

### 8.2 MASS CALCULATIONS: STOICHIOMETRY

**8 B**  Given a chemical equation, or a reaction for which the equation can be written, and the number of grams or moles of one species in the reaction, find the number of grams or moles of any other species. (Page 169)

### 8.3 PERCENTAGE YIELD

**8 C**  Given two of the following, or information from which two of the following may be determined, calculate the third: theoretical yield, actual yield and percentage yield. (Page 172)

### 8.4 LIMITING REAGENT PROBLEMS

**8 D**  Given a chemical equation, or information from which it may be determined, and initial quantities of two or more reactants, (a) identify the limiting reagent; (b) calculate the theoretical yield of a specified product, assuming complete consumption of the limiting reagent; and (c) calculate the quantity of the reactant initially in excess that remains unreacted. (Page 174)

---

## TERMS AND CONCEPTS

Stoichiometry (169)           Actual yield (172)
"Stoichiometry pattern" (169) Percentage yield (172)
Theoretical yield (172)       Limiting reagent (174)

---

## QUESTIONS AND PROBLEMS

*An asterisk (\*) identifies a question that is relatively difficult, or that extends beyond the performance goals in the chapter.*

Section 8.1

8.1)  The presence of carbon dioxide in a gas may be determined by the appearance of a cloudiness, calcium carbonate, when the gas is bubbled through limewater, which is a solution of calcium hydroxide: $Ca(OH)_2$ (aq) + $CO_2$ (g) → $CaCO_3$ (s) + $H_2O$ (l). If 3.91 moles of carbon dioxide pass through the solution, how many moles of calcium carbonate will form?

8.25)  When acetylene, $C_2H_2$, burns in an oxyacetylene torch, which is used in cutting steel, the equation is $2\,C_2H_2$ (g) + $5\,O_2$ (g) → $4\,CO_2$ (g) + $2\,H_2O$ (g). How many moles of oxygen are required for each mole of $C_2H_2$ used?

8.2) The possibility of a rock sample being limestone (calcium carbonate) may be tested with hydrochloric acid. The formation of bubbles indicates carbon dioxide: $CaCO_3$ (s) + 2 HCl (aq) → $CaCl_2$ (aq) + $CO_2$ (g) + $H_2O$ (l). How many moles of carbon dioxide will be released for the reaction of 0.284 mole of calcium carbonate?

### Section 8.2

8.3) In the reaction of problem 8.1, how many grams of $CO_2$ are required to produce 0.462 mole of $CaCO_3$?

8.4) Calculate the number of moles of $CaCl_2$ formed from 54.1 grams of $CaCO_3$ in the reaction of problem 8.2.

8.5) How many grams of sodium sulfate can be crystallized from the reaction of 95.2 grams of sodium chloride with excess sulfuric acid in the reaction 2 NaCl (s) + $H_2SO_4$ (aq) → $Na_2SO_4$ (s) + 2 HCl (g)?

8.6) Silver bromide that remains on photographic film after development is removed by reaction with "hypo," known chemically as sodium thiosulfate, $Na_2S_2O_3$. How many grams of sodium thiosulfate must react to remove 1.09 grams of silver bromide? The equation for the reaction is 2 $Na_2S_2O_3$ (aq) + AgBr (s) → $Na_3Ag(S_2O_3)_2$ (aq) + NaBr (aq).

8.7) Baking soda, $NaHCO_3$, reacts with cream of tartar, $KHC_4H_4O_6$, in the baking of a cake. A French recipe calls for 12 grams of cream of tartar. (A standard item in a French kitchen is a small scale, as quantities of solids appear in European recipes in grams rather than cups, teaspoons or other volume units.) How many grams of baking soda should be called for in the same recipe if all of both materials are to react completely? The equation is $KHC_4H_4O_6$ (s) + $NaHCO_3$ (s) → $KNaC_4H_4O_6$ (s) + $H_2O$ (l) + $CO_2$ (g).

8.8) How many grams of chlorine will be produced by the reaction of 2.69 grams of manganese dioxide, $MnO_2$, in the typical laboratory method of preparing chlorine:

$MnO_2$ (s) + 4 HCl (aq) → $MnCl_2$ (aq) + 2 $H_2O$ (l) + $Cl_2$ (g)?

8.9) Chloroform, $CHCl_3$, may be prepared in the laboratory by the reaction between chlorine and methane: 3 $Cl_2$ (g) + $CH_4$ (g) → $CHCl_3$ (l) + 3 HCl (g). Calculate the number of grams of chlorine that are required to produce 45.0 grams of $CHCl_3$.

8.26) The fermentation of molasses, $C_6H_{12}O_6$, in the presence of zymase, a catalyst derived from yeast, may be represented by the equation

$C_6H_{12}O_6$ (l) $\xrightarrow{\text{zymase}}$ 2 $C_2H_5OH$ (l) + 2 $CO_2$ (g).

How many moles of alcohol, $C_2H_5OH$, are produced from the decomposition of 2.80 moles of molasses?

8.27) Determine the number of grams of $C_2H_2$ that burned in producing 1.06 moles of $CO_2$ in the reaction of problem 8.25.

8.28) If 29.8 grams of $C_2H_5OH$ are formed in the reaction of problem 8.26, how many moles of $C_6H_{12}O_6$ fermented?

8.29) Stannous fluoride, $SnF_2$, the "fluoride" in many toothpastes, may be made by the reaction Sn (s) + 2 HF (g) → $SnF_2$ (s) + $H_2$ (g). Calculate the grams of HF that are required to prepare 0.750 gram of stannous fluoride.

8.30) A soda acid (sodium hydrogen carbonate) fire extinguisher makes carbon dioxide by the reaction 2 $NaHCO_3$ (s) + $H_2SO_4$ (aq) → $Na_2SO_4$ (aq) + 2 $H_2O$ (l) + $CO_2$ (g). How many grams of carbon dioxide will be produced by an extinguisher charged with 55.0 grams of $NaHCO_3$?

8.31) Hydrogen peroxide, $H_2O_2$, is capable of "cleaning" oil paintings of the black lead sulfide, PbS, formed by reaction of white lead in the paint with hydrogen sulfide in the air. The cleaning equation is PbS (s) + 4 $H_2O_2$ (aq) → $PbSO_4$ (s) + 4 $H_2O$ (l). How many grams of PbS can be removed by a solution containing 1.43 grams of $H_2O_2$, assuming all of the peroxide reacts?

8.32) The hydrate of copper sulfate, $CuSO_4 \cdot$ 5 $H_2O$, is deep blue. When heated, water is driven off, leaving a nearly white anhydrous salt: $CuSO_4 \cdot$ 5 $H_2O$ (s) → $CuSO_4$ (s) + 5 $H_2O$ (g). How many moles of the hydrate are required to release 48.6 grams of water?

8.33) Dichlorodifluoromethane is the chemical name for the common refrigerant called Freon. How many grams of Freon, $CCl_2F_2$, can be prepared from 26.8 grams of carbon tetrachloride, $CCl_4$, by the following catalytic reaction:

$CCl_4$ (l) + 2 HF (g) $\xrightarrow{\text{cat.}}$ $CCl_2F_2$ (g) + 2 HCl (g)?

8.10) One form of soap has the formula $C_{17}H_{35}COONa$. It is prepared by the reaction $C_3H_5(C_{17}H_{35}COO)_3$ (s) + 3 NaOH (aq) → 3 $C_{17}H_{35}COONa$ (s) + $C_3H_5(OH)_3$ (l). Calculate the number of grams of sodium hydroxide (lye) required to prepare 175 grams of soap.

8.11) The purification ability and bleaching action of chlorine both come from the release of oxygen in a reaction which may be summarized as 2 $H_2O$ (l) + 2 $Cl_2$ (aq) → 4 HCl (aq) + $O_2$ (g). How many grams of oxygen will be so liberated by 255 grams of chlorine?

8.12) Photosynthesis is the process by which carbon dioxide is converted into sugar, $C_6H_{12}O_6$, with the help of sunlight and the chlorophyll of green plants acting as a catalyst. It may be summarized in the equation 6 $CO_2$ (g) + 6 $H_2O$ (g) $\xrightarrow[\text{light energy}]{\text{chlorophyll}} C_6H_{12}O_6$ (aq) + 6 $O_2$ (g). How many grams of sugar may be produced in the reaction that consumes 6.90 grams of carbon dioxide?

8.13) Plaster of Paris is a compound whose formula may be written $CaSO_4 \cdot \frac{1}{2} H_2O$ or $(CaSO_4)_2 \cdot H_2O$. Casts made from this material are a higher hydrate of calcium sulfate, $CaSO_4 \cdot 2 H_2O$. The equation is $(CaSO_4)_2 \cdot H_2O$ (s) + 3 $H_2O$ (l) → 2 $CaSO_4 \cdot 2 H_2O$ (s). What mass of $CaSO_4 \cdot 2 H_2O$ will result from the reaction of 48.3 grams of $(CaSO_4)_2 \cdot H_2O$?

8.14) Dinitrogen oxide $N_2O$—known also as nitrous oxide and laughing gas—was one of the early anesthetics, and has recently regained popularity in this role in the dental field. It is made by the decomposition of ammonium nitrate: $NH_4NO_3$ (s) → $N_2O$ (g) + 2 $H_2O$ (l). How many grams of $NH_4NO_3$ are required to prepare 12,800 grams of $N_2O$?

8.15)* One of the early steps in the manufacture of sulfuric acid is to prepare sulfur dioxide from sulfide ores. How many tons of an ore that is effectively 12.1% $FeS_2$ must be processed to produce 5.00 tons of sulfur dioxide in the reaction 4 $FeS_2$ (s) + 11 $O_2$ (g) → 2 $Fe_2O_3$ (s) + 8 $SO_2$ (g)? (Hint: You can work with a *ton-mole*—the quantity of a substance containing the same number of units as there are atoms in exactly 12 tons of carbon-12—in the same manner as you have been using the "gram-mole." See also page 507 on the use of dimensional analysis for problems involving percent.)

8.34) The well known preservative from the biology laboratory, formaldehyde, HCHO, is prepared commercially in solution form by heating methanol, $CH_3OH$, with oxygen in the presence of copper:

$$2 \text{ CH}_3\text{OH (l)} + \text{O}_2 \text{ (g)} \xrightarrow{\text{Cu}} 2 \text{ HCHO (aq)} + 2 \text{ H}_2\text{O (l)}.$$

What mass of formaldehyde can be prepared from 14.8 grams of $CH_3OH$?

8.35) One part of the blast furnace operation in which iron is extracted from oxide ore is expressed in the equation $Fe_2O_3$ (s) + 3 CO (g) → 2 Fe (s) + 3 $CO_2$ (g). How many kg of CO are necessary to produce 1250 kg of iron?

8.36) Electrical energy is derived from chemical energy in starting an automobile engine with a storage battery. The chemical equation for the reaction that occurs is Pb (s) + $PbO_2$ (s) + 2 $H_2SO_4$ (aq) → 2 $PbSO_4$ (s) + 2 $H_2O$ (l). Calculate the number of grams of $H_2SO_4$ consumed in the reaction of 39.4 grams of lead dioxide, $PbO_2$.

8.37) Limestone caves, such as Mammoth Cave and Carlsbad Caverns, are made when carbon dioxide and water combine to dissolve the limestone, $CaCO_3$:

$$\text{CaCO}_3 \text{ (s)} + \text{H}_2\text{O (l)} + \text{CO}_2 \text{ (g)} \rightarrow \text{Ca(HCO}_3)_2 \text{ (aq).}$$

How many grams of carbon dioxide are required to dissolve 454 grams (1 pound) of limestone?

8.38) A chemical reaction that changed the history of the world is $N_2$ + 3 $H_2$ → 2 $NH_3$. Conducted at 500°C, up to 1000 atmospheres pressure, and in the presence of a catalyst, the so-called Haber process enabled Germany to manufacture explosives from the nitrogen of the air and thereby wage World War I. How many grams of ammonia, $NH_3$, can be made from 525 grams of nitrogen?

8.39)* One liter of air at normal atmospheric conditions contains 0.89 gram of nitrogen. What volume of air must be processed to make 100 kilograms (220 pounds) of ammonia by the Haber process if the reaction is $N_2$ (g) + 3 $H_2$ (g) → 2 $NH_3$ (g)?

8.16)* Calculate the volume of hydrogen that will result from the reaction of 3.94 grams of zinc in the reaction $Zn (s) + 2 HCl (aq) \rightarrow H_2 (g) + ZnCl_2 (aq)$, when conditions are such that 1 mole of hydrogen occupies 24.5 liters.

### Section 8.3

8.17) $3.81 \times 10^4$ grams of acetylene, $C_2H_2$, are produced when $5.19 \times 10^4$ grams of methane, $CH_4$, are treated in an electric arc in the commercial production of acetylene:

$$2 CH_4 (g) \xrightarrow[\text{arc}]{\text{electric}} C_2H_2 (g) + 3 H_2 (g).$$

Find the percentage yield.

8.18) In the extraction of copper from its sulfide ore, the overall process may be summarized by the equation $Cu_2S (s) + O_2 (g) \rightarrow 2 Cu (s) + SO_2 (g)$. If the percentage yield is 61.2%, how much copper will result from treatment of $7.00 \times 10^6$ grams of $Cu_2S$?

8.19) In a laboratory preparation of chloroform, $CHCl_3$, a student introduced 3.85 grams of methane, $CH_4$, and excess chlorine into a reaction chamber. On completion of the reaction, $CH_4 (g) + 3 Cl_2 (g) \rightarrow CHCl_3 (l) + 3 HCl (g)$, he recovered 23.0 grams of chloroform. What does this represent in percent?

8.20) The sea is a major industrial source of magnesium, where it exists as a very dilute solution of magnesium chloride. The hydroxide is precipitated from sea water by treatment with slaked lime (calcium hydroxide): $MgCl_2 (aq) + Ca(OH)_2 (s) \rightarrow Mg(OH)_2 (s) + CaCl_2 (aq)$. Calculate the percentage yield in the process if 489 kilograms of magnesium hydroxide are recovered from the addition of 891 kilograms of calcium hydroxide to an excess of sea water.

### Section 8.4

8.21) 40.0 grams of ammonia and 40.0 grams of oxygen are made to react until all of the limiting reagent is used. Calculate the grams of water vapor released if the reaction equation is $4 NH_3 (g) + 3 O_2 (g) \rightarrow 2 N_2 (g) + 6 H_2O (g)$. How many grams of which reactant will remain unchanged at the end of the reaction?

8.40)* Excess sulfuric acid is added to 125 milliliters of a solution in which the concentration of barium chloride is 0.750 mole per liter. Calculate the grams of barium sulfate that precipitate by the reaction $H_2SO_4 (aq) + BaCl_2 (aq) \rightarrow BaSO_4 (s) + 2 HCl (aq)$.

8.41) When $9.08 \times 10^5$ grams (one ton) of sodium chloride are electrolyzed in the commercial production of chlorine, $5.24 \times 10^5$ grams of chlorine are produced:

$$2 NaCl (aq) + 2 H_2O (l) \xrightarrow{\text{electrolysis}}$$

Calculate the percentage yield.

8.42) One step of the Ostwald process for manufacturing nitric acid involves making nitrogen monoxide by oxidizing ammonia in the presence of a platinum catalyst:

$$4 NH_3 (g) + 5 O_2 (g) \xrightarrow{Pt} 4 NO (g) + 6 H_2O (g).$$

If the percentage yield is 80.3%, how many grams of NO can be produced from $4.00 \times 10^3$ grams of $NH_3$?

8.43) One commercial method for preparing sodium hydroxide involves the reaction of sodium carbonate and calcium hydroxide in boiling water: $Na_2CO_3 (aq) + Ca(OH)_2 (aq) \rightarrow 2 NaOH (aq) + CaCO_3 (s)$. If the process is known to have a 92.0% yield, how many kilograms of sodium carbonate should be used to produce 325 kilograms of sodium hydroxide?

8.44) The metal chromium may be extracted from its oxide ore by reaction with carbon: $Cr_2O_3 (s) + 3 C (s) \rightarrow 2 Cr (s) + 3 CO (g)$. How many kilograms of chromium can be obtained from 402 kilograms of $Cr_2O_3$ if the percentage yield for the process is 86.4%?

8.45) A laboratory experiment calls for the use of 6.00 grams of manganese dioxide, $MnO_2$, and a solution containing 25.8 grams of HCl in the preparation of chlorine by the reaction $MnO_2 (s) + 4 HCl (aq) \rightarrow MnCl_2 (aq) + 2 H_2O (l) + Cl_2 (g)$. Assuming the limiting reagent reacts completely, how many grams of chlorine will result? Also, how many grams of the excess reagent will be unreacted at the end of the experiment?

8.22) Zinc may be extracted from its sulfide ore by first roasting to produce zinc oxide, and then heating with powdered coal. The second step is ZnO (s) + C (s) → Zn (s) + CO (g). In a laboratory run for the same reaction a chemist mixed 19.6 grams of carbon with 135 grams of zinc oxide. Calculate the grams of zinc metal that will be produced by complete conversion of the limiting species, and the grams of whatever other solid reactant that will remain.

8.23)* The solvent carbon tetrachloride—once widely used in dry cleaning and for extinguishing fires, but now considered too toxic (poisonous) to be used in a home, even as a spot remover—is prepared commercially by the reaction $CS_2$ (l) + $2 S_2Cl_2$ (l) → $CCl_4$ (l) + 6 S (s). How many tons of carbon tetrachloride will result from the reaction of all of the limiting reagent in a mixture of 3.00 tons of $S_2Cl_2$ and 1.00 ton of $CS_2$? Which reactant is in excess, and by how many tons? (The hint for Problem 8.15 applies here too.)

8.46) Though nearly 80% of the air is elemental nitrogen, it is very difficult to change this nitrogen into useful compounds. One successful process uses the reaction $Na_2CO_3$ (s) + 4 C (s) + $N_2$ (g) → 2 NaCN (s) + 3 CO (g). In a laboratory experiment with this reaction, 26.8 grams of carbon and 122 grams of sodium carbonate reacted with an abundance of air until one of the solids was completely consumed. Calculate the yield of sodium cyanide, NaCN, a highly poisonous salt used in electroplating. Also find the number of grams of the solid reactant that was left.

8.47)* One step in the extraction of copper from its ore involves the reaction of copper(I) sulfide with copper(I) oxide: $2 Cu_2O$ (s) + $Cu_2S$ (s) → 6 Cu (s) + $SO_2$ (g). To maximize this yield of copper it is evident that the producer must use two copper compounds in the proper proportion. Suppose he uses 1.00 ton of each compound in a production run. How many tons of elemental copper will he produce? How many *additional* tons of which compound should be added to balance the 1.00 ton of the limiting reagent? What then will be the total yield of copper in tons?

*At this point we close the chapter with a pair of 'challenge' questions that do not specify a chemical equation. The molar relationships between given and wanted species can be found, however, in the wording of the problems. With those relationships, the problems can be solved, using the principles developed in this chapter.*

8.24)* A mixed industrial product contains magnesium as a hexahydrate of magnesium chloride, $MgCl_2 \cdot 6 H_2O$. A 5.25 gram sample of the material is placed in water, dissolving the soluble components of the mixture, including the magnesium chloride. The chloride ion in the solution is then precipitated as silver chloride, AgCl, which is filtered, dried and weighed at 4.01 grams. Assuming magnesium chloride hexahydrate to be the only source of chloride ion, calculate (a) the grams of magnesium and (b) the percentage $MgCl_2 \cdot 6 H_2O$ in the sample.

8.48)* One of the methods of analyzing steel for its manganese content involves a series of chemical changes resulting in the complete conversion of the manganese to $Mn_3O_4$. From prior experience, however, it is known that the percentage yield is 96%. One such analysis of an 11.2 gram sample of steel resulted in the separation of 0.337 gram of $Mn_3O_4$. Find the percentage of manganese in the steel.

# 9

# CHEMICAL BONDING

For several chapters we have accepted Dalton's idea that atoms of different elements combine to form compounds. We've given them names and written their formulas. But never have we considered how these combinations come about—what causes atoms to stick together in molecular or ionic species. This is the kind of thing we will consider in the chapter on Chemical Bonding. Specifically we will examine two types of bonds, ionic bonds and covalent bonds. But first, let's introduce a tool that makes it easier to picture the role played by electrons in chemical bonding: the Lewis symbol.

## 9.1 LEWIS SYMBOLS OF THE ELEMENTS

PG 9 A Write the Lewis symbol of any atom in one of the A groups of the periodic table.

In Chapter 4 (page 84) it was stated that the similar chemical properties of members of the same family of elements can be attributed to their *valence electrons*, the s and p electrons of the highest occupied energy level when the atom is in its ground state. The ground state valence electron configuration of the alkali metals, for example, is $ns^1$, where n is the principal energy level of the valence electrons. For lithium this configuration is $2s^1$, and n = 2; for sodium, $3s^1$, and n = 3; and so forth. The ground state valence electron configurations of all A groups of the periodic table are summarized in Table 9.1.

The last line of Table 9.1 shows the **Lewis symbols,** or the **electron dot symbols,** of the elements of the third period. The Lewis symbol of an element is the chemical symbol with one or more dots distributed around it. The number of dots corresponds to the number of valence electrons in an atom of the element. Lewis symbols for alkali metals, represented by sodium in Table 9.1, have one dot for one valence electron, the $ns^1$ electron. The symbol for lithium is Li· ; for potassium, K· ; and so forth. Lewis symbols for the alkaline earth metals have two dots, matching the $ns^2$ configuration of the valence electrons, as shown for magnesium. For calcium the Lewis symbol is

TABLE 9.1   Lewis Symbols of the Elements

| Group | IA | IIA | IIIA | IVA | VA | VIA | VIIA | 0 |
|---|---|---|---|---|---|---|---|---|
| Highest Energy Electron Configuration | $ns^1$ | $ns^2$ | $ns^2np^1$ | $ns^2np^2$ | $ns^2np^3$ | $ns^2np^4$ | $ns^2np^5$ | $ns^2np^6$ |
| Number of Valence Electrons | 1 | 2 | 3 | 4 | 5 | 6 | 7 | 8 |
| Lewis Symbol Third Period Element | Na· | Mg· | ·Al· | ·Si· | ·P: | ·S: | :Cl: | :Ar: |

Ca:; and for barium, Ba:. Note that the exact location of the dots is not important. For magnesium, for example, the symbol may be Mg: or ·Mg·, whichever best serves the purpose for which the symbol is being used. Lewis symbols for the halogens have seven dots, as shown for chlorine, to match the $ns^2np^5$ (2 s electrons + 5 p electrons = 7 valence electrons) of their highest energy level. The noble gas family, Group 0, is characterized by a full set of 8 valence electrons, or an **octet,** as it is often called.

Lewis symbols are a great aid in accounting for what we believe to be taking place at the atomic level, which can never be observed directly, during the making and breaking of chemical bonds. We will use them for that purpose throughout this chapter.

## 9.2   MONATOMIC IONS WITH NOBLE GAS ELECTRON CONFIGURATIONS

PG   9  B   Identify by name and symbol the monatomic ions that are isoelectronic with a given noble gas atom, and write the electron configurations of those ions.

In Figure 4.15, page 83, we illustrated the formation of a sodium ion by the removal of an electron from a neutral sodium atom. This concept was presented a second time in the form of an equation on page 101 of Chapter 5. Table 9.2 offers the same idea, this time expanded to show the Lewis symbol for sodium and the electron configurations for the different species. A neutral sodium atom contains 11 protons and 11 electrons. Its electron configuration is $1s^22s^22p^63s^1$, and it has Na· as its Lewis symbol. If you remove the highest energy electron, the 3s electron, from the neutral sodium atom, the sodium ion that remains has the configuration $1s^22s^22p^6$. There are still 11 protons, leaving a net charge of +1 for the ion, which is represented by the symbol $Na^+$.

The formation of a magnesium ion is similar, except that the magnesium atom begins with 12 protons and 12 electrons, and produces an ion with a +2 charge by losing two electrons. Notice the electron configuration of the magnesium ion, $Mg^{2+}$; it is identical to that of the sodium ion. Two species having identical electron configurations are said to be **isoelectronic** with each other, *iso-* being a prefix suggesting *sameness*, as in *iso*tope.

What do you suppose is the electron configuration of the aluminum ion, $Al^{3+}$? The neutral atom must lose three electrons, two 3s and one 3p, to produce an ion with a +3 charge. Do you see that this will reduce the total electron population to 10, which will again yield the electron configuration $1s^2 2s^2 2p^6$? This is shown in Table 9.2.

It is significant to note that sodium, magnesium and aluminum are "active" metals. They do not occur in nature in their elemental states, nor are they completely stable in their elemental states when exposed to air or water. Freshly cut sodium is promptly covered with an oxide coating in which the sodium exists as the more stable $Na^+$ ion. Magnesium and aluminum are both used as structural metals, but it is possible to do so only because the elements cover themselves with a thin protective film when exposed to air. The film, a carbonate for magnesium and an oxide for aluminum, prevents further corrosion. The magnesium and aluminum are present as ions, $Mg^{2+}$ and $Al^{3+}$, in these films. It appears that sodium, magnesium and aluminum are more stable in nature as ions than as neutral atoms, and that these ions are isoelectronic with each other.

At this point ask yourself: Is there an *element* whose neutral atoms have this same $1s^2 2s^2 2p^6$ electron configuration? The answer, of course, is yes. Its neutral atom must have ten protons to match the ten electrons, and that identifies the element with atomic number 10, neon. And what are the chemical properties of neon? It is a noble gas, and the word *noble* means that it is unreactive, or stable, in its elemental state. The valence electron configuration of neon—of all noble gases—is the complete octet of highest energy s and p electrons, an $ns^2 np^6$ configuration, where n = 2 for neon. It appears that the noble gas configuration represents stability not only for the noble gases, but for many metal cations too.

An examination of nonmetal anions leads to the same conclusion. What must be done with neutral atoms of fluorine, oxygen and nitrogen to change

**TABLE 9.2** Formation of Monatomic Cations

| ELEMENT | ATOM | | MONATOMIC ION | | ELECTRON(S) |
|---|---|---|---|---|---|
| Sodium | 11 p⁺ / 11 e⁻ | → | 11 p⁺ / 10 e⁻ | + | $e^-$ |
| | Na· | → | Na⁺ | + | •(e⁻) |
| | $1s^2 2s^2 2p^6 3s^1$ | → | $1s^2 2s^2 2p^6$ | + | $e^-$ |
| Magnesium | 12 p⁺ / 12 e⁻ | → | 12 p⁺ / 10 e⁻ | + | $e^- + e^-$ |
| | Mg· | → | Mg²⁺ | + | •(e⁻) + •(e⁻) |
| | $1s^2 2s^2 2p^6 3s^2$ | → | $1s^2 2s^2 2p^6$ | + | 2 e⁻ |
| Aluminum | 13 p⁺ / 13 e⁻ | → | 13 p⁺ / 10 e⁻ | + | $e^- + e^- + e^-$ |
| | ·Al· | → | Al³⁺ | + | •(e⁻) + •(e⁻) + •(e⁻) |
| | $1s^2 2s^2 2p^6 3s^2 3p^1$ | → | $1s^2 2s^2 2p^6$ | + | 3 e⁻ |

TABLE 9.3  Formation of Monatomic Anions

| ELEMENT | ATOM | ELECTRON(S) | MONATOMIC ION |
|---|---|---|---|
| Fluorine | 9 p$^+$ / 9 e$^-$ | + e$^-$ → | 9 p$^+$ / 10 e$^-$ |
| | :Ḟ: | + •(e$^-$) → | [:Ḟ:]$^-$  F$^-$ |
| | $1s^2 2s^2 2p^5$ | + e$^-$ → | $1s^2 2s^2 2p^6$ |
| Oxygen | 8 p$^+$ / 8 e$^-$ | + e$^-$ + e$^-$ → | 8 p$^+$ / 10 e$^-$ |
| | •Ö: | + •(e$^-$) + •(e$^-$) → | [:Ö:]$^{2-}$  O$^{2-}$ |
| | $1s^2 2s^2 2p^4$ | + 2 e$^-$ → | $1s^2 2s^2 2p^6$ |
| Nitrogen | 7 p$^+$ / 7 e$^-$ | + e$^-$ + e$^-$ + e$^-$ → | 7 p$^+$ / 10 e$^-$ |
| | •Ṅ: | + •(e$^-$) + •(e$^-$) + •(e$^-$) → | [:Ṅ:]$^{3-}$  N$^{3-}$ |
| | $1s^2 2s^2 2p^3$ | + 3 e$^-$ → | $1s^2 2s^2 2p^6$ |

them into negatively charged ions, F$^-$, O$^{2-}$, and N$^{3-}$?* To make negatively charged ions from neutral atoms you must add one or more electrons to each atom. Lewis symbols and electron configurations again illustrate the process, which is summarized in Table 9.3. The fluoride, oxide and nitride anions are isoelectronic with each other—and with the stable noble gas neon.

A word of explanation is necessary, in case you suspect a flaw in the above comments about nitrogen and oxygen. These elements are stable in their elemental states, to be sure, but as diatomic molecules, not as individual atoms. As diatomic molecules the octet idea appears again in the sense of sharing electrons. We'll see this in greater detail later in this chapter.

The fact that all Group IA monatomic ions have a +1 charge, all Group IIA monatomic ions have a +2 charge, and all Group VIIA monatomic ions have a −1 charge, to name just three groups, suggests that the pattern built around the neon configuration is duplicated around other noble gas atoms. With a slight modification for lithium and hydrogen (see next page), this is the case.

---

*Nitride ion, N$^{3-}$, is not common, but does exist in a small number of compounds.

---

**Example 9.1**  Write the electron configurations for the calcium and chloride ions, Ca$^{2+}$ and Cl$^-$. With what noble gases are these ions isoelectronic?

To begin, note the locations of calcium and chlorine in the periodic table. Write their Lewis symbols and from them state the number of electrons that must be gained or lost to achieve complete octets at the highest energy level.

**9.1a**   Ca: must lose two electrons to achieve an octet;

:Cl· must gain one electron to achieve an octet.

You are now ready to answer the main questions: the electron configuration for $Ca^{2+}$ and $Cl^-$, please, and the noble gases with which they are isoelectronic.

---

**9.1b**  Both ions are isoelectronic with argon, $1s^22s^22p^63s^23p^6$. The calcium atom starts with the configuration $1s^22s^22p^63s^23p^64s^2$. In losing two electrons to yield $Ca^{2+}$ it reaches the electron configuration of argon. Chlorine, with configuration $1s^22s^22p^63s^23p^5$, must gain one electron to become $Cl^-$, which is isoelectronic with argon.

**Example 9.2**   Identify at least one more cation and one more anion that are isoelectronic with an atom of argon.

---

**9.2a**   $K^+$, $Sc^{3+}$, $S^{2-}$ and $P^{3-}$

The formation of $K^+$, $S^{2-}$ and $P^{3-}$ are identical to the formation of $Na^+$, $O^{2-}$ and $N^{3-}$, respectively, the ions immediately above them in the periodic table. You have not yet had information that would lead you to include the scandium ion, $Sc^{3+}$, in your answer to the question. If given the electron configuration of scandium, atomic number 21, $1s^22s^22p^63s^23p^64s^23d^1$, you might have guessed the charge on the ion would be 3+ because removal of the three highest energy electrons leaves the configuration of the noble gas, argon. Your guess would have been correct.

It was noted above that these ideas must be modified slightly when applied to monatomic ions formed by lithium and hydrogen. Both elements form monatomic ions that are isoelectronic with a noble gas atom, but they do not have a complete octet of electrons. The noble gas they duplicate is helium, which has the electron configuration of $1s^2$. Lithium, Li, with electron configuration $1s^22s^1$, loses its 2s electron to form the lithium ion, $Li^+$. Hydrogen, H, with configuration $1s^1$, gains an electron to reach the helium configuration and form the hydride ion, $H^-$.

In case you're wondering about the hydrogen ion, $H^+$, remember that this ion does not normally exist by itself (see page 104), but rather as a "hydrated" hydrogen ion, $H \cdot H_2O^+$, commonly called the hydronium ion, and written $H_3O^+$. This polyatomic ion exists in aqueous acid solutions, and is not properly a part of a consideration of monatomic ions.

Figure 9.1 is a periodic table showing most of the monatomic ions that are isoelectronic with noble gases, as well as the noble gases themselves. Referring to the table, this section may be summarized in the following generalization: **Metal atoms with 1, 2 or 3 electrons more than the preceding noble gas tend to acquire noble gas configurations by losing these electrons,**

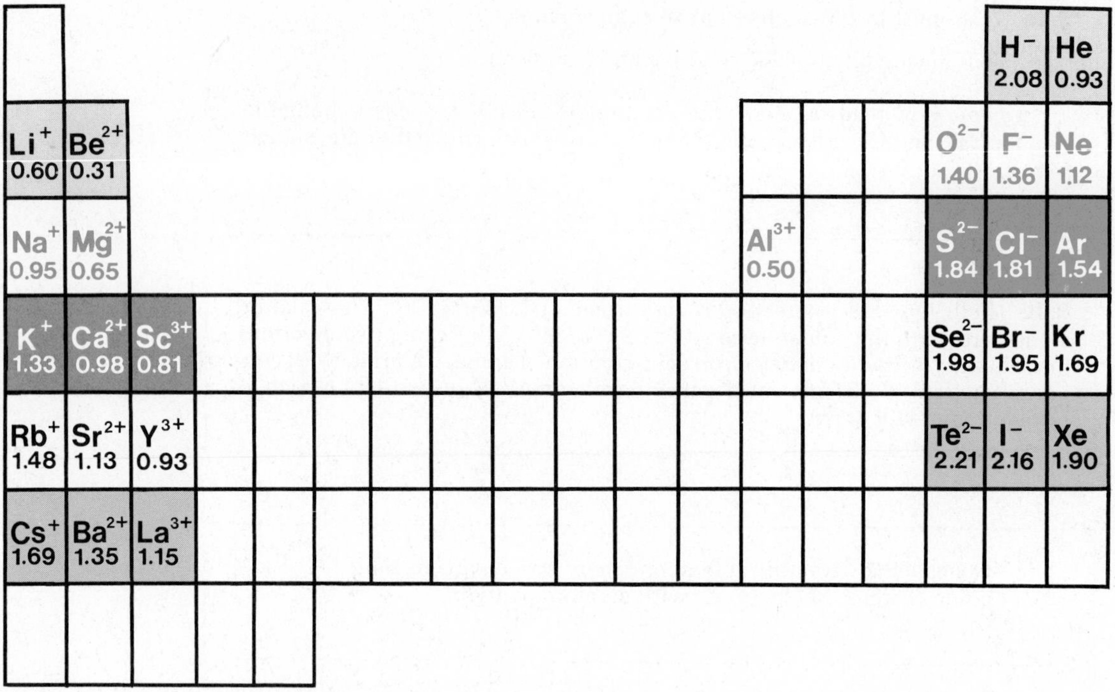

| Li+ 0.60 | Be2+ 0.31 | | | | | | | | | | | | | | | | | | | | | |

Figure 9.1 Monatomic ions having noble gas electron configurations. Each color group includes one noble gas atom and the monatomic ions that are isoelectronic with it. The numbers show the ionic or atomic radius in angstroms. For all isoelectronic species, size decreases as nuclear charge increases.

thereby forming cations with charges of +1, +2 and +3, respectively. Non-metal atoms with 1, 2 or 3 electrons fewer than the following noble gas reach their configurations by gaining these electrons, thereby forming anions with charges of −1, −2 and −3, respectively.

Not all monatomic cations are isoelectronic with the noble gases. The transition metals to the right of Group IIIB in the periodic table cannot form cations with noble gas structures because of the d electrons they contain. Yet in many compounds of these elements there is evidence of simple positive ions. We shall not, in this text, attempt to explain these ions in terms of their electron configurations, except to note that they are formed by the loss of various combinations of highest energy s and d electrons.

## 9.3   IONIC BONDS

We've been discussing the formation of monatomic ions as atoms gaining or losing electrons as if this is something they did quite freely. Actually it is a bit more complicated than that. One condition that absolutely must be satisfied is that, if one atom is going to lose one or more electrons, there must be one or more atoms around to gain them. Instead of simply being gained or lost, electrons are *transferred* from one atom to another. This underlies the general statement: **The formation of an ionic bond by two elements involves the transfer of one or more electrons from a metal to a nonmetal atom.**

The electron transfer idea for the formation of an ionic compound is

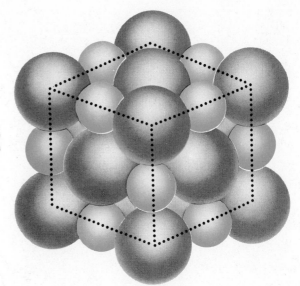

**Figure 9.2** Arrangement of ions in a sodium fluoride type crystal. The small spheres represent the sodium ions, the larger spheres the fluoride ions.

readily seen in the formation of sodium fluoride from sodium and fluorine atoms. Using Lewis structures,

$$Na\cdot \; + \; \cdot\ddot{F}\colon \; \rightarrow \; Na^+ \; + \; \left[\colon\!\ddot{F}\!\colon\right]^{-} \; \rightarrow \; NaF \; crystal$$

This diagram indicates that the electron is transferred from the sodium atom, leaving a sodium ion, to the fluorine atom, producing a fluoride ion.* When this occurs to a huge number of atoms, producing a huge number of ions, they are held together in a regular geometric crystal pattern by electrostatic forces of attraction that constitute ionic bonds. The sodium fluoride crystal is shown in Figure 9.2.

There are many kinds of ionic crystals, and the ions that make them up do not have to be in a 1:1 ratio, as with sodium fluoride. If the compound is calcium fluoride, for example, in which each calcium atom has two valence electrons to lose, there must be two fluorine atoms to receive them:

$$Ca\colon + \begin{matrix} \cdot\ddot{F}\colon \\ \\ \cdot\ddot{F}\colon \end{matrix} \; \rightarrow \; Ca^{2+} \; + \; 2\left[\colon\!\ddot{F}\!\colon\right]^{-} \; \rightarrow \; CaF_2 \; crystal$$

The 1:2 ratio of calcium ions to fluoride ions in calcium fluoride is reflected in the formula of the compound, $CaF_2$. Many combinations of numbers and charges enter into ionic crystals, but always in such proportions as to yield a compound that is electrically neutral. The crystal is not restricted to monatomic ions; polyatomic ions do exactly the same thing in compounds such as $(NH_4)_2SO_4$, although their structures are more complicated. One example is calcium carbonate, $CaCO_3$, pictured in Figure 9.3.

Ionic bonds are very strong. As a consequence, ionic compounds are

---

*The *overall* reaction is between metallic sodium and *diatomic fluorine molecules,* which are believed to separate into unstable atoms at one step in the process. It is after that step that the transfer of electrons probably occurs.

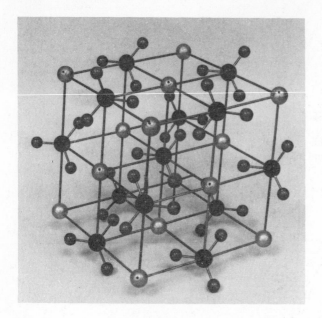

**Figure 9.3** Model of calcium carbonate. Each carbon atom is bonded to three oxygen atoms, making up the carbonate ion. There are equal numbers of calcium ions, $Ca^{2+}$, and carbonate ions, $CO_3^{2-}$, yielding a compound that is electrically neutral, $CaCO_3$.

typically solids at room temperature. Temperatures of 1000°C or greater are frequently required to furnish enough kinetic energy to overcome the electrostatic forces, free the ions from one another, and thereby melt the compound. In the molten state or in water solution, both being conditions in which electrically charged particles are free to move, ionic compounds are good conductors of electricity. They are ordinarily poor conductors in the solid state because the ions are held more or less rigidly in position in the crystal lattice, thereby preventing the general movement of electrical charge that constitutes conduction.

## 9.4 SIZES OF IONS

PG 9 C  Compare the radii of several given isoelectronic monatomic ions and suggest an explanation for the trend observed.

9 D  Compare the radii of several given ions formed by elements in the same group in the periodic table and suggest an explanation for the trend observed.

Returning to Figure 9.2 for a moment, notice the relative sizes of the two kinds of ions in the crystal. The smaller spheres are the $Na^+$ ions, and the larger spheres are the $F^-$ ions. This situation applies in most ionic compounds; the negative ion is usually the larger of the two. Occasionally, the effect is so great that anions in the crystal are actually "in contact" with each other, with the cations fitting into crevices between them. This happens with lithium iodide (see Fig. 9.4). Ionic size apparently plays a significant role in determining the properties of some species, particularly in regard to some of the smaller ions such as $Li^+$.

In Chapter 4, page 88, three influences on atomic size were identified: (1) number of occupied principal energy levels, (2) nuclear charge and (3) the "shielding" of part of the nuclear charge by inner electrons. These same

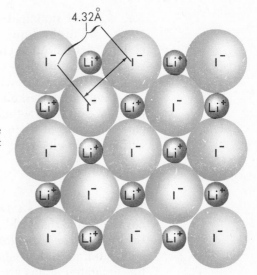

**Figure 9.4** Iodide ions in LiI appear to be in contact. Ionic radius is considered to be half the distance between adjacent iodide ions.

influences determine ionic sizes. The atomic size chart, Figure 4.18, page 87, also shows ionic sizes. Numerical values are given in Figure 9.1, page 188. As with atoms in the same group in the periodic table, the sizes of monatomic ions increase as you go down the group because the number of occupied principal energy levels increases, in accord with (1) above.

The difference in size between $Na^+$ and $F^-$ is explained in terms of nuclear charge, (2) above. Recall that $F^-$ and $Na^+$ are isoelectronic. The greater nuclear charge in $Na^+$ (11 protons vs. 9 in $F^-$) pulls the outer electrons in more tightly and hence makes the ion smaller. From Figure 9.1 we see that this effect is general. Compare, for example, the radii of the series of ions $O^{2-}$, $F^-$, $Na^+$, $Mg^{2+}$ and $Al^{3+}$. All of these ions have the neon structure with a total of 10 electrons. As the nuclear charge increases from 8 with $O^{2-}$ to 13 for $Al^{3+}$, the radius decreases steadily, from 1.40Å with $O^{2-}$ to 0.50Å for $Al^{3+}$. We note further that when the accepted size values of noble gas atoms are included, as they are in Figure 9.1, they fit perfectly into the size trend among isoelectronic species. While the regularity in sizes of isoelectronic species encourages us to believe the two measuring techniques—one for ions, the other for noble gas molecules—are compatible, we must remain cautious in reaching this conclusion.

## 9.5  COVALENT BONDS

PG  9 E  Distinguish between ionic and covalent bonds.

The type of bonding described in Section 9.3, where a transfer of electrons produces oppositely charged ions that are held together by electrostatic forces, explains quite satisfactorily the properties of many compounds, of which solid sodium fluoride, NaF, is typical. There are, however, a great many compounds whose properties are inconsistent with an ionic model. Consider, for example, hydrogen fluoride, HF, and methane, $CH_4$. Both HF and $CH_4$ are gases at room temperature and atmospheric pressure, while

NaF is a solid. Both hydrogen fluoride and methane are nonconductors when condensed to the liquid state, while sodium fluoride is a conductor when melted to the liquid state. Such striking contrasts in properties argue convincingly against the existence of ions in HF and $CH_4$.

In 1916, G. N. Lewis, professor of physical chemistry at the University of California at Berkeley, proposed that molecules are held together by **covalent bonds,** in which a **pair of electrons is shared by two atoms.** In the particular case of hydrogen, he considered that two hydrogen atoms, each with a single electron, could combine to form a molecule:

$$H\cdot\ +\ \cdot H \rightarrow H{:}H \text{ or } H{-}H$$

The two dots or the straight line drawn between the two atoms represents the covalent bond that holds the molecule together. In modern terms, we would say that the *electron cloud,* or *charge cloud,* formed by the two electrons is concentrated in the region between the two nuclei. The electron clouds of the separated atoms are said to *overlap,* as illustrated in Figure 9.5.

A similar approach may be taken to illustrate the formation of the covalent bond between two fluorine atoms to form a molecule of $F_2$, and between one hydrogen atom and one fluorine atom to form an HF molecule:

$$:\ddot{F}\cdot\ +\ \cdot\ddot{F}{:} \rightarrow :\ddot{F}{:}\ddot{F}{:} \text{ or } :\ddot{F}{-}\ddot{F}{:} \text{ or } F{-}F$$

$$H\cdot\ +\ \cdot\ddot{F}{:} \rightarrow H{:}\ddot{F}{:} \text{ or } H{-}\ddot{F}{:} \text{ or } H{-}F$$

Fluorine has seven valence electrons. The 2s orbital and two of the 2p orbitals are filled, but the remaining 2p orbital has only one electron. The bond is considered to be formed by the overlap of the half-filled 2p orbitals of two fluorine atoms. In the HF molecule the bond forms from the overlap of the 1s orbital of a hydrogen atom with the half-filled 2p orbital öf a fluorine atom.

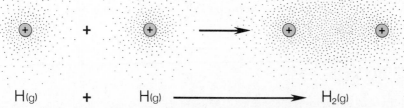

**Figure 9.5** The "electron cloud" picture of the hydrogen atom and the hydrogen molecule. Because nobody has ever seen an atom or the electrons that constitute chemical bonds between atoms, it is difficult to describe them accurately or to draw a satisfactory picture of them. The quantum mechanical concept, however, suggests that the nucleus of a hydrogen atom is, over a period of time, "surrounded" by an "electron cloud," or "charge cloud," which is represented as dots showing many instantaneous positions of the single electron. The uniform distribution of dots around the nucleus shows the spherical shape of the 1s orbital. When two hydrogen atoms form a covalent bond and become a molecule, their 1s orbitals are said to overlap. The bonding electrons are believed to spend more time between the two nuclei, as suggested by the heavier density of electron position dots between nuclei in the illustration, and lower dot density at the outer edges of the molecule.

When used to show the bonding arrangement between atoms in a molecule, electron-dot diagrams are commonly called **Lewis diagrams, Lewis formulas** or **Lewis structures.** Notice that the unshared electron pairs of fluorine are shown for two of the three Lewis diagrams for $F_2$ and HF above, but not for the third. Technically they should always be shown, but they are frequently omitted when not required by the context in which the diagrams appear.

The valence electron population around each atom in the Lewis diagrams of $H_2$, $F_2$ and HF is significant. Each hydrogen atom has two electrons, the same number as the noble gas helium. Each fluorine atom has eight valence electrons, or four electron pairs, corresponding to the noble gas neon. These and many similar observations lead us to believe that the stability of a noble gas electron configuration contributes to the formation of covalent bonds, just as it contributes to the formation of ions. This generalization is often referred to as the **octet rule,** or **rule of eight,** because of the eight electrons that characterize the highest occupied energy level of a noble gas atom.

Actually the tendency toward a completed octet of electrons in a bonded atom reflects one of the fundamental driving forces for all physical or chemical change: a natural tendency for a system to move to the lowest energy state possible. Every time you drop something, it falls spontaneously to the floor because its energy is lower on the floor than when you hold it in your hand. The formation of hydrogen molecules from hydrogen atoms illustrates this point. It may be represented by the equation

$$H\ (g)\ +\ H\ (g) \rightarrow H_2\ (g)\ +\ 104\ kcal$$

in which kcal is the abbreviation for kilocalorie, the unit in which reaction energies are usually measured. This equation indicates that the energy of a system involving two moles of hydrogen atoms is lower by 104 kilocalories when those atoms are bonded into one mole of hydrogen molecules than when they are separate atoms.

## 9.6 POLAR AND NONPOLAR COVALENT BONDS

PG   9 F   Distinguish between polar and nonpolar covalent bonds.

      9 G   Given a table of electronegativities, classify a given bond as nonpolar (or essentially nonpolar), polar or primarily ionic.

As we might expect, the two electrons joining the atoms in the $H_2$ molecule are shared equally by the two nuclei. Another way of saying this is that the electron cloud distribution is balanced in the molecule, with the greatest probability of locating the bonding electrons in the overlap region between the bonded atoms, as shown in Figure 9.5. **A bond in which the distribution of bonding electron charge is symmetrical is said to be nonpolar.** A bond between identical atoms, as in $H_2$ or $F_2$, is always nonpolar.

In the HF molecule, the distribution of the bonding electrons is somewhat different from that in $H_2$ or $F_2$. Here the density of the electron cloud favors the fluorine atom; the bonding electrons, on the average, are shifted

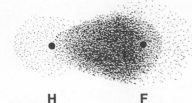

H                    F

**Figure 9.6** A polar bond. Fluorine in a molecule of HF has a higher electronegativity than hydrogen. The bonding electron pair is therefore shifted toward fluorine. The nonsymmetrical distribution of charge yields a polar bond.

towards fluorine and away from hydrogen (see Fig. 9.6). **A bond** such as this, **in which the electron distribution is unsymmetrical, or unbalanced, is referred to as polar.** In the hydrogen fluoride molecule, the fluorine atom acts as a negative pole and the hydrogen atom as a positive pole.

Bond polarity may be described in terms of the **electronegativities** of the bonded atoms. **Electronegativity is a measure of the relative ability of two atoms to attract the pair of electrons forming a single covalent bond between them.** High electronegativity indicates an element with a strong attraction for bonding electrons.

Electronegativity values are shown in periodic table form in Figure 9.7. Notice that electronegativities tend to rise among members of the same group from the bottom of the column to the top. This is because the bonding electrons are in a lower principal energy level in a smaller atom, and being closer to the nucleus, they are attracted by it more strongly. Electronegativities also increase from left to right across any row of the periodic table. This increasing attraction matches an increase in nuclear charge for atoms that are in the same principal energy level. Perhaps you recognize these two explanations as being identical with those given for atomic and ionic sizes, pages 88 and 191. In general, electronegativities are highest at the upper right region of the periodic table, and lowest in the lower left region.

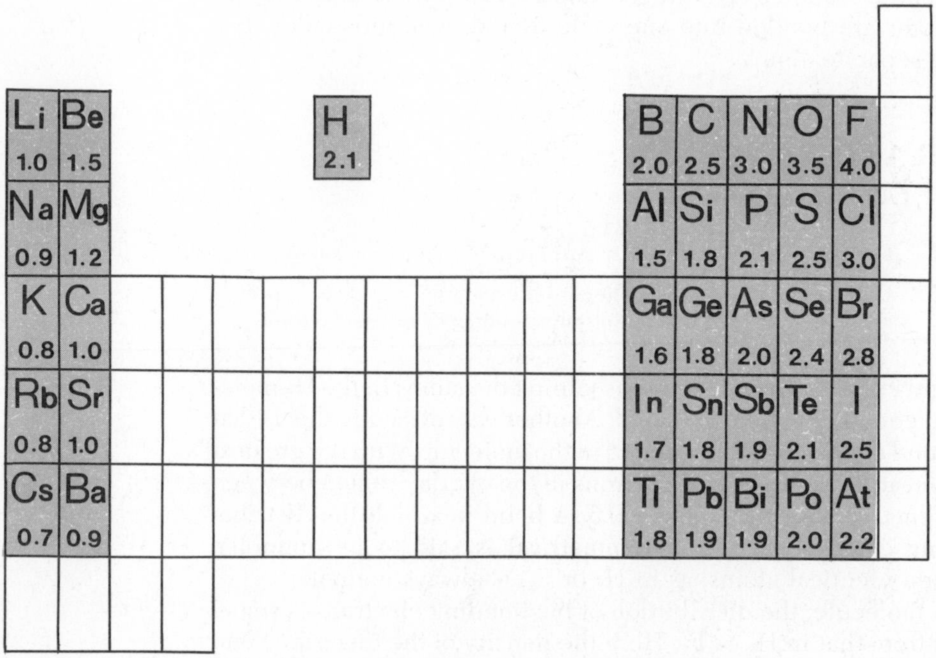

| Li | Be | | | | | | | | | | | | B | C | N | O | F | |
|---|---|---|---|---|---|---|---|---|---|---|---|---|---|---|---|---|---|---|
| 1.0 | 1.5 | | | | | | | | | H | | | 2.0 | 2.5 | 3.0 | 3.5 | 4.0 | |
| | | | | | | | | | | 2.1 | | | | | | | | |
| Na | Mg | | | | | | | | | | | | Al | Si | P | S | Cl | |
| 0.9 | 1.2 | | | | | | | | | | | | 1.5 | 1.8 | 2.1 | 2.5 | 3.0 | |
| K | Ca | | | | | | | | | | | | Ga | Ge | As | Se | Br | |
| 0.8 | 1.0 | | | | | | | | | | | | 1.6 | 1.8 | 2.0 | 2.4 | 2.8 | |
| Rb | Sr | | | | | | | | | | | | In | Sn | Sb | Te | I | |
| 0.8 | 1.0 | | | | | | | | | | | | 1.7 | 1.8 | 1.9 | 2.1 | 2.5 | |
| Cs | Ba | | | | | | | | | | | | Tl | Pb | Bi | Po | At | |
| 0.7 | 0.9 | | | | | | | | | | | | 1.8 | 1.9 | 1.9 | 2.0 | 2.2 | |
| | | | | | | | | | | | | | | | | | | |

**Figure 9.7** Periodic Table of Electronegativities.

You can estimate the polarity in a bond by calculating the difference between the electronegativity values for the two elements: the greater the difference, the more polar the bond. In nonpolar $H_2$ and $F_2$ molecules, where two atoms of the same element are bonded, the electronegativity difference is zero. In the polar HF molecule the electronegativity difference is 4.0 for fluorine minus 2.1 for hydrogen, or 1.9. A bond between carbon and chlorine, for example, with an electronegativity difference of $3.0 - 2.5 = 0.5$, is more polar than an H—H bond, but less polar than an H—F bond.

The more electronegative element toward which the bonding electrons are displaced acts as the "negative pole" in a polar bond. This is sometimes indicated by using an arrow rather than a simple dash, with the arrow pointing to the negative pole. In a bond between hydrogen and fluorine this would be H→F.

---

**Example 9.3** Using Figure 9.7, arrange the following bonds in order of increasing polarity: H—O, H—S, P—H, H—C. In each bond, state which atom would act as the negative pole.

Locate the atoms in the table and calculate the differences in electronegativity.

H—O_____; H—S_____; P—H_____; H—C_____

---

**9.3a** 1.4; 0.4; 0.0; 0.4

Now arrange the bonds in order from the least to the most polar.

---

**9.3b** P—H; H—S; H—C; H—O. The most polar bond arises when the difference in electronegativity is greatest (1.4 units in H—O). The H—C and H—S bonds are about equally polar since the difference in electronegativity is the same. The P—H bond is nonpolar (electronegativity difference = 0).

Now, decide which atom will act as the negative pole.

P—H_____; H—S_____; H—C_____; H—O_____

---

**9.3c** Neither; S; C; O

Since sulfur, carbon and oxygen are all more electronegative than hydrogen, the electron density in these three bonds will be shifted away from hydrogen toward S, C or O. In the P—H bond, where the electronegativities are identical, there will be no + and − poles. That is what is meant by a nonpolar bond.

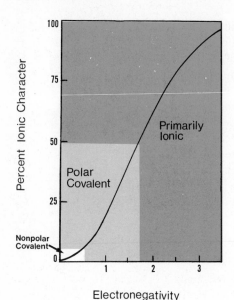

100

Percent Ionic Character

75

50

Primarily
Ionic

25

Polar
Covalent

Nonpolar
Covalent

0

1          2          3

Electronegativity
Difference

**Figure 9.8** Graph of nonpolar, polar and ionic character of a bond with difference in electronegativity of atoms forming the bond.

A polar covalent bond may be thought of as being intermediate between a pure (nonpolar) covalent bond, in which the electrons are essentially equally shared, and a pure ionic bond, in which there is theoretically a complete transfer of electrons from one atom to the other. In this sense, we sometimes express bond polarity in terms of **partial ionic character.** The greater the difference in electronegativity between two elements, the more ionic will be the bond between them. The relationship between these two variables is shown graphically in Figure 9.8.

A difference of electronegativity of 0.4 unit corresponds to a bond with only about 4 per cent ionic character. We find experimentally that bonds formed between atoms that differ in electronegativity by 0.4 unit or less behave very much like pure nonpolar covalent bonds. Such a bond might be classified as "essentially nonpolar." In contrast, a difference of about 1.7 to 1.9 electronegativity units corresponds to a bond with approximately 50 per cent ionic character. Such a bond might be described as being halfway between a pure covalent and a pure ionic bond. Bonds with greater than 1.7 to 1.9 electronegativity difference are usually regarded as ionic bonds. Bonds with electronegativity differences in the intermediate range—more than 0.5 but less than 1.7 to 1.9—are classified as polar covalent bonds.

---

**Example 9.4** Using Figure 9.7, classify each of the following bonds as essentially nonpolar, polar or primarily ionic: the H—C bond; the Li—F bond; the O—H bond.

First, we need the differences in electronegativity from Figure 9.7.

H—C_____;   Li—F_____;   O—H_____

---

**9.4a**  0.4; 3.0; 1.4

Now classify the bonds as essentially nonpolar, polar or primarily ionic.

-------------------------------------------------------------------------------

**9.4b**  H—C, essentially nonpolar; Li—F, ionic; O—H, polar. Electronegativity differences of 0.4 or less are characteristic of essentially nonpolar bonds; from 0.5 to about 1.7 to 1.9 the bonds are polar; above 1.9 they are primarily ionic.

## 9.7  WRITING LEWIS ELECTRON-DOT DIAGRAMS

PG  9 H   Write the Lewis diagram for any molecule or polyatomic ion made up of elements in the A groups of the periodic table.

### INSPECTION METHOD

Lewis diagrams of many simple molecules may be drawn by inspection, as illustrated in the formation of $H_2$, $F_2$ and HF molecules in Section 9.5. It was noted that a noble gas electron configuration characterizes each atom participating in a covalent bond. With the exception of hydrogen atoms, this means that a nonmetal atom surrounds itself with a full complement of eight valence electrons. Hydrogen is satisfied with two valence electrons, matching the electron configuration of the noble gas helium.

The bonds formed between hydrogen, on the one hand, and fluorine, oxygen, nitrogen or carbon, on the other, illustrate the inspection method of drawing Lewis structures. In the HF molecule it was noted that the unpaired electron in the hydrogen atom, H·, teamed with the unpaired electron in a fluorine atom, ·F̈:, to form the molecule, H:F̈:, or H—F̈:. The Lewis symbols for oxygen, nitrogen and carbon atoms are

$$·\ddot{O}: \qquad ·\ddot{N}· \qquad ·\dot{C}·$$

The two unpaired electrons of oxygen indicate it can form covalent bonds with two hydrogen atoms forming a water molecule, $H_2O$. Three unpaired electrons in a nitrogen atom suggest three covalent bonds with three hydrogen atoms, making up a molecule of ammonia, $NH_3$. Carbon has four valence electrons, so it forms covalent bonds with four hydrogen atoms to produce a molecule of methane, $CH_4$.* The Lewis diagrams for these compounds are shown:

$$H—\ddot{O}: \qquad\qquad H—\ddot{N}—H \qquad\qquad H—\overset{\overset{\textstyle H}{|}}{\underset{\underset{\textstyle H}{|}}{C}}—H$$
$$\,\,|\qquad\qquad\qquad\qquad |$$
$$H \qquad\qquad\qquad\qquad H$$
$$H_2O \qquad\qquad\qquad NH_3 \qquad\qquad\qquad CH_4$$

---

*A carbon atom, with its electron configuration $1s^2 2s^2 2p^2$, actually has only two unpaired electrons, the two 2p electrons. It is a fact, however, that carbon atoms form covalent bonds with four hydrogen atoms. One explanation for this involves "hybridized" orbitals, a topic beyond the scope of this text.

Let us pause for a moment to point out an important fact about Lewis diagrams. Though we call them Lewis *structures,* and they are used to *represent* and *suggest* physical structure, they do not and cannot show the true structure of a molecule. Actual molecular structure is three-dimensional, and cannot be displayed accurately on a two-dimensional page. In the diagrams above, for example, it appears that the angles formed by the bonds between the central atom and any two hydrogen atoms is a right angle. This is not so. As you will see shortly, all of the angles are about 109°. Even if the diagram for water were drawn with its proper angle it would not be possible to show the unshared electron pairs properly, as one pair would be above the plane of the paper and the other beneath it. Consequently, while the Lewis diagram for water is often drawn as shown to *suggest* that the three atoms are not in a straight line, the linear arrangement H—Ö—H is an equally correct Lewis diagram, drawn to conform to the octet rule. Lewis diagrams show what atoms in a molecule are bonded to each other, and by how many electron pairs they are bonded, but they do not show the shape of the molecule.

### MULTIPLE BONDS

Sometimes the total number of electrons and atoms in a molecule do not permit a Lewis diagram in which all atoms bonded to each other are connected by a single pair of electrons, and each atom has the configuration of a noble gas. The development of the diagram for ethene, $C_2H_4$, illustrates both the problem and the solution. To emphasize the contrast, this will be shown side by side with the Lewis diagram of ethane, $C_2H_6$, a compound where the problem is not present. In both molecules the carbon atoms are bonded to each other, leaving six electrons—three on each carbon—still available for bonding:

$$\cdot \overset{\cdot}{C} — \overset{\cdot}{C} \cdot$$

With ethane there are six hydrogen atoms to form covalent bonds with the six available electrons, but with ethene there are only four hydrogen atoms:

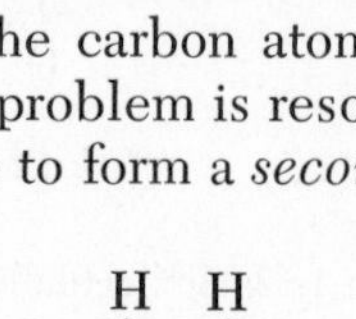

Ethane, $C_2H_6$ 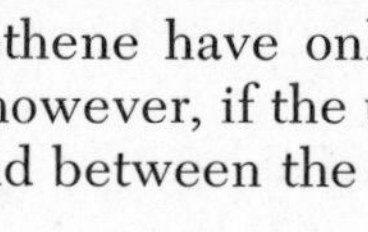          Ethene, $C_2H_4$

At this point each carbon atom in ethane is surrounded by a full octet of electrons, but the carbon atoms in ethene have only seven electrons around them. The problem is resolved, however, if the unpaired electrons in ethene combine to form a *second* bond between the two carbon atoms.

With two pairs of electrons between the carbon atoms, each carbon atom is now surrounded by eight electrons, satisfying the octet rule. When two atoms are bonded by *two* pairs of electrons, they are held together by a **double bond.**

The formation of a nitrogen molecule, $N_2$, requires the use of a **triple bond,** in which the atoms are held together by three pairs of electrons, so that each atom reaches the octet electron population found in noble gas atoms.

$$:\dot{N}\cdot \; + \; \cdot\dot{N}: \rightarrow :\dot{N} \rightleftharpoons \dot{N}: \rightarrow :N \equiv N:$$

There is experimental evidence to support the idea of *multiple* bonds, as double and triple bonds are referred to collectively. A triple bond is stronger and the bond length, the distance between bonded atoms, is shorter than the corresponding measurements for a double bond between the same atoms; and the double bond is shorter and stronger than a single bond. Bond strength is measured in the energy required to break the bond. The triple bond in nitrogen is among the strongest bonds known in chemistry. This is why elemental nitrogen is so stable and unreactive in the earth's atmosphere.

Multiple bonding is not limited to bonds between identical elements. For example, the Lewis structure of formaldehyde is H—C with double-bonded Ö: and H below , and for the cyanide ion, $[:C \equiv N:]^-$. Other examples will appear shortly.

## COMPLEX STRUCTURES

Lewis diagrams are not readily evident for some of the larger or more complex molecules and polyatomic ions. The procedure that follows may be used to sketch a Lewis structure for any species that obeys the octet rule:

1. **Count the number of valence electrons available.** For a molecule this can be done simply by adding the valence electrons contributed by each atom in the molecule. For a polyatomic ion this total must be adjusted to account for the charge on the ion. An ion with a −1 charge would have one more electron than the number of valence electrons in the neutral atoms; a −2 ion, two more; a +1 ion, one less; and so forth.

2. **Draw a tentative structure for the molecule or ion, joining atoms by single bonds. Place electron dots around each symbol so that the total number of electrons for each atom is eight.** Count both bonding electron pairs and unshared electrons in placing electron dots. In some cases, only one arrangement of atoms is possible. In others, two or more structures may be drawn. Ultimately chemical or physical evidence must be used to decide which of the possible structures is correct. A few general rules will help you in making diagrams that are most likely to be correct:

   a. A hydrogen atom always forms one bond; a carbon atom normally forms four bonds.

   b. When several carbon atoms appear in the same molecule, they are usually bonded to each other. In some compounds they are arranged in a closed loop; however, we will avoid such so-called cyclic compounds in this text.

   c. In compounds or ions having two or more oxygen atoms and one atom of another nonmetal, the oxygen atoms are usually arranged around the central nonmetal atom.

   d. In an oxyacid (hydrogen + oxygen + a nonmetallic element, as $H_2SO_4$ and $HNO_3$), hydrogen is usually bonded to an oxygen atom, which is then bonded to the nonmetallic atom: H—O—X, where X is a nonmetal.

  3. **Count the total number of electrons in your structure and compare it with the number available from Step 1.** If your count yields two electrons more than you have available, the molecule must have a double bond, usually between an oxygen and some other element, or between two carbon atoms. If your count is four more than there are available, a triple bond or two double bonds are indicated. Multiple bonds are used only when necessary, and then as few as possible are used in a given molecule.

  Although some of the examples that follow might be done simply and quickly by inspection, we shall nevertheless follow the procedure outlined above in order that you may become familiar with it, as well as see similarities and regularities that might otherwise escape unnoticed.

---

**Example 9.5**   Write Lewis structures for the ClF molecule and the $ClO^-$ ion.

  Following the rules given above, first determine the number of valence electrons in each species.

ClF_____; $ClO^-$_____.

- - - - - - - - - - - - - - - - - - - - - - - - - - - - - - - - - - - - - - - - - - - - - - - - - - - - - - -

**9.5a**   14; 14. In the ClF molecule, we simply add the valence electrons of the neutral atoms. Since both Cl and F are in Group VIIA and therefore have seven valence electrons (number of valence electrons equal to group number for A groups), we have: $7 + 7 = 14$. For the $ClO^-$ ion, we add the number of valence electrons in Cl (7) to the number contributed by O in Group VIA (6), and then add 1 to take care of the $-1$ charge: $7 + 6 + 1 = 14$.

  There is only one possible tentative structure in each case. Write it, and surround each atom with eight electrons.

- - - - - - - - - - - - - - - - - - - - - - - - - - - - - - - - - - - - - - - - - - - - - - - - - - - - - - -

**9.5b**   :C̈l—F̈:         [:C̈l—Ö:]⁻

The shared electron pair constituting the bond is counted for both atoms, so each atom requires six additional electrons.

  Now count the electrons in the above structure and compare it to the number available, determined in Step 1. Modify the above structures with multiple bonds to correct any differences in the two counts.

- - - - - - - - - - - - - - - - - - - - - - - - - - - - - - - - - - - - - - - - - - - - - - - - - - - - - - -

**9.5c**  Both structures have 14 electrons, the same as the number available. The Lewis diagrams are therefore correct as shown above.

The above example illustrates the fact that two species having the same number of electrons and the same number of atoms have similar Lewis diagrams, whether they are molecules or polyatomic ions.

**Example 9.6**  Draw the Lewis diagram for $SO_3^{2-}$, the sulfite ion.

Begin by counting up the number of valence electrons. Don't forget the ionic charge. . . .

------------------------------------------------------------

**9.6a**  26 electrons

Sulfur has six valence electrons; each oxygen atom has six electrons; and there are two extra electrons for the $-2$ charge: $6 + 3(6) + 2 = 26$

Now proceed to the tentative structure, with each atom surrounded by an octet of electrons.

------------------------------------------------------------

**9.6b**

$$\left[\ddot{\ddot{O}}-\overset{..}{\underset{\underset{\displaystyle :\overset{..}{\underset{..}{O}}:}{|}}{S}}-\ddot{\ddot{O}}:\right]^{2-}$$

Now count up the electrons in the tentative structure and compare it to the 26 valence electrons available. Modify the structure with multiple bonds as necessary.

------------------------------------------------------------

**9.6c**  The tentative structure has 26 electrons, equal to the valence electrons available. No modification is necessary.

**Example 9.7**  Derive the Lewis structure of $SO_2$.

First the number of valence electrons. . . .

------------------------------------------------------------

**9.7a**  18

Sulfur has six valence electrons, and oxygen six for each atom. $6 + 2(6) = 18$.

Now sketch the tentative structure, complete with an electron octet around each atom.

---

**9.7b**  $:\ddot{O}-\ddot{S}-\ddot{O}:$

As usual, the oxygen atoms are arranged around the nonmetal central atom.

Now count the electrons and compare to the 18 valence electrons available. Modify the structure with multiple bonds, if necessary, to correct for an unequal number.

---

**9.7c**  Tentative diagram has 20 electrons; only 18 electrons available.

$$:\ddot{O}-\ddot{S}=\ddot{O}:$$

This time the tentative structure has two electrons more than the number available. The total in the tentative structure must therefore be reduced by two. This is accomplished by replacing an unshared electron pair originally assigned to sulfur and an unshared electron pair originally assigned to one of the oxygens by an electron pair they share between them as a second bond:

$$:\ddot{O}-\ddot{S}-\ddot{O}: \qquad\qquad :\ddot{O}-\ddot{S}-\ddot{O}:$$

Replace                         New

Each atom is now surrounded by eight electrons, and the total number of electrons in the structure matches the number available.

You may wonder if it makes any difference on which side of the sulfur atom the double bond is placed. It does not, for the limited purpose of learning how to draw Lewis diagrams. It is a fact, however, that experimentally the bonds are identical, and have strengths and lengths that are between those normally associated with single bonds and double bonds connecting sulfur and oxygen atoms. This condition is known as *resonance,* and is frequently shown as

$$:\ddot{O}=\ddot{S}-\ddot{O}: \leftrightarrow :\ddot{O}-S=\ddot{O}:$$

In this text we will not be concerned with resonance structures beyond this point of information.

The rules we are following are readily applied to simple organic* molecules, which always contain carbon atoms, usually include hydrogen atoms, and may contain atoms of other elements, notably oxygen. If oxygen is present in an organic compound, it usually forms two bonds. If you remember

---

*Organic chemistry is the chemistry of compounds containing carbon, other than certain "inorganic" carbon compounds such as carbonates, CO and $CO_2$. A knowledge of bonding, including Lewis diagrams, is important in organic chemistry.

that carbon forms four bonds and hydrogen forms one, and that two or more carbon atoms usually bond to each other (Rules 2a and 2b, page 199), your tentative structures will probably be those that are found to be correct in the laboratory.

---

**Example 9.8**   Write the Lewis structure for propane, $C_3H_8$.

First the electron count. . . .

-----------------------------------------------------------------------------

**9.8a**   20

Four for each carbon atom, and one for each hydrogen: $3(4) + 8(1) = 20$.

The tentative structure follows readily from the rules mentioned just before this example.

-----------------------------------------------------------------------------

**9.8b**

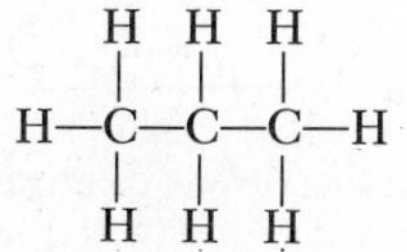

$$
\begin{array}{c}
\text{H} \quad \text{H} \quad \text{H} \\
| \quad\; | \quad\; | \\
\text{H--C--C--C--H} \\
| \quad\; | \quad\; | \\
\text{H} \quad \text{H} \quad \text{H}
\end{array}
$$

The last step is to check to see if the tentative structure contains the same number of electrons as the original count. Does it?

-----------------------------------------------------------------------------

**9.8c**   Yes, there are 20 electrons in the structure.

---

The Lewis diagrams of methane (page 197), ethane (page 198) and propane are drawn side by side for comparison:

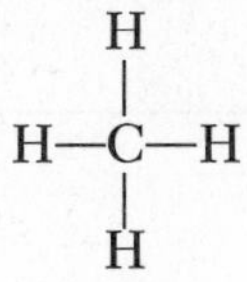

Methane, $CH_4$ 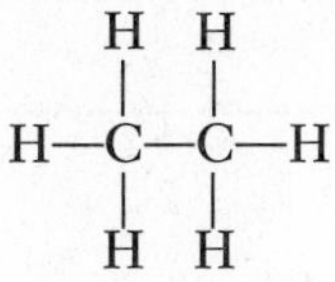       Ethane, $C_2H_6$ 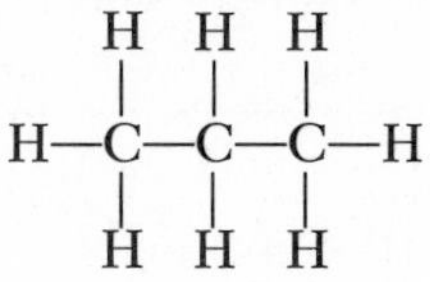       Propane, $C_3H_8$

Notice that each compound differs from the one before it by one $-\overset{\displaystyle H}{\underset{\displaystyle H}{C}}-$ unit.

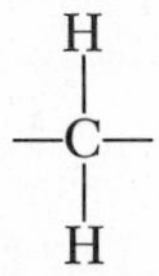

In fact, there is no end, theoretically, to the number of such units that might be inserted in the chain. The next member of the *alkane series,* as this is called, is $C_4H_{10}$, butane. It is possible to draw two structures of butane, and both are real but different compounds, each with its own set of physical properties. The structures are

H
|
H—C—H
|
H   H   H   H              H   |   H
|   |   |   |              |   |   |
H—C—C—C—C—H          H—C———C———C—H
|   |   |   |              |   |   |
H   H   H   H              H   H   H

Compounds having the same molecular formula but different structures are called **isomers** of each other. Among the alkanes the number of isomers increases dramatically as the number of carbon atoms increases. There are three pentanes, $C_5H_{12}$, five hexanes, $C_6H_{14}$, 35 nonanes, $C_9H_{20}$, over 300,000 with the formula $C_{20}H_{42}$, and over 100,000,000 having the formula $C_{30}H_{62}$! Needless to say, they have not all been isolated and studied. This gives you some idea why there are so many organic compounds—and we haven't even mentioned compounds that contain elements other than carbon and hydrogen. One kind of soap, for example, is $C_{18}H_{35}O_2Na$.

---

**Example 9.9**   Prepare the Lewis diagram of acetylene, $C_2H_2$.

As usual, start with the electron count.

----------------------------------------------------------------

**9.9a**   10        $2(4) + 2(1) = 10$

Now the tentative structure. . . .

----------------------------------------------------------------

**9.9b**   H—C̤—C̤—H

The electron count and structure modification, if necessary. . . .

----------------------------------------------------------------

**9.9c**   H—C≡C—H

The tentative structure has 14 electrons, four more than the number available. This requires a triple bond.

---

**Example 9.10**   Draw a Lewis diagram for $C_2H_6O$.

The electron count, please. . . .

----------------------------------------------------------------

**9.10a**   20          $2(4) + 6(1) + 6 = 20.$

It was mentioned earlier that oxygen atoms usually form two bonds in organic molecules. Try for the tentative structure, complete with unshared electrons. Bond the carbons to each other.

----------------------------------------------------------------------

**9.10b**

$$\begin{array}{c} \quad\text{H}\quad\text{H} \\ \quad|\quad\quad| \\ \text{H}-\text{C}-\text{C}-\ddot{\text{O}}-\text{H} \\ \quad|\quad\quad| \\ \quad\text{H}\quad\text{H} \end{array}$$

Now check the electrons in the diagram against the valence electron count, adjusting the structure with multiple bonds, if necessary.

----------------------------------------------------------------------

**9.10c**   20 electrons in both places. The Lewis diagram above is correct.

The compound in Example 9.10 is ethyl alcohol. If we had not insisted that the carbon atoms be bonded together you might have produced the diagram for an isomer of alcohol, another well known compound, dimethyl ether:

$$\begin{array}{c} \quad\text{H}\quad\quad\quad\text{H} \\ \quad|\quad\quad\quad\quad| \\ \text{H}-\text{C}-\ddot{\text{O}}-\text{C}-\text{H} \\ \quad|\quad\quad\quad\quad| \\ \quad\text{H}\quad\quad\quad\text{H} \end{array}$$

## 9.8   EXCEPTIONS TO THE OCTET RULE

While electron-dot diagrams derived from the octet rule are helpful in our understanding of properties of molecules, you should be aware that some substances do not "obey" the rule. Two common oxides of nitrogen, NO and $NO_2$, have an odd number of electrons. It is therefore impossible to write Lewis diagrams for these compounds in which each atom is surrounded by eight electrons. Phosphorus pentafluoride, $PF_5$, places five electron pair bonds around the phosphorus atom, and six pairs surround sulfur in $SF_6$.

Certain molecules whose Lewis diagrams obey the octet rule do not have the properties that would be predicted. Oxygen, $O_2$, was not used to introduce the double bond for that reason. On paper, $O_2$ appears to have an ideal double bond:

$$\ddot{\text{O}}=\ddot{\text{O}}$$

But liquid oxygen is *paramagnetic,* meaning that it is attracted by a magnetic field. This is characteristic of molecules that have unpaired electrons in their structure. This might suggest a Lewis diagram that has each oxygen surrounded by seven electrons:

$$:\ddot{O}—\ddot{O}:$$

But this is in conflict with other evidence that the oxygen atoms are connected by something other than a single bond. In essence, it is impossible to write a single Lewis diagram that satisfactorily explains all of the properties of molecular oxygen.

Two other species for which octet-rule diagrams are unsatisfactory are the fluorides of beryllium and boron. We might expect $BeF_2$ and $BF_3$ to be ionic compounds, but experimental evidence strongly supports covalent structures having the Lewis diagrams

$$:\ddot{F}—Be—\ddot{F}: \qquad \begin{array}{c} :\ddot{F} \qquad \ddot{F}: \\ \diagdown \quad \diagup \\ B \\ | \\ :\ddot{F}: \end{array}$$

in which the central atoms, Be and B, are surrounded by 2 and 3 pairs of electrons, respectively, rather than 4.

## 9.9    MOLECULAR GEOMETRY

The physical properties and, to some extent, the chemical properties of a molecular substance depend upon the shape of its molecules, that is, the way in which atoms within the molecule are arranged with respect to one another. We have already noted the inadequacy of Lewis diagrams in describing this **molecular geometry,** as it is called (page 198). If the molecule has only two atoms, the geometry is very simple. Two atoms joined by a covalent bond must be located on a straight line (2 points determine a straight line). We can safely predict that molecules such as $H_2$ and HCl are linear:

$$\text{H—H} \qquad \text{H—}\ddot{\text{C}}\text{l}:$$

When a molecule contains two or more covalent bonds, its molecular geometry includes the **bond angle, the angle formed by the bonds and the atom between them.** In the case of the $H_2O$ molecule, for example, there are two possibilities. The H—O—H bond angle might be 180°, in which case the three atoms would be located on a straight line. If, on the other hand, the bond angle is less than 180°, the three atoms are arranged in a bent pattern. It was mentioned on page 198 that the bond angle is "about 109°" (actually closer to 105°), so the molecule is indeed bent. The properties of the $H_2O$ molecule and hence of water itself depend on this bent geometry.

Chemists employ several theories or models in attempting to explain and understand molecular geometry. No single model is completely satisfactory, and each has its unique advantages. For the purpose of this text we have chosen to examine molecular geometry according to the *electron pair repulsion principle.*

## MOLECULAR GEOMETRY BASED ON ELECTRON PAIR REPULSION

PG 9 I  Given or having derived the Lewis diagram of a molecule in which a second period central atom is surrounded by 2, 3 or 4 pairs of electrons, predict its electron pair geometry and its molecular geometry.

The major features of molecular geometry can be predicted by the **electron pair repulsion principle,** which states that because of electrostatic repulsion, the **electron pairs surrounding an atom are located as far from each other as possible.** To illustrate this principle, we will consider cases in which a central atom is surrounded by 2, 3 or 4 electron pairs. In doing so, we will distinguish between two kinds of geometry: **electron pair geometry** describes the arrangement of *all* pairs of valence electrons around the central atom, whether they are shared with other atoms or not; and molecular geometry describes the arrangement of atoms bonded to the central atom.

TWO PAIRS OF ELECTRONS.  Example, $BeF_2$, Figure 9.9A. Two electron pairs around a central atom are as far apart as possible when they are on opposite sides of the atom, forming a straight line that includes the atom. *The electron pair geometry is thus* **linear.** Each electron pair bonds a fluorine atom to the central atom, creating a *linear molecular geometry* as well. The bond angle is 180°.

THREE PAIRS OF ELECTRONS.  Example $BF_3$, Figure 9.9B. Three electron pairs are as far apart as possible when they are located at the corners of an equilateral triangle with the central atom at the center and in the same plane. The *electron pair geometry* is described as **planar triangular.** Each electron pair bonds a fluorine atom to the beryllium atom, so the *molecular geometry is also planar triangular.* Each F-B-F bond angle is 120°.

FOUR PAIRS OF ELECTRONS.  Examples are $CH_4$, $NH_3$, $H_2O$. In these compounds carbon, nitrogen and oxygen all obey the octet rule. The *electron pair geometry* that places these four pairs as far from each other as possible is described as **tetrahedral.** A tetrahedral electron pair geometry has the central atom at the center of a tetrahedron with one electron pair at each of the four corners. A tetrahedron is a four sided solid, all four sides being identical equilateral triangles. It has the form of a pyramid with a triangular base (see Fig. 9.10). In a molecule of methane, $CH_4$, each electron pair bonds a hydrogen atom to the central carbon atom, so the *molecular geometry is*

**Figure 9.9**  Shapes of molecules with 2 and 3 pairs of bonding electrons. A, Two pairs of bonding electrons arrange themselves at 180° bond angle around the central atom. Both electron pair and molecular geometries are *linear.* Example, $BeF_2$. B, Three pairs of bonding electrons arrange themselves at 120° bond angles around the central atom. Both electron pair and molecular geometries are planar triangular. Example, $BF_3$.

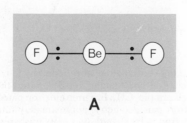

**A**

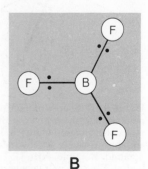

**B**

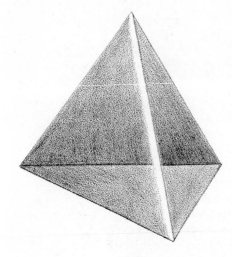

**Figure 9.10** Tetrahedron. The orientation of four equal pairs of bonding electrons around a central atom assumes the shape of a regular *tetrahedron*, a four sided solid with identical equilateral triangular faces. The central atom is in the center, forming a *bond angle of 109°28'* between atoms at any two corners.

*also tetrahedral,* as in Figure 9.11A. Each H-C-H bond angle is 109.5°, which is often described as the **tetrahedral angle.**

In ammonia, $NH_3$, the *electron pair geometry* remains approximately tetrahedral, with four pairs of electrons, but only three of these electron pairs are bonded to hydrogen atoms. The resulting *molecular geometry* is a low **trigonal pyramid,** or **pyramidal,** a pyramid with an equilateral triangle as a base, as in Figure 9.11B. The nitrogen atom is at the top of the pyramid, and the three hydrogen atoms form the base. Each H-N-H bond angle is slightly less than the tetrahedral angle, about 107°.

With water there are still four electron pairs oriented about the oxygen atom at approximately tetrahedral angles, but only two of those form bonds to other atoms. The result is a **bent** *molecular geometry* with a bond angle of about 105°, still quite close to the tetrahedral angle (see Fig. 9.11C).

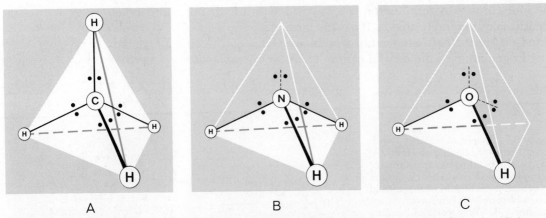

A                    B                    C

**Figure 9.11** Shapes of molecules having central atom surrounded by four electron pairs. A, Four shared electron pairs arrange themselves at the tetrahedral bond angle, 109.5°, around the central atom. Both electron pair and molecular geometries are tetrahedral. Example, $CH_4$. B, Three electron pairs shared, one pair unshared. Bond angles are approximately tetrahedral. Electron pair geometry, approximately tetrahedral; molecular geometry, trigonal pyramid. Example, $NH_3$. C, Two electron pairs shared, two pairs unshared. Bond angles are approximately tetrahedral. Electron pair geometry, approximately tetrahedral; molecular geometry, bent. Example, $H_2O$.

TABLE 9.4   Electron Pair and Molecular Geometries

| ELECTRON PAIRS | BONDED ATOMS | ELECTRON PAIR GEOMETRY | MOLECULAR GEOMETRY | EXAMPLE |
|:---:|:---:|:---|:---|:---:|
| 2 | 2 | Linear | Linear | $BeF_2$ |
| 3 | 3 | Planar triangular | Planar triangular | $BF_3$ |
| 4 | 4 | Tetrahedral | Tetrahedral | $CH_4$ |
| 4 | 3 | Tetrahedral | Trigonal pyramid | $NH_3$ |
| 4 | 2 | Tetrahedral | Bent | $H_2O$ |

The comparisons between electron pair and molecular geometries about a central atom are summarized in Table 9.4. The shapes described are for molecules containing single bonds only. Geometries of molecules containing multiple bonds are discussed on page 210.

---

**Example 9.11**   Predict the electron pair and molecular geometries of carbon tetrachloride, $CCl_4$.

It is always helpful in predicting electron pair and molecular geometries to have the Lewis diagram. Draw this for $CCl_4$.

----------------------------------------------------------------------

**9.11a**

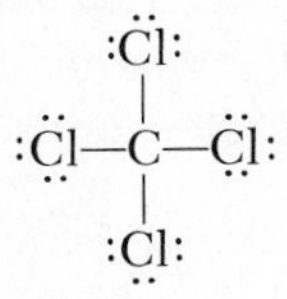

$$:\ddot{C}l:$$
$$|$$
$$:\ddot{C}l-C-\ddot{C}l:$$
$$|$$
$$:\ddot{C}l:$$

From the Lewis diagram you should establish the number of electron pairs around the central atom and the number of atoms bonded to the central atom. Both geometries follow. . . .

----------------------------------------------------------------------

**9.11b**   With four electron pairs around carbon, all bonded to other atoms, both geometries are tetrahedral.

---

**Example 9.12**   Describe the shape of a molecule of boron trihydride, $BH_3$.

First draw the Lewis structure. Remember that boron has only three valence electrons to contribute to covalent bonds. From the structure answer the question.

----------------------------------------------------------------------

**9.12a**
$$\begin{array}{c} H \\ \diagdown \\ \hspace{1em} B\!-\!H \hspace{2em} \text{Planar triangular} \\ \diagup \\ H \end{array}$$

Three electron pairs yield both an electron pair geometry and a molecular geometry that are planar triangular with 120° bond angles.

---

**Example 9.13** Predict the electron pair geometry and shape of a molecule of dichlorine oxide, $Cl_2O$.

-------------------------------------------------------------------------------------

**9.13a** $:\!\ddot{C}l\!-\!\ddot{O}:$ Electron pair geometry: tetrahedral; molecular geometry:
$\hspace{3.5em}|$ bent
$\hspace{3em}:\!\ddot{C}l\!:$

Oxygen has four electron pairs around it, yielding an electron pair geometry that is approximately tetrahedral. Only two of the electron pairs are bonded to other atoms, so the molecule is bent. The structure is similar to that of water.

## THE GEOMETRY OF THE MULTIPLE BOND

Experimental evidence shows that bond angles involving atoms connected by multiple bonds are the same as those that would be predicted for atoms connected only by single bonds. For example, in carbon dioxide the Lewis structure is

$$:\!\ddot{O}\!=\!C\!=\!\ddot{O}:$$

The carbon atom is connected to two oxygen atoms by two double bonds. Where might these bonds best locate themselves to be as far apart as possible? The answer is the same as it was for 2 pairs of electrons in beryllium fluoride on page 207: at 180° from each other. The carbon dioxide molecule is therefore linear.

The reasoning is not quite so clear when we consider mixtures of single and multiple bonds around the central atom. The Lewis structure for formaldehyde was given on page 199 as

$$\begin{array}{c} \ddot{O}: \\ \diagup\!\!\diagup \\ H\!-\!C \\ \diagdown \\ H \end{array}$$

As in $BF_3$, there are three atoms surrounding the central atom, two bonded by electron pairs and the third by two pairs of electrons. Finding maximum distance from each other again suggests a planar molecule, but the bond angles might be questionable. Laboratory data, however, show that the bond angles surrounding the carbon atom are 120°—just like the three bond angles surrounding the boron atom in $BF_3$.

## 9.10  POLARITY OF MOLECULES

> PG  9 J  Given or having determined the Lewis structure of a molecule, predict whether the molecule is polar or nonpolar.

We previously considered the polarity of covalent bonds. Now that we have some idea about how atoms are arranged in molecules, we are ready to discuss the polarity of molecules themselves. **A polar molecule is one in which there is an unsymmetrical distribution of charge,** resulting in + and − poles. A simple example is the HF molecule. The fact that the bonding electrons are somewhat closer to the fluorine atom gives it a partial negative charge, while the hydrogen atom acts as a positive pole. In general, any diatomic molecule in which the two atoms differ from each other will be at least slightly polar. Examples, in addition to HF, include HCl and BrCl; in both of these molecules, the chlorine atom acts as the negative pole.

When a molecule contains more than two atoms, we must know the bond angles in order to decide whether the molecule is polar or nonpolar. Consider, for example, the two triatomic molecules, $BeF_2$ and $H_2O$. Despite the presence of two strongly polar bonds, the linear $BeF_2$ *molecule* is nonpolar; since the fluorine atoms are symmetrically arranged around Be, the two polar Be–F bonds cancel each other. This may be shown as

$$:\ddot{F} \leftarrow Be \rightarrow \ddot{F}:$$

in which the arrows point to the more electronegative atoms. From a slightly different point of view, we can deduce that this molecule is nonpolar because the centers of positive and negative charge coincide at the center of the molecule.

In contrast, the bent water molecule is polar; the two polar bonds do not cancel each other because the molecule is not symmetrical around a horizontal axis.

$$\begin{array}{c} (-) \\ \ddot{O} \\ \nearrow \quad \nwarrow \\ H \quad (+) \quad H \end{array}$$

The negative pole is located at the more electronegative oxygen atom; the positive pole is midway between the two hydrogen atoms. In an electrical field, water molecules would tend to line up with the hydrogen atoms pointing toward the plate with the negative charge and the oxygen atoms toward the plate with the positive charge (see Fig. 9.12).

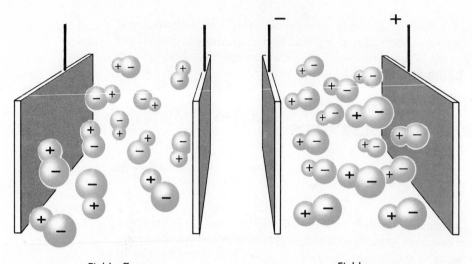

Field off                     Field on

**Figure 9.12** Orientation of polar molecules in an electric field. Two plates, immersed in a liquid whose molecules are polar, are connected through a switch to a source of an electric field. With the switch open the orientation of the molecules is random. When the switch is closed the molecules tend to line up with the positive end toward the negative plate, and the negative pole toward the positive plate.

Another molecule which is nonpolar despite the presence of polar bonds is $CCl_4$. The four C—Cl bonds are themselves polar, with the bonding electrons slightly displaced toward the chlorine atoms. However, since the four chlorines are symmetrically distributed about the central carbon atom (see Fig. 9.13), the polar bonds cancel each other. If one of the chlorine atoms in $CCl_4$ is replaced by hydrogen, the symmetry of the molecule is destroyed. The chloroform molecule, $CHCl_3$, is polar.

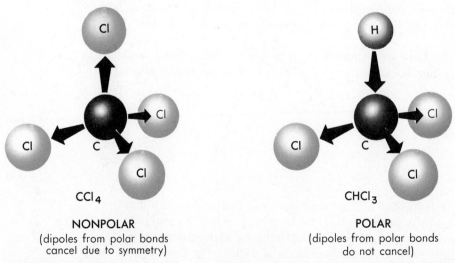

CCl$_4$                                CHCl$_3$

NONPOLAR                          POLAR
(dipoles from polar bonds            (dipoles from polar bonds
cancel due to symmetry)             do not cancel)

**Figure 9.13** Polar and nonpolar molecules. $CCl_4$ is nonpolar because polar bonds cancel owing to symmetry. $CHCl_3$ is polar because polar bonds do not cancel owing to lack of symmetry.

**Example 9.14** Is the $BF_3$ molecule polar? the $NH_3$ molecule?

First, the Lewis structures.

------------------------------------------------

**9.14a**

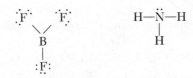

Now, the *shapes* of the molecules. . . .

------------------------------------------------

**9.14b** $BF_3$: planar, triangular; $NH_3$: trigonal pyramid.

Now, decide whether the molecules are polar or nonpolar.

------------------------------------------------

**9.14c** $BF_3$ is nonpolar; $NH_3$ is polar. Even though fluorine is more electronegative than boron, the three fluorine atoms are arranged symmetrically about the boron atom at the center. With ammonia, the bonding electrons are displaced toward the more electronegative nitrogen atom. The unsymmetrical pyramidal shape does not cancel the polar bonds, so the molecule is polar.

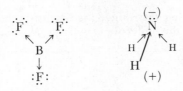

# CHAPTER 9 IN REVIEW

### 9.4   SIZES OF IONS

9 C   Compare the radii of several given isoelectronic monatomic ions and suggest an explanation for the trend observed. (Page 190)

9 D   Compare the radii of several given ions formed by elements in the same group in the periodic table and suggest an explanation for the trend observed. (Page 190)

### 9.5   COVALENT BONDS

9 E   Distinguish between ionic and covalent bonds. (Page 191)

### 9.6   POLAR AND NONPOLAR COVALENT BONDS

9 F   Distinguish between polar and nonpolar covalent bonds. (Page 193)

9 G   Given a table of electronegativities, classify a given bond as nonpolar (or essentially nonpolar), polar or primarily ionic. (Page 193)

### 9.7   WRITING LEWIS ELECTRON-DOT DIAGRAMS

9 H   Write a Lewis diagram for any molecule or polyatomic ion made up of elements in the A groups of the periodic table. (Page 197)

### 9.8   EXCEPTIONS TO THE OCTET RULE (Page 205)

### 9.9   MOLECULAR GEOMETRY

9 I   Given or having derived the Lewis structure of a molecule in which a second period central atom is surrounded by 2, 3 or 4 pairs of electrons, predict its electron pair geometry and its molecular geometry. (Page 207)

### 9.10   POLARITY OF MOLECULES

9 J   Given or having determined the Lewis structure of a molecule, predict whether the molecule is polar or nonpolar. (Page 211)

## TERMS AND CONCEPTS

Valence electrons (84, 183)
Lewis symbol (183)
Electron-dot symbol (183)
Octet (184)
Isoelectronic (184)
Ionic bond (188)
Covalent bond (192)
Electron (charge) cloud (192)
Overlap (of atomic orbitals) (192)
Lewis diagram, formula or structure (193)
Octet rule—rule of eight (193)
Nonpolar bond (193)
Polar bond (194)
Electronegativity (194)

Partial ionic character (196)
Multiple (double, triple) bond (198)
Isomer (204)
Molecular geometry (206)
Bond angle (206)
Electron Pair Repulsion Principle (207)
Electron pair geometry (207)
Linear (geometry) (207)
Planar triangular (geometry) (207)
Tetrahedral (geometry) (207)
Tetrahedral angle (208)
Trigonal pyramid, or pyramidal (208)
Bent (geometry) (208)
Polar molecule (211)
Nonpolar molecule (211)

# QUESTIONS AND PROBLEMS

*An asterisk (\*) identifies a question that is relatively difficult, or that extends beyond the performance goals in the chapter.*

## Section 9.1

9.1)   Write the Lewis symbols for atoms of potassium, phosphorus and bromine.

9.2)   Draw the electron-dot symbols for the elements whose atomic numbers are 31 and 82.

9.3)   Write the elemental symbols that might be represented by X if the Lewis symbol is $\cdot \ddot{X} \colon$.

9.37)   Write the electron-dot symbols for oxygen, calcium and boron.

9.38)   Write the Lewis symbols for the elements whose atomic numbers are 52 and 55.

9.39)   The electron-dot symbol M⋅ represents a group of elements. Write the symbols of the elements in that group.

## Section 9.2

9.4)   Identify those elements in the third period of the periodic table that form monatomic ions that are isoelectronic with a noble gas atom. Write the symbol for each such ion (example: $Ca^{2+}$ in the fourth period.)

9.5)   Identify two negatively charged monatomic ions that are isoelectronic with neon.

9.6)   Write the symbols of two ions that are isoelectronic with the chloride ion.

9.40)   Write the electron configuration of each third period monatomic ion identified in question 9.4. Also identify the noble gas atoms having the same configurations.

9.41)   Identify by symbol two positively charged monatomic ions that are isoelectronic with krypton, atomic number 36.

9.42)   Write the symbols of two ions that are isoelectronic with the barium ion.

## Section 9.3

9.7)\*   Using Lewis symbols, show how ionic bonds are formed by atoms of sulfur and potassium, leading to the correct formula of potassium sulfide.

9.43)\*   Aluminum oxide is an ionic compound. Sketch the transfer of electrons from aluminum atoms to oxygen atoms that accounts for the chemical formula of the compound.

## Section 9.4

9.8)   Identify the largest ion among the following: $Li^+$; $F^-$; $S^{2-}$.

9.9)   Arrange the following ions in order of their increasing size (smallest ion first): $Br^-$; $Ca^{2+}$; $Cl^-$; $K^+$.

9.10)   State the size trend among ions and noble gas atoms that have the same electron configuration. Suggest an explanation for this trend.

9.44)   Which among the following is the smallest ion: $Se^{2-}$; $Br^-$; $Te^{2-}$? (Se is at. no. 34; Te is at. no. 52.)

9.45)   List the following ions in order of their decreasing size (largest ion first): $Ba^{2+}$; $Cs^+$ (at. no. 55); $I^-$; $Rb^+$ (at. no. 37).

9.46)   State the size trend among monatomic ions derived from elements in the same group in the periodic table. Suggest an explanation for this trend.

## Section 9.5

9.11)   Explain why ionic bonds are called electron *transfer* bonds, and covalent bonds are known as electron *sharing* bonds.

9.47)   Show how atoms achieve the stability of noble gas atoms in forming covalent bonds.

9.12) Compare the bond between potassium and chlorine in potassium chloride with the bond between two chlorine atoms in chlorine gas. Which bond is ionic, and which is covalent? Describe how each bond is formed.

9.13) Show how a covalent bond forms between an atom of iodine and an atom of chlorine, yielding a molecule of ICl.

9.14)* "The bond between a metal atom and a nonmetal atom is most apt to be ionic, whereas the bond between two nonmetal atoms is most apt to be covalent." Explain why the foregoing statement is true.

9.15)* Does the energy of a system tend to increase, decrease or remain unchanged as two atoms form a covalent bond? Is the process exothermic or endothermic?

### Section 9.6

9.16) What is meant by classifying a bond as *polar* or *nonpolar*? What bonds are completely nonpolar?

9.17) Consider the following bonds: F—Cl; Cl—Cl; Br—Cl; I—Cl. Arrange these bonds in order of increasing polarity (lowest polarity first), based on Figure 9.8, page 196. Classify each bond as (a) nonpolar or essentially nonpolar, or (b) polar covalent.

9.18) For each bond in question 9.17 that exhibits any polarity at all, identify the atom that acts as the negative pole.

9.19) What is electronegativity? Why are the noble gases not included in the electronegativity table?

### Section 9.7

9.20)* What is a multiple bond?

*9.21–9.29 and 9.57–9.65: Write Lewis structures for the sets of molecules shown:*

9.21) HBr; $H_2S$; $PH_3$.

9.22) $OF_2$; CO; $SO_4^{2-}$.

9.23) $ClO^-$; $BrO_4^-$; $H_2SO_4$.

9.24) $C_4H_{10}$; $C_4H_8$; $C_4H_6$.

9.25)* $CH_3F$; $CH_2F_2$; $CF_4$.

9.26)* All possible isomers of $C_5H_{12}$.

9.27)* Two isomers of $C_5H_{10}$.

9.28)* Two isomers of $C_3H_6O$.

9.48) Considering bonds between the following pairs of elements, which are most apt to be ionic and which are most apt to be covalent: sodium and sulfur; fluorine and chlorine; oxygen and sulfur?

9.49) Sketch the formation of two covalent bonds by an atom of sulfur in making a molecule of hydrogen sulfide, $H_2S$.

9.50)* The bond between two metal atoms is neither ionic nor covalent. Explain, according to the octet rule, why this is so.
atoms and noble gas electron configurations appear to be related in forming covalent and ionic

9.51)* How do the energy and stability of bonded atoms and noble gas electron configurations appear to be related in forming covalent and ionic bonds?

9.52) What is electronegativity?

9.53) List the following bonds in order of decreasing polarity: K—Br; S—O; N—Cl; Li—F; C—C. Classify each bond as (a) nonpolar or essentially nonpolar, (b) polar covalent, or (c) primarily ionic.

9.54) For each bond in question 9.53, identify the positive pole, if any.

9.55) Identify the trends in electronegativities that may be observed in the periodic table.

9.56)* Double bonds and triple bonds conform to the octet rule. Could a quadruple (4) bond obey that rule? a quintuple (5) bond?

9.57) BrF; $SF_2$; $PF_3$.

9.58) $HS^-$; $ClO_3^-$; $NO_3^-$.

9.59) $IO_2^-$; $H_3PO_4$; $HSO_4^-$.

9.60)* $CH_4O$; $C_2H_4O$; $C_2H_6O_2$.

9.61) $CH_2ClF$; $CBr_2F_2$; $ClBrClF$.

9.62)* All possible isomers of $C_6H_{14}$.

9.63)* Two isomers of $C_3H_8O$, one of which, an ether, does not have all of the carbon atoms bonded to each other.

9.64)* Two isomers of $C_2H_2Cl_2$.

9.29)* Formic acid, HCOOH. (This compound is produced by ants.)

9.65)* Propionic acid, $C_2H_5COOH$.

### Section 9.8

9.30)* Why is it not possible to draw a Lewis diagram that conforms to the octet rule if the species contains an odd number of electrons?

9.66)* Two iodides of arsenic (at. no. 33) are $AsI_3$ and $AsI_5$. One of these iodides has a Lewis diagram that conforms to the octet rule, and one does not. Draw the octet-rule diagram that is possible, and explain why the other cannot be drawn.

### Section 9.9

*9.31, 9.32, 9.67 and 9.68: Describe (a) the electron pair geometry and (b) the molecular geometry predicted by the electron pair repulsion principle for each species listed.*

9.31) $BeH_2$; $CF_4$; $OF_2$.

9.67) $BH_3$; $NF_3$; $HF$.

9.32) $IO_4^-$; $ClO_2^-$; $CO_3^{2-}$.

9.68) $ClO^-$; $IO_3^-$; $NO_3^-$.

### Section 9.10

9.33) Explain how the carbon tetrafluoride molecule, $CF_4$, which contains four polar bonds (electronegativity difference 1.5), can be nonpolar.

9.69) The nitrogen-fluorine bond has an electronegativity difference of 1.0—less than the electronegativity difference between carbon and fluorine in $CF_4$. Yet the $NF_3$ molecule is polar, while $CF_4$ is nonpolar. How can this be?

9.34) Compare the polarities of the HCl and HI molecules. In each case identify the end of the molecule that is more negative.

9.70) Compare the polarities of the following molecules: $ClF$; $Cl_2$; $BrCl$; $ICl$. In each case identify the end of the molecule that is more negative.

9.35) Compare the polarities of the $H_2O$ and $H_2S$ molecules: which molecule is more polar? What would you predict about the polarity of the $H_2Te$ molecule (Te is at. no. 52)?

9.71) Draw Lewis diagrams of the $CF_4$ and $CH_2F_2$ molecules, using arrows pointing to the more electronegative element in each bond. From these diagrams show that $CH_2F_2$ is polar and $CF_4$ is not.

9.36) Sketch the water molecule, paying particular attention to the bond angle and using arrows to indicate the polarity of the individual bonds. Then sketch the methanol molecule, $HOCH_3$, again using arrows to show bond polarity and predicting the approximate shape of the molecule around the oxygen atom. Estimate the relative polarities of the water and methanol molecules, and explain your prediction.

9.72) As noted on page 205, there are two plausible Lewis structures for the compound with the molecular formula $C_2H_6O$, and both are real compounds: $C_2H_5OH$ is ethanol, or ethyl alcohol, and $CH_3OCH_3$ is diethyl ether, the anesthetic. Sketch these molecules with arrows to indicate the direction of bond polarity around the oxygen atom. Predict the relative polarities of these molecules. Predict also the relative polarity between $CH_3OH$ and $C_2H_5OH$. What would you expect of the polarity of $C_5H_{11}OH$?

# 10

# INORGANIC
# NOMENCLATURE

In Chapter 5 we presented a brief introduction to a formal system of chemical nomenclature. Included were monatomic ions from the A groups of the periodic table, certain common acids and the polyatomic anions derived from them, and two other very common and important polyatomic ions, the ammonium and hydroxide ions. In this chapter we will review that introduction and make it a part of a broader range of acids and ionic compounds. Covalent compounds of two elements will be included too. We will stop short of the interesting system that has been developed to identify the vast number of substances that are classified as organic compounds. That system will be introduced in Chapter 19.

The language of chemistry is like most languages in having many variations, many dialects, if you wish. Few are the people working with chemicals who speak a pure dialect. The standard for such a dialect, if one really does exist, is set by the International Union of Pure and Applied Chemistry (IUPAC), an international organization whose function it is to unify chemical terminology as it develops in the laboratories throughout the world. The system they propose is commonly called the Stock System. At the other extreme we have what are commonly known as trivial names, names derived by craftsmen who logically identified a substance by its use or some obvious physical or chemical property (see Appendix IV, page 512). In between are various levels of formality and sophistication.

We shall strive, in this chapter, to develop familiarity with the "professional" language of chemistry, the language used by chemists in the United States today. In areas where both old and new terms are employed, we shall mention both, but lean toward the more modern terminology in subsequent usage. This, after all, is the way a language grows.

## 10.1 OXIDATION STATE: OXIDATION NUMBER

We will open our consideration of chemical nomenclature by introducing the concept known as **oxidation number** or **oxidation state.** The terms are synonymous, and are used interchangeably. Oxidation numbers are a sort of electron bookkeeping system with which to keep track of electrons in what are called oxidation-reduction reactions. We will consider these reactions in some detail in Chapter 17. Of immediate interest and utility, however, are the rules by which oxidation numbers are assigned:

1. The oxidation number of any elemental substance is zero.
2. The oxidation number of a monatomic ion is the same as the charge on the ion.
3. The oxidation number of combined oxygen is $-2$, except in peroxides $(-1)$ and superoxides $(-\frac{1}{2})$. (We will not emphasize peroxides or superoxides in this text.)
4. The oxidation number of combined hydrogen is $+1$, except in hydrides $(-1)$.
5. In any molecular or ionic species, the sum of the oxidation numbers of all atoms in the species is equal to the charge on the species.

## 10.2 NAMES AND FORMULAS OF CATIONS AND MONATOMIC ANIONS

PG 10 A   Given the name (or formula), state the formula (or name) of (1) a monatomic ion which appears in Figure 10.1; (2) a monatomic cation of a transition metal, given the oxidation state; (3) the ammonium ion ($NH_4^+$, discussed in Section 5.6, page 106); (4) the mercury(I) ion ($Hg_2^{2+}$).

In studying the formation of monatomic cations in Chapter 5 (page 101), you learned that an atom of any Group IA element forms a monatomic cation with a $+1$ charge by losing a single electron. By Oxidation Rule Number 2 the oxidation number of alkali metal ion, such as $Na^+$ and $K^+$, is $+1$, the same as the ionic charge. An atom of a Group IIA element forms a monatomic ion with a $+2$ charge by losing two electrons. Thus the oxidation states of $Mg^{2+}$, $Ca^{2+}$ and $Ba^{2+}$ are all $+2$. The aluminum ion, $Al^{3+}$, the best known monatomic ion from Group IIIA, results from the loss of three electrons from a neutral atom. Its oxidation state is $+3$.

Some of the transition elements—elements in the B groups or Group VIII of the periodic table—are capable of forming monatomic cations with different charges. Iron is one example. If a neutral iron atom loses two electrons, the ion has a $+2$ charge, $Fe^{2+}$. A monatomic ion of iron may also be in the $+3$ oxidation state, $Fe^{3+}$. In this case the neutral atom has lost three electrons. Copper is another common element that is capable of more than one oxidation state as a monatomic ion. Its oxidation state may be +1, as in $Cu^+$, or it may be +2, as in $Cu^{2+}$.

For the purpose of naming chemical compounds it is necessary to distinguish between the two monatomic ions of iron. This is done in two ways. The older system applies either the *-ic* or the *-ous* suffix to the stem of the Latin name of the element: *ferrum* in the case of iron. The *-ic* suffix is always applied to the higher oxidation state, and *-ous* to the lower. Hence the name of the $Fe^{3+}$ ion is *ferric*; the *-ous* suffix applied to the $Fe^{2+}$ ion gives us *ferrous*. By this system the name of $FeCl_2$, made up of an $Fe^{2+}$ ion and two $Cl^-$ ions, is ferrous chloride. When $Fe^{3+}$ combines with three $Cl^-$ ions the resulting compound is ferric chloride, $FeCl_3$.

In this text we shall emphasize the newer Stock system whereby the ion is identified by its English name, followed immediately by the oxidation state, written in Roman numerals and enclosed in parentheses. Thus $Fe^{2+}$ is written iron(II), and $Fe^{3+}$ is written iron(III). In speaking, iron(II) becomes "iron two," and iron(III) is "iron three." Applied to the two chlorides, we have $FeCl_2$ as iron(II) chloride and $FeCl_3$ as iron(III) chloride.

Figure 10.1 is a partial periodic table showing most of the cations you are apt to encounter in your introductory course—perhaps all of them. The ammonium ion, $NH_4^+$, has been added beneath the alkali metal ions it so closely resembles. Notice that the mercury(I) ion, $Hg_2^{2+}$, is a *diatomic elemental* ion. It is the only polyatomic elemental ion of importance. Its +2

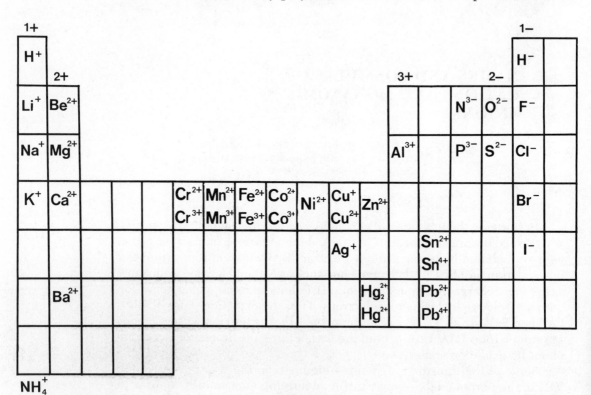

**Figure 10.1** Partial periodic table of common ions. Notes: (1) Tin (Sn) and lead (Pb) form monatomic ions in a +2 oxidation state. In their +4 oxidation states they are more accurately described as being covalently bonded, but such compounds are frequently named as if they were ionic compounds. (2) $Hg_2^{2+}$ is a diatomic elemental ion. Its name is mercury(I), indicating a +1 charge from each atom in the diatomic ion. (3) Ammonium ion, $NH_4^+$, is included as the only other common polyatomic cation, thereby completing this table as a minimum list of the cations you should be able to recall simply by referring to a full periodic table.

charge may be thought of as arising from each mercury atom losing one electron. Therefore the name is mercury(I), or *mercurous* by the older system.

Monatomic anions are formed when neutral atoms gain electrons (see page 101). Elements in Group VIIA acquire a $-1$ oxidation state by gaining one electron per atom, producing the $-1$ charge of the halide ions: fluoride, $F^-$, chloride, $Cl^-$, bromide, $Br^-$ and iodide, $I^-$. Group VIA elements form anions with a $-2$ charge by gaining two electrons per atom. The oxide, $O^{2-}$, and sulfide, $S^{2-}$, ions are common. Nitrogen and phosphorus form a few ionic compounds in which they are in the $-3$ oxidation state: $N^{3-}$ and $P^{3-}$. All of these monatomic anions are in Figure 10.1.

## 10.3 NAMES AND FORMULAS OF OXYACIDS AND THE POLYATOMIC ANIONS DERIVED FROM THEIR TOTAL IONIZATION

PG  10 B   Given the name (or formula) of an oxyacid of a Group VA, VIA or VIIA element, state its formula (or name).

   10 C   Given the name (or formula) of an oxyanion of a Group VA, VIA or VIIA element, state its formula (or name).

The term "acid" generally refers to one of a group of molecular compounds that contain hydrogen. When these compounds are placed in water they ionize, releasing hydrogen ions, $H^+$, and leaving behind negatively charged ions that balance the positive charge of the hydrogen ions lost. In Chapter 5 (page 105), *total ionization* was identified as the separation of all the hydrogen ions that *can* separate from an acid molecule, regardless of whether they actually separate or not. (In weak acids, which you will study

TABLE 10.1   Total Ionization of Acids

| ACID | TOTAL IONIZATION EQUATION | ANION NAME |
|---|---|---|
| Hydrochloric | $HCl\ (g) \xrightarrow{H_2O} H^+\ (aq) + Cl^-\ (aq)$ | Chloride |
| Chloric | $HClO_3\ (s) \xrightarrow{H_2O} H^+\ (aq) + ClO_3^-\ (aq)$ | Chlorate |
| Nitric | $HNO_3\ (l) \xrightarrow{H_2O} H^+\ (aq) + NO_3^-\ (aq)$ | Nitrate |
| Sulfuric | $H_2SO_4\ (l) \xrightarrow{H_2O} 2\ H^+\ (aq) + SO_4^{2-}\ (aq)$ | Sulfate |
| Carbonic* | $H_2CO_3\ (aq) \longrightarrow 2\ H^+\ (aq) + CO_3^{2-}\ (aq)$ | Carbonate |
| Phosphoric* | $H_3PO_4\ (s) \xrightarrow{H_2O} 3\ H^+\ (aq) + PO_4^{3-}\ (aq)$ | Phosphate |

*The reactions shown for carbonic and phosphoric acids occur to a very small extent in water solutions. They are shown to illustrate the system by which names of the carbonate and phosphate ions are derived. Both ions are commonly found from other sources.

in Chapter 16, only a small percentage of the hydrogen ions that *can* form are actually released.) As before, the assumed total ionization of an acid will be the basis for naming the polyatomic anions that result.

Equations representing the total ionization of several acids are shown in Table 10.1. Examining the names and formulas of the oxyacids and the ions derived from them, we note several regularities:

1. Each acid contains at least one hydrogen which can separate as $H^+$ (aq) in ionization.
2. Each acid contains a central atom which is a nonmetal, to which a varying number of oxygen atoms are covalently bonded.
3. The central atom and the oxygen atoms constitute the polyatomic anion.
4. The name of the acid is the name of the central atom, modified with an *-ic* suffix.
5. The name of the resulting polyatomic anion is the name of the central atom, modified with an *-ate* suffix.

The only acid in Table 10.1 that is new to you is chloric acid, $HClO_3$. It is included not so much because of its importance, but rather because it serves well to illustrate an even broader system of nomenclature for both acids and the polyatomic ions derived from them. There are three other oxyacids of chlorine that are a part of this system. They are perchloric acid, $HClO_4$, chlorous acid, $HClO_2$, and hypochlorous acid, $HClO$. All of these, as well as hydrochloric acid, which contains no oxygen, are listed in Table 10.2 in order of decreasing number of oxygen atoms. The Lewis diagrams, all of which conform to the octet rule, show structurally how chlorine is capable of forming this variety of acids. Notice also the oxidation states of chlorine in each molecule. Assigning oxidation states of $-2$ to oxygen and $+1$ to hydrogen (Rules 3 and 4 on page 219) and then applying Rule 5 to find the oxidation state of chlorine yield the sequence $+7$, $+5$, $+3$, $+1$ and $-1$ as the number of oxygens decreases from four to zero.

Our principal interest in Table 10.2 is the general oxyacid nomenclature system it illustrates, a system that extends beyond compounds of chlorine. In this system prefixes and suffixes are used to indicate the number of oxygen atoms present in the acid and corresponding anion, compared to the number of oxygens in the *-ic* acid and *-ate* anion. Table 10.3 identifies these prefixes and suffixes. The most direct application of this system occurs with the acids of other halogens. Bromine and iodine form acids and anions coresponding to those of chlorine, with a few exceptions that have not yet been isolated in the laboratory. Accordingly, $HBrO_3$ is bromic acid, and iodic acid is $HIO_3$; the corresponding anions are bromate, $BrO_3^-$, and iodate, $IO_3^-$. One additional oxygen yields periodic acid, $HIO_4$, and the periodate ion, $IO_4^-$. One less oxygen gives bromous acid, $HBrO_2$, and the bromite ion, $BrO_2^-$, the existence of which has been detected only in solution. Hypobromous acid, $HBrO$, and the hypobromite ion, $BrO^-$, have two oxygens fewer than the *-ic* acid and *-ate* ion.

Fluorine is omitted from the above consideration because it forms no oxyacids. Hydrofluoric acid does exist, however, as a water solution of hydrogen fluoride gas. The formula for both is HF, with state designations used to distinguish between them when necessary: HF (g) for the gas and HF (aq)

TABLE 10.2   Acids of Chlorine

| ACID | | LEWIS DIAGRAM | OXIDATION STATES | ANION | |
|---|---|---|---|---|---|
| $HClO_4$ | Perchloric | :Ö:<br>  &#124;<br>H—Ö—Cl—Ö:<br>  &#124;<br>:Ö: | H  Cl  4 O<br>+1  +7  4(−2) | $ClO_4^-$ | Perchlorate |
| $HClO_3$ | Chloric | H—Ö—Cl—Ö:<br>    &#124;<br>    :Ö: | H  Cl  3 O<br>+1  +5  3(−2) | $ClO_3^-$ | Chlorate |
| $HClO_2$ | Chlorous | H—Ö—Cl:<br>    &#124;<br>   :Ö: | H  Cl  2 O<br>+1  +3  2(−2) | $ClO_2^-$ | Chlorite |
| HClO<br>HOCl | Hypochlorous | H—Ö—Cl: | H  Cl  O<br>+1  +1  −2 | $ClO^-$<br>$OCl^-$ | Hypochlorite |
| HCl | Hydrochloric | H—Cl: | H  Cl<br>+1  −1 | $Cl^-$ | Chloride |

for the acid. Corresponding compounds exist for the other halogens; you are already familiar with hydrochloric acid. Note this important generalization: *hydro-ic* acids contain *no oxygen*. Anions derived from these acids are monatomic anions, and are named by modifying the elemental name with an *-ide* suffix in the usual manner.

TABLE 10.3   Prefixes and Suffixes in Acid and Anion Nomenclature

| OXYGEN ATOMS COMPARED TO *-IC* ACID AND *-ATE* ANION | ACID PREFIX AND/OR SUFFIX (EXAMPLE) | ANION PREFIX AND/OR SUFFIX (EXAMPLE) |
|---|---|---|
| One more | *per–ic* (perchloric) | *per–ate* (perchlorate) |
| Equal | *–ic* (chloric) | *–ate* (chlorate) |
| One less | *–ous* (chlorous) | *–ite* (chlorite) |
| Two less | *hypo–ous* (hypochlorous) | *hypo–ite* (hypochlorite) |
| No oxygen | *hydro–ic* (hydrochloric) | *–ide* (chloride) |

Variation in the number of oxygen atoms in oxyacids and oxyanions is not restricted to halogen compounds. Of particular importance are the compounds of sulfur and nitrogen that have one less oxygen than their -*ic* acids and -*ate* anions. $HNO_2$, for example, is nitr*ous* acid, and $SO_3^{2-}$ is the sulf*ite* ion. Furthermore, selenium and tellurium, atomic numbers 34 and 52, both in the same family as sulfur, form -*ic* and -*ous* acids that correspond with sulfuric and sulfurous acids, as well as the corresponding anions. The family characteristics extend to the *hydro-ic* acids too.

It is suggested that, in learning this material, you emphasize Table 10.3 in your study. By memorizing the formulas of a few -*ic* acids and the prefixes and suffixes used to indicate variations in oxygen atoms, you will be able to figure out the names and/or formulas of a large number of compounds, including some you may never have heard of before. This is more easily accomplished and stays with you longer than attempting to memorize all the names and formulas individually.

Let's try the system. In Examples 10.1 and 10.2 use the formulas of the -*ic* acids you already know, including chloric acid, $HClO_3$. Use Table 10.3 for the prefixes and suffixes for the moment. Also use the periodic table, assuming substitutions among members of the same chemical family are permissible, which they are for the substances in the example. Avoid looking at any of the foregoing text material; it will interfere with the thought processes we're trying to illustrate.

---

**Example 10.1**   Write the formula of the bromite ion.

The bromite ion is an oxyanion from a halogen acid. Your starting point for halogen compounds is chloric acid, $HClO_3$. What is the name and formula of the anion derived from chloric acid?

-------------------------------------------------------------------------------

**10.1a**   $ClO_3^-$, the chlorate ion

The complete ionization of an -*ic* acid always yields an -*ate* anion.

Now compare suffixes, using Table 10.3. How does the number of oxygen atoms in an -*ite* anion compare with the number of oxygen atoms in an -*ate* anion? From your answer and the formula of chlorate ion, $ClO_3^-$, write the formula of the chlorite ion.

-------------------------------------------------------------------------------

**10.1b**   $ClO_2^-$

An -*ite* anion has one less oxygen atom than an -*ate* anion. Removing one oxygen from chlorate ion, $ClO_3^-$, yields chlorite, $ClO_2^-$.

Now that you have the chlorite ion formula, changing it to bromite ion is simply a switch from one member of a chemical family to another. The bromite ion formula, please. . . .

-------------------------------------------------------------------------------

**10.1c** $BrO_2^-$

There are other reasoning paths by which you might have reached the same conclusion. For example, you might have made the halogen switch from chloric acid to bromic acid, $HBrO_3$, first, and then taken it to bromate ion, $BrO_3^-$, and then one less oxygen to bromite, $BrO_2^-$. The bromite ion, incidentally, is the one that is known in solution only. It doesn't form compounds, such as $NaBrO_2$.

Now that you've seen the technique, try it on three other substances. As before, limit your reference material to the periodic table and Table 10.3.

**Example 10.2** (a) Write the formula of selenic acid (selenium is at. no. 34); write the names of (b) HIO and (c) $AsO_3^-$ (As is arsenic, at. no. 33).

---

**10.2a** (a) $H_2SeO_4$; (b) hypoiodous acid; (c) arsenate ion

Explanations: (a) Selenium is in the same family as sulfur. If sulfuric acid is $H_2SO_4$, then selenic acid must be $H_2SeO_4$. (b) Substituting the halogen chlorine for the halogen iodine, HClO is hypochlorous acid. HIO must therefore be hypoiodous acid. (c) Arsenic is in the same family as nitrogen, which forms the corresponding $NO_3^-$ ion, nitrate. $AsO_3^-$ must therefore be an arsenate ion. (There are, in fact, two "arsenate" ions, the other being $AsO_4^{3-}$, corresponding to phosphate, $PO_4^{3-}$. Their names are distinguished by prefixes that are beyond the scope of our consideration.)

At this point we suggest you study Table 10.3 and fix it thoroughly in your thought. Then try the next two examples with reference to a periodic table only.

**Example 10.3** For each name given below, write the formula; for each formula, write the name.

| hydrofluoric acid | $HBrO_2$ |
| periodic acid | $ClO_2^-$ |

---

**10.3a** hydrofluoric acid: HF    $HBrO_2$: bromous acid
periodic acid: $HIO_4$    $ClO_2^-$: chlorite ion

**Example 10.4** For each name, write the formula; for each formula, write the name.

| sulfurous acid | $NO_2^-$ |
| carbonate ion | $SeO_3^{2-}$ (Se is selenium, at. no. 34) |

---

---

**10.4a**   sulfurous acid: $H_2SO_3$          $NO_2^-$: nitrite ion
carbonate ion: $CO_3^{2-}$          $SeO_3^{2-}$: selenite ion

---

## 10.4   NAMES AND FORMULAS OF ANIONS DERIVED FROM THE STEPWISE IONIZATION OF POLYPROTIC ACIDS

PG  10 D   Given the name (or formula) of an ion formed by the stepwise ionization of hydrosulfuric, sulfuric, phosphoric or carbonic acid, state the formula (or name).

A hydrogen atom consists of one proton and one electron. To form what we have been calling a hydrogen ion, $H^+$, the neutral atom must lose its electron. That leaves only the proton; a hydrogen ion is simply a proton. (In aqueous solution, the only place hydrogen ions commonly exist, the proton is hydrated. Recall the discussion of the hydronium ion on page 104.) Acids are sometimes classified by the number of hydrogen ions, or protons, released by a single molecule. **Acids** like hydrochloric, HCl, and nitric, $HNO_3$, **that yield only one proton per molecule, are called monoprotic acids. Acids that yield more than one proton per molecule are called polyprotic acids.** Sulfuric acid, $H_2SO_4$, and phosphoric acid, $H_3PO_4$, are polyprotic acids capable of releasing two and three hydrogen ions per molecule, respectively.

When polyprotic acids ionize, they do so by steps, releasing one hydrogen ion with each step. The intermediate anions produced are stable chemical species, constituting the negative ions in many ionic compounds. The theoretical ionization of carbonic acid, $H_2CO_3$, is an example. Examined from the Lewis structure, removal of one hydrogen ion yields $HCO_3^-$, the well-known **bicarbonate** ion:

$$H-\ddot{O}-C-\ddot{O}-H \quad \underset{\overset{\|}{\underset{:\ddot{O}:}{}}}{} \quad \rightarrow H^+ + \left[ :\ddot{O}-C-\ddot{O}-H \atop \overset{\|}{\underset{:O:}{}} \right]^-$$

*Bicarbonate* is a poor choice for the name of this ion—and one that appears again in principle in bisulfate and bisulfite ions resulting from partial ionization of sulfuric and sulfurous acids. A better name now being used by chemists is the more descriptive **hydrogen carbonate** ion. The second step in the ionization of carbonic acid yields the carbonate ion:

$$\left[ :\ddot{O}-C-\ddot{O}-H \atop \overset{\|}{\underset{:O:}{}} \right]^- \rightarrow H^+ + \left[ :\ddot{O}-C-\ddot{O}: \atop \overset{\|}{\underset{:O:}{}} \right]^{2-}$$

Phosphoric acid, $H_3PO_4$, has three steps in its ionization process, yielding $H_2PO_4^-$, *di*hydrogen phosphate ion, and $HPO_4^{2-}$, *mono*hydrogen phosphate ion, or simply hydrogen phosphate ion, as intermediates. All of these intermediate ions and their names are summarized in Table 10.4.

TABLE 10.4   Names and Formulas of Anions Derived from the Stepwise
Ionization of Acids

| ACID | ION | NAMES OF IONS | |
|---|---|---|---|
| | | *Preferred* | *Other* |
| $H_2CO_3$ | $HCO_3^-$ | Hydrogen carbonate | Bicarbonate Acid carbonate |
| $H_2S$ | $HS^-$ | Hydrogen sulfide | Bisulfide Acid sulfide |
| $H_2SO_4$ | $HSO_4^-$ | Hydrogen sulfate | Bisulfate Acid sulfate |
| $H_2SO_3$ | $HSO_3^-$ | Hydrogen sulfite | Bisulfite Acid sulfite |
| $H_3PO_4$ | $H_2PO_4^-$ | Dihydrogen phosphate | Monobasic phosphate |
| $H_2PO_4^-$ | $HPO_4^{2-}$ | Monohydrogen phosphate | Dibasic phosphate |

## 10.5   NAMES OF OTHER POLYATOMIC ANIONS

**ANIONS DERIVED FROM OTHER ACIDS.**   Organic acids also ionize slightly in water solution and yield anions. Only one such acid is significant at this point, namely, **acetic acid,** the component of vinegar that is responsible for its taste and odor. The Lewis structures for acetic acid and the **acetate** ion it yields are

The usual form for writing the acetate ion is $C_2H_3O_2^-$; some authors employ $CH_3COO^-$, or sometimes a "made-up" abbreviation, $OAc^-$.

**Hydrocyanic acid,** HCN, yields the **cyanide** ion, $CN^-$, in ionization. This ion and the hydroxide ion are the only two common polyatomic anions having an -*ide* ending, a suffix otherwise reserved for monatomic anions and covalent binary compounds.

**POLYATOMIC ANIONS FROM TRANSITION ELEMENTS.**   Chromium and manganese form some polyatomic anions that theoretically may be traced to acids, but only the ions are important. Their names and formulas are the chromate ion, $CrO_4^{2-}$, the dichromate ion, $Cr_2O_7^{2-}$, and the permanganate ion, $MnO_4^-$.

## 10.6 NAMES OF IONIC COMPOUNDS

> PG 10 E   Given the name (or formula) of an ionic compound composed of ions that are known or capable of being derived, state the formula (or name) of the compound.

The name of an ionic compound is the name of the positive ion followed by the name of the negative ion. The formula of the compound is the formula of the positive ion followed by the formula of the negative ion, plus whatever subscripts are required to adjust the total ionic charge to zero. This also, incidentally, brings the sum of the oxidation numbers to zero, in conformity with Oxidation Number Rule 5.

Compounds in which a monatomic cation has more than one possible oxidation state require special comment. For example, if asked to write the formula for iron(II) chloride, you know from the name that it is a compound made up of $Fe^{2+}$ and $Cl^-$ ions. The formula must therefore be $FeCl_2$. But what about the other direction? What, for example, is the name of $Fe_2O_3$? "Iron oxide" is not an adequate answer; it fails to distinguish between the two possible oxidation states of iron. Is it iron(II) oxide, or iron(III) oxide? To reach this decision you must use a combination of Oxidation Number Rules 3 and 5. Rule 3 states that oxygen has an oxidation number of $-2$. $Fe_2O_3$ has three oxygen atoms in the formula unit; hence oxygen contributes $3(-2) = -6$ to the summation of oxidation numbers in the compound. Rule 5 requires that the total oxidation number of the formula unit

**TABLE 10.5**   Cations

| IONIC CHARGE: +1 | IONIC CHARGE: +2 | IONIC CHARGE: +3 |
|---|---|---|
| *Alkali Metals:*<br>  *Group IA* | *Alkaline Earths:*<br>  *Group IIA* | *Group IIIA* |
| $Li^+$   Lithium<br>$Na^+$   Sodium<br>$K^+$   Potassium<br>$Rb^+$   Rubidium<br>$Cs^+$   Cesium | $Be^{2+}$   Beryllium<br>$Mg^{2+}$   Magnesium<br>$Ca^{2+}$   Calcium<br>$Sr^{2+}$   Strontium<br>$Ba^{2+}$   Barium | $Al^{3+}$   Aluminum<br>$Ga^{3+}$   Gallium |
| *Transition Elements* | *Transition Elements* | *Transition Elements* |
| $Cu^+$   Copper(I)<br>$Ag^+$   Silver | $Cr^{2+}$   Chromium(II)<br>$Mn^{2+}$   Manganese(II)<br>$Fe^{2+}$   Iron(II)<br>$Co^{2+}$   Cobalt(II)<br>$Ni^{2+}$   Nickel(II)<br>$Cu^{2+}$   Copper(II)<br>$Zn^{2+}$   Zinc<br>$Cd^{2+}$   Cadmium<br>$Hg_2^{2+}$   Mercury(I)<br>$Hg^{2+}$   Mercury(II) | $Cr^{3+}$   Chromium(III)<br>$Mn^{3+}$   Manganese(III)<br>$Fe^{3+}$   Iron(III)<br>$Co^{3+}$   Cobalt(III) |
| *Polyatomic Ions*<br><br>$NH_4^+$   Ammonium | | |
| *Others*<br><br>$H^+$   Hydrogen<br>  OR<br>$H_3O^+$   Hydronium | *Others*<br><br>$Sn^{2+}$   Tin(II)<br>$Pb^{2+}$   Lead(II) | |

TABLE 10.6   Anions

| IONIC CHARGE: −1 | | IONIC CHARGE: −2 | IONIC CHARGE: −3 |
|---|---|---|---|
| *Halogens:* *Group VIIA* | *Oxyanions* | *Group VIA* | *Group VA* |
| $F^-$ Fluoride | $ClO_4^-$ Perchlorate | $O^{2-}$ Oxide | $N^{3-}$ Nitride |
| $Cl^-$ Chloride | $ClO_3^-$ Chlorate | $S^{2-}$ Sulfide | $P^{3-}$ Phosphide |
| $Br^-$ Bromide | $ClO_2^-$ Chlorite | | |
| $I^-$ Iodide | $ClO^-$ Hypochlorite | *Oxyanions* | *Oxyanion* |
| | | $CO_3^{2-}$ Carbonate | $PO_4^{3-}$ Phosphate |
| *Acidic Anions* | $BrO_3^-$ Bromate | $SO_4^{2-}$ Sulfate | |
| | $BrO_2^-$ Bromite | $SO_3^{2-}$ Sulfite | |
| $HCO_3^-$ Hydrogen carbonate | $BrO^-$ Hypobromite | $C_2O_4^{2-}$ Oxalate | |
| $HS^-$ Hydrogen sulfide | $IO_4^-$ Periodate | $CrO_4^{2-}$ Chromate | |
| | $IO_3^-$ Iodate | $Cr_2O_7^{2-}$ Dichromate | |
| $HSO_4^-$ Hydrogen sulfate | | | |
| $HSO_3^-$ Hydrogen sulfite | $NO_3^-$ Nitrate | *Acidic Anion* | |
| | $NO_2^-$ Nitrite | $HPO_4^{2-}$ Monohydrogen phosphate | |
| $H_2PO_4^-$ Dihydrogen phosphate | $OH^-$ Hydroxide | | |
| | $C_2H_3O_2^-$ Acetate | *Diatomic Elemental* | |
| *Other Anions* | $MnO_4^-$ Permanganate | $O_2^{2-}$ Peroxide | |
| $SCN^-$ Thiocyanate | | | |
| $CN^-$ Cyanide | | | |
| $H^-$ Hydride | | | |

is the same as the charge on the unit, which, in the case of a compound, is zero. This means that a total of +6 must come from the two atoms of iron in the formula, or +3 from each atom. The compound is therefore iron(III) oxide. FeO is iron(II) oxide, a conclusion that would be reached by recognizing that the −2 of a single oxide ion would be balanced by a +2 from a single iron(II) ion.

In writing or speaking the name of an ionic compound in which a monatomic cation has more than one possible oxidation state, it is essential that the name include the oxidation state to avoid confusion.

Table 10.5 summarizes all the cations mentioned in this chapter, and Table 10.6 lists all the anions, plus a few more. If you have mastered the performance goals of this chapter, you know all of the cations and most of the anions. In completing the following exercises, therefore, try to use a periodic table as your only reference. Tables 10.5 and 10.6 should be used only as a last resort.

---

**Example 10.5**   For each formula below, write the name of the compound; for each name, write the formula.

$FeBr_3$          aluminum carbonate

$NaHSO_4$          manganese(II) chloride

---

**10.5a**

FeBr₃, iron(III) bromide      aluminum carbonate, $Al_2(CO_3)_3$

NaHSO₄, sodium hydrogen sulfate      manganese(II) chloride, $MnCl_2$

We reason that FeBr₃ is iron(III) bromide because the bromide ion, $Br^-$, has a single negative charge. There are three $Br^-$ ions, accounting for a $-3$ charge that must be balanced by a $+3$ charge. The iron ion must therefore be $Fe^{3+}$, or iron(III).

---

**Example 10.6** For each formula below, write the name of the compound; for each name, write the formula. Refer only to the periodic table.

$Zn_3N_2$      silver sulfate

HgS      copper(I) chloride

$Mg(NO_2)_2$      potassium dihydrogen phosphate

$NH_4BrO_3$      sodium periodate

-------------------------------------------------------------------------------

**10.6a**

$Zn_3N_2$, zinc nitride      silver sulfate, $Ag_2SO_4$

HgS, mercury(II) sulfide      copper(I) chloride, CuCl

$Mg(NO_2)_2$, magnesium nitrite      potassium dihydrogen phosphate, $KH_2PO_4$

$NH_4BrO_3$, ammonium bromate      sodium periodate, $NaIO_4$

## 10.7 COVALENT BINARY COMPOUNDS

PG 10 F   Given the name (or formula) of a covalent binary compound, state the formula (or name) of the compound.

**A binary compound is a compound that consists of two elements.** You have been using and writing the names and formulas of ionic binary compounds—compounds of two monatomic ions—for quite some time. We have used the names and formulas of covalent binary compounds throughout the text, but up to now we have not formalized a nomenclature system for them. This is the purpose of the present section.

Like ionic compounds, covalent binary compounds have first and last names. The first name is the name of the element that appears first in the chemical formula. The second name is the name of the second element in the formula, modified so it has an -*ide* suffix. With a few notable exceptions, the order of appearance of elements in the formula is that of increasing electronegativity. Thus the formulas of the oxides of fluorine and chlorine are $OF_2$ and $Cl_2O$, the electronegativity of oxygen being lower than that of fluorine, but higher than chlorine.

Straightforward application of these rules, unfortunately, is not free from uncertainty. For example, the formula CO indicates a covalent binary compound composed of carbon and oxygen. Carbon oxide would be the name, according to the rules. But then what would be the name of $CO_2$, also a covalent binary compound of carbon and oxygen? In establishing names that distinguish between two or more covalent binary compounds of the same elements, the chemist uses prefixes to identify the number of atoms of each element in the chemical formula. By this system, CO is named carbon monoxide, and $CO_2$ is called carbon dioxide. The prefix *mono-* indicates 1, and the prefix *di-* indicates 2. Strict adherence to the nomenclature rules would place prefixes before the names of both elements, such as monocarbon monoxide and monocarbon dioxide. As in previous situations, however, the number 1 is assumed in the absence of a prefix to indicate otherwise. The *mono-* is therefore not necessary for carbon, which is 1 in both compounds. It is used for a single oxygen atom in carbon monoxide because of possible uncertainty and confusion with carbon dioxide. A list of the prefixes up to ten is shown in Appendix V, page 513.

The oxides of nitrogen provide an excellent opportunity for you to practice the nomenclature of covalent binary compounds:

**Example 10.7**  For each formula below, write the name of the compound; for each name, write the formula.

| | |
|---|---|
| $NO_2$ | nitrogen monoxide |
| $N_2O_3$ | dinitrogen oxide |
| $N_2O_4$ | dinitrogen pentoxide |

------------------------------------------------

**10.7a**  

| | |
|---|---|
| $NO_2$, nitrogen dioxide | nitrogen monoxide, NO |
| $N_2O_3$, dinitrogen trioxide | dinitrogen oxide, $N_2O$ |
| $N_2O_4$, dinitrogen tetroxide | dinitrogen pentoxide, $N_2O_5$ |

The formulas of the named compounds come readily from the names. Nitrogen dioxide could be correctly identified as mononitrogen dioxide, but, as noted, *mono-* is normally omitted when the molecule has only one atom of an element, unless it is required to avoid ambiguity. Dinitrogen tetroxide might be called dinitrogen tetraoxide; however it is customary to drop the vowel at the end of a prefix if the element begins with a vowel, *if* the "sound" of the term permits.

# CHAPTER 10 IN REVIEW

**10.1  OXIDATION STATE: OXIDATION NUMBER (Page 219)**

**10.2  NAMES AND FORMULAS OF CATIONS AND MONATOMIC ANIONS**

10 A  Given the name (or formula), state the formula (or name) of (1) a monatomic ion which appears in Figure 10.1; (2) a monatomic cation of a transition metal, given the oxidation state; (3) the ammonium ion ($NH_4^+$); (4) the mercury(I) ion ($Hg_2^{2+}$). (Page 219)

10.3  NAMES AND FORMULAS OF OXYACIDS AND THE POLYATOMIC
ANIONS DERIVED FROM THEIR TOTAL IONIZATION

10 B  Given the name (or formula) of an oxyacid of a Group VA, VIA or
VIIA element, state its formula (or name). (Page 221)

10 C  Given the name (or formula) of an oxyanion of a Group VA, VIA or
VIIA element, state its formula (or name). (Page 221)

10.4  NAMES AND FORMULAS OF ANIONS DERIVED FROM THE
STEPWISE IONIZATION OF POLYPROTIC ACIDS

10 D  Given the name (or formula) of an ion formed by the incomplete
ionization of hydrosulfuric, sulfuric, phosphoric or carbonic acid,
state the formula (or name). (Page 226)

10.5  NAMES OF OTHER POLYATOMIC ANIONS (Page 227)

10.6  NAMES OF IONIC COMPOUNDS

10 E  Given the name (or formula) of an ionic compound composed of
ions that are known or capable of being derived, state the formula
(or name) of the compound. (Page 228)

10.7  COVALENT BINARY COMPOUNDS

10 F  Given the name (or formula) of a covalent binary compound, state
the formula (or name of the compound). (Page 230)

# TERMS AND CONCEPTS

Stock system (nomenclature) (218)
Oxidation number or state (219)
Monoprotic, polyprotic acids (226)

Stepwise ionization (226)
Binary compound (230)

# QUESTIONS AND PROBLEMS

*To answer all questions in this nomenclature chapter, write the name of the species for which the
formula is given, or the formula if the name is given. In naming ions or ionic compounds, use oxidation
numbers when necessary to avoid uncertainty, but not otherwise. For reference materials, try to limit
yourself to a periodic table giving no more information than the one inside the front cover of this book.
Atomic numbers are included in questions involving elements that are likely to be unfamiliar to you.*

Section 10.2

10.1)  $Ca^{2+}$; $Cr^{3+}$; $Zn^{2+}$; $P^{3-}$; $Br^-$.

10.2)  Lithium ion; ammonium ion; nitride ion;
fluoride ion; mercury(II) ion.

Section 10.3

10.3)  $HNO_3$ (aq); $H_2SO_3$ (aq); perchloric acid;
selenic acid (selenium is at. no. 34).

10.4)  Sulfate ion; chlorite ion; $IO_3^-$; $BrO^-$.

10.23)  $Cu^+$; $I^-$; $K^+$; $Hg_2^{2+}$; $S^{2-}$.

10.24)  Iron(III) ion; hydride ion; oxide ion;
aluminum ion; barium ion.

10.25)  HClO (aq); $H_2TeO_3$ (aq) (Te is tellurium,
at. no. 52); bromic acid; phosphoric acid.

10.26)  Selenite ion (selenium is at. no. 34);
periodate ion; $BrO_2^-$; $NO_2^-$.

Section 10.4

10.5) Hydrogen carbonate ion; dihydrogen phosphate ion; $HSO_4^-$.

10.27) $HPO_4^{2-}$; $HS^-$; hydrogen sulfite ion.

Section 10.6

10.6) Potassium sulfide; copper(II) nitrate; sodium hydrogen carbonate.

10.7) $MgSO_3$; $AlF_3$; $PbCO_3$.

10.28) Barium sulfate; chromium(III) oxide; calcium monohydrogen phosphate.

10.29) $CuSO_4$; $Ba(OH)_2$; $Hg_2I_2$.

Section 10.7

10.8) $SO_2$; $N_2O$; phosphorus tribromide; hydrogen iodide.

10.30) Dichlorine oxide; uranium hexafluoride (uranium is at. no. 92); HBr (g); $P_2O_3$.

*From this point items in the nomenclature exercise are selected at random from any section of the chapter. Unless marked with an asterisk (\*), all names and formulas are included in the performance goals, and should be found with reference to no more than the periodic table. Ions in compounds marked with an asterisk are included in Table 10.6, page 229; or if the unfamiliar ion is monatomic, the atomic number of the element is given.*

10.9) Hydrogen sulfite ion; potassium nitrate; $MnSO_4$; $SO_3$.

10.10) $BrO_3^-$; $Ni(OH)_2$; silver chloride; silicon hexafluoride.

10.11) Tellurate ion (tellurium is at. no. 52); iron(III) phosphate; $NaC_2H_3O_2$\*; $H_2S$ (g).

10.12) $HPO_4^{2-}$; CuO; sodium oxalate\*; ammonia.

10.13) Hypochlorous acid; chromium(II) bromide; $KHCO_3$; $Na_2Cr_2O_7$\*.

10.14) $Co_2O_3$; $Na_2SO_3$; mercury(II) iodide; aluminum hydroxide.

10.15) Calcium dihydrogen phosphate; potassium permanganate\*; $NH_4IO_3$; $H_2SeO_4$ (Se is selenium, at. no. 34).

10.16) $Hg_2Cl_2$; $HIO_4$ (aq); cobalt(II) sulfate; lead(II) nitrate.

10.17) Uranium trifluoride (uranium is at. no. 92); barium peroxide\*; $MnCl_2$; $NaClO_2$.

10.18) $K_2TeO_4$ (Te is tellurium, at. no. 52); $ZnCO_3$; chromium(II) chloride; acetic acid\*.

10.19) Barium chromate\*; calcium sulfite; CuCl; $AgNO_3$.

10.20) $Na_2O_2$\*; $NiCO_3$; iron(II) oxide; hydrosulfuric acid.

10.21) Zinc phosphide; cesium nitrate (cesium is at. no. 55); $NH_4CN$\*; $S_2F_{10}$.

10.22) $N_2O_3$; $LiMnO_4$\*; indium selenide (indium is at. no. 49; selenium is at. no. 34); mercury(I) thiocyanate\*.

10.31) Perchlorate ion; barium carbonate; $NH_4I$; $PCl_3$.

10.32) $HS^-$; $MgSO_3$; aluminum nitrate; oxygen difluoride.

10.33) Mercury(I) ion; cobalt(II) chloride; $SiO_2$; $LiNO_2$.

10.34) $N^{3-}$; $Ca(ClO_3)_2$; iron(III) sulfate; phosphorus pentabromide.

10.35) Tin(II) fluoride; potassium chromate\*; LiH; $FeCO_3$.

10.36) $HNO_2$; $Zn(HSO_4)_2$; potassium cyanide\*; copper(I) fluoride.

10.37) Magnesium nitride; lithium bromate; $NaHSO_3$; KSCN\*.

10.38) $Ni(HCO_3)_2$; CuS; chromium(III) iodide; potassium hydrogen phosphate.

10.39) Selenium dioxide (selenium is at. no. 34); magnesium nitrite; $FeBr_2$; $Ag_2O$.

10.40) SnO; $(NH_4)_2Cr_2O_7$\*; sodium hydride; oxalic acid\*.

10.41) Cobalt(III) sulfate; iron(III) iodide; $Cu_3(PO_4)_2$; $Mn(OH)_2$.

10.42) $Al_2Se_3$ (Se is selenium, at. no. 34); $MgHPO_4$; potassium perchlorate; bromous acid.

10.43) Strontium iodate (strontium is at. no. 38); sodium hypochlorite; $Rb_2SO_4$ (Rb is rubidium, at. no. 37); $P_2O_5$.

10.44) ICl; $AgC_2H_3O_2$\*; lead(II) dihydrogen phosphate; gallium fluoride (gallium is at. no. 31).

# 11

# THE GASEOUS STATE

## 11.1  PROPERTIES OF GASES

The air that surrounds us is a sea of mixed gases, called the atmosphere. It is not necessary, then, to search very far to find a gas whose properties we may study. Among the familiar characteristics of air which are, in fact, properties of all gases, we list the following:

**1. GASES MAY BE COMPRESSED.**  A fixed quantity of air may be made to occupy a smaller volume by applying pressure. Figure 11.1A shows a

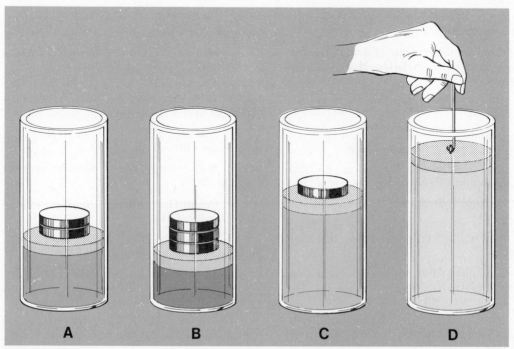

**Figure 11.1**  Properties of gases. Piston and cylinder show that gases may be compressed, and that they expand to fill uniformly the volume available to them.

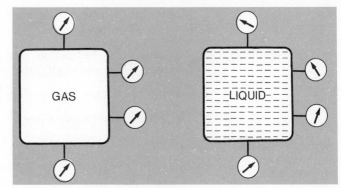

**Figure 11.2** Gas pressures are exerted uniformly in all directions; liquid pressures depend upon depth of liquid.

quantity of air in a cylinder having a leak-proof piston which can be moved to change the volume occupied by the air. Push the piston down by applying more force and the volume of air is reduced (see Fig. 11.1B).

2. **GASES EXPAND TO FILL THEIR CONTAINERS UNIFORMLY.** If less force were applied to the piston, as shown in Figure 11.1C, air would respond immediately, pushing the piston upward, expanding to fill uniformly the larger volume. If the piston were pulled up (Fig. 11.1D), air would again expand to fill the additional space.

3. **ALL GASES HAVE LOW DENSITY.** The density of air is 0.0013 g/cm³. The air in an empty 1 gallon (3.785 liter) bottle weighs 4.9 grams (about ⅙ ounce); one gallon of water has a mass of 3785 grams (8.3 pounds), about 770 times greater. All of the air in a typical bedroom weighs about 35 kg (77 pounds).

4. **GASES MAY BE MIXED.** "There's always room for more," is a phrase that may be applied to gases. You may add the same or a different gas to that gas already occupying a rigid container of fixed volume, provided there is no chemical reaction between them.

5. **A CONFINED GAS EXERTS CONSTANT PRESSURE ON THE WALLS OF ITS CONTAINER UNIFORMLY IN ALL DIRECTIONS.** This pressure, illustrated in Figure 11.2, is a unique property of the gas, independent of external factors such as gravitational forces.

## 11.2 THE KINETIC THEORY OF GASES AND THE IDEAL GAS MODEL

PG 11 A Explain physical properties of gases, or physical phenomena relating to gases, in terms of the ideal gas model.

In trying to account for the properties of gases, scientists have devised the **kinetic theory of gases.** This theory is actually the best understood portion of the *kinetic molecular theory* referred to in Chapter 2 (page 10). The theory describes an *"ideal gas model"* by which we visualize the nature of a gas by comparing it with a physical system that can be seen, or at least readily imagined. The main features of the ideal gas model are:

1. Gases consist of molecular particles moving at any given instant in straight lines.

2. Molecules collide with each other and with the container walls without loss of energy.

3. Gas molecules behave as independent particles; attractive forces between them are negligible.

4. Gas molecules are very widely spaced.

5. The actual volume of molecules is negligible compared to the space they occupy.

Particle motion explains why gases fill their containers. It also suggests how they exert pressure. When an individual particle strikes a container wall it exerts a force at the point of collision. When this is added to billions upon billions of similar collisions occurring continuously, the total effect is the steady force that is responsible for gas pressure.

There can be no loss of energy as a result of these collisions. If the particles lost energy, or slowed down, the combined forces would become smaller and the pressure would gradually decrease. Furthermore, because of the relationship between temperature and average molecular speed (see page 243), temperature would drop if energy was lost in collisions. But these things do not happen, so we conclude that energy is not lost in molecular collisions, either with the walls or between molecules.

Gas molecules must be widely spaced; otherwise the density of a gas would not be so low. One gram of liquid water at the boiling point occupies 1.04 cm$^3$. When changed to steam at the same temperature the same number of molecules fills 1670 cm$^3$, an expansion of 1600 times. If the molecules were touching each other in the liquid state, they must be widely separated in the vapor state. The compressibility and mixing ability of gases are also attributable to the open space between the molecules. Finally, it is because of the large intermolecular distance that attractions between molecules are negligible.

In summary, the ideal gas model pictures a gas as consisting of a large number of independent and widely spaced molecules, moving in random and chaotic fashion at high speeds. Individual molecules move in straight lines until they collide with other molecules or the wall of the container without loss of energy.

## 11.3  GAS MEASUREMENTS

### MEASURABLE PROPERTIES

PG  11 B  List the measurable properties of a gas.

11 C  Explain the meaning of "one atmosphere" of pressure and identify its equivalent in millimeters of mercury, centimeters of mercury, inches of mercury, torr and pounds per square inch.

Experiments involving gases ordinarily require that one or more of the following be measured:

1. QUANTITY.  The quantity of gas in a sample may be determined by weighing, although the procedure is somewhat more involved than the

weighing of solids or liquids. The amount of gas in a sample is also frequently expressed in terms of moles.

**2. TEMPERATURE.** Far more than solids or liquids, gases expand when heated, contract when cooled. Temperature is therefore an important variable in experiments with gases. Temperature is generally measured with a thermometer.

**3. VOLUME.** The volume occupied by a gas is the full volume of its container. It may be measured in the usual ways.

**4. PRESSURE.** The chaotic movement of molecules makes measurement of the pressure of a gas of special importance. Pressure may be measured with mechanical gauges or with instruments called manometers.

By definition, pressure is the force exerted on a unit area. Typical English units of pressure emerge from the definition: if force is measured in pounds and area in square inches, the pressure unit is pounds per square inch. The SI unit of pressure is the **pascal,** which is **one newton per square meter.** (The *newton* is the SI unit of force.) One pascal is a very small pressure; the kilopascal is a more practical unit. The **millimeter of mercury,** or its equivalent the **torr,** and the **atmosphere** are the common units for expressing pressure.

Weather bureaus report *barometric* pressure, the pressure exerted by the atmosphere at a given weather station, generally in inches or centimeters of mercury. It is measured by a device known as a barometer, developed by Evangelista Torricelli in the 17th century. Torricelli found that if a tube, closed at one end and filled with mercury, is inverted in a dish of mercury (see Fig. 11.3), the liquid level inside the tube will fall until the pressure of the atmosphere on the surface of the liquid is exactly balanced by the pressure of the mercury column, whose height may be measured. This height thereby becomes a measure of atmospheric pressure. On a day when the mercury column in a barometer is 752 mm high, we say that atmospheric pressure is 752 mm Hg.

At sea level on an average day the height of the mercury column in a barometer will be 760.0 mm or 29.92 inches. This pressure is arbitrarily called **one standard atmosphere** of pressure. The atmosphere unit is par-

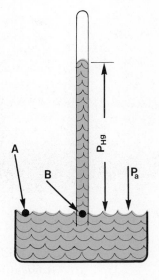

**Figure 11.3** Mercury barometer. Two operational principles govern the mercury barometer. (1) The total pressure at any point in a *liquid* system is the sum of the pressures of each gas or liquid phase above that point. (2) The total pressures at any two points at the *same level* in a *liquid* system are always equal. Point A at the liquid surface outside the tube is at the same level as Point B inside the tube. The only thing exerting downward pressure at A is the atmosphere; $P_a$ represents atmospheric pressure. The only thing exerting downward pressure at point B is the mercury above that point, designated $P_{Hg}$. A and B being at the same level, the pressures at these points are equal: $P_a = P_{Hg}$.

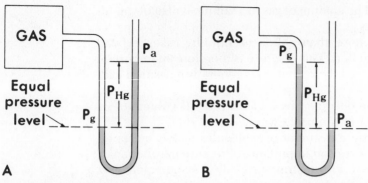

**Figure 11.4** Open end manometers. Open-end manometers are governed by the same principles as mercury barometers (see Fig. 11.3). The pressure of the gas, $P_g$, is exerted on the mercury surface in the closed (left) leg of the manometer. Atmospheric pressure, $P_a$, is exerted on the mercury surface in the open (right) leg. Using a meter stick, the *difference* between these two pressures, $P_{Hg}$, may be measured *directly in millimeters of mercury* (torr). Gas pressure is determined by equating the total pressures at the *lower liquid level.* In A, the pressure in the left leg is the gas pressure, $P_g$. Total pressure at the same level in the right leg is the pressure of the atmosphere, $P_a$, plus the pressure difference, $P_{Hg}$. Equating the pressures, $P_g = P_a + P_{Hg}$. In B, total pressure in the closed leg is $P_g + P_{Hg}$, which is equal to the atmospheric pressure, $P_a$. Equating and solving for $P_g$ yields $P_g = P_a - P_{Hg}$. In effect, the pressure of a gas, as measured by a manometer, may be found by adding the pressure difference to, or subtracting the pressure difference from, atmospheric pressure: $P_g = P_a \pm P_{Hg}$.

ticularly useful in referring to very high pressures. The English unit that is equal to one atmosphere is 14.69 pounds per square inch. In summary, the different pressure units and their relationships to each other are:

$$1.000 \text{ atm} = 14.69 \text{ lbs/in}^2 = 29.92 \text{ inches Hg} = 76.00 \text{ cm Hg} =$$
$$760.0 \text{ mm Hg} = 760.0 \text{ torr} = 1.013 \times 10^5 \text{ Pa} = 101.3 \text{ kPa,}$$

$$(11.1)$$

where the Pa and kPa are the pascal and kilopascal, respectively.

The *torr* is the recently introduced substitute for the *millimeter of mercury,* honoring the work of Torricelli. Both units are widely used, and the choice between them is one of personal preference. The advantage of *millimeter of mercury* is that it has physical meaning; it may be read by direct observation on an open-end manometer, the instrument by which pressure is most commonly measured in the laboratory (see Fig. 11.4). *Torr,* however, is both easier to say and write. We will use *torr* hereafter in this text.

## 11.4 BOYLE'S LAW

> PG 11 D Given the initial volume (or pressure) and initial and final pressures (or volumes) of a fixed quantity of gas at constant temperature, calculate the final volume (or pressure).

Robert Boyle, in the 17th century, investigated the quantitative relationship between pressure and volume of a fixed amount of gas at constant temperature. A modern laboratory experiment similar to that performed by Boyle employs a mercury-filled manometer shown in Fig. 11.5A. The closed and stationary left leg of the manometer is a gas measuring tube, calibrated in milliliters. Trapped above the mercury is a constant quantity of gas. The open right leg of the manometer is connected to the left leg by a flexible tube, making it possible to alter the pressure on the confined gas simply by moving the right leg up or down. An actual student experiment with this apparatus yielded the data in the table, Fig. 11.5B.

A plot of pressure versus volume for these data is shown in Figure 11.5C.

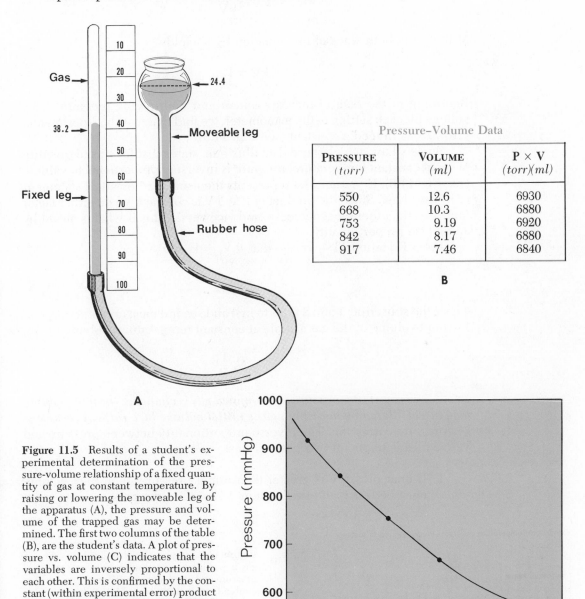

Pressure–Volume Data

| PRESSURE (torr) | VOLUME (ml) | P × V (torr)(ml) |
|---|---|---|
| 550 | 12.6 | 6930 |
| 668 | 10.3 | 6880 |
| 753 | 9.19 | 6920 |
| 842 | 8.17 | 6880 |
| 917 | 7.46 | 6840 |

B

**Figure 11.5** Results of a student's experimental determination of the pressure-volume relationship of a fixed quantity of gas at constant temperature. By raising or lowering the moveable leg of the apparatus (A), the pressure and volume of the trapped gas may be determined. The first two columns of the table (B), are the student's data. A plot of pressure vs. volume (C) indicates that the variables are inversely proportional to each other. This is confirmed by the constant (within experimental error) product of pressure × volume, shown in the third column of the table.

To a mathematician, the shape of the curve suggests an inverse proportionality between pressure and volume. Expressed mathematically,

$$P \propto \frac{1}{V} \qquad (11.2)$$

Introducing a proportionality constant (see Appendix I, page 499), we have

$$P = k_1 \frac{1}{V} \qquad (11.3)$$

Multiplying both sides of the equation by V yields

$$PV = k_1 \qquad (11.4)$$

Returning to the data of the experiment and multiplying pressure times volume for each setting of the manometer, we find that, within experimental error, PV is indeed a constant (see Fig. 11.5B).

Boyle's Law, which these data illustrate, states that **for a fixed quantity of gas at constant temperature, pressure is inversely proportional to volume** (see Fig. 11.6). Equation 11.4 represents the usual mathematical statement of Boyle's Law. Since the product of P and V is constant, when either factor increases the other must decrease, and vice versa. This is what is meant by an inverse proportionality.

From Equation 11.4 we see that $P_1V_1 = k_1 = P_2V_2$, or

$$P_1V_1 = P_2V_2. \qquad (11.5)$$

where the subscripts 1 and 2 refer to first and second measurements of pressure and volume of the gas sample at constant temperature. Solving for $V_2$,

$$V_2 = V_1 \frac{P_1}{P_2} \qquad (11.6)$$

In other words, *if the pressure of a confined gas is changed, the final volume may be calculated by multiplying the initial volume by a ratio of pressures* —a pressure correction. The inverse proportionality between pressure and volume leads to one of two possibilities:

1. If final pressure is greater than initial pressure, then final volume must be less than initial volume. Therefore the pressure correction

Condition: Temperature constant—no gas gained or lost

A       B       C

Figure 11.6 Boyle's Law. A fixed quantity of gas is confined to a cylinder, as in A, at a given pressure and constant temperature. If pressure is doubled, as in B, the volume is reduced to one half its original value. If pressure is doubled again, now four times the original pressure, the volume is reduced to one fourth of its original value.

must be a ratio less than 1—the numerator must be smaller than the denominator.

2. If final pressure is less than initial pressure, then final volume must be more than initial volume. Therefore the pressure correction must be a ratio more than 1—the numerator must be larger than the denominator.

The above statements are the keys to the "reasoning" method of solving pressure-volume problems. By knowing the inverse character of the pressure-volume relationship you may *reason* whether final volume is more or less than initial volume, and thereby choose a pressure correction greater or less than one.

Solve the following example by the reasoning method:

---

**Example 11.1**  A certain gas sample occupies 3.25 liters at 740 torr. Find the volume of the gas sample if the pressure is changed to 790 torr. Temperature remains constant.

From the statement of the problem, does pressure increase or decrease?

------------------------------------------------------------

**11.1a**  It increases—from 740 to 790 torr.

Will this cause an increase or decrease in volume?

------------------------------------------------------------

**11.1b**  Decrease. Pressure and volume are inversely related: as one goes up, the other goes down.

The new volume will be found by applying a pressure correction to the initial volume—by multiplying the initial volume by a ratio of initial and final pressures. The ratio may be either $\frac{740 \text{ torr}}{790 \text{ torr}}$ or $\frac{790 \text{ torr}}{740 \text{ torr}}$. Bearing in mind that the final volume must be less than the initial volume, which ratio is correct?

------------------------------------------------------------

**11.1c**  $\frac{740 \text{ torr}}{790 \text{ torr}}$. If the final volume, the product of the multiplication, is to be less than the initial volume, the initial volume must be multiplied by a ratio less than 1.

Now complete the problem.

------------------------------------------------------------

**11.1d**  3.04 liters

$$3.25 \text{ liters} \times \frac{740 \ \cancel{\text{torr}}}{790 \ \cancel{\text{torr}}} = 3.04 \text{ liters}$$

---

Example 11.1 may also be solved by substitution into Equation 11.5 or 11.6. This "formula" method is equally valid and preferred by some. These alternatives will be mentioned as they arise. Which method is "better" is, of course, a matter of opinion. It is recommended that you use the method presented in your chemistry class. In this text we shall generally use the reasoning approach; but the formula method will be mentioned with each example for those who may prefer it.

---

**Example 11.2**   1.44 liters of gas at 0.935 atmosphere are compressed to a volume of 0.275 liter. Find the new pressure in atmospheres.

Qualitatively, will the reduction in volume cause an increase or decrease in pressure?

---

**11.2a**   Increase. Pressure and volume are inversely related.

The correction factor this time is a ratio of volumes. Complete the problem.

---

**11.2b**   4.90 atmospheres

$$0.935 \text{ atm} \times \frac{1.44 \text{ liters}}{0.275 \text{ liter}} = 4.90 \text{ atm}$$

This problem might also be solved by substitution into Equation 11.5.

---

## 11.5   ABSOLUTE TEMPERATURE

PG   11 E   Given a temperature in degrees Celsius (or degrees Kelvin), convert it to degrees Kelvin (or degrees Celsius).

We know that the temperature of things can be reduced until they become very cold. But how cold? Is there a bottom limit to temperature? Experiments suggest there is. Two such experiments performed by students in college laboratories are described in Figures 11.7 and 11.8. These experiments, as well as many far more sophisticated investigations of low temperature phenomena, predict an *absolute zero* temperature at −273°C.

The temperature scale adopted for the SI system of units is the **Kelvin temperature scale** (see page 43), which has its zero at −273°C. All temperatures on the Kelvin scale therefore have positive values. The size of the Kelvin degree is the same as the size of the Celsius degree. This leads to the equation

$$\text{temperature (°K)} = \text{temperature (°C)} + 273, \qquad (11.7)$$

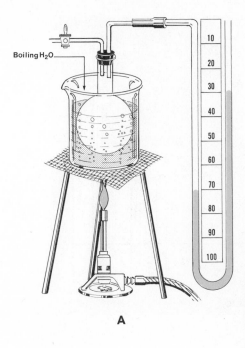

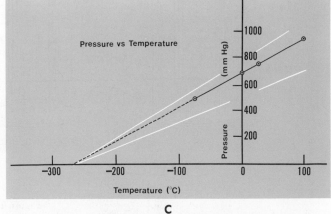

Pressure–Temperature Data

| TEMPERATURE (°C) | PRESSURE (torr) |
|---|---|
| 100 | 936 |
| 25 | 761 |
| 0 | 691 |
| −79 | 497 |

B

**Figure 11.7** Results of a student's experimental determination of the pressure-temperature relationship of a fixed quantity of gas at constant volume. The flask is immersed in a liquid bath at different temperatures (A). The pressures are measured at each temperature and tabulated (B). A graph of the data (C) is plotted in black. The white lines in C show what the graph would be if the experiment was repeated with larger or smaller quantities of gas. Extrapolation to zero pressure suggests an "absolute zero" temperature at −273°C.

by which temperatures expressed in either scale may be converted to the other (see Fig. 11.9). From Equation 11.7 the freezing point of water, 0°C, is 273°K, or "two hundred seventy three degrees Kelvin," as it traditionally is written and spoken. Under SI, the word "degree" and its symbol are dropped, leading to 273K and "two hundred seventy three Kelvin." We will continue to use "degree" in this text.

**Example 11.3** Tomorrow's weather forecast is for a high temperature of 77°F, or 25°C. Express this temperature in degrees Kelvin.

-------------------------------------------------------------------------------------

**11.3a** 298°K

$$25° + 273 = 298°$$

To understand why there is an absolute zero in temperature it is necessary to recognize what is measured by temperature. Experiments indicate that temperature is a measure of the average kinetic energy of the particles

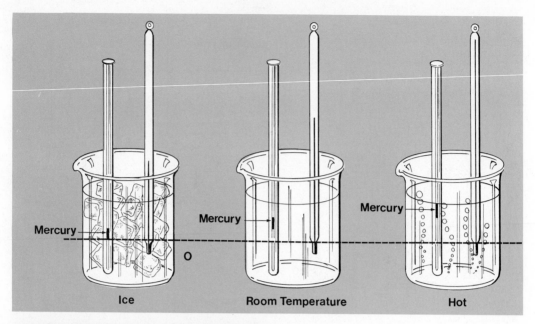

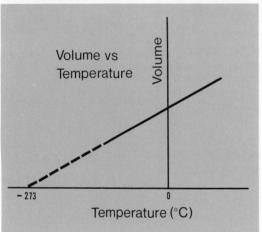

Volume vs Temperature

**Figure 11.8** Student experiment to determine the temperature-volume relationship of a fixed quantity of gas at constant pressure. A plug of mercury is drawn into a fine bore glass tube. The tube is sealed at one end, trapping a fixed quantity of air. When the tube is held vertically, open end up, the gas pressure is constant at atmospheric pressure plus the height of the mercury plug. The tube is then immersed in water at different temperatures, as shown. The mercury plug moves up or down as it will to maintain constant gas pressure at the different gas temperatures. Gas volume is measured at each temperature. A graph of volume vs. temperature is similar to the graph in Figure 11.7, again predicting an absolute zero at −273°C.

in a sample of matter. Kinetic energy is energy of motion, and is expressed mathematically as $\frac{1}{2}mv^2$, where m is the mass of the particle and v is its velocity. There is no reason to believe that the mass of a particle changes as temperature is reduced, so we conclude that particle velocity becomes less at lower temperatures. At absolute zero we imagine that all molecular movement stops.

Research with gases at absolute zero is impossible because the gases condense to liquids before absolute zero is reached. Other experiments support the concept of absolute zero, however, and fix the temperature somewhat more accurately at −273.16°K. No attempt to penetrate this bottom temperature barrier has ever been successful, although some researchers claim to have reached within 0.0014°K of the theoretical zero.

## 11.6  CHARLES' LAW

> PG 11 F  Given the initial volume, and initial and final temperatures of a fixed quantity of gas at constant pressure, calculate the final volume.

A graph of volume or pressure vs. *absolute* temperature may be obtained by shifting the vertical axes of the graphs in Figures 11.7 and 11.8 to −273°C. This is shown in Figure 11.10. A straight line passing through the origin is the graph of a direct proportionality. We conclude, therefore, that $V \propto T$ and $P \propto T$, where V is volume, T is absolute temperature and P is pressure. Applying a proportionality constant to the volume-temperature relationship yields

$$V = k_2 \, T \qquad (11.8)$$

Equation 11.8 is the mathematical expression of what is known as Charles' Law: **the volume of a fixed quantity of gas at constant pressure is proportional to absolute temperature.** The physical significance of Charles' Law is illustrated in Figure 11.11.

Dividing both sides of Equation 11.8 by T yields

$$\frac{V}{T} = k_2 \qquad (11.9)$$

From this it follows that $\dfrac{V_1}{T_1} = k_2 = \dfrac{V_2}{T_2}$, or

$$\frac{V_1}{T_1} = \frac{V_2}{T_2} \qquad (11.10)$$

where the subscripts 1 and 2 again refer to first and second measurements of the two variables. Solving for $V_2$,

$$V_2 = V_1 \times \frac{T_2}{T_1} \qquad (11.11)$$

Thus, if the temperature of a confined gas is changed at constant pressure, the final volume may be found by multiplying the initial volume by a *ratio of absolute temperatures*—a temperature correction. Again there are two possibilities, this time dictated by the *direct* proportionality between temperature and volume:

1. If final temperature is greater than initial temperature, then final volume must be greater than initial volume. Therefore the temperature correction must be a ratio more than 1—the numerator must be larger than the denominator.

2. If final temperature is less than initial temperature, then final volume

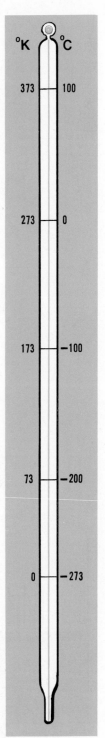

**Figure 11.9**  Kelvin (absolute) and Celsius temperature scales compared.

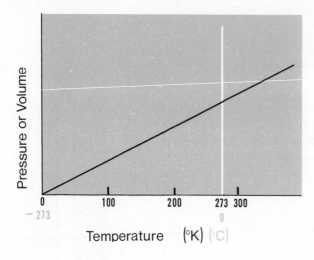

**Figure 11.10** Graph of pressure or volume vs. absolute temperature, with other variables held constant. This graph is produced from graphs in Figures 11.7 and 11.8, with vertical axis moved to absolute zero at −273°K. The straight line through the origin shows the pressure to be directly proportional to the absolute temperature at constant volume and quantity, and volume to be directly proportional to absolute temperature at constant pressure and quantity.

must be less than initial volume. Therefore the temperature correction must be a ratio less than 1—the numerator must be smaller than the denominator.

One extremely important fact must be remembered in working gas law problems involving temperature: the proportional relationships apply to *absolute* temperatures rather than to Celsius temperatures, which are frequently given in the statement of the problem. Before solving gas law problems, Celsius temperatures must be converted to °K.

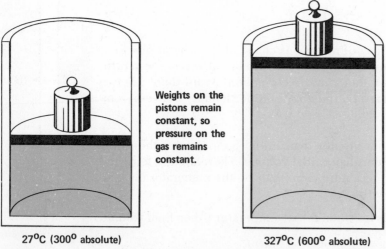

Weights on the pistons remain constant, so pressure on the gas remains constant.

27°C (300° absolute)    327°C (600° absolute)

**Figure 11.11** Illustration of Charles' Law. If the absolute temperature is doubled, the volume is doubled at constant pressure and amount of gas.

**Example 11.4**  2.42 liters of a gas, measured at 22°C, are heated to 45°C at constant pressure. What will be the new volume of the gas?

From the conditions of the problem, will the final volume be more or less than 2.42 liters?

---

**11.4a**  More. Gas volume varies directly with temperature; if temperature increases, volume must also increase.

What temperature ratio will you use as a multiplier to find the new volume?

---

**11.4b**  $\dfrac{318°K}{295°K}$.  $\dfrac{(45 + 273)°K}{(22 + 273)°K} = \dfrac{318°K}{295°K}$.

Always be sure to convert to °K for gas law problems.

Complete the problem: find the final volume.

---

**11.4c**  2.61 liters

$$2.42 \text{ liters} \times \frac{318°\cancel{K}}{295°\cancel{K}} = 2.61 \text{ liters}$$

Direct substitution into Equation 11.11 would yield the same result.

## 11.7  COMBINED GAS LAWS

PG   11  G   For a fixed quantity of a confined gas, given the initial volume, pressure and temperature, and the final pressure and temperature, calculate the final volume.

The pressure-volume-temperature relationships we have been considering can be combined to yield

$$\frac{PV}{T} = K^* \tag{11.12}$$

---

*As $\dfrac{V}{T} = k_2$ was derived from $V \propto T$, so $\dfrac{P}{T} = k_3$ may be derived from $P \propto T$. If the left sides of this equation and Equations 11.4 and 11.9 are multiplied and equated to the similar product of their right sides,

$$\frac{P^2V^2}{T^2} = K^2$$

The product $(k_1)(k_2)(k_3)$ is a constant. Let it be $K^2$. Then

$$\frac{P^2V^2}{T^2} = (k_1)(k_2)(k_3)$$

Taking the square root of both sides yields Equation 11.12.

and

$$\frac{P_1 V_1}{T_1} = \frac{P_2 V_2}{T_2} \qquad (11.13)$$

Solving this equation for $V_2$,

$$V_2 = V_1 \times \frac{P_1}{P_2} \times \frac{T_2}{T_1} \qquad (11.14)$$

From Equation 11.14, it appears that a final volume may be found from initial volume by applying a pressure correction and a temperature correction, both of which may be reasoned out just as we have done them in the previous examples. To illustrate . . . .

---

**Example 11.5**  A certain gas occupies 3.40 liters at 65°C and 680 torr. What volume will it occupy if it is cooled to room temperature, 21°C, and compressed to 800 torr?

We shall approach this problem by first setting up for the volume change caused by a change in pressure, holding temperature constant; and then determining the further volume change caused by a change in temperature, holding pressure constant. By steps, what will happen to the volume because of the change in pressure from 680 torr to 800 torr, considering temperature constant? Will volume increase or decrease?

-------------------------------------------------------------------------------

**11.5a**  Volume will *decrease* if pressure increases: they are inversely proportional.

Now begin the setup of the problem with the initial volume multiplied by the proper pressure correction. Do not solve.

-------------------------------------------------------------------------------

**11.5b**  $3.40 \text{ liters} \times \frac{680 \text{ torr}}{800 \text{ torr}}$

3.40 liters is the volume at 680 torr and 65°C. Solving the setup as far as it is written would give the volume at 800 torr and 65°C. Now extend the setup by applying the proper temperature correction to get the volume at 800 torr and 21°C. Solve for the answer.

-------------------------------------------------------------------------------

**11.5c**  2.51 liters

$$3.40 \text{ liters} \times \frac{680 \text{ torr}}{800 \text{ torr}} \times \frac{294°\text{K}}{338°\text{K}} = 2.51 \text{ liters}$$

Reducing temperature reduces volume; they are directly proportional. The temperature correction is therefore less than 1.

Direct substitution into Equation 11.14 would reproduce the above setup and yield the same result.

You now have two ways to solve volume problems by the gas laws: you may *reason* your way through temperature and pressure corrections, or you may solve the problems algebraically by Equation 11.13. If your instructor states a preference, by all means adopt it, at least for the present. If you must memorize an equation, either by your own choice or by teacher direction, Equation 11.13 is recommended. With it you may solve for any variable, given the other five.

## 11.8 STANDARD TEMPERATURE AND PRESSURE

PG  11 H    Given the volume of a gas at one temperature and pressure (or at STP), find the volume it would occupy at STP (or at a stated temperature and pressure).

Because of the interdependence of temperature, pressure and volume of a confined gas, it is not possible to specify the *quantity* of gas in volume units without also specifying the temperature and pressure. Volume may be used as a quantity unit, however, if it is referred to arbitrarily established standards of temperature and pressure. The commonly accepted "standard conditions," or "standard temperature and pressure, STP," as they are called and frequently abbreviated, are 0°C (273°K) and 1.00 atmosphere (760 torr). Many gas law applications require conversion of one of the variables, usually volume, to or from STP. The problems are solved in the same manner as those just discussed.

---

**Example 11.6**  What would be the volume at STP of 4.06 liters of nitrogen, measured at 712 torr and 28°C?

Set up the problem in its entirety and solve.

- - - - - - - - - - - - - - - - - - - - - - - - - - - - - - - - - - - - - - - - - - - - - - - - - - - - - - - - - - - - - - - - - - - - - - - - - - - - - -

**11.6a**  3.45 liters

$$4.06 \text{ liters} \times \frac{712 \text{ torr}}{760 \text{ torr}} \times \frac{273°\text{K}}{301°\text{K}} = 3.45 \text{ liters}$$

---

## 11.9  THE IDEAL GAS EQUATION

### *DEVELOPMENT OF THE IDEAL GAS EQUATION*

PG  11 I    Write the equation that relates the measurable properties of an ideal gas to each other (the ideal gas equation).

    11 J    Write the variation of the ideal gas equation that includes the mass of the gas sample and its molar weight.

To this point our quantitative consideration of gases has been restricted to a fixed quantity. Figure 11.12 describes an experiment in which the rela-

.**A**

Mass–Pressure Data

Volume = 0.050 liter                                    Temperature = 23°C = 296°K

| DATA | | | RESULTS | |
|---|---|---|---|---|
| WEIGHT *(g)* | PRESSURE *(atm)* | OXYGEN *(g)* | OXYGEN *(moles)* | $R$ $\left(\dfrac{liter \times atm}{°K \times moles}\right)$ |
| 1412.07 | 0.00 | 0.00 | 0.00 | — |
| 1412.40 | 0.50 | 0.33 | 0.010 | 0.084 |
| 1412.72 | 1.00 | 0.65 | 0.020 | 0.084 |
| 1413.07 | 1.50 | 1.00 | 0.0313 | 0.081 |
| 1413.41 | 2.00 | 1.34 | 0.0419 | 0.081 |
| 1413.72 | 2.50 | 1.65 | 0.0515 | 0.082 |

**B**

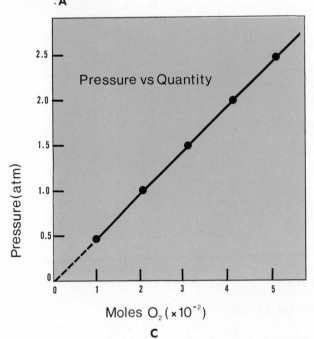

**C**

**Figure 11.12** The relationship between quantity and pressure of a gas at constant volume and constant temperature. A metallic gas weighing bottle (A) with a pressure gauge attached has a fixed volume of 0.50 liter at room temperature, 23°C. When evacuated (pressure = 0.00 atm), its mass is measured at 1412.07 grams and recorded in the table (B). Oxygen is introduced until the pressure reaches 0.50 atmosphere at room temperature. The mass is measured and recorded. The procedure is repeated at 0.50 atmosphere intervals to a total pressure of 2.50 atmospheres. Subtracting the mass of the empty bottle from each recorded weighing gives the mass of the oxygen in the bottle. The mass may be converted to moles by dividing by the molecular weight of oxygen, 32.0 g/mole. The straight line graph of pressure vs. moles (C) indicates a direct proportionality between these variables. Data from the experiment can be used to find an experimental value of R, the ideal gas constant (see accompanying text).

tionship between pressure and quantity is determined. While the experiment is pictured as being conducted with a single gas, it is a fact that the same experiment performed with any other gas whose behavior over the temperature and pressure range approximates ideal gas behavior would yield identical results. The conclusion drawn from such experiments is that pressure is directly proportional to the number of moles of gas present when temperature and volume are held constant. If n is the number of moles, the proportionality is

$$P \propto n \qquad (11.15)$$

We have seen previously that pressure is proportional to $\frac{1}{V}$ (page 240) and also to T (page 245). A general proportionality that combines all these (see Appendix I, page 499) is

$$P \propto \frac{nT}{V} \qquad (11.16)$$

Introducing a proportionality constant, R, gives what is known as the **ideal gas equation:**

$$P = \frac{nRT}{V} \qquad (11.17)$$

which is more frequently written in the form

$$PV = nRT \qquad (11.18)$$

The constant, R, is referred to as the **universal gas constant.** Solving Equation 11.18 for R gives

$$R = \frac{PV}{nT} \qquad (11.19)$$

It is an experimental fact that at STP 1.00 mole of any gas approximating ideal gas behavior occupies a volume of 22.4 liters. Substituting these values of pressure, volume, moles and absolute temperature into Equation 11.19 gives both the values and units of R:

$$R = \frac{1.00 \text{ atm} \times 22.4\ell}{1.00 \text{ mole} \times 273°K} = 0.0821 \frac{(\ell)\,(\text{atm})}{(\text{mole})\,(°K)} \qquad (11.20)$$

$$R = \frac{760 \text{ torr} \times 22.4\ell}{1.00 \text{ mole} \times 273°K} = 62.4 \frac{(\ell)\,(\text{torr})}{(\text{mole})\,(°K)} \qquad (11.21)$$

The choice between these values of R for a given problem is dictated by the units in the problem. *It is essential that the measurement units of pressure, volume, temperature and quantity correspond with the units of R.*

A useful variation of the ideal gas equation may be derived as follows: If the weight of any chemical species, g, is divided by the molar weight, MW, the quotient is the number of moles: $\frac{\text{grams}}{\text{grams/mole}}$ = moles. Therefore $\frac{g}{MW}$ may be substituted for its equivalent, n, in Equation 11.18:

$$PV = \frac{g}{MW} RT \qquad (11.22)$$

A gas that conforms to the ideal gas equation is called, logically enough, an ideal gas. Real gases, however, deviate somewhat from ideal behavior. These deviations are significant at the relatively high pressures and/or low temperatures which cause a gas to condense to a liquid. Under these circumstances intermolecular attractions are large enough to violate the conceptual model of a gas. Fortunately deviations from the ideal gas equation are negligible for many common gases over broad ranges of temperature and pressure. The equation therefore provides a satisfactory basis for most quantitative work with gases.

## APPLICATIONS OF THE IDEAL GAS EQUATION

PG 11 K   Given the pressure, temperature and volume (or number of moles) of a gas, calculate the number of moles (or volume).

11 L   Given the pressure, temperature and density of a gas—or the mass of a known volume—calculate the molar weight.

In using the ideal gas equation, Equation 11.18 or 11.22, the recommended procedure is to solve the equation algebraically for the desired quantity, substitute known values of the other variables, and solve. Include and cancel units in the usual way; it is your best check on the correctness of your setup. While the equations may be used to determine the value of any unknown when the others are given, we shall limit our consideration to the calculation of moles, volume and molar weight.

---

**Example 11.7**   What volume will be occupied by 0.393 mole of nitrogen at 738 torr and 24°C?

*Solution.*   We begin by solving Equation 11.18 for the required volume:

$$V = \frac{nRT}{P}$$

Notice that pressure is given in torr. We therefore use the value of R in which torr is the pressure unit, Equation 11.21. Substituting this and other given data,

$$V = \frac{nRT}{P} = \frac{0.393 \text{ mole} \times \frac{62.4 \text{ (liter) (torr)}}{\text{(mole) (°K)}} \times (273 + 24)°K}{738 \text{ torr}}$$

$$= \frac{0.393 \text{ mole}}{738 \text{ torr}} \times \frac{62.4 \text{ (liter) (torr)}}{\text{(mole) (°K)}} \times (273 + 24)°K = 9.87 \text{ liters}$$

---

**Example 11.8**   How many moles of ammonia are in a 5.00 liter gas cylinder at 18°C if they exert a pressure of 8.65 atmospheres?

The method is again direct: solve Equation 11.18 for n, substitute and compute the answer.

**11.8a**  1.81 moles $NH_3$

$$n = \frac{PV}{RT} = \frac{8.65 \text{ atm} \times 5.00 \text{ liters}}{0.0821 \frac{(\text{liter})(\text{atm})}{(\text{mole})(°K)} \times (273 + 18)°K}$$

$$= 8.65 \text{ atm} \times \frac{(\text{mole})(°K)}{0.0821 (\text{liter})(\text{atm})} \times \frac{5.00 \text{ liters}}{291°K} = 1.81 \text{ moles}$$

One of the most useful applications of the ideal gas equation is to determine the molar weight of an unknown substance in the vapor state. The following example illustrates the method:

**Example 11.9**  1.67 grams of an unknown liquid are vaporized at a temperature of 125°C. Its volume is measured as 0.421 liter at 749 torr. Calculate the molar weight.

Using Equation 11.22 we can solve for the molar weight, substitute and calculate the answer. Complete the problem.

------------------------------------------------

**11.9a**  132 g/mole

$$MW = \frac{gRT}{PV} = \frac{1.67 \text{ g} \times \frac{62.4 (\text{liter})(\text{torr})}{(\text{mole})(°K)} \times (125 + 273)°K}{749 \text{ torr} \times 0.421 \text{ liter}}$$

$$= \frac{1.67 \text{ g}}{749 \text{ torr}} \times \frac{398°K}{0.421 \text{ liter}} \times \frac{62.4 (\text{liter})(\text{torr})}{(\text{mole})(°K)} = 132 \text{ g/mole}$$

**Example 11.10**  Find the molar weight of an unknown gas if its density is 1.45 grams/liter at 25°C and 756 torr.

Density does not appear in the ideal gas equation as such, but its component units, grams/liter, do. The problem is solved just as the last one, interpreting density as the mass of 1.45 grams and the volume as 1.00 liter. Complete the problem.

------------------------------------------------

**11.10a**  35.7 grams/mole

$$MW = \frac{gRT}{PV} = \frac{1.45 \text{ g}}{1.00 \text{ liter}} \times \frac{62.4 (\text{liter})(\text{torr})}{(\text{mole})(°K)} \times \frac{298°K}{756 \text{ torr}} = 35.7 \text{ g/mole}$$

## 11.10 MOLAR VOLUME AT STANDARD TEMPERATURE AND PRESSURE

PG  11 M   Define molar volume. State the molar volume of any gas at STP.

11 N   Given the volume (or number of moles) of any gas at STP, find the number of moles (or volume).

11 O   Given two of the following for any gas at STP, find the third: grams, volume, molar weight.

11 P   Given gas density at STP (or molar weight), find molar weight (or gas density at STP).

**Molar volume** is comparable to molar weight: as molar weight represents grams per mole, molar volume is liters per mole. An expression for molar volume may be found by solving Equation 11.18 for $\frac{V}{n}$:

$$\frac{V}{n} = \frac{RT}{P} \qquad (11.23)$$

From this it is readily apparent that the volume occupied by one mole of *any gas* depends upon the gas pressure and temperature. The value of molar volume at standard temperature and pressure, 0°C (273°K) and 1 atmosphere, may be found by substituting appropriate figures into Equation 11.23 and solving:

$$\frac{V}{n} = \frac{0.0821 \text{ (liter) (atm)}}{\text{(mole) (°K)}} \times \frac{273°\text{K}}{1 \text{ atm}} = 22.4 \text{ liters/mole}$$

The resulting equivalence, 22.4 $\ell$ (gas at STP) ≃ 1 mole, provides a unit path from $\ell$ (gas at STP) → moles, or moles → $\ell$ (gas at STP).

Because standard temperature and pressure are so frequently used as reference conditions for gases, the molar volume at these conditions is a useful quantity. You must be aware, however, of the restrictions placed on the value, 22.4 liters per mole. It is the molar volume of a *gas*—never a solid or a liquid—if *that volume is measured at standard temperature and pressure,* not some other combination of temperature and pressure. Do not use 22.4 liters per mole unless these conditions are satisfied.

Three quantities are usually involved in molar volume problems. They are density, measured in $\frac{\text{grams}}{\text{liter}}$; molar weight, $\frac{\text{grams}}{\text{mole}}$; and molar volume, 22.4 $\frac{\text{liters}}{\text{mole}}$ at STP. While density and molar weight are unique for each gas, molar volume at STP is 22.4 liters/mole for all gases. The use of molar volume is illustrated by the following examples:

**Example 11.11**   Find the volume of 0.350 mole of helium at STP.

At 22.4 liters per mole, the calculation for this problem should be apparent. Solve completely.

------------------------------------------------------------------------

**11.11a**   7.84 liters

$$\boxed{0.350 \; \cancel{mole}} \times \frac{22.4 \text{ liters}}{1 \; \cancel{mole}} = 7.84 \text{ liters}$$

**Example 11.12**   Find the volume of 12.0 grams of oxygen at STP.

The only difference between this example and the one before it is that the quantity is given in grams rather than moles. Conversion of 12.0 grams of oxygen to moles is straightforward, and converting moles to liters as in the previous example completes the problem.

------------------------------------------------------------------------

**11.12a**   8.40 liters $O_2$

$$\boxed{12.0 \; \cancel{g \; O_2}} \times \frac{\cancel{\text{moles } O_2}}{32.0 \; \cancel{g \; O_2}} \times \frac{22.4 \text{ liters}}{1 \; \cancel{mole}} = 8.40 \text{ liters}$$

**Example 11.13**   Find the density of ammonia, $NH_3$, at STP.

This time there is no "given quantity." But we do know, or can find, two things about ammonia. Its molar volume at STP is 22.4 liters/mole, and its molar weight is 17.0 grams per mole. The mole is the connecting link. One mole weighs 17.0 grams, and one mole—the same quantity—occupies 22.4 liters. We seek the density, grams/liter. With that information you can complete the problem.

------------------------------------------------------------------------

**11.13a**   0.759 gram/liter

$$\frac{17.0 \text{ grams}}{22.4 \text{ liters}} = 0.759 \text{ gram/liter}$$

Without the reasoning shown above, the problem may be solved strictly from the units. Knowing grams are in the numerator of the answer, it is reasonable to assume grams will be in the numerator of the setup of the problem. Starting with molar weight, $\frac{\text{grams}}{\text{mole}}$, by what must we multiply to get $\frac{\text{grams}}{\text{liter}}$?

Apparently moles in the denominator of $\frac{\text{grams}}{\text{mole}}$ must be replaced by liters—and we know the relationship between liters and moles of gas at STP. Therefore

$$\frac{17.0 \text{ grams}}{1 \text{ mole}} \times \frac{1 \text{ mole}}{22.4 \text{ liters}} = 0.759 \text{ gram/liter}$$

---

**Example 11.14**  The density of an unknown gas at STP is 1.25 grams per liter. Estimate the molar weight of the gas.

If 1 liter weighs 1.25 grams, and there are 22.4 liters in 1 mole, what is the weight of 1 mole? Set up and solve.

-----------------------------------------------------------------------------------

**11.14a**  28.0 grams/mole

$$\frac{1.25 \text{ grams}}{1 \text{ liter}} \times \frac{22.4 \text{ liters}}{1 \text{ mole}} = 28.0 \text{ grams/mole}$$

## 11.11  GAS STOICHIOMETRY

PG  11  Q    For a chemical reaction for which the equation may be written, given the grams of any species OR the volume of any gaseous species at specified temperature and pressure, calculate the grams of any other species OR the volume of any gaseous species at specified temperature and pressure.

In Section 8.2, page 169, a three-step pattern was established for the solving of stoichiometry problems. It is repeated here for your ready reference:

1. Convert the quantity of the given species to moles.
2. Convert the moles of given species to moles of wanted species.
3. Convert the moles of wanted species to the quantity units required.

In Chapter 8 the only measurable quantity unit considered was grams. Molar weight—grams per mole—was the vehicle by which conversion was made between moles and grams in completing the first or third step in the stoichiometric pattern. We are now ready to apply the pattern to a second quantity unit, volume of gas. *If* the gas is measured at STP, molar volume—22.4 liters per mole for *any* gas—becomes a second vehicle by which conversion may be made between a quantity measurement and moles. The thought process is identical. Let's see how it works . . . .

**Example 11.15**  Hydrogen gas is released when sodium reacts with water: $2 \text{ Na (s)} + 2 \text{ HOH (l)} \rightarrow \text{H}_2 \text{ (g)} + 2 \text{ NaOH (aq)}$. What volume of $\text{H}_2$, measured at STP, will be liberated by 8.62 grams of sodium?

*Solution.* The first step is to convert the given quantity of moles:

$$\boxed{8.62 \ \cancel{g \ Na}} \times \frac{1 \ \text{mole Na}}{23.0 \ \cancel{g \ Na}}$$

Using equation coefficients, we now extend the setup to moles of $H_2$ (Step 2):

$$\boxed{8.62 \ \cancel{g \ Na}} \times \frac{1 \ \cancel{\text{mole Na}}}{23.0 \ \cancel{g \ Na}} \times \frac{1 \ \text{mole } H_2}{2 \ \cancel{\text{moles Na}}}$$

Finally we convert the moles of wanted species to grams (Step 3), using molar volume at STP, as in Example 11.11, page 255:

$$\boxed{8.62 \ \cancel{g \ Na}} \times \frac{1 \ \cancel{\text{mole Na}}}{23.0 \ \cancel{g \ Na}} \times \frac{1 \ \cancel{\text{mole } H_2}}{2 \ \cancel{\text{moles Na}}} \times \frac{22.4 \ \text{liters}}{1 \ \cancel{\text{mole}}} = 4.20 \ \text{liters } H_2$$

---

**Example 11.16** Calculate the number of grams of oxygen that are required to react with sulfur dioxide to yield 18.6 liters of sulfur trioxide, measured at STP: $2 \ SO_2 \ (g) + O_2 \ (g) \rightarrow 2 \ SO_3 \ (g)$.

Begin with the given quantity and set up the three conversions of the stoichiometric pattern. Solve completely.

---

**11.16a** 13.3 grams $O_2$

$$\boxed{18.6 \ \cancel{\text{liters } SO_3}} \times \frac{1 \ \cancel{\text{mole } SO_3}}{22.4 \ \cancel{\text{liters}}} \times \frac{1 \ \cancel{\text{mole } O_2}}{2 \ \cancel{\text{moles } SO_3}} \times \frac{32.0 \ \text{g } O_2}{1 \ \cancel{\text{mole } O_2}} = 13.3 \ \text{g } O_2$$

In this example your gas conversion was from liters to moles, dividing by the molar volume of 22.4 liters/mole.

Other techniques are required if the gas volume is measured at non-STP conditions. The method we will consider will divide the stoichiometric pattern into two parts. In one part the volume-mole conversion will be accomplished by means of the ideal gas law. The other two steps of the problem will be completed in the usual manner.

**Example 11.17** How many grams of ammonia can be produced by the reaction of 3.85 liters of hydrogen, measured at 15.0 atmospheres and 85°C? The equation is $N_2 \ (g) + 3 \ H_2 \ (g) \rightarrow 2 \ NH_3 \ (g)$.

*Solution.* Step 1 of the stoichiometric pattern calls for the conversion of the given quantity, a volume of hydrogen at non-STP conditions, to moles. The ideal gas equation may be used for this purpose, as it was in Example 11.8, page 252:

$$n = \frac{PV}{RT} = \frac{(\text{mole}) \ (°K)}{0.0821 \ (\cancel{\ell}) \ (\cancel{\text{atm}})} \times \frac{15.0 \ \cancel{\text{atm}}}{(273 + 85) \ (\cancel{°K})} \times 3.85 \ \cancel{\ell} = 1.96 \ \text{moles}$$

The remaining steps of the stoichiometric pattern are conversion of moles of hydrogen to moles of ammonia via the equation quantities, and then moles of ammonia to grams by molar weight:

$$1.96 \text{ moles } H_2 \times \frac{2 \text{ moles } NH_3}{3 \text{ moles } H_2} \times \frac{17.0 \text{ g } NH_3}{1 \text{ mole } NH_3} = 22.2 \text{ g } NH_3$$

---

**Example 11.18**   How many liters of $CO_2$, measured at 740 torr and 130°C, will be produced by the complete burning of 16.2 grams of butane, $C_4H_{10}$?

To begin, you need an equation . . . .

-------------------------------------------------------------------------------

**11.18a**   $C_4H_{10} \text{ (g)} + {}^{13}/_2 \, O_2 \text{ (g)} \rightarrow 4 \, CO_2 \text{ (g)} + 5 \, H_2O \text{ (g)}$

The stoichiometric pattern takes you from grams of given species to moles of wanted species. The gas law equation guides you from moles of wanted species to volume. Complete the problem.

-------------------------------------------------------------------------------

**11.18b**   38.1 liters $CO_2$

$$16.2 \text{ g } C_4H_{10} \times \frac{1 \text{ mole } C_4H_{10}}{58.0 \text{ g } C_4H_{10}} \times \frac{4 \text{ moles } CO_2}{1 \text{ mole } C_4H_{10}} = 1.12 \text{ moles } CO_2$$

$$V = \frac{nRT}{P} = \frac{1.12 \text{ moles } CO_2}{740 \text{ torr}} \times \frac{62.4 \text{ (liter) (torr)}}{(°K) \text{ (mole)}} \times (273 + 130)°K$$

$$= 38.1 \text{ liters } CO_2$$

There is an alternative approach to Examples 11.17 and 11.18 that is preferred by some teachers and students. Applied to Example 11.17, instead of using the ideal gas equation to convert a given volume of gas at non-STP conditions to moles, the given volume is first converted to STP by pressure and temperature corrections, as in Section 11.7 (page 248):

P Correction      T Correction

$$3.85 \text{ liters } H_2 \times \frac{15.0 \text{ atm}}{1.00 \text{ atm}} \times \frac{273°K}{(273 + 85)°K}$$

Starting with this volume at STP, the remainder of the problem carries through the three steps of the stoichiometric pattern:

$$3.85 \text{ liters } H_2 \times \frac{15.0 \text{ atm}}{1.00 \text{ atm}} \times \frac{273°K}{(273 + 85)°K} \times \frac{1 \text{ mole } H_2}{22.4 \text{ liters } H_2}$$

P Correction      T Correction            Step 1

$$\times \frac{2 \text{ moles } NH_3}{3 \text{ moles } H_2} \times \frac{17.0 \text{ g } NH_3}{1 \text{ mole } NH_3} = 22.3 \text{ grams } NH_3$$

Step 2                Step 3

Applying this method to Example 11.18, the volume of product gas at STP is first set up as in Example 11.15 and then converted to the required pressure and temperature by appropriate corrections:

$$16.2 \text{ g } \cancel{C_4H_{10}} \times \frac{1 \text{ mole } \cancel{C_4H_{10}}}{58.0 \text{ g } \cancel{C_4H_{10}}} \times \frac{4 \text{ moles } \cancel{CO_2}}{1 \text{ mole } \cancel{C_4H_{10}}} \times \frac{22.4 \text{ liters } CO_2}{1 \text{ mole } \cancel{CO_2}}$$

$$\qquad\qquad\qquad\text{Step 1}\qquad\qquad\text{Step 2}\qquad\qquad\text{Step 3}$$

$$\times \frac{760 \cancel{\text{ torr}}}{740 \cancel{\text{ torr}}} \times \frac{(273 + 130)°\cancel{K}}{273°\cancel{K}} = 37.9 \text{ liters } CO_2$$

$$\quad\text{P Correction}\qquad\text{T Correction}$$

## AVOGADRO'S HYPOTHESIS: VOLUME-VOLUME PROBLEMS

PG 11 R State Avogadro's Hypothesis regarding gas volumes and number of molecules.

Solving the ideal gas equation, Equation 11.18, page 251, for n, we obtain

$$n = \frac{PV}{RT} \qquad\qquad (11.24)$$

For a given value of pressure and temperature, $\frac{P}{RT}$ is a constant.

Therefore $\qquad\qquad n = \text{constant} \times V \qquad\qquad (11.25)$

The ideal gas equation holds for *all* gases. It follows that the "constant" in the equation will be the same for all gases at the same temperature and pressure. We therefore conclude that **equal volumes of all gases at the same temperature and pressure contain the same number of molecules.** This statement is known as Avogadro's Hypothesis.

Avogadro's Hypothesis leads us to realize that the volume ratios of the *gaseous* species in a chemical reaction, *provided they are measured at the same temperature and pressure,* are identical to the mole ratios appearing as coefficients in the chemical equation. In the burning of ethene, for example, it takes three moles of oxygen to burn one mole of ethene, according to the equation $C_2H_4 \text{ (g)} + 3 \text{ } O_2 \text{ (g)} \rightarrow 2 \text{ } CO_2 \text{ (g)} + 2 \text{ } H_2O \text{ (l)}$. At STP three moles of oxygen would occupy $3 \times 22.4 = 67.2$ liters, while the single mole of ethene would occupy $1 \times 22.4 = 22.4$ liters. The volume ratio would be

$$\frac{67.2}{22.4} = \frac{22.4 \times 3}{22.4 \times 1} = \frac{3}{1},$$

the same as the mole ratio. Consequently, as 1 mole $C_2H_4 \simeq 3$ moles $O_2$, we can say 1 liter $C_2H_4 \simeq 3$ liters $O_2$. This provides a short cut in solving stoichiometry problems involving volumes of two gases measured at the same temperature and pressure.

---

**Example 11.19**  1.30 liters of ethene, $C_2H_4$, are burned completely. What volume of oxygen is required, if both gas volumes are measured at STP? The equation is

$$C_2H_4 \text{ (g)} + 3 \ O_2 \text{ (g)} \rightarrow 2 \ CO_2 \text{ (g)} + 2 \ H_2O \text{ (l)}$$

The equivalence 1 liter $C_2H_4 \simeq 3$ liters $O_2$ tells us the volume of oxygen is three times the volume of ethene. Set up and solve.

---

**11.19a**  3.90 liters $O_2$

$$1.30 \ \text{liters } C_2H_4 \times \frac{3 \text{ liters } O_2}{1 \text{ liter } C_2H_4} = 3.90 \text{ liters } O_2$$

This result is confirmed if the problem is solved by the full three steps of the stoichiometric pattern:

$$1.30 \ \text{liters } C_2H_4 \times \frac{1 \text{ mole } C_2H_4}{22.4 \text{ liters } C_2H_4} \times \frac{3 \text{ moles } O_2}{1 \text{ mole } C_2H_4} \times \frac{22.4 \text{ liters } O_2}{1 \text{ mole } O_2}$$

$$= 3.90 \text{ liters } O_2$$

---

Be sure to recognize the restriction on this simplified solution process: both gas volumes *must be measured at the same temperature and pressure*, but not necessarily STP as in the above example.

---

**Example 11.20**  When oxygen comes into contact with nitrogen monoxide, nitrogen dioxide is produced:

$$2 \ NO \text{ (g)} + O_2 \text{ (g)} \rightarrow 2 \ NO_2 \text{ (g)}$$

How many liters of nitrogen dioxide will form by the reaction of 4.30 liters of oxygen if both volumes are measured at the same temperature and pressure?

---

**11.20a**  8.60 liters $NO_2$

$$4.30 \ \text{liters } O_2 \times \frac{2 \text{ liters } NO_2}{1 \text{ liter } O_2} = 8.60 \text{ liters } NO_2$$

Sometimes the given and wanted gas volumes are at different temperatures and pressures. The procedure for this kind of problem is to convert the volume of the *given* species from its temperature and pressure to the volume it *would occupy* at the temperature and pressure specified for the *wanted* species, using temperature and pressure correction ratios. From that point the solution is as in the last two examples. To illustrate . . . .

---

**Example 11.21** 1.75 liters of oxygen, measured at 24°C and 755 torr, are consumed in burning sulfur. At one point in the exhaust hood the sulfur dioxide produced is at 165°C and a pressure of 785 torr. Find the volume of the sulfur dioxide at those conditions. The equation is

$$S \text{ (s)} + O_2 \text{ (g)} \rightarrow SO_2 \text{ (g)}$$

First, we find the volume that would be occupied by the oxygen at 165°C and 785 torr. Example 11.5, page 248, is similar, so you may find it helpful to look back. Set up that far, but do not solve.

---

**11.21a**     $\boxed{1.75 \text{ liters O}_2} \times \dfrac{755 \text{ torr}}{785 \text{ torr}} \times \dfrac{438°K}{297°K}$

The pressure change is from 755 torr to 785 torr. An increase in pressure reduces volume, so the pressure correction is less than 1. Temperature increases from 297°K to 438°K. A temperature rise increases volume, so the temperature correction is more than 1.

From here the problem is of the form, "How many liters of $SO_2$ are equivalent to the volume of $O_2$ in the above setup, both gases measured at the same temperature and pressure?" Extend the setup and complete the problem.

---

**11.21b**   2.48 liters $SO_2$

$\boxed{1.75 \text{ liters O}_2} \times \dfrac{755 \text{ torr}}{785 \text{ torr}} \times \dfrac{438°K}{297°K} \times \dfrac{1 \text{ liter SO}_2}{1 \text{ liter O}_2} = 2.48 \text{ liters SO}_2$

---

Avogadro's Hypothesis was not based originally on the ideal gas equation, as we have presented it here. Rather it was suggested by Avogadro to explain the experimental observation that *when gases react with each other, the reacting volumes, measured at the same temperature and pressure, are in the ratio of small, whole numbers.* This relationship is known as the law of combining volumes, and is demonstrable in the laboratory by these and other reactions between gases:

$$H_2 \text{ (g)} + Cl_2 \text{ (g)} \rightarrow 2 \text{ HCl (g)}$$
1 liter    1 liter

$$2 H_2 \text{ (g)} + O_2 \text{ (g)} \rightarrow 2 H_2O \text{ (g)}$$
2 liters    1 liter

$$N_2 \text{ (g)} + 3 H_2 \text{ (g)} \rightarrow 2 NH_3 \text{ (g)}$$
1 liter    3 liters

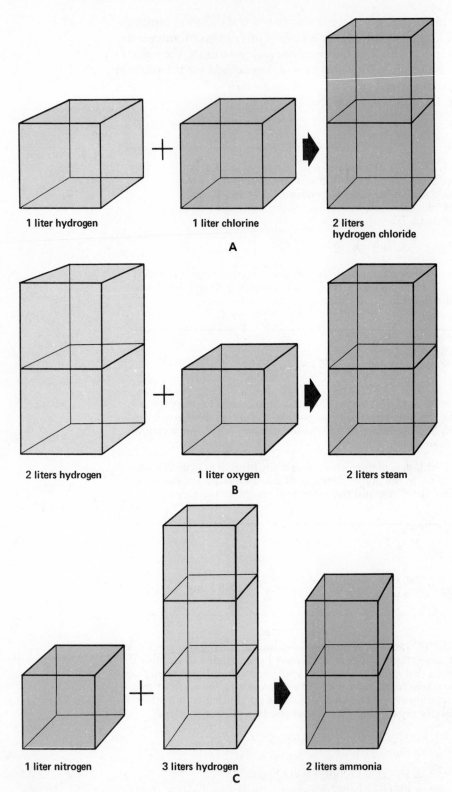

**Figure 11.13** A, Volume ratios in the reaction of hydrogen and chlorine to form hydrogen chloride. B, Volume ratios in the reaction of hydrogen and oxygen to form water (steam). C, Volume ratios in the reaction of nitrogen and hydrogen to form ammonia.

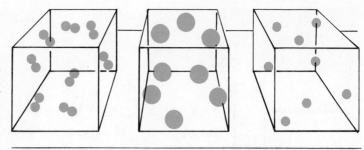

**Figure 11.14** Avogadro's Hypothesis: equal volumes of gases, measured at the same temperature and pressure, contain the same number of molecules.

**Same volume, temperature, and pressure**

The combining volume relationships are illustrated in Figure 11.13, which suggests correctly that the principle extends to product gases.

Avogadro reasoned that experimental observations such as these could be explained if equal volumes of different gases, measured at the same temperature and pressure, contain the same number of molecules, as shown in Figure 11.14.

## 11.12 DALTON'S LAW OF PARTIAL PRESSURES

PG 11 S Show that Dalton's Law of Partial Pressures conforms to the model of an ideal gas.

11 T Given the partial pressure of each component in a mixture of gases, find the total pressure.

11 U Given the total pressure of a gas saturated with water vapor, and the temperature of the system, find the partial pressure of the dry gas.

Two gas cylinders are connected by a short pipe of negligible volume, as shown in Figure 11.15. The cylinder on the right has a fixed volume, whereas the volume of the cylinder on the left may be adjusted by the position of the piston. In a certain experiment, during which the temperature of the entire system is held constant, different quantities of two gases, A and B, are placed into the cylinders. The piston is positioned so the volumes occupied by the two gases are *exactly equal* (see I, Fig. 11.15). The pressures exerted by the individual gas samples, $P_a$ and $P_b$, are recorded. The valve separating the two compartments is then opened and the piston moved to the right, combining the two gas samples by forcing all the gas originally in the left cylinder into the right cylinder (see II, Fig. 11.15). The final pressure in the right cylinder, P, is recorded. The final pressure at the end of the experiment is equal to the sum of the original pressures: $P = P_a + P_b$.

Experimental results such as these may be interpreted in terms of the model of an ideal gas. Initially, $n_a$ moles of gas A occupy volume V at temperature T. According to ideal gas Equation 11.18 solved for pressure, they will exert pressure $P_a = \dfrac{n_a RT}{V}$. Similarly, in the other chamber $n_b$ moles of gas B exert pressure $P_b = \dfrac{n_b RT}{V}$. When the gases are combined in the same volume

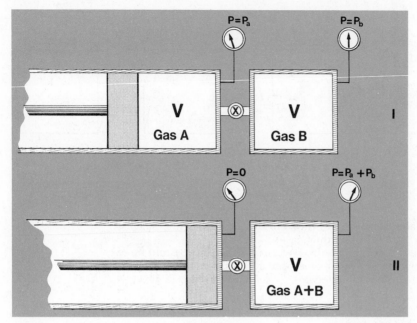

**Figure 11.15** Experimental apparatus to demonstrate Dalton's Law of Partial Pressures.

they each occupied initially—which is possible, according to the gas model, because of the large amount of space between the individual gas molecules —the total number of moles of molecules, $n_{tot}$, is equal to $n_a + n_b$. The gas laws are independent of the identity of the gas or gases involved, so we would anticipate that if $P_{tot}$ is total pressure, $P_{tot} = \dfrac{n_{tot}RT}{V}$. This expression agrees with the additivity of individual pressures:

$$P_{tot} = \frac{n_{tot}RT}{V} = \frac{(n_a + n_b)RT}{V} = \frac{n_aRT}{V} + \frac{n_bRT}{V} = P_a + P_b \quad (11.26)$$

In the mixture of gases A and B, $P_a$ and $P_b$ are referred to as the **partial pressures** of gases A and B respectively.

These observations are summed up in **Dalton's Law of Partial Pressures,** which states that **the total pressure exerted by a mixture of gases is the sum of the partial pressures of the components. The partial pressure of a component is the pressure that component would exert if it alone occupied the total volume at the same temperature** (see Fig. 11.16). Expressed mathematically, Dalton's Law is

$$P = p_1 + p_2 + p_3 + \ldots \quad (11.27)$$

where P is the total pressure and $p_1$, $p_2$, $p_3$ ... are the partial pressures of components 1, 2, 3, . . . .

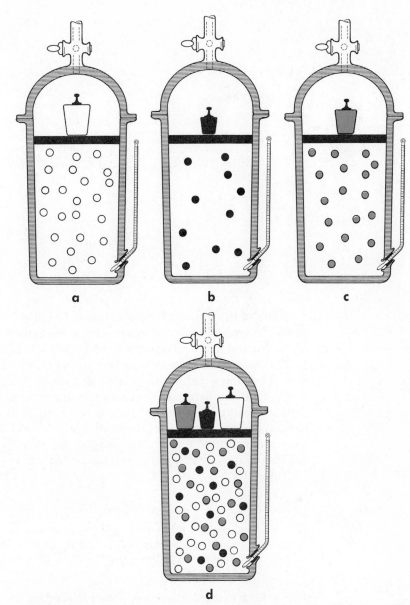

**Figure 11.16** If gases in *a*, *b* and *c*, which exert individual pressures illustrated by the weights shown, are combined as in *d*, the total pressure will be the sum of the individual (partial) pressures.

**Example 11.22** In a gas mixture the partial pressure of methane is 150 torr; of ethane, 180 torr, and of propane, 450 torr. Find the total pressure exerted by the mixture.

This is a straightforward application of Equation 11.27.

-----

**11.22a** 780 torr      P = 150 + 180 + 450 = 780 torr

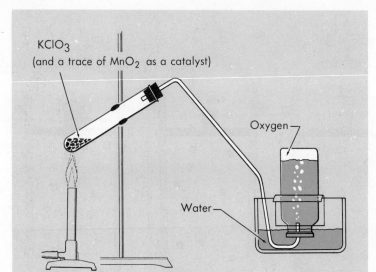

KClO₃
(and a trace of MnO₂ as a catalyst)

Oxygen

Water

**Figure 11.17** Laboratory preparation of oxygen.

One of the important applications of Dalton's Law of Partial Pressures has to do with mixtures of gases with water vapor. Gases may be prepared and collected in the laboratory by an apparatus such as that shown in Figure 11.17. The oxygen formed in the test tube is bubbled through water, at which time it becomes "saturated" with water vapor. The "gas" collected is therefore actually a mixture of the oxygen being generated and water vapor. The pressure exerted by the mixture is the sum of the partial pressure of the oxygen and the partial pressure of the water vapor. The latter is the equilibrium vapor pressure of water (Chapter 12, page 284) which depends only on temperature; values for this quantity at various temperatures are given in Appendix VI, page 514.

**Example 11.23** If the total gas pressure in an oxygen generator such as that shown in Fig. 11.18 is 755 torr, and the temperature of the system is 22°C, find the partial pressure of the oxygen.

First, to see clearly where you are headed in this problem, write the partial pressure equation (Equation 11.27) as it applies specifically to this problem.

---

**11.23a** $P = p_{O_2} + p_{H_2O}$

The water vapor pressure, $p_{H_2O}$, may be found from Appendix V. It is . . . .

---

**11.23b** 19.8 torr

From the total pressure given in the problem and the partial pressure of the water vapor, the partial pressure of oxygen follows readily . . . .

---

**11.23c**   $p_{O_2}$ = 735 torr

$755 = p_{O_2} + 19.8$
$p_{O_2} = 755 - 19.8 = 735$ torr

# CHAPTER 11 IN REVIEW

**11.1  PROPERTIES OF GASES (Page 234)**

**11.2  THE KINETIC THEORY OF GASES AND THE IDEAL GAS MODEL**

11 A   Explain physical properties of gases, or physical phenomena related to gases, in terms of the ideal gas model. (Page 235)

**11.3  GAS MEASUREMENTS**

11 B   List the measurable properties of a gas. (Page 236)

11 C   Explain the meaning of "one atmosphere" of pressure and identify its equivalent in millimeters of mercury, centimeters of mercury, inches of mercury, torr and pounds per square inch. (Page 236)

**11.4  BOYLE'S LAW**

11 D   Given the initial volume (or pressure) and initial and final pressures (or volumes) of a fixed quantity of gas at constant temperature, calculate the final volume (or pressure). (Page 238)

**11.5  ABSOLUTE TEMPERATURE**

11 E   Given a temperature in degrees Celsius (or degrees Kelvin), convert it to degrees Kelvin (or degrees Celsius). (Page 242)

**11.6  CHARLES' LAW: GAY-LUSSAC'S LAW**

11 F   Given the initial volume, and initial and final temperatures of a fixed quantity of gas at constant pressure, calculate the final volume. (Page 245)

**11.7  COMBINED GAS LAWS**

11 G   For a fixed quantity of a confined gas, given the initial volume, pressure and temperature, and the final pressure and temperature, calculate the final volume. (Page 247)

**11.8  STANDARD TEMPERATURE AND PRESSURE**

11 H   Given the volume of a gas at one temperature and pressure (or at STP), find the volume it would occupy at STP (or at a stated temperature and pressure). (Page 249)

**11.9  THE IDEAL GAS EQUATION**

11 I   Write the equation that relates the measurable properties of an ideal gas to each other (the ideal gas equation). (Page 249)

11 J   Write the variation of the ideal gas equation that includes the mass of the gas sample and its molar weight. (Page 249)

11 K   Given the pressure, temperature and volume (or number of moles) of a gas, calculate the number of moles (or volume). (Page 252)

11 L    Given the pressure, temperature and density of a gas—or the mass of a known volume—calculate the molar weight. (Page 252)

## 11.10   MOLAR VOLUME AT STANDARD TEMPERATURE AND PRESSURE

11 M    Define molar volume. State the molar volume of any gas at STP. (Page 254)

11 N    Given the volume (or number of moles) of any gas at STP, find the number of moles (or volume). (Page 254)

11 O    Given two of the following for any gas at STP, find the third: grams, volume, molar weight. (Page 254)

11 P    Given gas density at STP (or molar weight), find molar weight (or gas density at STP). (Page 254)

## 11.11   GAS STOICHIOMETRY

11 Q    For a chemical reaction for which the equation may be written, given the grams of any species OR the volume of any gaseous species at specified temperature and pressure, calculate the grams of any other species OR the volume of any gaseous species at specified temperature and pressure. (Page 256)

11 R    State Avogadro's Hypothesis regarding gas volumes and number of molecules. (Page 259)

## 11.12   DALTON'S LAW OF PARTIAL PRESSURES

11 S    Show that Dalton's Law of Partial Pressures conforms to the model of an ideal gas. (Page 263)

11 T    Given the partial pressure of each component in a mixture of gases, find the total pressure. (Page 263)

11 U    Given the total pressure of a gas saturated with water vapor, and the temperature of the system, find the partial pressure of the dry gas. (Page 263)

# TERMS AND CONCEPTS

Kinetic theory of gases (235)
Kinetic molecular theory (KMT) (235)
Ideal gas model (235)
Pressure (237)
Pascal (237)
Millimeter of mercury (unit of pressure) (237)
Torr (237)
Atmosphere (unit of pressure) (237)
Barometer (237)
Barometric (atmospheric) pressure (237)
Manometer (238)

Boyle's Law (238)
Absolute zero (242)
Kelvin temperature scale (242)
Charles' Law (245)
Standard temperature and pressure, (STP) (249)
Ideal gas equation (251)
Universal gas constant, R (251)
Molar volume (254)
Avogadro's Hypothesis (259)
Dalton's Law of Partial Pressures (263)

# QUESTIONS AND PROBLEMS

*An asterisk (\*) identifies a question that is relatively difficult, or that extends beyond the performance goals of the chapter.*

Section 11.2

11.1) What is the kinetic theory of gases? How does it describe an "ideal gas"?

11.42) What is kinetic energy? What "properties" of gases are involved in their kinetic energies?

*11.2–11.6 and 11.43–11.47: Explain how each of the gaseous phenomena described below is related to one or more of the features of the model of an ideal gas.*

11.2) Gases with a distinctive odor can be detected some distance from their source.

11.3) Under proper conditions, air pressure in an automobile tire remains constant over extended periods of time.

11.4) The density of liquid oxygen is about 1.4 grams/cm³. Vaporized at 0°C and 760 torr, this same 1.4 grams occupies 980 cm³, an expansion of nearly 1000 times.

11.5) Properties of gases become less "ideal"— the substance adopts behavior patterns not typical of gases—when subjected to very high pressures such that the individual molecules are close to each other.

11.6) Any container, regardless of size, will be completely filled by one gram of hydrogen.

11.43) Pressure is exerted on the top of a tank holding a gas, as well as on its sides and bottom.

11.44) Balloons expand in all directions when blown up, not just at the bottom as when filled with water.

11.45) Even though an automobile tire is "filled" with air, more air can always be added without increasing the volume of the tire significantly.

11.46) Very small dust particles, seen in a beam of light passing through a darkened room, appear to be moving about erratically.

11.47) Gas bubbles always rise through a liquid.

Section 11.3

11.7) Four properties of gases may be measured. Name them.

11.8) Define and/or distinguish between the atmosphere, millimeter of mercury, and the torr as units of pressure.

11.9)\* An open-end manometer (see Fig. 11.18) is used to measure the pressure of a confined gas. Calculate this pressure if $P_a = 747$ torr and the mercury levels in the manometer are as shown.

11.48) What is the meaning of *pressure?*

11.49)\* Distinguish between a barometer and a manometer. Explain how each measures the pressure of a gas.

11.50)\* Figure 11.19 shows an open-end manometer attached to an aspirator, a device commonly used to draw air through laboratory equipment. With mercury levels as shown and $P_a = 758$ torr, calculate the pressure in the hose.

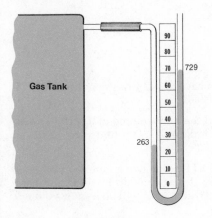

Figure 11.18

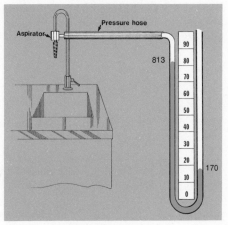

Figure 11.19

11.10) 3.15 liters of gas at 0.940 atmospheres are compressed to 6.26 atmospheres. Calculate the new volume.

11.11) What does the term "absolute zero" mean? What physical condition theoretically exists at absolute zero?

11.12) Lead melts at 328°C. Express this in absolute temperature, °K.

11.13) Liquid oxygen "boils" at 90°K. What is its boiling point in °C?

11.14) An industrial gas storage tank has a "floating top" that maintains constant pressure. If, on a day during which there is no gas consumption, the volume is 85,600 liters at 5 AM, when the temperature is 15°C, what will be the volume at 3 PM when it has warmed to 25°C?

11.15)* In a laboratory experiment, gas collected at 68°C exerts a pressure of 912 torr. What will the pressure be after cooling to 27°C, assuming no volume change?

11.16) 73.4 ml of oxygen, measured at 43°C and 824 torr, are collected in a laboratory experiment. What will be the volume when the gas cools to 23°C and the pressure is adjusted to 749 torr?

11.17) A variable volume marine research device is designed to hold a fixed quantity of gas at a pressure that balances the pressure outside the unit. At the surface of the ocean its volume is 1.62 liters at 23°C and 1.02 atm. Find its volume when lowered to a depth where the pressure is 4.86 atm and the temperature is 3°C.

11.18) What is the meaning of "STP?"

11.19) 47.9 ml hydrogen are collected at 26°C and 718 torr. Find the volume occupied at STP.

11.20) If a gas occupies 46.9 ml at STP, what volume will it fill at 24°C and 738 torr?

11.51) To what volume must 127 milliliters of a confined gas at 749 torr be expanded to reduce pressure to 506 torr?

11.52)* Describe an experiment by which the Celsius equivalent of absolute zero may be estimated.

11.53) Find the Celsius boiling point of sulfur, which boils at 718°K.

11.54) Antifreeze may be added to water to reduce the freezing point to −29°C. What is this temperature in °K?

11.55) Only 620 milliliters of a collapsible plastic bag are filled with air at a temperature of −4°C. What will the volume be when the temperature rises to 21°C?

11.56)* Beside the ideal gas storage tank of Problem 11.14 is a tank of compressed air, which is also unused. In the morning, when the temperature is 15°C, its pressure gauge records 1650 pounds per square inch. What pressure will it reach in the afternoon when the temperature is 25°C?

11.57) 7.92 liters of nitrogen, measured at 1.28 atm and 17°C, are in a piston-fitted variable volume cylinder. The nitrogen is warmed to 39°C, and the piston moved to adjust the pressure to 1.39 atm. Calculate the new volume.

11.58) A weather balloon is partly filled with 34.9 liters of helium, measured at 752 torr and 19°C. What volume will the balloon occupy after rising to an altitude where the pressure is 512 torr and temperature is −45°C?

11.59) Why have the arbitrary conditions of STP been established?

11.60) Find the STP volume of a certain quantity of air if it occupies 48.6 liters at −12°C and 1.72 atm.

11.61) The STP volume of a sample of nitrogen is 1.46 liters. If it is cooled to −44°C and compressed to 2.06 atm, what will be its new volume?

Section 11.9

11.21)  What volume will be occupied by 0.16 mole of hydrogen at 751 torr and 22°C?

11.22)  How many moles of methane, $CH_4$, are in a 2.55 liter cylinder if the temperature is 20°C and the pressure is 9.40 atmospheres?

11.23)  2.68 liters of an unknown gas, measured at 22°C and 745 torr, weigh 5.89 grams. Calculate the molar weight.

11.24)*  How many atmospheres of pressure will be exerted by 25.0 grams of sulfur dioxide when confined in a 2.15 liter cylinder at 20°C?

11.25)*  A 21.2 liter oxygen tank is installed in an unventilated corner of an industrial plant. The tank is fitted with a relief valve that opens at 24.0 atmospheres of pressure. On a very hot day, when the tank happened to contain 624 grams of oxygen, the valve opened. What was the Celsius temperature of the oxygen in the tank?

11.26)*  Calculate the number of kilograms of helium in a 1.75 cubic meter balloon at 9°C and 798 torr.

11.27)*  The density of an unknown hydrocarbon gas at 25°C and 750 torr is 1.69 grams/liter. It consists of 85.6% carbon and 14.4% hydrogen. (a) What is the empirical formula of the gas? (b) Calculate the molar weight of the gas. (c) Write the molecular formula of the gas.

11.62)  1.26 moles of oxygen are compressed to 4.36 atmospheres at a temperature of 36°C. Calculate the volume.

11.63)  Calculate the number of moles of ammonia in a 24.0 liter tank when the pressure is 894 torr and the temperature is 29°C.

11.64)  1.06 grams of an unknown compound are placed in a 0.500 liter testing vessel and vaporized by partial evacuation and heating. When the temperature is 86°C, the pressure is 0.390 atmosphere. Calculate the molar weight.

11.65)*  At what Celsius temperature will 0.200 mole of nitrogen exert a pressure of 545 torr in a 4.12 liter cylinder?

11.66)*  How many grams of chlorine are in a 0.716 liter cylinder if the pressure is 10.9 atm at 30°C?

11.67)*  How much pressure (torr) will be exerted by 190 kilograms of methane, $CH_4$, when stored in a 269 cubic meter tank at 22°C?

11.68)*  Analysis of 1.19 grams of hydrazine, a rocket fuel, shows that it contains 1.04 grams of nitrogen, and the balance is hydrogen. If the same quantity of hydrazine is vaporized at 96°C in a 1.47 liter chamber, it exerts a pressure of 0.766 atmosphere. Determine the molar weight, empirical formula and molecular formula of hydrazine.

Section 11.10

11.28)  What is the meaning of "molar volume?"

11.29)  How many moles of ethane, $C_2H_6$, are in 16.9 liters at STP?

11.30)  Estimate the molar weight of a gas having a density of 1.83 g/liter at STP.

11.31)  0.937 gram of an unknown gas occupy 0.744 liter at STP. Find the molar weight.

11.69)  Explain the restrictions placed on the statement that 22.4 liters per mole is the molar volume of any gas.

11.70)  Find the STP volume of 4.62 grams of ammonia.

11.71)  What is the mass of 3.25 liters of neon (at. no. 10) at STP?

11.72)  Find the density of sulfur dioxide at STP.

Section 11.11

11.32)  How many liters of $O_2$, measured at STP, will be released on the decomposition of 2.65 g of mercury(II) oxide: $2\ HgO \rightarrow 2\ Hg + O_2$?

11.73)  What quantity of potassium chlorate must be decomposed to produce 1.50 liters of oxygen at STP? $2\ KClO_3\ (s) \rightarrow 2\ KCl\ (s) + 3\ O_2\ (g)$.

11.33) What quantity of magnesium must a student react with excess hydrochloric acid to produce 85.0 ml hydrogen, measured at STP? $Mg$ (s) + 2 $HCl$ (aq) → $H_2$ (g) + $MgCl_2$ (aq).

11.34) What volume of oxygen, measured at 25°C and 752 torr, is required to "burn" 3.26 grams of magnesium by the reaction 2 $Mg$ (s) + $O_2$ (g) → 2 $MgO$ (s)?

11.35) Copper(II) oxide may be "reduced" to copper by heating in a stream of hydrogen: $CuO$ (s) + $H_2$ (g) → $Cu$ (s) + $H_2O$ (g). How many grams of CuO will be reduced by 4.16 liters of hydrogen, measured at 1.65 atm and 243°C?

11.36) What is Avogadro's Hypothesis? Explain why the volume of gaseous reactants and products in a reaction, provided they are measured at the same temperature and pressure, are in the same ratio as the coefficients of the equation for the reaction.

11.37) How many liters of hydrogen will combine directly with 1.28 liters of oxygen to produce steam if both volumes are measured at 200°C and 1.00 atm? 2 $H_2$ (g) + $O_2$ (g) → 2 $H_2O$ (g).

11.38) In an all gas phase reaction 28.3 liters of steam at 540°C and 1.46 atm react with carbon monoxide to produce hydrogen and carbon dioxide: $CO$ (g) + $H_2O$ (g) → $CO_2$ (g) + $H_2$ (g). How many liters of CO, measured at 460°C and 2.19 atm, will be used?

## Section 11.12

11.39) State Dalton's Law of Partial Pressures, either in words or as an equation. Explain how this law "fits" the ideal gas model.

11.40) Helium, neon and argon are mixed in such a fashion that their respective partial pressures are 0.364 atm, 0.108 atm and 0.529 atm. What is the total pressure of the system?

11.41) A laboratory hydrogen generator collects the gas produced by bubbling it through water. The total pressure of the gas collected is 751 torr. The temperature is 34°C, at which water vapor pressure is 40.0 torr. Calculate the partial pressure of hydrogen.

11.74) When sodium hydrogen carbonate is treated with sulfuric acid, carbon dioxide bubbles off: $H_2SO_4$ (aq) + 2 $NaHCO_3$ (s) → $Na_2SO_4$ (aq) + 2 $H_2O$ (l) + 2 $CO_2$ (g). What volume of $CO_2$, measured at STP, is available from 8.58 g $NaHCO_3$?

11.75) Chlorine combines directly with sodium to form sodium chloride: 2 $Na$ (s) + $Cl_2$ (g) → 2 $NaCl$ (s). How many grams of NaCl will result from the reaction of 0.745 liter $Cl_2$, measured at 1.21 atm and 22°C?

11.76) Calculate the volume of ammonia, measured at 746 torr and 26°C, released by the complete reaction of 15.8 grams of lime, CaO, with excess $NH_4Cl$ solution: 2 $NH_4Cl$ (aq) + $CaO$ (s) → 2 $NH_3$ (g) + $CaCl_2$ (aq) + $H_2O$ (l).

11.77)* Suggest a reason why Avogadro's Hypothesis is acceptable for gases, but not for solids or liquids.

11.78) How many liters of gaseous hydrogen chloride, measured at 1.10 atm and 31°C, will result from the combination of 0.734 liter of chlorine, also measured at 1.10 atm and 31°C, with excess hydrogen? $H_2$ (g) + $Cl_2$ (g) → 2 $HCl$ (g).

11.79) Determine the volume of hydrogen, measured at 768 torr and 35°C, that is required to produce 4.80 liters of methane, $CH_4$, measured at 220°C and 1800 torr, by the reaction 4 $H_2$ (g) + $CS_2$ (g) → $CH_4$ (g) + 2 $H_2S$ (g).

11.80)* A gaseous mixture contains only nitrogen and hydrogen at a total pressure of 1.00 atm. If the partial pressure of $H_2$ is 0.50 atm, find $p_{N_2}$. Which gas, if either, is present in the greatest mass? Explain your answer.

11.81) Find the partial pressure of oxygen in clean air if the total pressure is 757.0 torr, and partial pressures of other gases are (1) nitrogen, 590 torr; (2) argon, 7 torr; (3) all others, 0.2 torr.

11.82) If the total volume of gas collected in Problem 11.41 was 50.3 ml, what volume would be occupied by the dry hydrogen at STP?

# LIQUIDS AND SOLIDS

The general differences between the three states of matter, gases, liquids and solids, were first mentioned in Section 2.2, page 10. These were also related to the kinetic theory of matter, pointing out that all three states are believed to consist of particles in constant motion, with decreasing degrees of freedom in the order named. These distinctions are reviewed in Figure 12.1. In Chapter 11 we examined the gaseous state in some detail. In this chapter we turn our attention to the so-called condensed states of matter, liquids and solids.

## 12.1  THE NATURE OF THE LIQUID STATE

PG  12 A  In terms of the relative distances between molecules and the effect of intermolecular forces, explain the differences in physical behavior between liquids and gases.

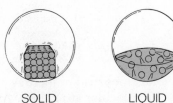

SOLID          LIQUID          GAS

**Figure 12.1** Characteristics of solids, liquids and gases. Solid has definite volume and definite shape. Liquid has definite volume, but assumes the shape of the bottom of its container, up to that total volume. Gas fills its container, acquiring both its shape and volume. Particles are in constant motion in all states: vibrating in fixed positions in a solid, moving randomly within the volume occupied by a liquid, and moving randomly within the entire volume of the container as a gas. Particles are very close to each other in solid and liquid, but widely separated in the gaseous state.

The behavior of gases is neatly summarized in the ideal gas equation that may be applied quite satisfactorily to nearly all gases. No similar relationship is available for liquids. But we do know two things about liquids that are quite different from gases. First, liquid molecules are very close to each other, instead of being widely separated, as in a gas. Second, intermolecular attractions are of major importance in determining the behavior of a liquid, whereas the properties of gases are explainable only if we assume that these attractions are negligible, if they exist at all.

The fact that gases can be compressed and liquids, for all practical purposes, cannot, is explained by these two differences. To compress a gas you need only to push the molecules closer to each other. In a liquid the molecules are already "touchingly close," so to compress a liquid you would have to crush or distort the actual molecules, which does not appear to occur. If you heat a gas at constant pressure, or reduce the pressure on a gas at constant temperature, the gas expands readily. Neither an increase in temperature nor a reduction in pressure causes a significant increase in the volume of a liquid, because the intermolecular attractions apparently hold the molecules together in an almost fixed total volume.

The strong intermolecular attractions in a liquid are the direct result of the closeness of the molecules to each other. These attractive forces are electrostatic in character, and their strength is inversely related to the distance between molecules; that is, the smaller the distance the larger the force. A gas may be condensed to a liquid by reducing its temperature, by compressing it, or by a combination of these acts.* A temperature reduction slows the molecules in their random movement. Intermolecular collisions are no longer elastic; some become so weak the particles no longer bounce off each other. Instead the attractions between them while in contact cause them to stay together, leading eventually to the formation of a liquid drop, and ultimately to nearly total condensation. When gas molecules are pushed close to each other in compression, the intermolecular attractions become significant, and cause the molecules to stick together, leading to condensation.

## 12.2 PHYSICAL PROPERTIES OF LIQUIDS

PG 12 B  For two liquids, given comparative values of physical properties that depend on intermolecular attractions, predict the relative strengths of those attractions; or, given a comparison of the strengths of intermolecular attractions, predict the relative values of physical properties that depend upon them.

Many physical properties of liquids are directly related to the strength of intermolecular attractions. Among them are the following:

**VAPOR PRESSURE.**  Because of evaporation, the open space above any liquid contains some of the liquid molecules in the gaseous, or vapor, state. The partial pressure exerted by these gaseous molecules is called **vapor pressure.** If the gas space above the liquid is closed, the vapor pressure will increase to a definite value, referred to as the **equilibrium vapor pressure.** (See Section 12.4, page 282.) Equilibrium vapor pressures of different

---

*There is a temperature, called *critical temperature*, above which a gas cannot be liquefied by compression alone.

liquids are related to the strength of intermolecular forces. *Liquids with relatively weak intermolecular attractions evaporate more readily, yielding higher concentrations in the vapor state, and therefore higher vapor pressures.*

**BOILING POINT.** Liquids may be changed to gases by boiling. A liquid must be heated to make it boil. When the temperature reaches the **boiling point** the average kinetic energy of the liquid particles is sufficient to overcome the forces of attraction that hold molecules in the liquid state, and it becomes a gas. *Liquids with stronger intermolecular forces require higher temperatures for boiling.*

**MOLAR HEAT OF VAPORIZATION.** Even after a liquid has been raised to its boiling point, additional energy is required to separate the liquid molecules from each other and keep them apart. As you continue to heat a boiling liquid, the temperature of both the liquid and the vapor remain at the boiling point until all of the liquid has been converted to the gaseous state. The energy required to vaporize one mole of liquid at its boiling point is called the **molar heat of vaporization.** *A mole of a liquid with strong intermolecular attractions requires more energy to vaporize it than a mole of a liquid with weak intermolecular attractions.*

The trends in vapor pressure, boiling point and molar heat of vaporization are correlated with strength of intermolecular attractions in Table 12.1.

**VISCOSITY.** One of the characteristics of liquids, according to the kinetic molecular theory, is that the liquid molecules are free to move about relative to each other within the body of the liquid. That freedom to move about, however, is not the same in all liquids. You are aware, for example, that water may be poured much more freely than syrup, and syrup more readily than honey. The unique pouring characteristic of each liquid is the result of its **viscosity.** Viscosity may be thought of as an internal resistance to flow. *Liquids with strong intermolecular forces are generally more viscous than liquids with weak intermolecular attractions.*

**SURFACE TENSION.** When a liquid is broken into "small pieces" it forms drops. Ideally, isolated drops are perfect spheres. This is not readily apparent when we look at water drops because we usually see them in the act of falling, at which time they take on a "tear-drop" form. A sphere has the smallest surface area possible for a drop of any given volume. This tendency toward a minimum surface is the result of **surface tension,** the unbalanced attraction of molecules at the surface of a liquid by those molecules beneath the surface. Within a liquid each molecule is attracted in all directions by the molecules that surround it. At the surface, however, the attraction is nearly all downward, pulling the surface molecules into a sort of tight skin over the liquid. The effect of surface tension in water may be seen when a needle

TABLE 12.1  Physical Properties of Liquids

| SUBSTANCE | VAPOR PRESSURE AT 20°C | NORMAL BOILING POINT | HEAT OF VAPORIZATION | INTERMOLECULAR FORCES |
|---|---|---|---|---|
| Mercury | 0.0012 torr | 357°C | 14.2 kcal/mole | Strongest |
| Water | 17.5 | 100 | 9.7 | ↑ |
| Benzene | 75 | 80 | 7.3 | |
| Ether | 442 | 35 | 6.2 | |
| Ethane | 27,000 | −89 | 3.7 | Weakest |

**Figure 12.2** Surface tension. Unbalanced downward attractive forces at the surface of a liquid pull molecules into a difficult-to-penetrate skin capable of supporting small bugs or thin pieces of steel, such as a needle or razor blade. Bug literally runs *on* the water; it does not float *in* it. Molecules within the water are attracted in all directions, as shown.

floats if placed gently on a still water surface, or when small bugs run across the surface of a quiet pond (see Fig. 12.2). *Liquids with strong intermolecular attractions have higher surface tension than liquids in which intermolecular forces are weak.*

We now summarize these relationships between intermolecular attractions and physical properties. If you think in terms of the "stick togetherness" of molecules with strong attractive forces, you can usually reason to correct conclusions. With all other influences being equal, strong intermolecular attractions generally lead to:

(1) Low vapor pressure. This is an "inverse" relationship because the vapor pressure is a result of the *gas* released by evaporation when the intermolecular attractions between liquid molecules are overcome. If the *stick togetherness* is high between liquid molecules, not much gas escapes, so the vapor pressure is low.

(2) High boiling point. When *stick togetherness* is high, it takes a lot of agitation (high temperature) before the molecules can even *begin* to tear loose from each other within the liquid, where boiling occurs.

(3) High heat of vaporization. Again, if *stick togetherness* is high, it takes still more energy to separate the molecules, even after they are shaking vigorously enough to boil.

(4) High viscosity. High *stick togetherness* means more "gooey-ness."

(5) High surface tension. High *stick togetherness* at the surface means more resistance to anything that would break through or stretch that surface (high surface tension).

## 12.3 TYPES OF INTERMOLECULAR FORCES

PG  12 C  Identify and describe or explain dipole forces, dispersion forces and hydrogen bonds.

12 D   Given the structure of a molecule, or information from which it may be determined, identify the significant intermolecular forces present.

12 E   Given the molecular structures of two substances, or information from which they may be obtained, compare or predict relative values of physical properties that are related to them.

As a group, the intermolecular forces that are responsible for so many physical properties of liquids are called **van der Waals forces.** These are the same forces that account for the nonideal behavior of many real gases, particularly when temperatures and/or pressures are near the values at which the gas would condense to a liquid. It is logical to ask, "Where do these forces come from? What causes them to exist?" It has already been stated that they are electrostatic in character; the attractions are between negative and positive charges. But molecules, you may protest, are electrically neutral. How, therefore, can there be electrostatic attractions between them? To find an answer to that question you must recall that, even though the molecule may be neutral as a whole, the *distribution* of electrical charge within the molecule may be either symmetrical or nonsymmetrical—that is, balanced or unbalanced. Molecules with a symmetrical distribution of charge are nonpolar; if the charge distribution is unbalanced, the molecule is polar. This concept was introduced in Chapter 9, page 211.

There are three kinds of intermolecular forces that can be traced to electrostatic attractions:

**1. DIPOLE FORCES.**  Polar molecules are sometimes described as "dipoles," meaning that the molecule has two "poles," positive in one region and negative in another. Depending on their size and shape, dipoles can interact with each other in a number of ways, as shown in Figure 12.3. Iodine chloride is an example. In the solid state the molecules are oriented

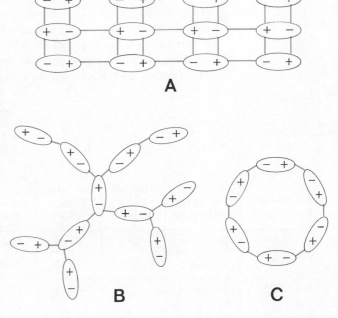

**Figure 12.3**  Dipole forces. In A, the polar molecules are arranged in a regular geometric pattern, as they might be found in a molecular solid. B shows similar polar molecules arranged in a random fashion, such they might be found in a liquid. In C, the molecules have grouped themselves into a larger unit, evidence of which is known in all three states of matter. In all cases the positive region of one polar molecule is attracted to the negative region of the second polar molecule.

as in Figure 12.3A. The bonds between molecules are weak, however. At 27°C or above they cannot withstand the vigorous kinetic energy of the molecular motion so the substance melts. In the liquid state there continue to be intermolecular attractions leading to constantly changing groups of molecules that are temporarily attracted to each other. These attractions also are not very strong: they are overcome at 97°C, the boiling point of iodine chloride. By contrast, nonpolar bromine, a substance of comparable size, shape and molecular weight, melts under 0°C, and boils at 59°C. Comparisons between boiling points of polar and nonpolar compounds of similar molecular weight are listed in Table 12.2. The consistently higher boiling points of polar compounds is evidence of dipole forces between the molecules.

2. DISPERSION (LONDON) FORCES. Returning to bromine, $Br_2$, what type of intermolecular attractions hold it in the liquid state at temperatures up to 59°C? Why is it not, like chlorine, a gas at room temperature? The forces responsible for bromine being a liquid are called **dispersion forces,** or **London forces.**

Dispersion forces are believed to be the result of "temporary dipoles" that are formed by the shifting electron clouds within molecules that are, over a period of time, nonpolar. (See Fig. 12.4.) These dipoles attract or repel the electron clouds of nearby nonpolar molecules, thereby inducing them to become dipoles temporarily. As long as these dipoles exist—a very small fraction of a second in each individual case—there is an attraction between them. The strength of dispersion forces depends upon the ease with which electron distributions can be distorted, or "polarized." Large molecules, with many electrons, or electrons far removed from the nucleus, are more easily polarized than small, compact molecules in which the nuclei hold electrons in position more firmly. Larger molecules are generally heavier. As a consequence intermolecular forces tend to increase with increasing molecular weight among otherwise similar substances, as indicated in Figure 12.5.

Dispersion forces exist between *all* kinds of molecules, polar or nonpolar. Among small, low molecular weight molecules they are weak, and significant only when not overshadowed by dipole forces. But among compounds with large molecules and higher molecular weight, these forces can become quite strong—strong enough to be responsible for room-temperature liquids and solids among nonpolar substances, as bromine illustrates.

TABLE 12.2   Boiling Points of Polar vs. Nonpolar Substances

| FORMULAS | POLAR OR NONPOLAR | MOLECULAR WEIGHT | BOILING POINT (°C) | FORMULAS | POLAR OR NONPOLAR | MOLECULAR WEIGHT | BOILING POINT (°C) |
|---|---|---|---|---|---|---|---|
| $N_2$ | Nonpolar | 28 | −196 | $GeH_4$ | Nonpolar | 77 | −90 |
| CO | Polar | 28 | −192 | $AsH_3$ | Polar | 78 | −55 |
| $SiH_4$ | Nonpolar | 32 | −112 | $Br_2$ | Nonpolar | 160 | 59 |
| $PH_3$ | Polar | 34 | −85 | ICl | Polar | 162 | 97 |

**Molecules**

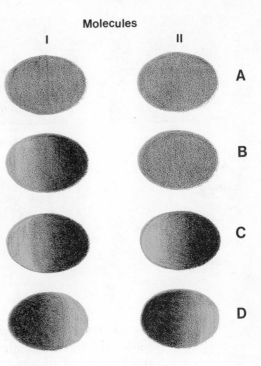

**Figure 12.4** The origin of dispersion forces. A, Molecules I and II are temporarily close to each other. Uniform shading indicates overall uniform distribution of electron cloud in molecule. B, Electron cloud in Molecule I has shifted to right, forming a temporary dipole with concentration of negative charge to right, positive charge to left. C, Negative charge concentration at the right of Molecule I repels electron cloud in Molecule II to right side of molecule, inducing it to become a temporary dipole similar to Molecule I. At this time there is a weak dipole-dipole attraction between the two "instantaneous dipoles." D, A small fraction of a second later the electron clouds may shift to opposite sides of the molecules. The instantaneous dipoles still attract each other weakly. Again a small fraction of a second later the molecules interact similarly with outer nearby molecules, forming new weak intermolecular attractions.

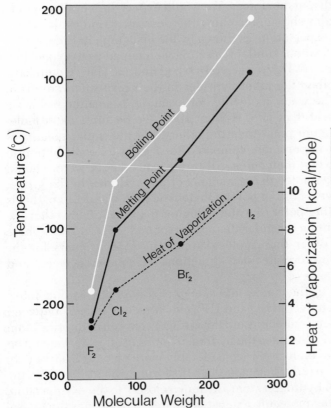

Boiling Points of Normal Alkanes

| FORMULA | BOILING POINT (°C) |
|---|---|
| $CH_4$ | −162 |
| $C_2H_6$ | −88 |
| $C_3H_8$ | −42 |
| $C_4H_{10}$ | 0 |
| $C_5H_{12}$ | 36 |
| $C_6H_{14}$ | 69 |
| $C_7H_{16}$ | 98 |
| $C_8H_{18}$ | 126 |
| $C_9H_{20}$ | 151 |
| $C_{10}H_{22}$ | 174 |

**Figure 12.5** Physical properties as a function of molecular size. The graph at the left shows the boiling points, melting points and heats of vaporization of the halogens. The table at the right lists the boiling points of normal alkane hydrocarbons, which have the general formula $C_nH_{2n+2}$. As the molecular sizes increase, as they do with increasing molecular weight, intermolecular attractions due to dispersion forces increase. This causes an increase in the magnitude of physical properties related to these attractions.

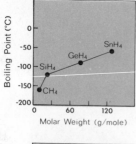

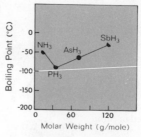

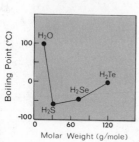

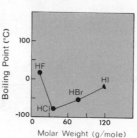

**Figure 12.6** Boiling points of different hydrides. In the graph at the upper left, there is no hydrogen bonding in the compound with the smallest molecule (lowest molecular weight), so it has the lowest boiling point in its family. In the other three graphs the compound with the smallest molecule exhibits hydrogen bonding between molecules. These strong intermolecular forces cause the boiling points to be abnormally high compared to other members of their respective families.

**3. HYDROGEN BONDS.** Four hydrides of the Group IVA elements, in order of decreasing molecular weight, are $SnH_4$, $GeH_4$, $SiH_4$ and $CH_4$. These molecules are all tetrahedral in shape; they are nonpolar. The only intermolecular forces are dispersion forces. We would expect their boiling points to decrease in the order shown, matching the decreasing molecular sizes. They conform to this expectation, as shown by the first graph in Figure 12.6.

With some modification, similar reasoning would lead us to expect that the hydrides of nitrogen ($NH_3$), oxygen ($H_2O$) and fluorine (HF) would have lower boiling points than the other hydrides in their respective groups in the periodic table. The water molecule, for example, is smaller and has a lower molecular weight than $H_2S$, $H_2Se$ or $H_2Te$. To be sure, water molecules are somewhat more polar than the others, but molecular polarity is usually of secondary importance, compared to size, in its contribution to attractive forces between molecules. All things considered, water should have the smallest intermolecular attractions of the hydrides in the group, and therefore the lowest boiling point. Obviously, from Figure 12.6, this is not the case; the boiling point of water is higher than that of any other hydride in the group. Similar irregularities appear for both $NH_3$ and HF. There must be something other than molecular polarity that accounts for the unexpectedly strong intermolecular attractions in ammonia, water and hydrogen fluoride.

**Hydrogen bond** is the name given to the abnormally strong intermolecular attractions in $NH_3$, $H_2O$ and HF. The attractive force is between the hydrogen atom of one molecule and the nitrogen, oxygen or fluorine atom of another. Figure 12.7 illustrates the hydrogen bonding effect in water. The oxygen atom in a water molecule is considerably more electronegative (3.5) than the hydrogen atom (2.1). The electron pairs forming bonds between the oxygen atom and each hydrogen atom are drawn closely to the oxygen atom, giving that region of the molecule a negative charge. This leaves the hydrogen nucleus—nothing more than a proton—as a small and therefore highly concentrated region of positive charge. The negatively charged oxygen region of one water molecule can get quite close to the small positively

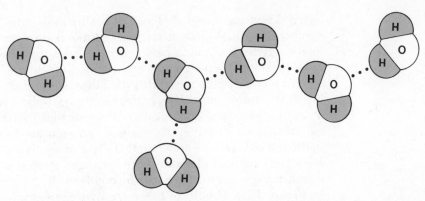

**Figure 12.7** Hydrogen bonding in water. Intermolecular hydrogen bonds are present between electronegative oxygen region of one molecule and electropositive hydrogen region of second molecule.

| ELECTRONEGATIVE ELEMENT | LEWIS DIAGRAM | EXAMPLE |
|---|---|---|
| Nitrogen | H—N̈— | 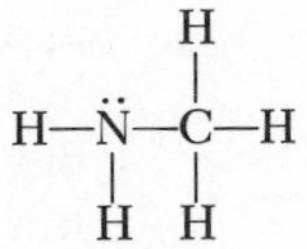 H—N̈—C—H (Methylamine) |
| Oxygen | :Ö— | 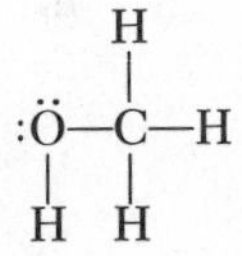 :Ö—C—H (Ethanol) |
| Fluorine | H—F̈: | 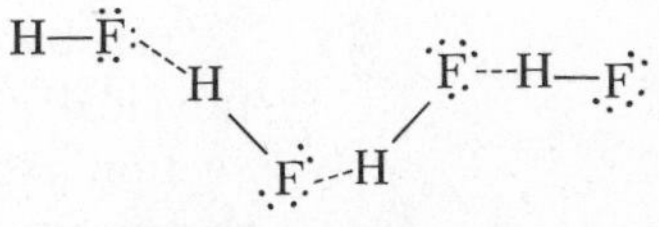 Hydrogen fluoride |

**Figure 12.8** Recognizing hydrogen bonding. Hydrogen bonding most frequently occurs when hydrogen is covalently bonded to nitrogen, oxygen or fluorine in a molecule. The temporary intermolecular bonds are established between one of these strongly electronegative atoms in one molecule and a hydrogen atom of a *different* molecule. Ammonia (not shown) and methylamine are examples of compounds of nitrogen that exhibit hydrogen bonding; water (not shown) and ethanol are oxygen compounds that are hydrogen bonded. In hydrogen fluoride there is evidence that molecules form chains of variable length, as shown.

charged hydrogen region of a neighboring molecule. The result is a hydrogen bond, a dipole-like attraction, but much stronger than ordinary dipole forces. Hydrogen bonds between molecules are roughly $1/10$ as strong as covalent bonds between atoms within a molecule.

To recognize the possibility of hydrogen bonding in a substance, examine its structure. Normally a hydrogen atom will be covalently bonded to an atom of a highly electronegative element that has one or more unshared electron pairs; fluorine, oxygen and nitrogen are the most common. The hydrogen bond forms between that electronegative atom and the hydrogen atom in a nearby molecule. If the hydrogen atom is covalently bonded to a nitrogen atom, the nitrogen may be bonded to two other species, as indicated in Figure 12.8. If both of these are hydrogen atoms, the compound is, of course, ammonia, $NH_3$. If one is a hydrogen atom and the other is a $-CH_3$ group, methyl amine is formed. If the hydrogen atom is bonded to oxygen, the oxygen will be bonded to one other species, which may be another hydrogen atom, producing water, $H_2O$. Another example is methyl alcohol, as shown.

Hydrogen bonding involving fluorine is unique because both hydrogen and fluorine form only one covalent bond. This sets up the possibility of a linkage of indefinite length between several HF molecules, as indicated in Figure 12.8. Such chains are also known to form closed rings. Finally the hydrogen difluoride ion present in such compounds as $KHF_2$ is considered to be the result of a hydrogen bond in which two $F^-$ ions are linked by a $H^+$ ion, forming $HF_2{}^-$, $\left[ :\ddot{F}-H-\ddot{F}: \right]^-$.

The abnormal boiling point of water—abnormal in the sense that it departs from what would be expected when compared to other Group VIA hydrides—is but one example of a long list of "abnormal" physical properties of water. Water is so common a substance we take it much for granted. But in relative trends of physical properties predicted from the periodic table, water is one of the most "uncommon" substances known. Nearly all of its unique properties are at least partly the result of hydrogen bonding.

## 12.4 LIQUID-VAPOR EQUILIBRIUM

PG 12 F Describe or explain the equilibrium between a liquid and its own vapor, and the process by which the equilibrium is reached.

In Section 12.2, vapor pressure and boiling point were identified as physical properties that are related to intermolecular attractions. In this section and the next we will examine these properties in greater detail, and thereby learn a bit more about the liquid state.

In Chapter 11, page 243, temperature was described as a measure of the average kinetic energy of the particles in a sample of matter. The word *average* suggests that at a given temperature the particles in the sample do not all have the same kinetic energy, but rather a range of energies. Kinetic energy is expressed mathematically as $\frac{1}{2} mv^2$, where m is the mass of the particle and v is its velocity. If the sample is a pure substance, and disregarding the minor variations introduced by isotopes, all particles have the same mass. Their different kinetic energies therefore indicate that they have different velocities.

We will now consider what happens at the molecular level when a liquid evaporates. Intermolecular attractions tend to keep the substance in the liquid state. But a few of the faster moving molecules near the surface have enough kinetic energy to overcome these attractions and escape (evaporate) from the liquid. At a given temperature the fraction, or percentage, of the total sample that is capable of evaporating is constant. As a consequence the rate of evaporation per unit of surface area is also constant at that temperature. If the temperature rises, a larger portion of the sample has sufficient kinetic energy to evaporate, and the evaporation rate is greater.

If a volatile liquid such as benzene is placed in an Erlenmeyer flask, which is then stoppered as in Figure 12.9, the benzene will begin to evaporate. If we assume the entire system maintains a constant temperature by absorbing sufficient heat from the surroundings to offset the cooling effect of evaporation, the *rate* of evaporation will remain constant throughout the experiment. The constant evaporation rate is represented by the fixed length of black arrows pointing upward from the liquid in each view of the flask, and also as the horizontal black line in a graph of evaporation rate versus time in Figure 12.9.

At first (Time 0) the movement of molecules is entirely in one direction, from the liquid to the vapor. However, as the concentration of molecules in the vapor builds up, an occasional molecule collides with the surface and reenters the liquid. The change of state from a gas to a liquid is called **condensation.** The *rate* of condensation, or return from vapor to liquid, depends upon the concentration of molecules in the vapor state. At Time 1

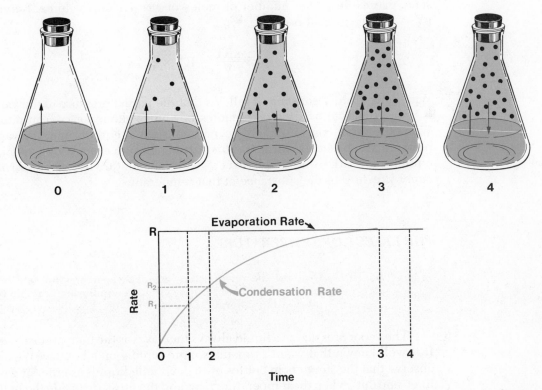

**Figure 12.9** The development of a liquid-vapor equilibrium. Depth of shading in vapor space indicates vapor concentration.

there will be a small accumulation of molecules in the vapor state, so the condensation rate will be more than zero, but considerably less than the evaporation rate. This is shown by the arrow lengths in the Time 1 flask of Figure 12.9.

As long as the rate of evaporation is greater than the rate of condensation, the vapor concentration will rise over the next interval of time. Therefore the rate of return from vapor to liquid rises with time (Time 2). Eventually the rates of vaporization and condensation become equal (Time 3); the number of molecules moving from vapor to liquid in unit time just balances the number moving in the opposite direction. We describe this situation, **when opposing rates of change are equal,** as a condition of **dynamic equilibrium** between liquid and vapor. Once equilibrium is reached, the concentration of molecules in the vapor has a certain fixed value which does not change. The rates therefore remain equal (Time 4).

Changes that occur in either direction, such as the change from a liquid to a vapor and the opposite change from a vapor to a liquid, are called **reversible changes;** if the change is chemical it is a **reversible reaction.** Chemists write equations describing reversible changes with a double arrow, one pointing in each direction. For example, the reversible change between liquid benzene, $C_6H_6$ (l), and benzene vapor, $C_6H_6$ (g), is represented by the equation

$$C_6H_6 \text{ (l)} \rightleftarrows C_6H_6 \text{ (g)} \tag{12.1}$$

The ideal gas law predicts that the partial pressure exerted by the vapor at any time will depend upon the concentration of molecules in the vapor state, expressed as the number of moles of gas per unit volume. Solving $pV = nRT$ for partial pressure,

$$p = \frac{nRT}{V} = RT \times \frac{n}{V}$$

With both R and T constant, it follows that the partial pressure of the vapor is proportional to n/V, which is moles per unit volume, or concentration. When the vapor concentration becomes constant at equilibrium, the vapor pressure also becomes constant. This partial pressure exerted by a vapor in equilibrium with its liquid phase at a given temperature is the *equilibrium vapor pressure* of the substance at that temperature.

## THE EFFECT OF TEMPERATURE

PG 12 G Describe the relationship between vapor pressure and temperature for a liquid-vapor system in equilibrium; explain this relationship in terms of the kinetic molecular theory.

The vapor pressure of a liquid always increases as the temperature rises. It is well known that water evaporates more rapidly on a hot day. We also observe that the stoppers in bottles of such volatile liquids as ether or gasoline "pop out" when the temperature rises and the pressure inside the bottle increases. Laboratory experiments have measured the equilibrium vapor pressures of many liquids at different temperatures. Some of these are plotted in Figure 12.10.

The rapid increase in vapor pressure with temperature is understood in terms of how equilibrium is reached in a liquid-vapor system. It was pointed out earlier in this section (page 283) that the percentage of molecules having sufficient kinetic energy to evaporate increases with temperature, resulting in a higher evaporation rate. This means that the condensation rate must also be greater if equilibrium is present at the higher temperature. This takes a higher equilibrium vapor concentration, which in turn exerts a larger vapor pressure.

The extent to which vapor pressure increases with temperature may be surprising. The vapor pressure of water at 25°C, for example, is 24 torr, but at 60°C it is 149 torr, an increase of about 500%. The absolute temperature increase from 298°K to 333°K is only 35°, or about 12%. Clearly the vapor pressure change is not a gas law pressure increase, in which pressure is proportional to absolute temperature. It is a fact, however, that the percentage

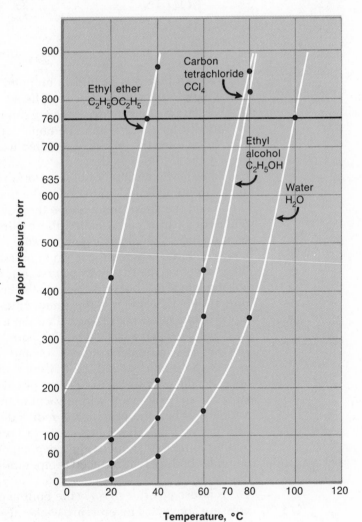

**Figure 12.10** Vapor pressures of some common liquids at different temperatures.

of water molecules with sufficient kinetic energy to evaporate increases by about 600% over the same temperature range, matching the percentage increase in vapor pressure. It is the *number of particles with enough energy to evaporate* that is responsible for the vapor pressure dependence on temperature.

Perhaps this idea of percentage increases may be more easily understood by comparing it to the heights of one hundred 15-year-old students having an average height of 5'4", or 64 inches, with only five students (5%) that are six feet tall. Three years later they are measured again, and the average height has increased to 5'8", or 68 inches, and fifteen (15%) have passed the six foot mark. At both times the large majority are shorter than six feet, and the average height has increased only 4" out of 64", or about 6%, over the three year period; but the *number* of students having reached six feet has *tripled*, or increased by 200%, over the same period.

## 12.5  THE PHENOMENON OF BOILING

PG 12 H  Describe the process of boiling and the relationship between boiling point, vapor pressure and surrounding pressure.

When a liquid is heated in an open container, bubbles form, usually at the base of the container where heat is being applied. The first bubbles that we see are often air, driven out of solution by an increase in temperature. Eventually, however, when a certain temperature is reached, vapor bubbles form throughout the liquid, rise to the surface, and break. When this happens we say the liquid is boiling.

In order for a stable bubble to form in a boiling liquid, the vapor pressure within the bubble must be high enough to push back the surrounding liquid and the atmosphere above the liquid. The minimum temperature at which this can occur is called the **boiling point: the boiling point is that temperature at which the vapor pressure of the liquid is equal to the pressure above its surface.** Actually the vapor pressure within a bubble must be a tiny bit greater than the surrounding pressure, which suggests that bubbles probably form in local "hot spots" within the boiling liquid. The boiling temperature at one atmosphere—the temperature at which the vapor pressure is equal to one atmosphere—is called the **normal boiling point.** From Figure 12.10 we see that the normal boiling point of water is 100°C; of ethyl alcohol, 78°C; of carbon tetrachloride, 77°C; and of ethyl ether, 35°C.

As you might expect, the boiling point of a liquid can be reduced by lowering the pressure above it. It is possible to boil water at 25°C by evacuating the space above it with a vacuum pump or even with a simple water aspirator. When the pressure is reduced to 24 torr, the equilibrium vapor pressure at 25°C, the water starts to boil. Chemists often purify a high-boiling compound, which might decompose or oxidize at its normal boiling point, by boiling it at reduced temperature under vacuum and condensing the vapor.

It is also possible to *raise* the boiling point of a liquid by *increasing* the pressure above it. The pressure cooker used in the kitchen takes advantage of this effect. By allowing the pressure to build up within the cooker, it is possible to reach temperatures as high as 110°C without boiling off the water. At this temperature, foods cook in about half the time required at 100°C.

## 12.6 THE NATURE OF THE SOLID STATE

PG 12 I Distinguish between crystalline and amorphous solids.

Solids can be classified into two different categories, depending upon the degree of order in their structure. In a **crystalline solid,** the particles are arranged in a fixed geometric pattern called a **crystal lattice** which repeats itself over and over again in three dimensions. Each particle is restricted to a particular site in the crystal lattice. It can vibrate about that site but cannot move past its neighbors. The high degree of order often leads to large crystals which have a precise geometric shape. In ordinary table salt, we can distinguish small cubic crystals of sodium chloride. Large, beautifully formed crystals of such minerals as quartz ($SiO_2$) and fluorite ($CaF_2$) are found in nature. Figure 12.11 has photographs of three crystalline solids.

In an **amorphous solid** such as glass, rubber or polyethylene, there is no long-range order. Even though the arrangement around a particular site may resemble that in a crystal, the pattern does not repeat itself throughout the solid (Figure 12.12). From a structural standpoint, we may regard an amorphous solid as intermediate between the crystalline and the liquid states. In many amorphous solids, the particles have some freedom to move with respect to one another. The elasticity of rubber and the tendency of glass to flow when subjected to stress over a long period of time suggest that the particles in these materials are not rigidly fixed in position.

Crystalline solids have characteristic physical properties which can serve to identify them. Sodium chloride, for example, melts sharply at 800°C. This is in striking contrast to glass, which first softens and then slowly liquifies over a wide range of temperatures. The physical properties of a few crystalline solids are listed in Table 12.3.

**Figure 12.11**  Crystals of NaCl, $NH_4H_2PO_4$ and $CuSO_4 \cdot 5H_2O$.

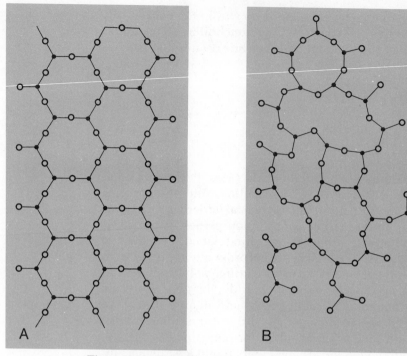

**Figure 12.12** Crystalline (A) and amorphous (B) solids.

## 12.7 TYPES OF CRYSTALLINE SOLIDS

PG 12 J Distinguish between the following types of crystalline solids: ionic; molecular; macromolecular; and metallic.

Solids such as those listed in Table 12.3 can be divided into four classes on the basis of their particle structure and the type of forces that hold these particles together in the crystal lattice:

TABLE 12.3  Physical Properties of Crystalline Solids

| SUBSTANCE | FOR-MULA | MELTING POINT (°C) | HEAT OF FUSION* | WATER SOLU-BILITY** | ELECTRICAL CONDUCTIVITY |
|---|---|---|---|---|---|
| Naphthalene | $C_{10}H_8$ | 80 | 4.61 | 0.03 | nonconductor |
| Iodine | $I_2$ | 114 | 4.02 | 0.3 | nonconductor |
| Potassium Nitrate | $KNO_3$ | 333 | 2.6 | 316 | nonconductor† |
| Sodium Chloride | NaCl | 800 | 6.9 | 360 | nonconductor† |
| Copper | Cu | 1083 | 3.1 | 0 | conductor |
| Iron | Fe | 1535 | 2.7 | 0 | conductor |
| Silicon Dioxide | $SiO_2$ | 1710 | — | 0 | nonconductor |
| Magnesium Oxide | MgO | 2800 | — | 0.006 | nonconductor† |
| Diamond | C | 3500 | — | 0 | nonconductor |

*kcal/mole
**g/1000 g water at room temperature
†conducts when melted

1. **Ionic crystals.** Examples: NaF, $CaCO_3$, AgCl, $NH_4Br$. Oppositely charged ions are held together by strong electrostatic forces (recall Figure 9.2, page 189). As pointed out earlier, ionic crystals are typically high-melting, frequently water-soluble, and have very low electrical conductivities. Their melts and water solutions, in which the ions are mobile, conduct electricity readily.

2. **Molecular crystals.** Examples: $I_2$, ICl. Small discrete molecules are held together by relatively weak intermolecular forces of the types discussed in Section 12.3. Molecular crystals are typically soft, low-melting, and generally (but not always) insoluble in water. They usually dissolve in nonpolar or slightly polar organic solvents such as carbon tetrachloride or chloroform. Molecular substances, with rare exceptions, are nonconductors when pure, even in the liquid state.

3. **Macromolecular crystals.** Examples: C (diamond), $SiO_2$ (quartz). Atoms are covalently bonded to each other to form one "huge molecule" making up the entire crystal.

There are no small, discrete molecules in macromolecular crystals. In diamonds each carbon atom is covalently bonded to four other carbon atoms to give a network structure that extends throughout the entire crystal (see Fig. 12.13A). The structure of silicon dioxide resembles that of diamond in that the atoms are held together by a continuous series of covalent bonds. Each silicon atom is bonded to four oxygen atoms, each oxygen to two silicons, as shown in Figure 12.13B.

Macromolecular solids, like ionic crystals, are high-melting. One has to go to very high temperatures to supply enough energy to break the covalent bonds holding the atoms together in crystals of quartz (mp = 1710°C) or diamond (mp = 3500°C). Unlike ionic solids, macromolecular crystals are almost always insoluble in water and, indeed, in any common solvent. They are generally poor conductors of electricity either in the solid or in the molten state.

A        B

**Figure 12.13** A, Model of diamond (carbon) crystal. B, Model of quartz ($SiO_2$) crystal. (Diamond photograph by Judy Serface; quartz photograph, courtesy of Klinger Scientific Apparatus Corporation, Jamaica, N.Y.)

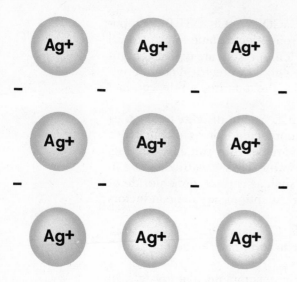

**Figure 12.14** Electron sea model of metallic bond.

4. **Metallic crystals.** A simple model of the bonding in a metal pictures it as consisting of an array of positive ions immersed in a "sea" of relatively mobile valence electrons. This so-called "electron sea" model of metals is illustrated in Figure 12.14 for silver. The $Ag^+$ ions form the backbone of the crystal; the electrons surrounding these ions are not tied down to any particular ion and hence are not restricted to a particular location. It is because of these freely moving electrons that metals are excellent conductors of electricity.

The general properties of the four kinds of crystalline solids are summarized in Table 12.4.

TABLE 12.4    Types of Crystals

| TYPE | EXAMPLES | PROPERTIES |
|---|---|---|
| Ionic | $KNO_3$, NaCl, MgO | High-melting; generally water-soluble; brittle; conduct only when melted or dissolved in water. |
| Molecular | $C_{10}H_8$, $I_2$ | Low-melting; usually more soluble in organic solvents than in water; non-conductors in pure state. |
| Macromolecular | $SiO_2$, C | Very high-melting; insoluble in all common solvents; brittle; non- or semi-conductors. |
| Metallic | Cu, Fe | Wide range of melting points; insoluble in all common solvents; malleable; ductile; good electrical conductors. |

# CHAPTER 12 IN REVIEW

## 12.1 THE NATURE OF THE LIQUID STATE

12 A  In terms of the relative distances between molecules and the effect of intermolecular forces, explain the differences in physical behavior between liquids and gases. (Page 273)

## 12.2 PHYSICAL PROPERTIES OF LIQUIDS

12 B  For two liquids, given comparative values of physical properties that depend on intermolecular attractions, predict the relative strengths of those attractions; or, given a comparison of the strengths of intermolecular attractions, predict the relative values of physical properties that depend on them. (Page 274)

## 12.3 TYPES OF INTERMOLECULAR FORCES

12 C  Identify and describe or explain dipole forces, dispersion forces and hydrogen bonds. (Page 276)

12 D  Given the structure of a molecule, or information from which it may be determined, identify the significant intermolecular forces present. (Page 277)

12 E  Given the molecular structures of two substances, or information from which they may be obtained, compare or predict relative values of physical properties that are related to them. (Page 277)

## 12.4 LIQUID-VAPOR EQUILIBRIUM

12 F  Describe or explain the equilibrium between a liquid and its own vapor, and the process by which the equilibrium is reached. (Page 282)

12 G  Describe the relationship between vapor pressure and temperature for a liquid-vapor system in equilibrium; explain this relationship in terms of the kinetic molecular theory. (Page 284)

## 12.5 THE PHENOMENON OF BOILING

12 H  Describe the process of boiling and the relationship between boiling point, vapor pressure and surrounding pressure. (Page 286)

## 12.6 THE NATURE OF THE SOLID STATE

12 I  Distinguish between crystalline and amorphous solids. (Page 287)

## 12.7 TYPES OF CRYSTALLINE SOLIDS

12 J  Distinguish between the following types of crystalline solids: ionic; molecular, macromolecular; metallic. (Page 288)

## TERMS AND CONCEPTS

Vapor pressure (274)
Equilibrium vapor pressure (274)
Boiling point (275)
Molar heat of vaporization (275)
Viscosity (275)
Surface tension (275)
van der Waals forces (277)

Dipole (277)
Dipole forces (277)
Dispersion (London) forces (278)
Hydrogen bond (279)
Condensation (283)
Dynamic equilibrium (284)
Reversible change; reversible reaction (284)

## QUESTIONS AND PROBLEMS

*An asterisk (\*) identifies a question that is relatively difficult, or that extends beyond the performance goals of the chapter.*

### Section 12.1

12.1) Explain why gases are less dense than liquids.

12.2) Explain why water is less compressible than air.

### Section 12.2

12.3) Identify and explain the relationship between intermolecular attractions and equilibrium vapor pressure.

12.4) What is meant by molar heat of vaporization?

12.5) Which liquid is more viscous, water or motor oil? In which liquid do you suppose the intermolecular attractions are stronger? Explain.

12.6) A falling drop of any liquid tends to be spherical in shape. If the liquid is water there is a visible elongation of the drop into a "teardrop" form. A drop of mercury, however, remains more spherical. Compare the surface tension of water with that of mercury. What does this suggest about the strength of intermolecular attractions in water compared to the interatomic attractions in mercury?

12.7) One of the functions of soap is to alter the surface tension of water. Considering the purpose of laundry soap, do you think soap increases or decreases intermolecular attractions in water? Explain.

### Section 12.3

12.8) Explain the origin of the three major types of intermolecular forces.

12.28) Explain why two gases will mix with each other more rapidly than two liquids.

12.29) Explain why intermolecular attractions are stronger in the liquid state than in the gas state.

12.30) How do intermolecular attractions influence the boiling point of a pure substance?

12.31) Why is molar heat of vaporization dependent upon the strength of intermolecular attractions?

12.32) A tall glass cylinder is filled to a depth of 1 meter with water. Another tall glass cylinder is filled to a depth of 1 meter with syrup. Identical ball bearings are dropped into each tube at the same instant. In which tube will the ball bearing reach the bottom first? Explain your prediction in terms of viscosity and intermolecular attractions.

12.33) If water is spilled on a laboratory desk top it usually spreads over the surface, wetting any papers or books that may be in its path. If mercury is spilled, it neither spreads nor makes paper it contacts wet, but rather forms little drops that are easily combined into pools by pushing them together. Suggest an explanation for these facts in terms of the apparent surface tension and intermolecular attraction in mercury and in water.

12.34) The level at which a duck floats on water is determined more by the thin oil film that covers his feathers than by a lower density of his body compared to water. The water does not "mix" with the oil, and therefore does not penetrate the feathers. If, however, a few drops of "wetting agent" are placed in the water near the duck, the poor bird will sink. State the effect of wetting agent on surface tension and intermolecular attractions of water.

12.35) Under what circumstances are dispersion forces likely to produce stronger intermolecular attractions than dipole forces, and under what circumstances are the dispersion forces likely to be weaker?

12.9) Identify the principal intermolecular forces in each of the following compounds: HBr; $C_2H_2$; $NF_3$; $C_2H_5OH$.

12.10) Given that ionic compounds generally have higher melting points than molecular compounds of similar molar weight, compare dipole forces and forces between ions. How are they alike and how are they different?

12.36) Identify the principal intermolecular forces in each of the following compounds: $NH(CH_3)_2$; $CH_2F_2$; $C_3H_8$.

12.37) Compare dipole forces and hydrogen bonds. How are they different, and how are they similar?

*12.11, 12.12, 12.38 and 12.39: Predict, on the basis of molecular weight and polarity, which member of each of the following pairs has the higher boiling point, and state the reason for your choice:*

12.11) $CH_4$ and $CCl_4$.

12.12) $H_2S$ and $PH_3$.

12.38) $CH_4$ and $NH_3$.

12.39) Ar and Ne.

12.13) Hydrogen is usually covalently bonded to one of what three elements in order for hydrogen bonds to be present? What unique feature do these elements share that sets them apart from other elements?

12.14) Of the three types of intermolecular forces, which one(s) (a) operate in all molecular substances; (b) operate between all polar molecules?

12.15) In which of the following substances would you expect dipole forces to operate?
(a) H—C≡N   (b) O=C=O

12.40) What physical feature of the hydrogen atom, when covalently bonded to an appropriate second element, contributes significantly to the strength of hydrogen bonding between molecules?

12.41) Of the three types of intermolecular forces, which one(s) (a) account for the high melting point, boiling point and other abnormal properties of water; (b) increase with molecular size?

12.42) In which of the following substances would you expect hydrogen bonds to form?

*12.16 and 12.43: Predict which compound will have the higher melting and boiling points, and explain your prediction:*

12.16) $C_3H_8$ or $C_6H_{14}$

12.43) $CO_2$ or $CS_2$

## Section 12.4

12.17) What essential condition exists when we say a system is in a state of *dynamic* equilibrium?

12.44) Explain why the rate of evaporation from a liquid depends upon temperature. Explain why the rate of return from the vapor to the liquid state depends upon concentration in the vapor state.

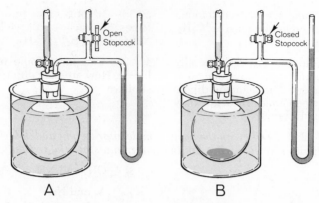

**Figure 12.15** Measurement of vapor pressure. A, The buret contains the liquid whose vapor pressure is to be measured. The flask, tubes and manometer above the mercury in the left leg are all at atmospheric pressure through the open stopcock. The mercury in the open right leg of the manometer is also at atmospheric pressure, so the mercury levels are the same in the two legs. B, To measure vapor pressure, stopcock is closed, trapping air in flask in space above the left mercury level. Liquid is introduced to flask from buret. Evaporation occurs until equilibrium is reached. Vapor causes increase in pressure, which is measured directly by the difference in mercury levels.

*Questions 12.18, 12.19, 12.45 and 12.46 are based on the apparatus shown in Figure 12.15. Study the caption that describes how vapor pressure is measured and then answer the questions.*

12.18)* A student uses the apparatus in Figure 12.15 to determine the equilibrium vapor pressure of a volatile liquid. He observes that the pressure increases rapidly at first, but more slowly as equilibrium is approached. Suggest a reason for this.

12.19) Suppose in making the vapor pressure measurement described in Figure 12.15, all of the liquid introduced into the flask evaporates. Explain what this means in terms of evaporation and condensation rates. How does the vapor pressure in the flask compare with the equilibrium vapor pressure at the existing temperature?

12.20) Using the ideal gas equation, show why the partial pressure of a gaseous component depends upon its vapor concentration at a given temperature.

12.21)* Three closed boxes have identical volumes. A beaker containing a small quantity of acetone, an easily vaporized liquid, is placed in one. Over a period of time it evaporates completely. A medium quantity of acetone is placed in the second. Eventually it evaporates until only a small portion of liquid remains. A larger quantity is placed in the third, and about half of it eventually evaporates. (a) Which box or boxes develop the greatest acetone vapor pressure? (b) Which probably has the least? (c) Explain both answers.

12.45)* Why would the apparatus in Figure 12.15 be of little or no value in determining the equilibrium vapor pressure of water if used as described? Under what conditions might the apparatus give acceptable values for water vapor pressure?

12.46) After the system has come to equilibrium, as in Figure 12.15B, an additional volume of liquid is introduced to the flask. Describe and explain what will happen to the pressure indicated by the manometer. Disregard any Boyle's Law effect; assume the change in gas volume resulting from increased liquid volume is negligible.

12.47) Using the ideal gas equation, show why equilibrium vapor pressure is dependent upon temperature.

12.48)* Three closed boxes have different volumes; one is small, one medium sized, and one large. Beakers containing equal quantities of acetone are placed in the boxes. Eventually all the acetone evaporates in one box, but equilibrium is reached in the other two. (a) In which box does complete evaporation occur? (b) Compare the eventual vapor pressures in the three boxes. (c) Explain both answers.

## Section 12.5

12.22)  Define boiling point. Draw a vapor pressure-temperature curve and locate the boiling point on it.

12.23)  The vapor pressure of a certain compound at 20°C is 906 torr. Is the substance a gas or a liquid at 760 torr? Explain.

12.24)  Explain why high-boiling liquids usually have low vapor pressures.

12.25)  The molar heat of vaporization of substance X is 8.1 kcal/mole; of substance Y, 6.4 kcal/mole. Which substance would be expected to have the higher normal boiling point; the higher vapor pressure at 25°C?

12.49)  An industrial process requires boiling a liquid whose boiling point is so high that maintenance costs on associated pumping equipment are prohibitive. Suggest a way this problem might be solved.

12.50)  Normally a gas may be condensed by cooling it. Suggest a second method, and explain why it will work.

12.51)  Explain why low-boiling liquids usually have low molar heats of vaporization.

12.52)  At 20°C the vapor pressure of substance M is 520 torr; of substance N, 634 torr. Predict which substance will have the lower boiling point; the lower molar heat of vaporization.

## Section 12.6

12.26)  List the properties that distinguish a crystalline solid from an amorphous solid.

12.53)  Is ice a crystalline solid or an amorphous solid? On what properties do you base your conclusion?

## Section 12.7

12.27 and 12.54: *For each of the solids having the physical properties tabulated below, indicate whether it is most likely to be ionic, molecular, macromolecular or metallic:*

|         | SOLID | MELTING POINT | WATER SOLUBILITY | CONDUCTIVITY (PURE) | TYPE OF SOLID |
|---------|-------|---------------|------------------|---------------------|---------------|
| 12.27)  | A     | 150°C         | Insoluble        | Nonconductor        | _____   |
|         | B     | 1450°C        | Insoluble        | Excellent           | _____   |
| 12.54)  | A     | 2000°C        | Insoluble        | Nonconductor        | _____   |
|         | B     | 1050°C        | Soluble          | Nonconductor        | _____   |

# 13

# ENERGY IN PHYSICAL
# AND CHEMICAL CHANGE

### 13.1 PHYSICAL AND CHEMICAL
### ENERGY—A REVIEW

When the subject of *energy* was introduced qualitatively in Chapter 2 (page 18), many of its basic ideas were described. Energy was defined as the ability to do *work*, or *to exert a force* over a distance. Two kinds of *mechanical energy* were identified. The energy possessed by an object in motion is *kinetic energy*, whereas energy possessed by an object because of its position in a force field is *potential energy. Temperature* is a measure of the average kinetic energy of the particles in a sample of matter (page 243), and *chemical energy* turns out to be associated with potential energy changes resulting from the rearrangement of electrons, protons, atoms, ions and molecules in electric fields.

A natural tendency for a system to reach its lowest energy state was identified as one of the major driving forces underlying the formation of chemical bonds (page 193). The energy of a chemical change is the combination of energies absorbed and released as old bonds are broken and new bonds are formed. In the last chapter (pages 276–282) we considered intermolecular forces, which have their origin in electrostatic attractions and repulsions. Molar heat of vaporization is one physical property that is directly concerned with energy; two others will be introduced in this chapter.

*The Law of Conservation of Energy* (page 19) reminds us that in ordinary (nonnuclear) physical and chemical changes energy may be converted from one form to another, but it is neither created nor destroyed. This suggests that chemical energy may be changed to such things as electrical energy in car batteries, kinetic energy in a moving automobile, sound in a firecracker, light in flames, and heat in the warming of our homes. A detailed study of energy and work related to chemical change takes nearly all of these into account. For our purposes, however, we will limit consideration to the most obvious of these forms of energy, heat. In fact, the terms *endothermic* and *exothermic* (page 18) refer respectively to the absorption or release of heat energy.

## 13.2  UNITS OF ENERGY

So far our descriptions of energy have been entirely qualitative. In this chapter we will approach the quantitative side of the subject. This means we will have to identify a unit of energy. The SI energy unit is the **joule,** which is assigned the symbol J. The joule is formally defined in terms of mechanical work, a force times a distance: one joule is a force of one newton (another SI unit) acting over a distance of one meter. The joule is a particularly convenient energy unit in electrochemistry, but heat energy is generally measured in a more familiar unit, the *calorie.* **One calorie is the quantity of heat required to raise the temperature of one gram of water one degree Celsius.*** The relationship between a calorie and a joule is

$$\text{1 calorie} = 4.184 \text{ joules} \qquad (13.1)$$

Both the calorie and the joule are small quantities of energy, so energy is often expressed in terms of the *kilo-* unit 1000 times larger, the kilocalorie (kcal) and kilojoule (kJ):

$$\text{1 kcal} = 1000 \text{ cal; 1 kJ} = 1000 \text{ J} \qquad (13.2)$$

All energy calculations will be expressed in calories or kilocalories in this text. Each example or problem answer, however, will be followed by its equivalent in joules or kilojoules, enclosed in square brackets.

The term *Calorie* used in discussion of nutrition is actually the kilocalorie. Capitalization of the first letter is used to distinguish the larger unit from the smaller. Caloric requirements from food vary considerably according to individual needs and activities. A small adult doing average physical work can probably derive sufficient energy from the intake of as little as 2000–2400 Calories per day. This requirement may increase to 5000 to 6000 Calories per day for a large man engaged in strenuous physical labor.

A precautionary note: when a person begins the quantitative consideration of energy in the form of heat, there is a strong tendency to think of the temperature degree as a unit of heat. It is not. The distinction between temperature and energy can probably be seen in a practical kitchen procedure. If you place a small quantity of water on a stove to bring it from room temperature to boiling for, say, two cups of tea, you must deliver a certain amount of energy from your electric or gas range. The final temperature is 100°C. If you perform the identical act that will provide 20 cups of tea for a party, you will heat to boiling ten times the quantity of water required for two cups. This will require ten times as much heat as is needed for two cups. Yet the 100° temperatures reached are the same. Clearly temperature degrees are not suitable units for quantity of heat energy.

---

*More exactly, the calorie is the amount of heat required to raise the temperature of one gram of water from 14.5°C to 15.5°C. The heat required for a 1° change varies slightly with temperature.

## 13.3 ENERGY, TEMPERATURE AND CHANGE OF STATE

PG 13 A Sketch, interpret and/or identify regions in a graph of temperature versus energy for a pure substance over a temperature range from below the melting point to above the boiling point.

When you remove a cube of ice from the refrigerator, it feels dry to the touch. This is because it *is* dry; it is all solid at a temperature below the melting point. Using Figure 13.1, which is drawn for one gram of ice, we will follow what happens as the ice absorbs energy through several changes in state and temperature. For purposes of our analysis we will assume that temperature changes and heat absorption occur simultaneously and uniformly throughout the sample.

1. The ice from the refrigerator is represented as Point A, at $-10°C$. As the ice absorbs energy, moving to the right on the graph, the temperature gradually increases to $0°C$, rising vertically on the graph. This change is represented by the line from A to B. The ice remains dry; it is entirely solid throughout the change.

2. As the ice continues to absorb energy from the surroundings it begins to melt, but *the temperature remains constant at 0°C*. The melting process is represented by the horizontal line (constant temperature) from B to C on the graph. Between B and C the sample is partly solid, partly liquid.

3. As energy is absorbed by liquid water at the freezing point, temperature increases once again. When room temperature is reached heat must be supplied by a burner or hotplate. Temperature increases until the boiling point is reached at D. Throughout the change from C to D the entire sample is in the liquid state.

4. Heat input beyond Point D causes boiling, another constant temperature process. The sample, partly liquid and partly gas between D and E, remains at $100°C$, until vaporization is complete at E.

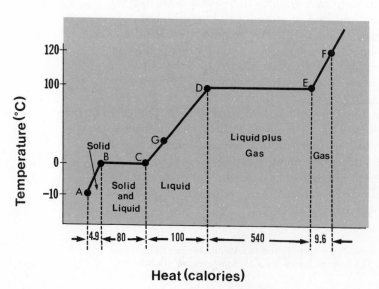

Figure 13.1 Temperature vs. energy absorbed as one gram of ice at $-10°C$ is warmed, melted to the liquid state, heated to the boiling point, boiled, and then heated to $120°C$. The curve is typical of most pure substances passing between the solid and vapor states. (Axes are not drawn to scale.)

5. As energy is absorbed beyond E, the gas temperature rises. At F the sample is superheated steam, a gas at a temperature higher than its boiling point, just as the oxygen in the air is a gas above its boiling point.

The temperature-energy curves for most pure substances are similar to Figure 13.1. The horizontal lines represent changes of state at constant temperature; the slanted lines represent temperature increases with a single state, solid, liquid or gas. Two *kinds* of calculations emerge from this analysis: (1) Quantitative consideration of temperature changes within a single state involve *specific heat*. (2) State changes involve what is sometimes called *latent heat*, but more frequently identified with the change in question, such as *heat of fusion, heat of solidification, heat of vaporization or heat of condensation*. We will now examine each of these separately.

## 13.4  ENERGY AND CHANGE OF TEMPERATURE: SPECIFIC HEAT

PG  13 B  Given (1) the mass of a pure substance, (2) its specific heat, and (3) its temperature change, or initial and final temperatures, calculate the heat flow.

13 C  Given the amount of heat flow to or from a known mass of a substance, and its temperature change, or initial and final temperatures, calculate the specific heat of the substance.

In order to raise the temperature of a substance you must heat it; heat must flow into the substance. If the substance is to be cooled, there must be heat flow from the substance to its surroundings. Experiments such as those illustrated in Figures 13.2 and 13.3 indicate that heat flow is proportional to both the mass of the sample and its temperature change. Combining these proportionalities and introducing a proportionality constant yields the equation

$$Q = m \times c \times \Delta T \qquad (13.3)$$

where Q is the heat flow in calories, m is the mass in grams, c is the proportionality constant and $\Delta T$ is the temperature change, or final temperature minus initial temperature, $T_f - T_i$.*

The proportionality constant, c, is a property of a pure substance called **specific heat. Specific heat is the heat flow required to change the temperature of one gram of a substance one degree Celsius.** The units of specific heat may be determined by solving Equation 13.3 for the proportionality constant:

$$\text{Specific heat} = c = \frac{Q}{m \times \Delta t} = \frac{\text{calories}}{(\text{grams})(°C)} \qquad (13.4)$$

---

*The Greek letter **delta**, $\Delta$, is used in scientific writing to designate change. It represents the final value of a measurement minus the initial value: $\Delta X = X(\text{final}) - X(\text{initial})$. Applied to temperature, for example, the initial temperature is subtracted from the final temperature, as warming from 20°C to 25°C yields $\Delta T = 25°C - 20°C = 5°C$. Note that change may be negative as well as positive. In *cooling* from 25°C to 20°C, $\Delta T$ is $20°C - 25°C = -5°C$. $\Delta X$ will be negative whenever the final value of X is less than the initial value.

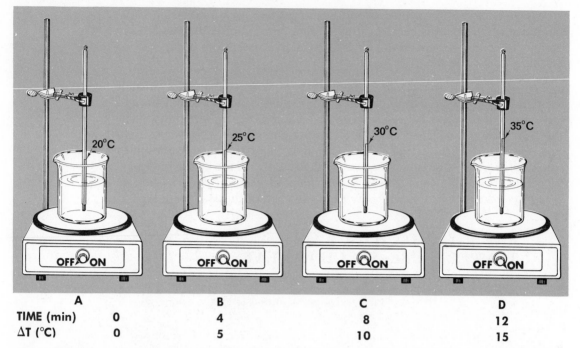

| | A | B | C | D |
|---|---|---|---|---|
| TIME (min) | 0 | 4 | 8 | 12 |
| ΔT (°C) | 0 | 5 | 10 | 15 |

**Figure 13.2** Experiment demonstrating relationship between heat absorbed by a sample of matter and its temperature change. A beaker containing 100 grams of water at 20°C is placed on a hot plate (A). Assume that, when the hot plate is turned on, water absorbs heat at a constant rate. This makes time proportional to heat absorbed; the longer the water is heated, the more heat it absorbs. Therefore time is a "measure" of heat flow to the water. After 4 minutes (B), the temperature has risen to 25°C. At 8 minutes (C) the temperature is 30°C, and at 12 minutes (D), 35°C. These data and the total change in temperature, ΔT, are tabulated beneath the illustrations. The data show that ΔT is proportional to time. At 4 minutes, ΔT is 5°C. At 8 minutes (twice 4), ΔT is 10°C (twice 5°); and at 12 minutes (3 times 4), ΔT is 15°C (3 × 5). Because time is a measure of heat, we conclude ΔT is proportional to heat absorbed for a fixed quantity of water.

These units are read "calories per gram degree." Recalling the definition of the calorie (page 297), it follows that the specific heat of water is 1 calorie per gram degree (or 4.18 joules per gram degree in SI units). Specific heats of other substances are given in Table 13.1.

Specific heat problems may be solved algebraically by direct substi-

TABLE 13.1   Selected Specific Heats

| SUBSTANCE | $\frac{cal}{(g)(°C)}$ | $\frac{J}{(g)(°C)}$ | SUBSTANCE | $\frac{cal}{(g)(°C)}$ | $\frac{J}{(g)(°C)}$ |
|---|---|---|---|---|---|
| *Elements* | | | | | |
| Aluminum | 0.21 | 0.88 | Sulfur | 0.175 | 0.732 |
| Carbon: | | | Zinc | 0.092 | 0.38 |
| Diamond | 0.12 | 0.50 | | | |
| Graphite | 0.17 | 0.71 | *Compounds* | | |
| Cobalt | 0.11 | 0.46 | Acetone | 0.51 | 2.1 |
| Copper | 0.092 | 0.38 | Benzene | 0.42 | 1.8 |
| Gold | 0.031 | 0.13 | Carbon | | |
| Iron | 0.106 | 0.444 | tetrachloride | 0.20 | 0.84 |
| Lead | 0.038 | 0.16 | Ethanol | 0.59 | 2.5 |
| Magnesium | 0.24 | 1.0 | Methanol | 0.61 | 2.6 |
| Silicon | 0.17 | 0.71 | Ice [$H_2O(s)$] | 0.49 | 2.1 |
| Silver | 0.057 | 0.24 | Steam [$H_2O(g)$] | 0.48 | 2.0 |

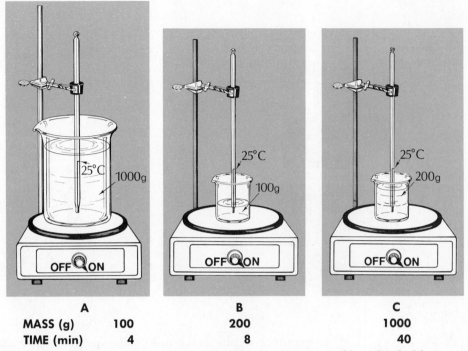

|  | **A** | **B** | **C** |
|---|---|---|---|
| **MASS (g)** | 100 | 200 | 1000 |
| **TIME (min)** | 4 | 8 | 40 |

**Figure 13.3** Experiment demonstrating relationship between mass and heat absorbed for a constant temperature change. All temperature changes are from 20°C to 25°C. Beaker A contains 100 grams of water, and requires 4 minutes. Beaker B contains 200 grams (twice as much as A) and requires 8 minutes (twice as long as A). Beaker C holds 1000 grams (10 times A) and requires 40 minutes (10 times A). Assuming heat absorbed is proportional to time, as in Figure 13.2, we conclude that heat absorbed is proportional to mass for a constant temperature change.

tution into Equation 13.3. The method is the same whether the substance is a solid, liquid or gas and there is no *change* of state. You must be sure, of course, to use the specific heat for that particular state the substance is in over the temperature range of the problem. Resulting answers correspond to changes from one point to another on one of the *slanted* lines in Figure 13.1, A–B, C–D or E–F.

---

**Example 13.1** How much heat is required to raise the temperature of 500 grams of water for a pot of tea from 14°C to 90°C? Answer in both calories and kilocalories.

Direct substitution into Equation 13.3 yields the heat flow in calories, which may be expressed in kilocalories too. For $\Delta T$, substitute T(final) − T(initial).

- - - - - - - - - - - - - - - - - - - - - - - - - - - - - - - - - - - - - - - - - - - - - - - - - - - - - - - - - - - - - - - - - -

**13.1a** 38,000, or $3.8 \times 10^4$ cal; 38 kcal $\quad$ [$1.6 \times 10^5$ J; $1.6 \times 10^2$ kJ]

$$Q = m \times c \times \Delta T = 500 \ \cancel{g} \times \frac{1 \ \text{cal}}{(g) \ (°\cancel{C})} \times (90 - 14)°\cancel{C} = 38,000 \ \text{cal} \quad [160,000 \ \text{J}]$$

$$38,000 \ \cancel{\text{cal}} \times \frac{1 \ \text{kcal}}{1000 \ \cancel{\text{cal}}} = 38 \ \text{kcal.} \quad [1.6 \times 10^2 \ \text{kJ}]$$

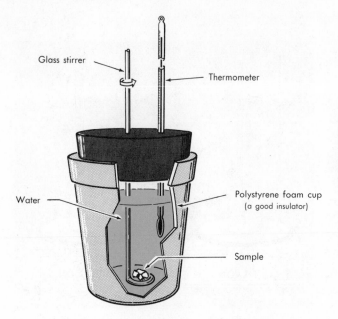

Glass stirrer

Thermometer

Water

Polystyrene foam cup
(a good insulator)

Sample

**Figure 13.4** Coffee cup calorimeter often used in college chemistry laboratory experiments. The heat flow from a physical or chemical change is determined by measuring the temperature change in a known mass of water.

As a quantitative physical property, the specific heat of a substance must be determined by experimental measurements. If in a calorimeter (see Figure 13.4) you determine the calories of heat gained or lost by a known mass of the substance as it passes through a measured temperature change, the specific heat of that substance may be calculated by direct substitution into Equation 13.4.

---

**Example 13.2**  In a college laboratory experiment a student observes an increase from 25.0°C to 31.7°C when 141 grams of aluminum absorb 192 calories of heat. Calculate the specific heat of aluminum *from these data.*

---

**13.2a**  0.20 cal/(g) (°C)    [0.84 J/(g) (°C)]

$$\text{Specific heat} = c = \frac{\text{calories}}{(\text{grams}) \, (°C)} = \frac{192 \text{ cal}}{(141 \text{ g}) \, (31.7{-}25.0)°C} = 0.20 \text{ cal/(g) (°C)}$$
$$[0.84 \text{ J/(g) (°C)}]$$

---

If any three of the four quantities, heat flow, mass, specific heat or temperature change, are known, the fourth may be determined algebraically by substitution into Equation 13.3. For substances other than water, specific heat values from Table 13.1 should be used. Notice that $\Delta T$ will have a negative value if final temperature is lower than initial temperature. The resulting negative heat flow indicates simply that the substance is cooling rather than heating.

# 13.5 ENERGY AND CHANGE OF STATE: LATENT HEAT

We will now consider heat flow as a substance changes state, known as **latent heat.** Change of state energies appear as horizontal lines on Figure 13.1, page 298. Line B–C represents the change between a solid and a liquid, and line D–E is for the change between a liquid and gas. We will examine the liquid-gas change first.

## *HEAT OF VAPORIZATION*

> PG 13 D Given the quantity of a pure substance changing between the liquid and vapor states, and the heat of vaporization, calculate the heat flow.

Any time you lift an object from the floor and place it on a table, you do work. There is a gravitational attraction between the object and the earth. You must expend energy—do work—against that attractive force when you increase the distance of separation between the objects. Similarly there is an attractive force between molecules in a liquid. To separate those molecules and convert the liquid to a gas requires energy. To boil water, you heat it. When a liquid evaporates at room temperature, heat is absorbed from the surroundings. This is why you feel cold when you come out of a shower or swimming pool; the evaporation of moisture draws heat from the body surface, lowering its temperature.

In Section 12.2 (page 275), *molar heat of vaporization* was defined as *the energy required to vaporize one mole of a liquid at its boiling point.* From the definition, molar heat of vaporization is measured in energy units per mole, usually kilocalories per mole. An alternative approach to change of state problems identifies simply **heat of vaporization as the energy required to vaporize one gram of a liquid at its boiling point.** According to this definition, heat of vaporization units are calories per gram. SI values are usually given in kilojoules per gram. Heats of vaporization for several substances are given in Table 13.2.

When a vapor condenses to a liquid at the boiling point the reverse energy change occurs. The heat flow may then be referred to as **heat of condensation.** Values are identical to heats of vaporization, except that they are negative, indicating that heat is flowing *from* the substance (an exothermic change), rather than into it.

The energy required to vaporize a given quantity of a liquid may be calculated from the equation

$$Q = m \times \Delta H_{vap} \qquad (13.5)$$

where m is mass in grams and $\Delta H_{vap}$ is heat of vaporization in calories per gram. Energy changes calculated from this equation correspond with the energy change from D to E in Figure 13.1. Applications are straightforward "substitute-and-solve" problems.

TABLE 13.2   Latent Heats of Selected Substances

| SUBSTANCE | MELTING POINT (°C) | BOILING POINT (°C) | HEAT OF FUSION | | HEAT OF VAPORIZATION | |
|---|---|---|---|---|---|---|
| | | | (cal/g) | (J/g) | (cal/g) | (kJ/g) |
| H₂O (s) H₂O (l) | 0 | 100 | 80 | 335 | 540 | 2.26 |
| Na | 98 | 892 | 27 | 113 | 1020 | 4.27 |
| NaCl | 801 | 1413 | 124 | 519 | | |
| Cu (s) Cu (l) | 1083 | 2595 | 49 | 205 | 1150 | 4.81 |
| Zn | 419 | 907 | 24 | 100 | 420 | 1.76 |
| Bi | 271 | 1560 | 13 | 54 | | |
| Pb | 327 | 1744 | 5.5 | 23 | | |
| Ni | 1453 | 2732 | 74 | 310 | | |

---

**Example 13.3**   Calculate the energy required to vaporize 250 grams of water at its boiling point. Express the answer in kilocalories.

*Solution.*   From Table 13.2, $\Delta H_{vap}$ = 540 cal/gram for water. Substituting into Equation 13.5,

$$Q = m \times \Delta H_{vap} = 250 \ g \times \frac{540 \ cal}{1 \ g} = 135{,}000 \ cal = 135 \ kcal \quad [565 \ kJ]$$

---

**Example 13.4**   If the molar heat of vaporization of benzene is 7.3 kcal/mole, how much energy is absorbed as 62.5 grams of benzene, $C_6H_6$, vaporize at the boiling point?

Again the problem is a direct substitution into Equation 13.5, but be sure to include units and make any conversions that may be required.

------------------------------------------------------------

**13.4a**   Q = 5.8 kcal   [24 kJ]

$$Q = m \times \Delta H_{vap} = 62.5 \ g \ C_6H_6 \times \frac{1 \ mole \ C_6H_6}{78.0 \ g \ C_6H_6} \times \frac{7.3 \ kcal}{1 \ mole \ C_6H_6} = 5.8 \ kcal \quad [24 \ kJ]$$

---

## HEAT OF FUSION

PG   13 E   Given the quantity of a pure substance changing between the solid and liquid states, and the heat of fusion, calculate the heat flow.

To melt a solid its crystal lattice must be broken down. This is an endothermic process; energy must be supplied. This is evident when you consider that ice must be warmed to change it to liquid water, and candle wax melts from the heat of the burning candle. Conceptually similar to heat of vaporization, the **heat of fusion, $\Delta H_{fus}$, of a substance is the energy required to melt one gram of a solid at its melting point.** Again the definition indicates units of calories per gram, or joules per gram in SI units. Heat of fusion may also be given as molar heat of fusion in kilocalories or kilojoules per mole. Heats of fusion are generally much smaller than heats of vaporization, as may be seen in Table 13.2, where both values are given for several substances.

Just as condensation is the opposite of vaporization, freezing is the opposite of melting. The quantity of heat released in freezing a sample is identical to the heat required to melt that sample. Accordingly, **heat of solidification** is numerically equal to heat of fusion, but the sign is negative. Melting and freezing processes are associated with changes occurring along horizontal line B–C in Figure 13.1.

The heat flow equation for the change between solid and liquid is

$$Q = m \times \Delta H_{fus} \qquad (13.6)$$

Calculation methods are identical to those in vaporization problems.

---

**Example 13.5** Calculate the energy required to melt 135 grams of sodium at its melting point. Express the answer in both calories and kilocalories.

The heat of fusion of sodium may be found in Table 13.2. That and equation 13.6 are all you need . . . .

-------------------------------------------------------------------------------------------------

**13.5a**

$$Q = m \times \Delta H_{fus} = 135 \ \cancel{g} \times \frac{27 \ cal}{1 \ \cancel{g}} = 3.6 \times 10^3 \ cal = 3.6 \ kcal \qquad [1.5 \times 10^4 \ J = 15 \ kJ]$$

---

## 13.6 CHANGE IN TEMPERATURE PLUS CHANGE OF STATE

PG 13 F  Given (1) the quantity of a pure substance, (2) $\Delta H_{vap}$ and/or $\Delta H_{fus}$ of the substance and (3) average specific heat of the substance in the solid, liquid and/or vapor state, calculate the total heat flow in going from one state and temperature to another state and temperature.

The temperature changes and changes of state we have been considering occur independently of each other, and sometimes one after the other. This can be illustrated by retracing the five steps of change from Point A to Point F in Figure 13.1, this time quantitatively for one gram of water:

1. *Warming of ice from −10°C to 0°C.* This corresponds with the change from Point A to Point B in Figure 13.1. From Table 13.1 the specific heat of $H_2O$ (s) is 0.49 cal/(g)(°C). From Equation 13.3,

$$Q_{A-B} = m \times c \times \Delta T = 1.00 \, g \times \frac{0.49 \text{ cal}}{(g)(°C)} \times [0 - (-10)]°C = 4.9 \text{ cal } [20 \text{ J}]$$

2. *Melting ice at 0°C.* This corresponds with the change from Point B to Point C in Figure 13.1. From Table 13.2, $\Delta H_{fus}$ is 80 calories per gram. Using Equation 13.6,

$$Q_{B-C} = m \times \Delta H_{fus} = 1.00 \, g \times \frac{80 \text{ cal}}{1 \, g} = 80 \text{ cal } [330 \text{ J}]$$

The total heat input to convert 1 gram of ice at −10°C to liquid at the melting point is the heat to raise the temperature of the ice to the freezing point plus the heat to melt the ice:

$$Q_{A-C} = Q_{A-B} + Q_{B-C} = 4.9 + 80 = 85 \text{ cal } [360 \text{ J}]$$

3. *Warming water from 0°C to 100°C.* This corresponds with the change from Point C to Point D in Figure 13.1.

$$Q_{C-D} = m \times c \times \Delta T = 1.00 \, g \times \frac{1.00 \text{ cal}}{(g)(°C)} \times (100 - 0)°C = 100 \text{ cal } [418 \text{ J}]$$

The total heat to convert 1 gram of ice at −10°C to liquid at the boiling point is the sum of the three steps:

$$Q_{A-D} = Q_{A-B} + Q_{B-C} + Q_{C-D} = 4.9 + 80 + 100 = 185 \text{ cal } [774 \text{ J}]$$

4. *Boiling water at 100°C.* This corresponds with the change from Point D to Point E in Figure 13.1.

$$Q_{D-E} = m \times \Delta H_{vap} = 1.00 \, g \times \frac{540 \text{ cal}}{1 \, g} = 540 \text{ cal } [2260 \text{ J}]$$

The total heat to convert 1 gram of ice at −10°C to steam at the boiling point is the sum of the four steps:

$$Q_{A-E} = Q_{A-B} + Q_{B-C} + Q_{C-D} + Q_{D-E}$$
$$= 4.9 + 80 + 100 + 540 = 725 \text{ cal } [3030 \text{ J}]$$

5. *Heating steam from 100°C to 120°C.* This corresponds with the change from Point E to Point F in Figure 13.1.

$$Q_{E-F} = m \times c \times \Delta T = 1.00 \, g \times \frac{0.48 \text{ cal}}{(g)(°C)} \times (120 - 100)°C = 9.6 \text{ cal } [40 \text{ J}]$$

The total heat to convert 1 gram of ice at −10°C to steam at 120°C is the sum of the five steps:

$$Q_{A-F} = Q_{A-B} + Q_{B-C} + Q_{C-D} + Q_{D-E} + Q_{E-F}$$
$$= 4.9 + 80 + 100 + 540 + 9.6 = 735 \text{ cal } [3070 \text{ J}]$$

As this example illustrates, heat flow from one state and temperature to another state and temperature may be found by calculating the energy of each step, and then adding these energies. It helps to sketch a graph like Figure 13.1 and identify each step on the graph.

---

**Example 13.6**  How many kcal are required to convert 45.0 grams of water, initially at 25.0°C, to steam at 100.0°C?

This conversion may be visualized on Figure 13.1 if Point G is at a temperature of 25.0°C. Two heat calculations are required: first, from G to D, warming the water to 100.0°C; and second, from D to E, vaporizing it to steam. Calculate first the energy consumed in warming the 45.0 grams of water from 25.0°C to 100.0°C.

-------------------------------------------------------

**13.6a**  3.37 kcal [14.1 kJ]

$$45.0 \text{ g} \times \frac{1.00 \text{ cal}}{\text{g·°C}} \times (100.0 - 25.0)\text{°C} = 3380 \text{ cal} = 3.38 \text{ kcal} \quad [14.1 \text{ kJ}]$$

The heat required to vaporize 45.0 grams of water may now be determined.

-------------------------------------------------------

**13.6b**  24.3 kcal [102 kJ]

$$45.0 \text{ g} \times \frac{540 \text{ cal}}{1 \text{ g}} = 24300 \text{ cal} = 24.3 \text{ kcal} \quad [102 \text{ kJ}]$$

You now have the energy required for the separate steps. What is the total energy required for the whole process?

-------------------------------------------------------

**13.6c**  27.7 kcal [116 kJ]

$$3.38 + 24.3 = 27.7 \text{ kcal} \quad [116 \text{ kJ}]$$

---

**Example 13.7**  Copper melts at 1083°C. Calculate the number of kcal that will be lost by 250 grams of molten copper, initially at 1343°C, as it cools to the melting point, solidifies, and then cools as a solid down to room temperature, 26°C.

The required data may be found in Tables 13.1 and 13.2. Complete the problem.

-------------------------------------------------------

**13.7a**  43 kcal [180 kJ]

Cooling liquid:

$$250 \; \cancel{g} \times \frac{0.10 \; \text{cal}}{\cancel{g} \cdot \cancel{°C}} \times (1083 - 1343)\cancel{°C} = -6500 \; \text{cal} = -6.5 \; \text{kcal} \; [-27.2 \; \text{kJ}]$$

(The minus sign indicates heat is being lost, not absorbed.)

Solidification:

$$250 \; \cancel{g} \times \frac{-49 \; \text{cal}}{1 \; \cancel{g}} = -12{,}000 \; \text{cal} = -12 \; \text{kcal} \qquad [-51 \; \text{kJ}]$$

Cooling solid:

$$250 \; \cancel{g} \times \frac{0.092 \; \text{cal}}{\cancel{g} \cdot \cancel{°C}} \times (26 - 1083)\cancel{°C} = -24{,}000 \; \text{cal} = -24 \; \text{kcal} \; [-100 \; \text{kJ}]$$

Total heat flow is $-6.5 + (-12) + (-24) = -43$ kcal $[-180$ kJ$]$. The total heat *lost* is 43 kcal [180 kJ].

## 13.7  THERMOCHEMICAL EQUATIONS

> PG  13  G  Given a chemical equation, or information from which it may be written, and the heat (enthalpy) of reaction, write the thermochemical equation in two forms.

At the beginning of Chapter 7 (page 143) we introduced the idea of chemical equations by examining the reaction when sodium is placed into water. It was noted that the solution formed became hot and released heat to the surroundings. The process was described by the equation

$$2 \; \text{Na (s)} + 2 \; H_2O \; \text{(l)} \rightarrow H_2 \; \text{(g)} + 2 \; \text{NaOH (aq)} + \text{heat}$$

It was then pointed out that nearly all chemical changes involve an energy transfer, usually in the form of heat. You have since learned that these heat effects result from the rearrangement of charged bodies in electric force fields as existing chemical bonds are broken and new bonds formed. We will now consider quantitatively the heat effects of a chemical change.

Heat flow resulting from a chemical change is referred to as **heat of reaction** or **enthalpy of reaction. Enthalpy, designated H, may be thought of as "heat content," the amount of heat possessed by a chemical substance at a given temperature and pressure.** The enthalpy of individual substances cannot be measured directly, but changes in enthalpy may be determined in calorimeters. "Heat of reaction" is therefore equivalent to "change of enthalpy," or $\Delta H$.

If you burn two moles of ethane, $C_2H_6$, 736 kilocalories are released. The $\Delta H$ of the reaction is $-736$ kcal, the negative sign indicating that energy has been lost to the surroundings. There are two ways to express this in a chemical equation. One is to write the equation in the usual way, including the energy term as a "product":

$$2 \ C_2H_6 \ (g) + 7 \ O_2 \ (g) \rightarrow 4 \ CO_2 \ (g) + 6 \ H_2O \ (l) + 736 \ kcal \quad (13.7)$$

Alternatively, the enthalpy change is written separately, to the right of the conventional equation:

$$2 \ C_2H_6 \ (g) + 7 \ O_2 \ (g) \rightarrow 4 \ CO_2 \ (g) + 6 \ H_2O \ (l); \ \Delta H = -736 \ kcal \quad (13.8)$$

When writing thermochemical equations, state designations, i.e., (g), (l) and (s), *must* be used. The equation is ambiguous without them because the magnitude of enthalpy change depends upon the state of the reactants and products. If Equation 13.8 is written with water in the gaseous state, for example, $2 \ C_2H_6 \ (g) + 7 \ O_2 \ (g) \rightarrow 4 \ CO_2 \ (g) + 6 \ H_2O \ (g)$, the value of $\Delta H$ is $-673$ kcal.

The burning of ethane is, as might be expected, an exothermic reaction: heat is given off. The law of conservation of energy requires that the total heat content of the reactants be greater by 736 kcal than the total heat content of the products. In other words,

$$H \ products = H \ reactants - 736 \ kcal$$

It therefore follows that

$$\Delta H = H_P - H_R = -736 \ kcal$$

A graphic representation of this relationship is helpful in understanding the sign of $\Delta H$. In this graph, Figure 13.5A, enthalpy is plotted on the vertical axis. The enthalpy of the reactants ($2$ moles $C_2H_6$ + $7$ moles $O_2$) is shown by the horizontal line at the left. The line at the right gives the enthalpy of the products ($4$ moles of $CO_2$ + $6$ moles of $H_2O$). The difference between them represents $\Delta H$, the enthalpy of reaction. For every exothermic reaction, $H_{products} < H_{reactants}$; total enthalpy has been reduced in the reaction. Therefore $\Delta H$ is always negative for an exothermic reaction. By similar reasoning $\Delta H$ is always positive for an endothermic reaction, as shown by the enthalpy-reaction graph in Figure 13.5B. The thermal decomposition of potassium

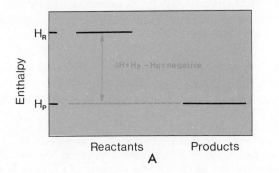
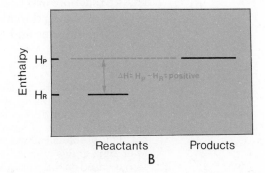

**Figure 13.5** Enthalpy-reaction graphs for an exothermic change (A) and an endothermic change (B).

chlorate illustrates an endothermic reaction. The thermochemical equations are

$$2 \ KClO_3 \ (s) \rightarrow 2 \ KCl \ (s) + 3 \ O_2 \ (g) \qquad \Delta H = +21.4 \ kcal \qquad (13.9)$$

$$2 \ KClO_3 \ (s) + 21.4 \ kcal \rightarrow 2 \ KCl \ (s) + 3 \ O_2 \ (g) \qquad (13.10)$$

---

**Example 13.8**  The thermal decomposition of $CaCO_3$ (s) to CaO (s) and $CO_2$ (g) is an endothermic reaction requiring 42 kcal per mole of $CaCO_3$ (s) decomposed. Write the thermochemical equation in two forms.

---

**13.8a**      $CaCO_3 \ (s) \rightarrow CaO \ (s) + CO_2 \ (g) \qquad \Delta H = +42 \ kcal$

$$CaCO_3 \ (s) + 42 \ kcal \rightarrow CaO \ (s) + CO_2 \ (g)$$

---

## 13.8  THERMOCHEMICAL STOICHIOMETRY

PG   13 H   Given a thermochemical equation, or information from which it may be written, calculate the amount of heat evolved or absorbed for a given amount of reactant or product; alternatively, calculate how many grams of a reactant are required to produce a given amount of heat.

The molar equivalence relationships that may be drawn from a chemical equation include energy in thermochemical equations. From

$$2 \ C_2H_6 \ (g) + 7 \ O_2 \ (g) \rightarrow 4 \ CO_2 \ (g) + 6 \ H_2O \ (l) + 736 \ kcal \qquad (13.7)$$

the following equivalences may be established:

2 moles $C_2H_6 \simeq$ 7 moles $O_2 \simeq$ 4 moles $CO_2 \simeq$ 6 moles $H_2O \simeq$ 736 kcal (13.11)

Heats of reaction may be related quantitatively to amounts of chemical species by changing the stoichiometry pattern (page 169) to convert between moles of one species and kcal of energy.

---

**Example 13.9**  How many kilocalories of heat are evolved by the burning of 84.0 grams of ethane, $C_2H_6$, according to Equation 13.7 (above)?

This problem is a two-step conversion. The unit path is from given quantity to energy: grams ethane $\rightarrow$ moles ethane $\rightarrow$ kilocalories. The second conversion is based on the equivalence that may be drawn from Equation 13.11. Go all the way to the answer.

---

**13.9a**  $1.03 \times 10^3$ kcal [$4.31 \times 10^3$ kJ]

$$84.0 \text{ g C}_2\text{H}_6 \times \frac{1 \text{ mole C}_2\text{H}_6}{30.0 \text{ g C}_2\text{H}_6} \times \frac{736 \text{ kcal}}{2 \text{ moles C}_2\text{H}_6} = 1.03 \times 10^3 \text{ kcal}$$

The final conversion comes from the fact that 2 moles $C_2H_6$ are equivalent to 736 kcal, according to Equations 13.7 and 13.11.

**Example 13.10**  How many grams of hexane, $C_6H_{14}$ (l), must be burned in order to provide $4.00 \times 10^3$ kcal of heat? The thermochemical equation is

$$2 \text{ C}_6\text{H}_{14} \text{ (l)} + 19 \text{ O}_2 \text{ (g)} \rightarrow 12 \text{ CO}_2 \text{ (g)} + 14 \text{ H}_2\text{O (l)}; \quad \Delta H = -1980 \text{ kcal}$$

or

$$2 \text{ C}_6\text{H}_{14} \text{ (l)} + 19 \text{ O}_2 \text{ (g)} \rightarrow 12 \text{ CO}_2 \text{ (g)} + 14 \text{ H}_2\text{O (l)} + 1980 \text{ kcal}$$

In this problem the given quantity is kilocalories. The path is the reverse of that in the previous example. Carry it all the way to an answer.

---

**13.10a**  347 grams $C_6H_{14}$

$$4000 \text{ kcal} \times \frac{2 \text{ moles C}_6\text{H}_{14}}{1980 \text{ kcal}} \times \frac{86.0 \text{ g C}_6\text{H}_{14}}{1 \text{ mole C}_6\text{H}_{14}} = 347 \text{ grams C}_6\text{H}_{14}$$

A word about algebraic signs: In these examples we have been able to disregard the sign of $\Delta H$ because of the way the questions were worded. The question, "How much heat . . .?" is answered simply with a quantity, a number, of kilocalories. The context of the question is sufficient specification as to whether the heat is gained or lost. In example 13.9, the question was "How many grams . . .?" Obviously that answer must be positive regardless of the sign of $\Delta H$. However, whenever a question is of the form, "What is the value of $\Delta H$?", the algebraic sign is an essential part of the answer.

# CHAPTER 13 IN REVIEW

**13.1  PHYSICAL AND CHEMICAL ENERGY—A REVIEW (Page 296)**

**13.2  UNITS OF ENERGY (Page 297)**

**13.3  ENERGY, TEMPERATURE AND CHANGE OF STATE**

13 A  Sketch, interpret and/or identify regions in a graph of temperature versus energy for a pure substance over a temperature range from below the melting point to above the boiling point. (Page 298)

**13.4  ENERGY AND CHANGE OF TEMPERATURE: SPECIFIC HEAT**

13 B  Given (1) the mass of a pure substance, (2) its specific heat and (3) its temperature change, or initial and final temperatures, calculate the heat flow. (Page 299)

13 C   Given the amount of heat flow to or from a known mass of a substance, and its temperature change, or initial and final temperatures,
calculate the specific heat of the substance. (Page 299)

**13.5   ENERGY AND CHANGE OF STATE: LATENT HEAT**

13 D   Given the quantity of a pure substance changing between the
liquid and vapor states, and the heat of vaporization, calculate the
heat flow. (Page 303)

13 E   Given the quantity of a pure substance changing between the solid
and liquid states, and the heat of fusion, calculate the heat flow.
(Page 304)

**13.6   CHANGE IN TEMPERATURE PLUS CHANGE IN STATE**

13 F   Given (1) the quantity of a pure substance, (2) $\Delta H_{vap}$ and/or $\Delta H_{fus}$
of the substance and (3) average specific heat of the substance in the
solid, liquid and/or vapor state, calculate the total heat flow in going
from one state and temperature to another state and temperature.
(Page 305)

**13.7   THERMOCHEMICAL EQUATIONS**

13 G   Given a chemical equation, or information from which it may be
written, and the heat (enthalpy) of reaction, write the thermochemical equation in two forms. (Page 308)

**13.8   THERMOCHEMICAL STOICHIOMETRY**

13 H   Given a thermochemical equation, or information from which it may
be written, calculate the amount of heat evolved or absorbed for a
given amount of reactant or product; alternatively, calculate how
many grams of a reactant are required to produce a given amount
of heat. (Page 310)

# TERMS AND CONCEPTS

Joule, kilojoule (297)                   Heat of fusion (305)
Calorie, kilocalorie (297)               Heat of solidification (305)
Delta, $\Delta$ (footnote, 299)          Heat of reaction (308)
Specific heat (299)                      Enthalpy of reaction, $\Delta H$ (308)
Latent heat (303)                        Enthalpy, H (308)
Heat of vaporization (303)               Thermochemical equation (309)
Heat of condensation (303)               Thermochemical stoichiometry (310)

# QUESTIONS AND PROBLEMS

*An asterisk (\*) identifies a question that is relatively difficult, or that extends beyond the performance
goals of the chapter.*

Section 13.3

*Note: The graph in Figure 13.6 shows the relationship between temperature and energy for a sample
of a pure substance. Assume letters J through P on the horizontal and vertical axes represent numbers,
and that expressions such as R − S or X + Y + Z represent arithmetic operations to be performed with
those numbers. Questions 13.1–13.5 and 13.24–13.28 all relate to Figure 13.6.*

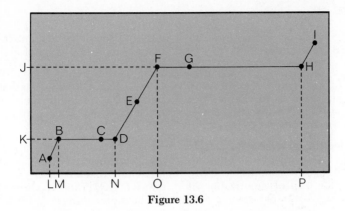

**Figure 13.6**

13.1) What values are plotted, both vertically and horizontally?

13.2) Identify in Figure 13.6 all points on the curve where the substance is entirely liquid.

13.3) Identify all points on the curve in Figure 13.6 where the substance is partly liquid and partly gas.

13.4) Describe what happens physically as the energy represented by N − M is added to the sample.

13.5) Using letters from the graph, write the expression for the energy required to raise the temperature of the liquid from the freezing point to the boiling point.

Section 13.4

13.6) A certain solution has a specific heat of 0.84 cal/(g)(°C). How many kilocalories must be removed from 325 grams of this solution to reduce its temperature from 22°C to −19°C?

13.7) In a carefully controlled calorimetry experiment 5.624 grams of an organic solid absorb 22.63 calories as the solid is warmed from 18.6°C to 32.4°C. Calculate the average specific heat over that temperature range.

13.8)* A piece of copper absorbs 58.6 calories while gaining 14.9°C in temperature. Calculate the mass of the copper.

13.9)* A student measures 100 grams of water into a calorimeter and finds that its temperature is 23.1°C. She then drops 39.0 grams of hot metal of unknown specific heat into the water, stirs, and records the maximum temperature reached, 29.6°C. If the metal was initially at 98.2°C, calculate its specific heat. Disregard any heat loss to the calorimeter or surroundings.

13.24) Identify by letter the boiling and freezing points in Figure 13.6.

13.25) Identify all points on the curve in Figure 13.6 where the substance is entirely gas.

13.25) Identify in Figure 13.6 all points on the curve where the substance is partly solid and partly liquid.

13.27) Describe the physical changes that occur as energy N − P is removed from the sample.

13.28) Using letters from the graph, show how you would calculate the energy required to boil the liquid at its boiling point.

13.29) How many kilocalories are required to warm 75 kilograms of an electroplating solution from 20°C to 35°C if its specific heat is 0.74 cal/(g)(°C)?

13.30) A 64.5 gram piece of a metal is cooled from 89.3°C to 31.8°C in a calorimeter that permits measurement of the number of calories lost. If the sample releases 519 calories, calculate the specific heat of the metal.

13.31)* What temperature will be reached by a 185 gram pressing surface of an electric iron, initially at 23°C, if it absorbs 2.81 kcal of energy and its specific heat is 0.110 cal/(g)(°C)?

13.32)* In an experiment similar to that described in Problem 13.9, a student uses toluene, $C_6H_5CH_3$, which has a specific heat of 0.43 cal/(g)(°C), instead of water in order to maximize temperature change of the liquid. 48.6 grams of a metal of unknown specific heat are heated to 91.9°C, and then dropped into 94.3 grams of toluene at 21.2°C. The entire system comes to equilibrium at 27.6°C. Calculate the specific heat of the metal.

Section 13.5

13.10) How many kilocalories are required to boil one gallon (3.785 kg) of water at 100°C?

13.11) Octane, $C_8H_{18}$, is a component of gasoline. Its heat of vaporization is 74 calories per gram. How much heat is absorbed in vaporizing 25.0 grams of octane in the process of ignition?

13.12)* Using an electric immersion heater it is found that 2.32 kcal of heat must be supplied to vaporize 50.0 grams of carbon tetrachloride, $CCl_4$, in a hood. Calculate the heat of vaporization of carbon tetrachloride in calories per gram.

13.13)* Experimental data indicate that 5.72 kcal of heat must be furnished in order to boil 65.8 grams of toluene, $C_7H_8$, one of the useful byproducts recovered from the destructive distillation of coal to produce coke, a fuel used in the steel industry. Determine the *molar* heat of vaporization of toluene in kcal/mole.

13.14)* Ammonia is a widely used refrigerant. Its cooling effect arises from the heat absorbed from the surroundings as liquid ammonia is allowed to vaporize. Calculate the mass of ammonia that changes state in the transfer of 12.5 kcal in a small refrigerating unit. The molar heat of vaporization of ammonia is 5.22 kcal/mole.

13.15) If solid benzene is removed from a refrigerator and permitted to warm to 5°C it begins to melt. How many kilocalories will be absorbed from the surroundings by 75.0 grams of benzene in melting if its heat of fusion is 32.7 cal/g?

13.16)* A student determines in the laboratory that 437 calories are required to melt 31.2 grams of tin at its melting point. Calculate the molar heat of fusion of tin in kilocalories per mole.

13.17)* How many grams of silver at its melting point can be liquified by 1.50 kcal if the heat of fusion is 21.0 cal/g?

Section 13.6

13.18) Assume the ice tray of your refrigerator holds 225 grams of water. How many calories must your refrigerator remove from one tray full of water initially at 16°C if the ultimate temperature of the ice is −12°C? Disregard the heat that must be removed from the tray.

13.33) Calculate the number of calories needed to vaporize 0.450 gram of metallic sodium at its boiling point.

13.34) Calculate the heat that must be removed from 50.0 grams of ethane, $C_2H_6$, when it is condensed at its normal boiling point of −87°C if $\Delta H_{vap} = 117$ cal/g.

13.35)* Sulfur dioxide, a gas at room temperature, boils at −10°C. In a laboratory experiment it is found that 2.23 kcal are absorbed when 42.0 grams of $SO_2$ liquid vaporize at −10°C. Calculate the heat of vaporization in calories per gram.

13.36)* The formula for ethyl ether may be written $(C_2H_5)_2O$. A student warms 32.5 grams of the liquid to its boiling point, 35°C, by immersing it in a water bath at that temperature. She then transfers it to 500 grams of water in a calorimeter, the temperature of which is 50.0°C. The ether is vaporized and escapes through the hood, absorbing sufficient heat from the water to reduce its temperature from 50.0°C to 44.7°C. Assuming all the heat lost by the water went to vaporizing the ether, calculate the *molar* heat of vaporization of ether in kcal/mole.

13.37) Components of gasoline are separated in the fractional distillation of petroleum in oil refineries. One such substance is hexane, $C_6H_{14}$, which has a molar heat of vaporization of 6.90 kcal/mole. How many grams of hexane condense when the process releases 8.35 kilocalories of heat?

13.38) Naphthalene, a compound commonly used for moth balls, melts at 80°C. Its heat of fusion is 36.0 cal/g. Calculate the number of kilocalories required to melt $2.00 \times 10^2$ grams of naphthalene.

13.39) The melting point of mercury is −39°C. At that temperature 209 calories must be removed from 74.8 grams of mercury to liquefy it. Calculate $\Delta H_{fus}$ in calories per gram.

13.40) 925 calories are released by a sample of pure gold (at. no. 79) when it solidifies at its freezing point. If the molar heat of fusion of gold is 3.15 kcal per mole, calculate the mass of the sample.

13.41) Some alloys, unlike most mixtures, have definite melting points. One such alloy melts at 262°C, and has a heat of fusion of 5.2 calories per gram. As a solid this alloy has a specific heat of 0.0287 cal/(g)(°C), and as a liquid 0.0349 cal/(g)(°C). Calculate the kilocalories of heat absorbed by an 8.55 kg slug of this metal as it comes to the 355°C temperature of a melting pot if the slug is dropped into the pot at room temperature, 30°C.

*13.19–13.21 and 13.42–13.44: Thermochemical equations may be written in two ways, one with a heat term as a part of the equation, and alternatively with ΔH set apart from the regular equation. In the examples that follow, write the equations for the reactions described in both forms. Recall that state designations of all substances are essential in thermochemical equations.*

13.19)   When propane, $C_3H_8$ (g)—a primary constituent in bottled gas used to heat trailers and rural homes—is burned to form gaseous carbon dioxide and liquid water, 531 kilocalories of heat energy are released for every mole of butane consumed.

13.20)   In "slaking" lime, CaO (s), by converting it to solid calcium hydroxide through reaction with water, 15.6 kilocalories of heat are released for each mole of calcium hydroxide formed.

13.21)   The extraction of elemental aluminum from aluminum oxide is a highly endothermic electrolytic process. The reaction may be summarized by an equation in which the oxide reacts with carbon to yield aluminum and carbon dioxide gas. The energy requirement is 129 kcal/mole of aluminum metal produced. (Caution on the value of ΔH.)

13.42)   673 kilocalories of energy are absorbed from sunlight in the photosynthesis reaction in which carbon dioxide and water combine to produce sugar, $C_6H_{12}O_6$, and release oxygen.

13.43)   The electrolysis of water is an endothermic reaction, absorbing 68.3 kcal for each mole of liquid water decomposed to its elements.

13.44)   The reaction in an oxyacetylene torch is highly exothermic, releasing 312 kilocalories of heat for every mole of acetylene, $C_2H_2$ (g), burned. The end products are gaseous carbon dioxide and liquid water.

13.22)   How much heat energy will be released by burning $4.00 \times 10^3$ grams of butane, $C_4H_{10}$, one of the two principal fuels used in bottled gas? The thermochemical equation is 2 $C_4H_{10}$ (g) + 13 $O_2$ (g) → 8 $CO_2$ (g) + 10 $H_2O$ (l) + 1380 kcal.

13.23)   Calculate the number of grams of hexane, a component of gasoline, required to release $5.5 \times 10^4$ kcal if ΔH = −990 kcal for the reaction $C_6H_{14}$ (l) + $^{19}/_2$ $O_2$ (g) → 6 $CO_2$ (g) + 7 $H_2O$ (l).

13.45)   The direct combination of powdered zinc with powdered sulfur is a spectacular reaction—though not one to be tried by students!—that yields bright light, flame and smoke: Zn (s) + S (s) → ZnS (s). Calculate the energy released by the reaction of 9.63 grams of zinc if ΔH = −148.5 kcal/mole.

13.46)   Carbon monoxide is used as a fuel in many industrial processes: 2 CO (g) + $O_2$ (g) → 2 $CO_2$ (g) + 135 kcal. How much carbon monoxide must be burned in a process requiring 1830 kcal of energy?

# 14

# SOLUTIONS

## 14.1 THE CHARACTERISTICS OF A SOLUTION

PG 14 A Distinguish between a pure substance, a solution, a suspension and a colloid.

Solutions abound in nature. We are surrounded by the gaseous solution known as air. The oceans are a water solution of sodium chloride and other dissolved substances in lower concentration. Some of these substances are in sufficient concentration to make it profitable to extract them from sea water, magnesium being a notable example. Even what we call "fresh" water is a solution, although the concentrations are so low we tend to think of the water we drink as "pure." "Hard" water may be sufficiently pure for human consumption, but there are enough calcium and magnesium salts present to form solid deposits in hot water pipes and boilers. Even rain water is a solution, containing dissolved gases. Oxygen is not very soluble in water, but what little there is in solution is mighty important to fish, who cannot survive without it.

A **solution** is classified as a **homogeneous mixture.** This implies uniform distribution of solution components, so that a sample taken from any part of the solution will have the same composition. Two solutions made up of the same components, however, may have different compositions. This characteristic, variable composition, is an important distinguishing feature between solutions and compounds. A solution of ammonia in water, for example, may contain 1% ammonia by weight, 2%, 5%, 20.3% . . . up to the 29% solution we call "concentrated ammonia." In contrast, the composition of pure compounds is always the same (the Law of Constant Composition, page 15); for ammonia the composition is always 82.4% nitrogen and 17.6% hydrogen.

Solutions may exist in any of the three states, gas, liquid or solid. Air is a gaseous solution, made up of nitrogen, oxygen, carbon dioxide and other gases in small amounts. Odors are usually the result of liquids (perfume, benzene) or solids (moth balls) that vaporize and become a part of the air solution. In addition to oxygen in water, dissolved carbon dioxide in carbonated beverages is a familiar liquid solution of a gas. Alcohol in water is

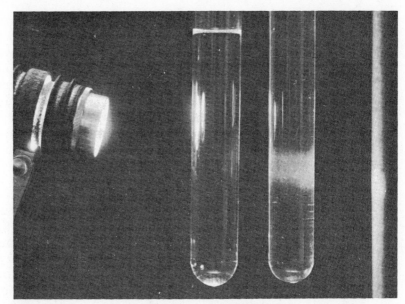

**Figure 14.1** Reflection of light by a colloidal dispersion. A beam of light is invisible as it passes through a solution (left tube) because the dissolved particles are too small to reflect light. Colloidal particles (right tube) are too small to be seen with the naked eye, but they are large enough to reflect light and produce a visible beam. The effect is similar to the beam of a search light, which is visible only because the light is reflected by the dust particles in the air.

an example of the solution of two liquids, and the oceans of the world are natural liquid solutions of solids. Solid state solutions are common in the form of metal alloys.

Particle size distinguishes solutions from other mixtures. Dispersed particles in solutions, which may be atoms, ions or molecules, are very small —generally less than $5 \times 10^{-7}$ cm (50 Å) in diameter. Particles of this size do not settle on standing, and they are too small to be seen.

In contrast, a **suspension** is a heterogeneous distribution of larger particles, usually $10^{-4}$ cm or larger in diameter, in a second medium. In the liquid or gaseous state, suspended particles will settle over a period of time. Homogeneous dispersions of particles between solution and suspension size are called **colloids.** Milk, fog and Jello are common examples of colloids. Colloidal particles, which do not settle on standing, may consist of a single huge molecule, or more commonly an aggregate of smaller molecules. Colloidal particles, like dissolved particles, are too small to be seen, but they are large enough to reflect light and produce a visible beam, as shown in Figure 14.1.

## 14.2   SOLUTION TERMINOLOGY

PG   14  B   Distinguish between terms in the following groups:

Solute and solvent

Concentrated and dilute

Solubility, saturated, unsaturated and supersaturated

Miscible and immiscible

In discussing solutions one uses a language of closely related and sometimes overlapping terms. We will now identify and define these terms.

**Solute and Solvent.**   When solids or gases are dissolved in liquids, the solid or gas is said to be the **solute** and the liquid the **solvent.** More generally, the solute is taken to be the substance present in relatively small quantity. The medium in which the solute is dissolved is generally relatively large in quantity, and is called the solvent. The distinction is not precise, however. Water is capable of dissolving more than its own weight of some solids, but the water continues to be called the solvent. In alcohol-water solutions, either liquid may be the more abundant and, in a given context, either might be called the solute or solvent.

**Concentrated and Dilute.**   A **concentrated** solution has a *relatively* large quantity of a specific solute per unit amount of solution, and a **dilute** solution has a *relatively* small quantity of the same solute per unit amount of solution. The terms compare concentrations of two solutions of the *same solute and solvent.* They carry no other quantitative significance.

**Solubility, Saturated and Unsaturated.**   **Solubility** is a measure of how much solute will dissolve in a given amount of solvent at a given temperature. It is sometimes expressed by giving the number of grams of solute that will dissolve in 100 grams of solvent. A solution which can exist in equilibrium with undissolved solute is a **saturated** solution. A solution whose concentration corresponds to the solubility limit is therefore saturated. If the concentration of a solute is less than the solubility limit it is **unsaturated.**

**Supersaturated Solutions.**   Under carefully controlled conditions, a solution can be produced in which the concentration of solute is greater than the normal solubility limit. Such a solution is said to be **supersaturated.** A supersaturated solution of sodium acetate, for example, may be prepared by dissolving 80 grams of the salt in 100 grams of water at about 50°C. If the solution is then cooled to 20°C without stirring, shaking, or other disturbance, all 80 grams of solute will remain in solution even though the solubility at 20°C is only 46.5 grams/100 grams of water. The supersaturated solution can be maintained indefinitely so long as there are no tiny particles upon which crystallization can start. If a small seed crystal of sodium acetate is added, crystallization takes place until equilibrium is attained by the formation of a saturated solution.

**Miscible and Immiscible.**   Miscible and immiscible are terms customarily limited to solutions of liquids in liquids. If two liquids dissolve in each other in all proportions they are said to be **miscible** in each other. Alcohol and water, for example, are miscible liquids. Liquids that are insoluble in each other, as oil and water, are **immiscible.** Some liquid pairs will mix appreciably with each other, but in limited proportions; these are said to be *partially miscible.*

## 14.3   THE FORMATION OF A SOLUTION

PG   14 C   Describe the formation of a saturated solution from the time excess solid solute is first placed into a liquid solvent.

14 D   Identify and explain the factors that determine the time required to dissolve a given amount of solute, or to reach equilibrium.

We will now examine the mechanism of solution formation at the particle level. Consider what happens when an ionic crystal is placed in water (Fig. 14.2). The negatively charged ions at the surface of the crystal are attracted by the positive end of the polar water molecule. A "tug-of-war" for the negative ions begins: water molecules tend to pull them from the crystal, while neighboring positive ions tend to hold them in the crystal. In a similar fashion, positive ions at the surface are attacked by the negative portion of the water molecule and torn from the crystal.

Once the charged ions are released from the crystal the electrostatic forces of attraction between solute and solvent particles continue to exert themselves. At least in dilute solutions, ions are surrounded by polar water molecules, drawn there by the attraction of opposite charges. Such ions are said to be "hydrated." Ion hydration is frequently indicated by the symbol (aq), as in $Na^+$ (aq) and $Cl^-$ (aq).

The dissolving process is reversible. As the dissolved solute particles move randomly through the solution, they come into contact with the undissolved solute and crystallize—return to the solid state. The rate at which crystallization occurs depends upon the concentration of solute at the surface of the undissolved solid. If, before all the solute is dissolved, the concentration increases to the point that the crystallization rate is equal to the rate at which the solid is dissolving, an equilibrium is established. For NaCl, this equilibrium may be represented by the "reversible reaction" equation (see page 284).

$$NaCl \ (s) \rightleftharpoons Na^+ \ (aq) + Cl^- \ (aq) \qquad (14.1)$$

The concentration at equilibrium identifies a saturated solution, as noted in the previous section; a numerical expression of this concentration represents the solubility at the existing temperature (see Fig. 14.3).

The time required to dissolve a given amount of solute—or to reach

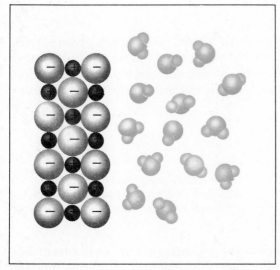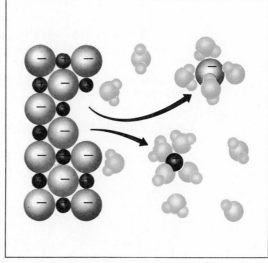

**Figure 14.2**   Dissolving of an ionic solute in water.

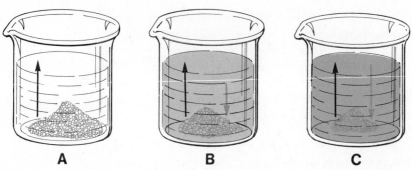

**A**  **B**  **C**

**Figure 14.3** Development of equilibrium in producing a saturated solution. The solution dissolves at a constant rate, shown by the black arrows. In A, when dissolving has just begun, the solute concentration in the solvent is zero, so no crystallization can occur. In B the solute concentration has risen to yield a crystallization rate indicated by the colored arrow—still less than the dissolving rate. Eventually, in C, the concentration has increased to the point that the dissolving and crystallization rates are equal. Equilibrium has been reached, and the solution is saturated.

equilibrium concentration, if excess solute is present—depends upon several factors:

First, the dissolving process being a surface phenomenon, a finely divided solid, which offers more surface per unit of mass than a coarsely divided solid, will dissolve more rapidly.

Second, stirring or agitating the solution prevents a localized buildup of concentration to near saturation level at the solute surface, thereby minimizing the rate of crystallization. The "net" dissolving rate is therefore maximized.

Third, at higher temperatures particle movement is more rapid, thereby speeding up all physical processes.

## 14.4 FACTORS THAT DETERMINE SOLUBILITY

The extent to which a particular solute will dissolve in a given solvent may be correlated to three factors: the magnitude of intermolecular forces between solvent molecules and solute molecules; the temperature; and, in cases of gases dissolved in liquids, the partial pressure of the solute gas.

### INTERMOLECULAR FORCES

PG 14 E   Given the structural formulas of two molecular substances, or other information from which the strength of their intermolecular forces may be estimated, predict if they will dissolve appreciably in each other, and state the criteria on which your prediction is based.

You saw in Chapter 12 that physical properties of a molecular substance usually may be associated with the intermolecular forces resulting from the geometry of its molecules. Solubility is among these properties. Generally speaking, *if the forces between molecules of substance A are roughly equal to the intermolecular forces of substance B, substances A and B will probably dissolve in each other.* From the standpoint of these forces, the molecules might be said to be able to replace each other. On the other hand, if the intermolecular forces between solute molecules are quite different from the forces between solvent molecules, it is unlikely that the substances will dissolve in each other.

If we consider, for example, intermolecular attractions between such substances as hexane, $C_6H_{14}$, and decane, $C_{10}H_{22}$, we find that each substance has intermolecular forces attributable to relatively weak dispersion forces. These forces are roughly equal for the two substances, which are mutually soluble in each other. Neither, however, is soluble in water or methanol, $CH_3OH$, two liquids that exhibit strong hydrogen bonding intermolecular attractions. But water and methanol are soluble in each other, again supporting the correlation between solubility and similarity of intermolecular forces.

This generalization is sometimes summarized by the saying "like dissolves like," implying that polar substances dissolve in other polar substances, nonpolars dissolve in nonpolars, but polar and nonpolar combinations do not dissolve. Molecular polarity does indeed contribute to solubility, but other factors frequently outweigh polarity in importance, and the generalization has many exceptions. Predictions based on molecular polarity alone must be verified in the laboratory before they may be relied upon.

## TEMPERATURE

Temperature exerts a major influence on most chemical equilibria, solution equilibria included. Consequently solubility is temperature-dependent. Figure 14.4 indicates that solubility of most solids increases with rising temperature, but there are notable exceptions. The solubilities of gases in liquids, on the other hand, are generally lower at higher temperatures. The explanation of the interrelationship between temperature and solubility involves energy changes in the solution process, as well as other factors. We shall look into this matter qualitatively in Chapter 15.

## PRESSURE

PG  14  F   Predict how the solubility of a gas in a liquid will be affected by a change in the partial pressure of that gas over the liquid.

Changes in partial pressure of a solute gas over a liquid solution have a pronounced effect on the solubility of the gas (see Fig. 14.5). This is sometimes startlingly apparent on opening a bottle or can of a carbonated beverage. Such beverages are bottled or canned under carbon dioxide partial pressure slightly greater than one atmosphere, which increases the solubility of the gas. This is what is meant by "carbonated." As the pressure is released on opening, solubility decreases, resulting in bubbles of carbon dioxide escaping from the solution. In an "ideal" solution, solubility of a gas

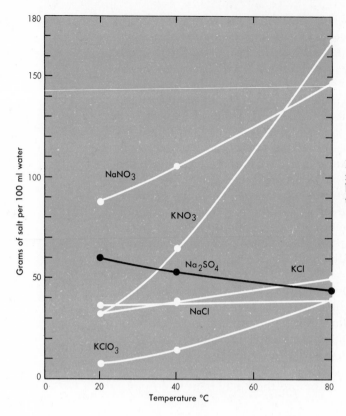

Figure 14.4 Temperature-solubility curves for various salts in water. Note that the solubility of most salts increases with temperature, but there are exceptions.

in a liquid is directly proportional to the partial pressure of the gas. On the other hand, pressure has little or no effect on the solubility of solids or liquids in a liquid solvent.

## 14.5 SOLUTION CONCENTRATION: PERCENTAGE BY WEIGHT

The concentration of a solution is ordinarily expressed in terms of quantity of solute present in a given quantity of solvent or of total solution. Quantity of solute may be given in grams or moles; quantity of solvent or solution

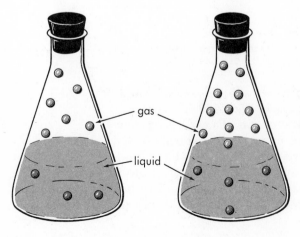

Figure 14.5 Effect of partial pressure of a gas on its solubility in a liquid. Solute gas concentration, and therefore its partial pressure, is lower in the left flask than in the right flask. Consequently the solute concentration in solution is lower in the left flask.

may be stated in mass or volume units. In this and the next three sections we will examine four ways to express solution concentration. We begin with percentage by weight.

PG 14 G Given grams of solute and grams of solvent or solution, calculate percentage concentration.

14 H Given grams of solution and percentage concentration, calculate grams of solute and grams of solvent.

If the concentration of a solute in a solution is given in per cent, you may assume it to be per cent by weight unless specifically stated otherwise. As always, per cent is $\frac{\text{part quantity}}{\text{total quantity}} \times 100$. Accordingly, the percentage by weight of solute is defined by the following equations:

$$\% \text{ by weight of solute} = \frac{\text{grams solute}}{\text{grams solution}} \times 100 \qquad (14.2)$$

$$= \frac{\text{grams solute}}{\text{grams solute + grams solvent}} \times 100 \qquad (14.3)$$

---

**Example 14.1** 125 grams of solution, when evaporated to dryness, was found to contain 42.3 grams of solute. What was the percentage by weight of the solute?

This involves only a direct substitution into one of the defining equations.

--------------------------------------------------------------------------------

**14.1a** 33.8%

We have 42.3 grams of solute; the total quantity is 125 grams of solution. Substituting into Equation 14.2,

$$\frac{\text{g solute}}{\text{g solution}} \times 100 = \frac{42.3}{125} \times 100 = 33.8\%$$

---

**Example 14.2** 3.50 grams of potassium nitrate are dissolved in 25.0 grams of water. Calculate the percentage concentration of potassium nitrate.

Again one of the defining equations may be used . . . .

--------------------------------------------------------------------------------

**14.2a** 12.3% potassium nitrate

By Equation 12.2,

$$\frac{\text{g solute}}{\text{g solute + g solvent}} \times 100 = \frac{3.50}{3.50 + 25.0} \times 100 = 12.3\%$$

**Example 14.3** You are to prepare 250 grams of 7.00% sodium carbonate solution. How many grams of sodium carbonate and how many milliliters of water do you use? (Recall that the density of water is 1.00 g/ml.)

You may approach this as a typical percentage problem, or from a dimensional analysis viewpoint, where percentage is grams of solute per 100 grams of solution. On this basis, a 7.00% solution means 7.00 grams of solute ≃ 100 grams of solution. (See page 507.)

---

**14.3a** 17.5 grams sodium carbonate, 232 ml of water

$$\boxed{250 \text{ g solution}} \times \frac{7.00 \text{ g Na}_2\text{CO}_3}{100 \text{ g solution}} = 17.5 \text{ g Na}_2\text{CO}_3; \text{ or}$$

$$7.00\% \text{ of } 250 = 0.0700 \times 250 = 17.5 \text{ g Na}_2\text{CO}_3$$

$$\text{grams water} = \text{grams solution} - \text{grams solute}$$

$$= 250 - 17.5 = 232 \text{ g water}$$

At 1.00 g/ml, 232 grams of water = 232 milliliters.

## 14.6 SOLUTION CONCENTRATION: MOLALITY (OPTIONAL)

Many physical properties of solutions are related to the solution concentration expressed as **molality. Molality is the number of moles of solute dissolved in one kilogram of solvent.** Mathematically,

$$m = \frac{\text{moles of solute}}{\text{kilograms of solvent}} \qquad (14.4)$$

where m is the symbol of molality. In essence, the molality of a solution is *the number of moles of solute that are equivalent to one kilogram of solvent.* In a 1.5 molal solution, for example, 1.5 moles of solute ≃ 1.0 kilogram of solvent.

The quantity of solvent in molality problems is usually given in *grams*, or in volume, which may be converted to grams by multiplying by density. Molality, however, is based on *kilograms* of solvent. To convert grams to kilograms you divide by 1000—or more simply, move the decimal three places to the left. It is convenient to do this mentally in the initial setup of the problem.

**Example 14.4** Calculate the molality of a solution prepared by dissolving 15.0 grams of sugar, $C_{12}H_{22}O_{11}$ (MW = 342 g/mole), in 350 milliliters of water.

*Solution.* The data of the problem may be interpreted as stating the concentration of the solution in grams of solute per milliliter of solvent: 15.0 grams solute per 350 ml $H_2O$. Because the density of water is 1.00 g/ml, the concentration is 15.0 grams solute per 350 g $H_2O$, or

$$\frac{15.0 \text{ g } C_{12}H_{22}O_{11}}{0.350 \text{ kg } H_2O}$$

Molality is moles of solute per kilogram of solvent. To convert the above expression to molality it is necessary only to change 15.0 grams of $C_{12}H_{22}O_{11}$ to moles:

$$\frac{15.0 \text{ g } C_{12}H_{22}O_{11}}{0.350 \text{ kg } H_2O} \times \frac{1 \text{ mole } C_{12}H_{22}O_{11}}{342 \text{ g } C_{12}H_{22}O_{11}} = 0.125 \text{ m}$$

**Example 14.5**  How many grams of KCl (MW = 74.6 g/mole) must be dissolved in 250 grams of water to make a 0.400 m solution?

*Solution.*  Molality provides a unit conversion between moles of solute and kilograms of solvent. Beginning with the given quantity, 250 grams of water, or 0.250 kilogram, the unit path becomes kilograms water → moles solute → grams solute, the second conversion being via molar weight. The entire setup is

$$0.250 \text{ kg } H_2O \times \frac{0.400 \text{ mole KCl}}{1 \text{ kg } H_2O} \times \frac{74.6 \text{ g KCl}}{1 \text{ mole KCl}} = 7.46 \text{ g KCl}$$

## 14.7  SOLUTION CONCENTRATION: MOLARITY

PG  14 I  Given two of the following, calculate the third: volume of solution, molarity, and moles (or grams with known or calculable molar weight) of solute.

Percentage and molality concentrations have both been based on mass of solute and mass of solvent, each considered separately. In using liquids, volume is much more easily measured than mass. Therefore solution concentrations are frequently expressed in terms of their **molarity,** which relates to a certain volume of solution. Abbreviated M, molarity is defined as **moles of solute per liter of solution.** In the form of an equation,

$$M = \frac{\text{moles solute}}{\text{liter solution}} \tag{14.5}$$

In essence, the molarity of a solution is *the number of moles of solute that are equivalent to one liter of solution.* In a 1.5 molar solution, for example, 1.5 moles of solute ≃ 1.0 liter of solution.

The preparation of a solution of known molarity is a frequent procedure in a laboratory. The volume of solution to be prepared is often expressed in milliliters. Because molarity is defined in liters, volume must be changed to express the larger unit. Conversion is simply division by 1000, which requires moving the decimal three places to the left. Thus 750 milliliters is 0.750 liter:

$$750 \text{ ml} \times \frac{1 \ell}{1000 \text{ ml}} = 0.750 \ell$$

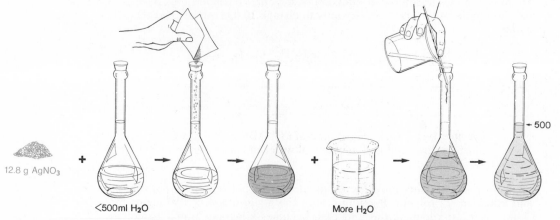

**Figure 14.6**   Preparation of 500 ml 0.150 M AgNO₃.

This milliliters-to-liters conversion will be performed without comment in the examples that follow.

---

**Example 14.6**   How many grams of silver nitrate, AgNO₃ (MW = 170 g/mole), must be dissolved to prepare 500 milliliters of 0.150 M AgNO₃?

The given quantity is 500 milliliters. If this is expressed in liters, the unit path becomes liters → moles AgNO₃ → grams AgNO₃. Complete the problem.

- - - - - - - - - - - - - - - - - - - - - - - - - - - - - - - - - - - - - - - - - - - - - - - - - - - - - - - - - - - - - - - - -

**14.6a**   $0.500 \, \ell \times \dfrac{0.150 \, \text{mole AgNO}_3}{1 \, \ell} \times \dfrac{170 \text{ g AgNO}_3}{1 \text{ mole AgNO}_3} = 12.8 \text{ g AgNO}_3$

---

A clearer idea of the meaning of molarity may be gained by imagining the preparation of a solution of known molarity. Building on Example 14.6, the first step in the process is to weigh out 12.8 grams of silver nitrate (see Fig. 14.6). This is transferred to a 500 ml volumetric flask containing *less than* 500 ml water. After dissolving the solute, water is added to the 500 milliliter mark on the neck of the flask. Notice the definition of molarity is based on the volume of *solution,* not the volume of *solvent.* This is why the solute is dissolved in less than 500 milliliters of water and then diluted to that volume, rather than being dissolved in 500 milliliters of water, which would yield slightly more than 500 milliliters of solution.

---

**Example 14.7**   15.8 grams of sodium hydroxide (MW = 40.0 g/mole) are dissolved in water and diluted to 100 milliliters. Calculate the molarity.

From the data presented, the concentration may be expressed in grams of solute per liter of solution. Write the concentration in these units.

- - - - - - - - - - - - - - - - - - - - - - - - - - - - - - - - - - - - - - - - - - - - - - - - - - - - - - - - - - - - - - - - -

**14.7a** $\dfrac{15.8 \text{ g NaOH}}{100 \text{ ml solution}} = \dfrac{15.8 \text{ g NaOH}}{0.100 \,\ell \text{ solution}}$

The only requirement for converting the above expression to molarity, or moles per liter, is to convert grams of solute to moles. Complete the problem.

-------------------------------------------------------------------------------

**14.7b** 3.95 M NaOH

$$\dfrac{15.8 \cancel{\text{ g NaOH}}}{0.100 \,\ell} \times \dfrac{1 \text{ mole NaOH}}{40.0 \cancel{\text{ g NaOH}}} = 3.95 \text{ moles NaOH}/\ell = 3.95 \text{ M}$$

One of the more important functions of the molarity concept is the conversion between volume of solution and moles of solute.

---

**Example 14.8**  How many moles of solute are there in 45.3 ml of 0.550 M solution?

From the given quantity it is a one-step conversion to moles. Set up and solve.

-------------------------------------------------------------------------------

**14.8a**  0.0249 mole

$$\boxed{0.0453 \cancel{\,\ell\,}} \times \dfrac{0.550 \text{ mole}}{1 \cancel{\,\ell}} = 0.0249 \text{ mole}$$

The reverse problem is also significant . . . .

---

**Example 14.9**  Find the number of milliliters of 1.40 M solution that contain 0.287 mole of solute.

Converting the given quantity to volume is easy enough, but be sure you answer the question in the units required.

-------------------------------------------------------------------------------

**14.9a**  205 milliliters

$$\boxed{0.287 \cancel{\text{ mole}}} \times \dfrac{1 \cancel{\,\ell}}{1.40 \cancel{\text{ moles}}} \times \dfrac{1000 \text{ ml}}{1 \cancel{\,\ell}} = 205 \text{ milliliters}$$

This setup might be shortened by one step by recognizing that moles/liter is the same as moles/1000 ml:

$$0.287 \cancel{\text{ mole}} \times \dfrac{1000 \text{ ml}}{1.40 \cancel{\text{ moles}}} = 205 \text{ milliliters}$$

## 14.8   SOLUTION CONCENTRATION: NORMALITY (OPTIONAL)

*Normality* is a convenient way to express solution concentration in quantitative analytical work. **The normality of a solution is the number of equivalents of solute per liter.** Designated N, normality may be expressed mathematically as

$$N = \frac{\text{equivalents solute}}{\text{liter solution}} \tag{14.6}$$

The concept of normality leads immediately to the meaning of the term *equivalent.* In an acid-base neutralization reaction, the only type of normality problem we will discuss here, **one equivalent of an acid is that quantity that yields one mole of hydrogen ions in a chemical reaction; one equivalent of a base is that quantity that reacts with one mole of hydrogen ions.** Because hydrogen and hydroxide ions combine on a one-to-one basis, one mole of hydroxide ions is one equivalent. According to these definitions, both one mole of HCl and one mole of NaOH are one equivalent because they yield, respectively, one mole of hydrogen ion and one mole of hydroxide ion in a reaction. $H_2SO_4$, on the other hand, has two equivalents per mole because each mole of the compound can release two moles of hydrogen ions. Similarly one mole of $Al(OH)_3$ may represent three equivalents because three moles of hydroxide ion may react.

Notice that the number of equivalents in a mole of an acid or base depends on a specific reaction, not simply on the number of moles of hydrogen or hydroxide ion present in one mole of solute. The number of moles of these ions that actually react may or may not be the same as those present. For example, we might expect that phosphoric acid, $H_3PO_4$, has three equivalents per mole—and indeed it does in the reaction

$$H_3PO_4 \text{ (aq)} + 3 \text{ NaOH (aq)} \rightarrow Na_3PO_4 \text{ (aq)} + 3 \text{ HOH (l)} \tag{14.7}$$

For reasons beyond the scope of this discussion, this reaction is difficult to perform quantitatively in the laboratory. It is possible, however, to control the reaction in a titration experiment (see page 331) so 1.00 mole of $H_3PO_4$ reacts with 1.00 mole of NaOH in

$$H_3PO_4 \text{ (aq)} + \text{NaOH (aq)} \rightarrow NaH_2PO_4 \text{ (aq)} + \text{HOH (l)} \tag{14.8}$$

In equation 14.8, one mole of phosphoric acid yields one mole of hydrogen ions to react with one mole of sodium hydroxide. This is one equivalent of sodium hydroxide. Phosphoric acid has one equivalent per mole in Equation 14.8.

Under other conditions the phosphoric acid-sodium hydroxide reaction can be controlled to produce a one-to-two mole ratio between the acid and the base:

$$H_3PO_4 \text{ (aq)} + 2 \text{ NaOH (aq)} \rightarrow Na_2HPO_4 \text{ (aq)} + 2 \text{ HOH (l)} \tag{14.9}$$

In this reaction one mole of phosphoric acid releases two moles of hydrogen ions, so there are two equivalents per mole. This reaction, combined with Equations 14.7 and 14.8, shows that the number of equivalents per mole depends on a specific reaction, not just on how many hydrogen or hydroxide ions are in a mole of an acid or base.

---

**Example 14.10**   Calculate the normality of a solution prepared by dissolving 2.50 grams of NaOH in 500 milliliters of water.

*Solution.*   The concentration of the solution may be expressed in grams of solute per liter and converted to moles per liter (molarity) in the usual way:

$$\frac{2.50 \text{ g NaOH}}{0.500 \text{ } \ell} \times \frac{1 \text{ mole NaOH}}{40.0 \text{ g NaOH}}$$

Because one mole of NaOH contains only one mole of hydroxide ions, the only possibility is one mole NaOH = 1 equivalent NaOH. Extending the setup to convert moles of NaOH to equivalents,

$$\frac{2.50 \text{ g NaOH}}{0.500 \ell} \times \frac{1 \text{ mole NaOH}}{40.0 \text{ g NaOH}} \times \frac{1 \text{ eq NaOH}}{1 \text{ mole NaOH}} = 0.125 \text{ N NaOH}$$

It is apparent from this setup that the normality of a sodium hydroxide solution is equal to its molarity. This is because there is one equivalent per mole.

---

**Example 14.11**   A sulfuric acid solution contains 10.5 grams of $H_2SO_4$ per 250 milliliters. Calculate its normality for the reaction

$$H_2SO_4 \text{ (aq)} + 2 \text{ NaOH (aq)} \rightarrow Na_2SO_4 \text{ (aq)} + 2 \text{ HOH (l)}$$

*Solution.*   Again we begin by expressing the acid concentration in moles per liter. The equation indicates that two moles of hydrogen ion are released by one mole of acid, so there are two equivalents per mole. The complete setup is

$$\frac{10.5 \text{ g } H_2SO_4}{0.250 \ell} \times \frac{1 \text{ mole } H_2SO_4}{98.1 \text{ g } H_2SO_4} \times \frac{2 \text{ eq } H_2SO_4}{1 \text{ mole } H_2SO_4} = 0.856 \text{ N}$$

---

The usefulness of normality as a concentration unit will become clear in the next section. It depends on the ease by which conversion may be made between solution volume and number of equivalents. Normality states the number of equivalents that are equivalent to one liter of solution. For example, in a 0.750 N solution, 0.750 equivalent solute $\simeq$ 1 liter solution.

---

**Example 14.12**   How many equivalents are present in 18.6 milliliters of 0.856 N $H_2SO_4$?

*Solution.*   Converting milliliters to liters the problem is solved in a single step:

$$0.0186 \ell \times \frac{0.856 \text{ eq } H_2SO_4}{\ell} = 0.0159 \text{ eq } H_2SO_4$$

---

Example 14.12 illustrates an important relationship involving normality:

$$V \times N = \text{number of equivalents} \tag{14.10}$$

## 14.9   SOLUTION STOICHIOMETRY

PG   14 J   Given the quantity of any species participating in a chemical reaction for which the equation may be written, find the quantity of any other species, either quantity being measured in (a) grams; (b) volume of a gas at specified temperature and pressure; or (c) volume of solution at specified molarity.

**14 K** Given the volumes of two solutions that react with each other in a titration experiment, and the molarity of one solution, calculate the molarity of the other.

In Examples 14.8 and 14.9 you learned how to convert in either direction between volume of solution of known molarity and moles of solute. These operations provide another method for the first and third steps in the stoichiometric pattern (see Section 8.2, page 169). Solution stoichiometry calculations are essentially the same as in the other stoichiometry problems you have solved. They will be illustrated in the examples that follow.

---

**Example 14.13**  How many grams of lead(II) iodide will be precipitated from the addition of excess potassium iodide solution to 50.0 ml of 1.22 M $Pb(NO_3)_2$?

First, the equation . . . .

---

**14.13a**  $Pb(NO_3)_2$ (aq) + 2 KI (aq) $\rightarrow$ $PbI_2$ (s) + 2 $KNO_3$ (aq)

Begin with the given quantity from the statement of the problem. The unit path is $\ell$ given $\rightarrow$ moles given (see Ex. 14.8, page 327) $\rightarrow$ moles wanted $\rightarrow$ grams wanted. Set up the first step to moles of given species, but do not solve.

---

**14.13b**

$$\boxed{0.0500\,\ell} \times \frac{1.22 \text{ moles } Pb(NO_3)_2}{1\,\ell}$$

The remainder of the problem is like the first stoichiometry problems: convert moles of given to moles of wanted, and then to grams. Complete the setup and the solution.

---

**14.13c**  28.1 g $PbI_2$

$$\boxed{0.0500\,\ell} \times \frac{1.22 \text{ moles } Pb(NO_3)_2}{1\,\ell} \times \frac{1 \text{ mole } PbI_2}{1 \text{ mole } Pb(NO_3)_2} \times \frac{461 \text{ g } PbI_2}{1 \text{ mole } PbI_2}$$

$$= 28.1 \text{ g } PbI_2$$

---

**Example 14.14**  Calculate the number of milliliters of 0.842 M NaOH that will be required to precipitate as $Cu(OH)_2$ all the copper in 30.0 ml 0.635 M $CuSO_4$. The equation is

$$2 \text{ NaOH (aq)} + CuSO_4 \text{ (aq)} \rightarrow Cu(OH)_2 \text{ (s)} + Na_2SO_4 \text{ (aq)}$$

The first two steps in the unit path this time are as in the last example: $\ell$ given $\rightarrow$ moles given $\rightarrow$ moles wanted. Set up that far, but do not solve.

**14.14a**
$$\boxed{0.0300\,\ell} \times \frac{0.635 \text{ mole CuSO}_4}{1\,\ell} \times \frac{2 \text{ moles NaOH}}{1 \text{ mole CuSO}_4}$$

At this point you have the number of moles of sodium hydroxide required.
It is "packaged" at $0.842 \dfrac{\text{mole}}{\ell}$, or $0.842 \dfrac{\text{mole}}{1000 \text{ ml}}$.

How many such "packages" are required to obtain the number of moles indicated by the above setup? Recall Example 14.9, in which you performed the identical operation. Complete the setup and solve the problem—remembering, of course, that the answer is required in milliliters.

---

**14.14b** 45.2 ml NaOH

$$\boxed{0.0300\,\ell} \times \frac{0.635 \text{ mole CuSO}_4}{1\,\ell} \times \frac{2 \text{ moles NaOH}}{1 \text{ mole CuSO}_4}$$

$$\times \frac{1\,\ell \text{ NaOH}}{0.842 \text{ mole NaOH}} \times \frac{1000 \text{ ml}}{1\,\ell} = 45.2 \text{ ml NaOH}$$

Alternatively, using the direct 0.842 mole NaOH ≐ 1000 ml NaOH conversion, as in Example 14.9,

$$\boxed{0.0300\,\ell} \times \frac{0.635 \text{ mole CuSO}_4}{1\,\ell} \times \frac{2 \text{ moles NaOH}}{1 \text{ mole CuSO}_4} \times \frac{1000 \text{ ml NaOH}}{0.842 \text{ mole NaOH}}$$

$$= 45.2 \text{ ml NaOH}$$

---

**Example 14.15**   How many liters of dry hydrogen, measured at STP, will be released by the complete reaction of 45.0 ml 0.486 M $H_2SO_4$ with excess granular zinc? The equation is
$$Zn\,(s) + H_2SO_4\,(aq) \rightarrow ZnSO_4\,(aq) + H_2\,(g).$$

Recalling that molar volume of any gas at STP is 22.4 liters/mole, set up and solve this problem completely.

---

**14.15a**   0.490 liter $H_2$

$$\boxed{0.0450\,\ell} \times \frac{0.486 \text{ mole H}_2\text{SO}_4}{1\,\ell} \times \frac{1 \text{ mole H}_2}{1 \text{ mole H}_2\text{SO}_4} \times \frac{22.4\,\ell \text{ H}_2}{1 \text{ mole H}_2}$$

$$= 0.490\,\ell \text{ H}_2$$

One of the more important laboratory operations in analytical chemistry is called **titration**. Titration is the very careful addition of one solution into another by means of a buret, which measures the volume of solution added. Figure 14.7 shows a buret in use.

Suppose you have a sodium chloride solution and you wish to determine the weight of sodium chloride it contains. This can be done very accu-

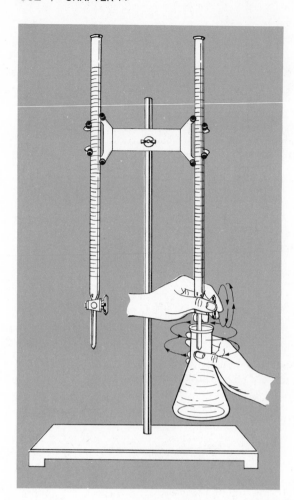

**Figure 14.7**   Titration from a buret into a flask.

rately by precipitating all of the chloride ion as silver chloride, using a solution of silver nitrate for this purpose:

$$AgNO_3 \text{ (aq)} + NaCl \text{ (aq)} \rightarrow AgCl \text{ (s)} + NaNO_3 \text{ (aq)}$$

The silver nitrate solution is carefully prepared so its concentration is known. It is then introduced to the sodium chloride solution until *just the right volume required for precipitation* has been added and *no more*. In other words, the quantities of the reactants must be *chemically equal*. The volume of silver nitrate solution is measured. The product of volume and molarity gives us moles of silver nitrate. From moles of silver nitrate the path to moles of sodium chloride and grams of sodium chloride is routine.

The one remaining question is how to tell when "just the right volume" of silver nitrate solution has been added. For this purpose a reagent is used that gives a visible signal just as soon as all the chloride ion has been precipitated. This reagent is called an **indicator.** In the reaction at hand, sodium chromate solution is the indicator. When chromate ion precipitates with silver ion it forms deep red silver chromate, $Ag_2CrO_4$, which is detected immediately when it first appears mixed with the white silver chloride.

There is one other important requirement for an indicator: it must not change color until the reaction for which it is being used is completed. In other words, no red silver chromate may form until all the chloride ion has been precipitated. The chromate ion satisfies this requirement. If silver ion is added to a solution containing both chloride and chromate ion, it will react first with the chloride ion to form white AgCl. Only when the chloride ion is gone will it combine with chromate ion to impart the red color that signals completion of the main reaction.

Let's make our problem specific with an example . . . .

---

**Example 14.16** Determine the weight of sodium chloride present in a solution if 38.6 ml of 0.813 M $AgNO_3$ are required to precipitate all the chloride present.

The foregoing description of the reaction should enable you to set up and complete this problem. Your unit path is from $\ell$ $AgNO_3$ → moles $AgNO_3$ → moles NaCl → g NaCl.

------------------------------------------------------------

**14.16a** 1.84 g NaCl

$$0.0386\,\ell \times \frac{0.813\text{ mole }AgNO_3}{1\ell} \times \frac{1\text{ mole NaCl}}{1\text{ mole }AgNO_3} \times \frac{58.5\text{ g NaCl}}{1\text{ mole NaCl}}$$

$$= 1.84\text{ g NaCl}$$

---

Titration is quite frequently the process by which the concentrations of acids or bases are determined. These titrations are neutralization reactions. Organic substances are usually used as indicators. They have one color in a solution that is acidic, and another color in a solution that is basic. Phenol-phthalein and litmus are two of the better known acid-base indicators, the latter most familiar as litmus paper that turns pink when dipped into an acid and blue if dipped into a base.

---

**Example 14.17** 25.0 ml of a NaOH solution of unknown concentration are titrated with 0.439 M $H_2SO_4$. 18.6 milliliters of the sulfuric acid are required for the reaction. Calculate the molarity of the sodium hydroxide solution. The equation is

$$2\text{ NaOH (aq)} + H_2SO_4\text{ (aq)} \rightarrow Na_2SO_4\text{ (aq)} + 2\text{ HOH (l)}$$

The procedure with this example begins in the same way as the last. This time, however, we are not interested in the number of *grams* of NaOH in the 25.0 ml sample, but rather the *concentration* in moles per liter. The first two steps of the stoichiometry pattern take us to moles of NaOH. If we divide the number of moles by the volume in liters, we have molarity.

As a first step, set up the problem to the determination of moles of NaOH, but do not solve.

------------------------------------------------------------

**14.17a**    $\boxed{0.0186 \quad \ell\,\text{H}_2\text{SO}_4} \times \dfrac{0.439 \,\text{mole H}_2\text{SO}_4}{1\,\ell\,\text{H}_2\text{SO}_4} \times \dfrac{2 \text{ moles NaOH}}{1 \,\text{mole H}_2\text{SO}_4}$

The setup above gives you the number of moles of NaOH in 25.0 ml of solution. Complete the problem: dividing by volume in liters—or multiplying by the reciprocal of volume—gives the concentration of the solution in moles per liter.

---

**14.17b**    0.653 mole NaOH/liter

$$\boxed{0.0186\,\ell\,\text{H}_2\text{SO}_4} \times \frac{0.439 \,\text{mole H}_2\text{SO}_4}{1\,\ell\,\text{H}_2\text{SO}_4} \times \frac{2 \text{ moles NaOH}}{1 \,\text{mole H}_2\text{SO}_4} \times \frac{1}{0.0250\,\ell\,\text{NaOH}}$$

$$= 0.653 \text{ mole NaOH/liter}$$

**Optional**

Example 14.17 furnishes an excellent illustration of the use of normality as a concentration unit. The reaction equation shows that two moles of sodium hydroxide react with one mole of sulfuric acid. But notice how the numbers of equivalents compare. There is one equivalent per mole of NaOH, so two moles represent two equivalents. One mole of $\text{H}_2\text{SO}_4$ contains two equivalents of $\text{H}_2\text{SO}_4$, the same as the number of equivalents of sodium hydroxide. Because of the way *equivalent* has been defined, it follows that, in any reaction, *the number of equivalents of* all *reactants is always the same.*

Equation 14.10 indicates that the number of equivalents of a species in a reaction may be found by multiplying the volume of a solution by its normality. Because the numbers of equivalents of all reactants are equal, the volume times normality products of the reactants may be equated:

$$V_a N_a = V_b N_b \tag{14.11}$$

where $V_a$ and $V_b$ are the volumes of the acid and base, respectively, and $N_a$ and $N_b$ are their normalities. Equation 14.11 and the idea of normality simplify greatly many calculations that are done over and over again in routine control and research operations in industry, as the following example shows.

---

**Example 14.18**   The concentration of an alkaline (basic) solution in an industrial process is checked periodically by titrating a 25.0 ml sample with 0.878 N $\text{H}_2\text{SO}_4$. One such titration required 18.6 ml of the acid to neutralize the base. Calculate the normality of the base.

*Solution.*   If Equation 14.8 is solved for the normality of the base, direct substitution of the given information yields the answer:

$$N_b = \frac{V_a N_a}{V_b} = \frac{18.6 \times 0.878}{25.0} = 0.653 \text{ N}$$

---

The simplification of the normality calculation compared to molarity is apparent when you compare Examples 14.17 and 14.18. They are *exactly the same problem.* With two equivalents per mole, 0.439 M $\text{H}_2\text{SO}_4$ is the same solution as 0.878 N $\text{H}_2\text{SO}_4$.

Normality calculations are not limited to acid-base reactions. They are particularly useful in oxidation-reduction reactions in which molar relationships are quite complex. But the mole concept remains the essential connecting link between substances in any stoichiometric problem, even though it may be hidden in the relationship between moles and equivalents. For this reason some chemists prefer to disregard normality as a concentration unit and work exclusively in moles and molarity.

## 14.10 COLLIGATIVE PROPERTIES OF SOLUTIONS (OPTIONAL)

A pure solvent has certain distinct, definite physical properties, as does any pure substance. The introduction of a solute into the solvent affects these properties. The properties of the solution, a mixture, depend upon the relative quantities of solvent and solute. It has been found experimentally that, in *dilute* solutions of certain solutes, the *change* in some of these properties is proportional to the molal concentration of the solute particles. The proportionality constant is independent of the solute; it is a property of the solvent. Solution properties that are determined solely by the *number* of solute particles dissolved in a fixed quantity of solvent are called **colligative properties.**

Freezing and boiling points of solutions are colligative properties. Perhaps the most common example is the antifreeze used in the cooling systems of automobiles. The solute dissolved in the radiator water reduces the freezing temperature to a level well below the normal freezing point of pure water, and also raises the boiling point above the normal boiling point.

The mathematical relationship for the change in freezing point between a solution and a pure solvent is

$$\Delta T_f = K_f\, m \tag{14.12}$$

where $\Delta T_f$ is the **freezing point depression,** as it is called, $K_f$ is the proportionality constant known as a **molal freezing point constant,** and m is the molality of the solution. Similarly,

$$\Delta T_b = K_b\, m \tag{14.13}$$

in which $\Delta T_b$ is the **boiling point elevation** and $K_b$ is the **molal boiling point constant.** For water $K_f = -1.86$, and $K_b = 0.52$. Boiling and freezing point problems may be solved algebraically by direct substitution into one or the other of the above equations.

---

**Example 14.19** Determine the freezing point of a solution of 12.0 grams of urea, $CO(NH_2)_2$, in 250 grams of water.

*Solution.* To use Equation 14.12 it is necessary to express the solution concentration in molality:

$$\frac{12.0 \text{ g } CO(NH_2)_2}{0.250 \text{ kg } H_2O} \times \frac{1 \text{ mole } CO(NH_2)_2}{60.0 \text{ g } CO(NH_2)_2} = 0.800 \text{ m}$$

If $K_f = -1.86$ for water,

$$\Delta T_f = -1.86 \times 0.800 = -1.49°C$$

This indicates the freezing point of the solution is 1.49°C below the normal freezing point of water, 0°C. The solution freezes at $-1.49°C$.

---

Notice that Equations 14.12 and 14.13 do not yield freezing or boiling points, but *changes* in these properties, $\Delta T_f$ and the freezing point of the solution are the same in Example 14.18 only because the solvent happens to freeze at 0°C, a convenience that will not be true for any solvent other than water. Notice also that we have not used units in the usual way. They can be included, but they are awkward and tend to confuse the calculation rather than aid it.

## 14.11 ELECTROLYTES AND SOLUTION CONDUCTIVITY

### STRONG ELECTROLYTES, WEAK ELECTROLYTES AND NONELECTROLYTES

PG   14 L   Distinguish between strong electrolytes, weak electrolytes and nonelectrolytes.

14 M   Explain why the solution of an ionic compound is always a good conductor of electricity.

14 N   Explain why the solution of a molecular compound may be a good conductor of electricity, a poor conductor, or a nonconductor.

When an electric light bulb is assembled as shown in Figure 14.8, current must pass from one metal strip to another in order to complete the circuit. If the beaker contains pure water, no current passes; the bulb does not light up. However, if the metal strips are immersed in a solution of sodium

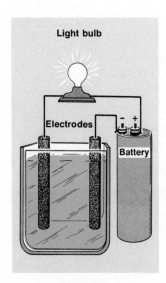
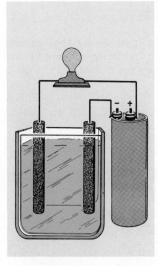
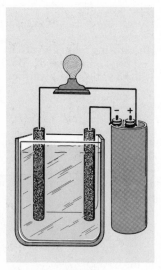

| a. Solution of table salt (an electrolytic solution) | b. Solution of table sugar (a nonelectrolytic solution) | c. Pure water (a nonelectrolyte) |

**Figure 14.8** Electrolytes and nonelectrolytes. (a) NaCl solution conducts electricity, as shown by the glowing light bulb. NaCl is an electrolyte. (b) Sugar solution does not conduct electricity, so the bulb does not glow. Sugar is a nonelectrolyte. (c) Pure water does not conduct electricity; it is a nonelectrolyte. Therefore the light bulb does not glow.

chloride, the bulb glows brightly—almost as much as if the metal strips were "shorted" with a screw driver. A solution of sugar produces no such effect; like pure water, a sugar solution is a nonconductor of electricity.

Passage of electricity through a liquid is referred to generally as **electrolysis.** The metal strips through which the "current" of electricity enters and leaves the solution are called **electrodes.** A solute which, when dissolved in water, yields a solution that is a good conductor of electricity is called a **strong electrolyte.** Sugar, whose solution is a nonconductor, is called a **nonelectrolyte.** Some solutes, such as acetic acid, are **weak electrolytes.** Their solutions conduct electricity, but poorly, permitting only a dim glow in the lamp of Figure 14.8. The term *electrolyte* is also used to refer to the *solution* through which electric current passes. Accordingly the acid solution in an automobile storage battery is an electrolyte. In commercial electroplating, *electrolyte* is almost always used in the solution sense.

An electric "current" is a movement of electric charge. In a metal wire, a mass movement of negatively charged electrons constitutes an electric current. In a water solution the charge is carried by ions rather than electrons. Once released from their crystalline lattice, positive and negative ions in solution are free to move. When a pair of electrodes is connected to a source of direct current, such as a storage battery, one electrode acquires a positive charge, the other a negative charge. The positively charged electrode attracts the negatively charged ions, while the positive ions move toward the negative electrode, as shown in Figure 14.9. This counter-current movement of ions through the solution constitutes a flow of electric current. *Solution conductivity is conclusive evidence of the presence of ions.*

When sugar dissolves, the crystal breaks into individual sugar molecules. These molecules are electrically neutral, and therefore show no tendency to move in either direction. Even if there were movement of these molecules through the solution it would not constitute an electric current because the molecules have no charge.

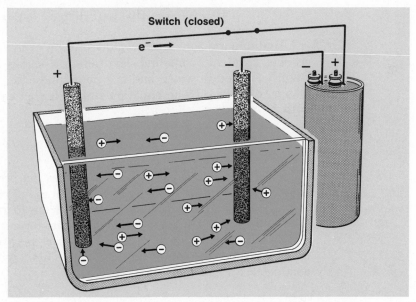

**Figure 14.9** Conductivity in an ionic solution. The conductivity of a solution is positive evidence that mobile ions are present.

In general, whenever an ionic compound dissolves, the ions are free to move and the solution is a good conductor. The compound is a strong electrolyte. When a molecular compound dissolves, it may be a nonelectrolyte, a weak electrolyte or a strong electrolyte, depending on the degree of ionization. Many molecular solutes do not ionize at all in solution; they are the nonelectrolytes. Weak electrolytes are those molecular compounds that ionize slightly when dissolved, and the few molecular compounds that ionize almost completely are the strong electrolytes. The terms *strong* and *weak*, when used in this sense, refer to the degree of ionization of a solute.

## STRONG ACIDS AND WEAK ACIDS

It is important in the next section that, given the name or formula of an acid, you first recognize it as an acid, and then be able to classify it as a strong acid or a weak acid. An acid, as we have used the term so far, is a hydrogen-bearing compound that releases hydrogen ions in water solution.* Its formula is generally in the form $HX$, $H_2X$ or $H_3X$, where X represents anything that can become a negatively charged ion when the acid ionizes, such as $Cl^-$ from $HCl$ or $SO_4^{2-}$ from $H_2SO_4$. Notice that the negative ion may or may not contain oxygen. Even un-ionizable hydrogen may be present; acetic acid, $HC_2H_3O_2$, for example, yields the acetate ion, $C_2H_3O_2^-$. Do not be concerned if the negative ion produced by the acid ionization happens to be unfamiliar. It will behave in exactly the same way as the ion from a familiar acid, and should be treated accordingly.

Hydrochloric acid is the water solution of hydrogen chloride, a gaseous covalent compound. Hydrochloric acid is a strong acid. It is almost completely ionized in water, and is an excellent conductor. The ionization process may be described by the equation

$$HCl\ (g) \xrightarrow{H_2O} HCl\ (aq) \longrightarrow H^+\ (aq) + Cl^-\ (aq) \qquad (14.14)$$

Hydrofluoric acid is the water solution of hydrogen fluoride, another gaseous covalent compound. Hydrofluoric acid is a weak acid. It is only slightly ionized in water, and is a poor conductor. The ionization process may be described by an equation similar to that for hydrochloric acid, but with an important difference:

$$HF\ (g) \xrightarrow{H_2O} HF\ (aq) \xleftarrow{\quad\longrightarrow} H^+\ (aq) + F^-\ (aq) \qquad (14.15)$$

The double arrow indicates that the ionization process is reversible (see page 284), and its greater length from right to left shows that the tendency is much stronger for hydrogen and fluoride ions to form dissolved hydrogen fluoride molecules than for the molecules to form the ions. In other words, the dissolved particles in hydrofluoric acid are primarily un-ionized HF molecules, and very few (relatively) ions. This is why the solution is a poor conductor, a weak acid.

---

*You will become familiar with broader concepts of acids in Chapter 16.

Now, how do you determine if an acid is strong or weak? The answer is, by knowing the ones that are strong. We will consider only six: nitric, $HNO_3$; sulfuric, $H_2SO_4$; hydrochloric, HCl; hydrobromic, HBr; hydroiodic, HI; and perchloric, $HClO_4$. The first three are the best known acids. The fourth and fifth are similar to hydrochloric, containing two other halogens; and the sixth, not widely discussed in an introductory course, simply must be remembered as a strong acid. In classifying an acid, you first check to see if it is one of the strong acids. If not, it must be weak.*

## "SOLUTION INVENTORY"

We have seen that when an ionic solute dissolves, the ions separate and are free to move. In a solution of sodium chloride, for example, there are no sodium chloride "molecules" having the formula NaCl. The solute particles *actually present* in the solution are sodium ions, $Na^+$ (aq) and chloride ions, $Cl^-$ (aq). The term **solution inventory** will be used in this text to identify the solute particles actually present in a solution. Thus the solution inventory of potassium hydroxide solution, KOH (aq), is potassium ion, $K^+$ (aq) and hydroxide ion, $OH^-$ (aq); and the solution inventory of a solution of ammonium nitrate, $NH_4NO_3$ (aq), is ammonium ion, $NH_4^+$ (aq), and nitrate ion, $NO_3^-$ (aq).

Because of the nearly complete ionization of strong acids in water, their solution inventories are also ions. The solution inventory of hydrochloric acid is $H^+$ (aq) and $Cl^-$ (aq); and the solution inventory of sulfuric acid is $H^+$ (aq) and $SO_4^{2-}$ (aq). Weak acids do not ionize completely, however. We have seen that hydrofluoric acid, for example, contains relatively few ions and mostly un-ionized HF molecules. The solution inventory of hydrofluoric acid is therefore primarily HF (aq). In general, the solution inventory of a weak acid is written as the un-ionized acid molecule.

\*    \*    \*    \*    \*

The concepts in this section are the foundation for writing net ionic equations in the next section. They are therefore summarized for you here.

1. *An electrolyte is a solute that, in water solution, conducts electricity by means of ions. Its solution inventory consists of ions.*
2. *Ionic compounds in solution always act as electrolytes. The solutions are good conductors; their solution inventories consist of ions.*
3. *Unless they ionize by reaction with water, molecular compounds in solution usually behave as nonelectrolytes. They are nonconductors; their solution inventories consist of neutral molecules.*
4. *Strong acids (there are six you should know) are strong electrolytes because they ionize almost completely in water. The solutions are good conductors; the solution inventories consist of ions.*

---

*The dividing line between strong and weak acids is arbitrary, and at least two acids are marginal in their classifications. Both oxalic acid, $H_2C_2O_4$, and phosphoric acid, $H_3PO_4$, may logically be considered as strong acids in their first ionization steps, but weak in the second and, in the case of phosphoric acid, third. For our purposes we will regard them as weak, and then avoid questions in which they might have to be classified.

5. *Weak acids (all acids except those that are strong) are weak electro-lytes because they ionize only slightly in water. The solutions are poor conductors; the solution inventories are written as neutral molecules.*

## 14.12 NET IONIC EQUATIONS

In Chapter 7, we wrote chemical equations for some types of reactions occurring in water solution. At that time you were promised a description of these reactions in the form of net ionic equations. Building now on your knowledge of electrolytes in solution, we shall develop net ionic equations for three kinds of aqueous solution reactions: (1) reactions that release a gas to the atmosphere; (2) reactions that yield a precipitate; and (3) reactions that yield a molecular product. These reactions are sometimes referred to as reactions that "go to completion," because they continue until essentially all of at least one reactant is used up.

Your skill in writing net ionic equations will depend largely on your understanding of the ideas introduced in the last section. If there has been a time lapse since you studied it, you may wish to review the summary with which it ends. Items 2, 4 and 5 are particularly important.

### REACTIONS THAT RELEASE A GAS TO THE ATMOSPHERE

> PG 14 O   Given reactants that yield an identified gaseous product, write the net ionic equation.

If a piece of zinc is dropped into sulfuric acid, a vigorous evolution of hydrogen gas occurs. Remaining in the reaction vessel will be a solution of zinc sulfate. The question we now propose is, what happened?

Heretofore this question has been answered by a chemical equation —a "conventional" equation, as we shall call it in this chapter to distinguish it from ionic equations. The conventional equation for the reaction between zinc and sulfuric acid is

$$Zn \ (s) + H_2SO_4 \ (aq) \rightarrow H_2 \ (g) + ZnSO_4 \ (aq) \tag{14.16}$$

This equation, however, does not reveal *specifically* what happened, because the equation includes the formulas for two chemical species that are not physically present, namely, $H_2SO_4$ (aq) and $ZnSO_4$ (aq). Sulfuric acid is a strong acid. Its solution inventory, $H^+$ (aq) and $SO_4^{2-}$ (aq), identifies the species actually present. Further, zinc sulfate being a soluble ionic compound, the solution inventory of $ZnSO_4$ (aq) is $Zn^{2+}$ (aq) and $SO_4^{2-}$ (aq).

If the dissolved substances in Equation 14.16 are replaced by their solution inventory species, an **ionic equation** results:

$$Zn \ (s) + H^+ \ (aq) + SO_4^{2-} \ (aq) \rightarrow H_2 \ (g) + Zn^{2+} \ (aq) + SO_4^{2-} \ (aq) \tag{14.17}$$
$$\text{(Unbalanced)}$$

You will observe that the ionic equation is not balanced: there are two hydrogen atoms on the right, and only one on the left. Normally it is not necessary to balance either the conventional equation or the ionic equation to find the **net ionic equation,** which answers fully and accurately the question, "What happened?"*

What did happen in Equation 14.17? Zinc started out as a solid, and wound up as a dissolved ion. Zinc changed chemically. Hydrogen went into the reaction as an aqueous ion, but came out as gaseous molecules. Hydrogen experienced a chemical change. Sulfate ion entered the reaction as a dissolved ion, and ended up as a dissolved ion. It did *not* change: it appears in exactly the same form on both sides of the equation. Such an ion—an ion that is present at the scene of a reaction, but not a participant in the reaction—is called a **spectator ion.** Spectator ions do not appear in a net ionic equation, which is restricted to those species that actually experience a chemical change. The ionic equation is therefore converted to a net ionic equation by eliminating the spectators and balancing:

$$\text{Zn (s)} + \text{H}^+ \text{ (aq)} + \cancel{\text{SO}_4^{2-} \text{ (aq)}} \rightarrow \text{H}_2 \text{ (g)} + \text{Zn}^{2+} \text{ (aq)} + \cancel{\text{SO}_4^{2-} \text{ (aq)}} \quad (14.18)$$
$$\text{(Unbalanced)}$$

$$\text{Zn (s)} + \text{H}^+ \text{ (aq)} \rightarrow \text{H}_2 \text{ (g)} + \text{Zn}^{2+} \text{ (aq) (Unbalanced)} \quad (14.19)$$

$$\text{Zn (s)} + 2 \text{ H}^+ \text{ (aq)} \rightarrow \text{H}_2 \text{ (g)} + \text{Zn}^{2+} \text{ (aq)} \quad (14.20)$$

In summary, a net ionic equation includes only those substances that actually undergo chemical change in the reaction; it tells what happened. Net ionic equations may be written by a three step procedure:

1. Write the conventional equation, including designation of state, (g), (l) or (s), or aqueous solution, (aq).
    State designations are important guides for the next step in the procedure.
    The conventional equation may be unbalanced.*
2. Write the ionic equation by replacing all *dissolved* substances in the conventional equation by their solution inventory species.
    Only the dissolved substances, designated (aq) in the conventional equation, can be changed—and then only if they yield ions in solution (weak acids and some other molecular solutes do not). *Never change solids (s), liquids (l) or gases (g) in this step.*
    The ionic equation may be unbalanced.*
3. Write the net ionic equation by eliminating all spectators from the ionic equation and balancing both atoms and electric charge.

---

*Many instructors prefer that each equation be balanced. If this is the wish of your instructor, you should comply with it. Balancing the conventional equation is routine. If you have balanced the conventional equation, and then take into account the *number* of ions produced by each dissolved species in writing your solution inventory (e.g., $\text{H}_2\text{SO}_4 \rightarrow 2 \text{ H}^+ + \text{SO}_4^{2-}$), the ionic equation will be balanced automatically; but you should check to be sure. Elimination of the spectators will then yield a balanced net ionic equation, which again should be checked. Sometimes that equation will contain coefficients all divisible by a common factor, as $2 \text{ Ag}^+ + 2 \text{ Cl}^- \rightarrow 2 \text{ AgCl}$. Ordinarily such coefficients should be divided by the common factor to yield a net ionic equation having the lowest possible whole-number coefficients.

**Example 14.20**  If calcium metal is placed into hydrochloric acid, hydrogen gas evolves and a solution of calcium chloride remains. Write the net ionic equation.

First, the conventional equation should be written. It may remain unbalanced. Include (g), (l), (s) or (aq) designations for each species.

-------------------------------------------------------------------------------

**14.20a**  $Ca$ (s) + $HCl$ (aq) → $H_2$ (g) + $CaCl_2$ (aq) (Unbalanced)

Write next the ionic equation, in which the dissolved solutes, designated (aq), are shown in solution inventory form. Do not change items designated (s), (l) or (g).

-------------------------------------------------------------------------------

**14.20b**  $Ca$ (s) + $H^+$ (aq) + $Cl^-$ (aq) → $H_2$ (g) + $Ca^{2+}$ (aq) + $Cl^-$ (aq)
                                    (Unbalanced)

Finally, identify and eliminate any spectators, write and balance the net ionic equation.

-------------------------------------------------------------------------------

**14.20c**  $Ca$ (s) + 2 $H^+$ (aq) → $H_2$ (g) + $Ca^{2+}$ (aq)

The chloride ion is a spectator in the ionic equation.

---

**Example 14.21**  If solid potassium is placed in water, hydrogen gas and a solution of potassium hydroxide result. Write the net ionic equation for the reaction.

The unbalanced conventional equation is straightforward . . . .

-------------------------------------------------------------------------------

**14.21a**  $K$ (s) + $H_2O$ (l) → $H_2$ (g) + $KOH$ (aq) (Unbalanced)

In the ionic equation, show the solution inventories of all dissolved substances. Do not change items designated (s), (l) or (g).

-------------------------------------------------------------------------------

**14.21b**  $K$ (s) + $H_2O$ (l) → $H_2$ (g) + $K^+$ (aq) + $OH^-$ (aq) (Unbalanced)

Now eliminate the spectators and balance for the net ionic equation. Careful! This one's tricky.

-------------------------------------------------------------------------------

**14.21c**  2 $K$ (s) + 2 $H_2O$ (l) → $H_2$ (g) + 2 $K^+$ (aq) + 2 $OH^-$ (aq)

This reaction has no spectators; the balanced ionic equation is the net ionic equation. But the *balanced* ionic equation is not

$$K \text{ (s)} + 2 \text{ H}_2\text{O (l)} \rightarrow \text{H}_2 \text{ (g)} + \text{K}^+ \text{ (aq)} + 2 \text{ OH}^- \text{ (aq) (Unbalanced)}$$

This equation balances out the atoms all right, but not the charges. Net charge is zero on the left, but $-1$ on the right. Two $\text{K}^+$ ions on the right balance the charges, but then the second K is necessary on the left to restore atomic balance. Always be sure to check out charge balance.

## REACTIONS THAT FORM PRECIPITATES

PG  14 P   Given the product of a precipitation reaction, write the net ionic equation.

14 Q   Given a table of solubilities, predict whether or not a precipitate will form when two solutions are mixed; if a precipitate does form, write the net ionic equation for the reaction.

Net ionic equations are particularly useful for describing reactions in aqueous solution that yield products formed by the combination of ions from different sources. They are sometimes called *ion combination reactions*. One kind of ion combination reaction yields a precipitate as one product.

---

**Example 14.22**  When hydrochloric acid and silver nitrate solutions are mixed, a heavy, white precipitate of silver chloride is produced. Write the net ionic equation for the reaction.

As an opener, write the conventional equation.

---

**14.22a**  $\text{HCl (aq)} + \text{AgNO}_3 \text{ (aq)} \rightarrow \text{AgCl (s)} + \text{HNO}_3 \text{ (aq)}$

In writing the ionic equation, how will you handle $\text{HNO}_3$ (aq)? What kind of compound is it? Is it strong or weak? What is its solution inventory? If you know these answers, proceed with the ionic equation; if not, look back to Section 14.11.

---

**14.22b**  $\text{H}^+ \text{ (aq)} + \text{Cl}^- \text{ (aq)} + \text{Ag}^+ \text{ (aq)} + \text{NO}_3^- \text{ (aq)} \rightarrow$
$$\text{AgCl (s)} + \text{H}^+ \text{ (aq)} + \text{NO}_3^- \text{ (aq)}$$

$\text{HNO}_3$ (aq) is a strong acid; its solution inventory consists of the $\text{H}^+$ (aq) and $\text{NO}_3^-$ (aq) ions.

Elimination of the spectators yields the net ionic equation.

---

**14.22c**  $\text{Ag}^+ \text{ (aq)} + \text{Cl}^- \text{ (aq)} \rightarrow \text{AgCl (s)}.$

$\text{H}^+$ (aq) and $\text{NO}_3^-$ (aq) are spectators in the ionic equation.

Let's try another that has an interesting twist to it . . . .

---

**Example 14.23** When solutions of silver sulfate and aluminum chloride are combined, silver chloride precipitates. Write the net ionic equation.

By now you know the three step sequence: conventional equation, ionic equation, net ionic equation. Go all the way to the net ionic equation on this one.

---

**14.23a** $Ag^+$ (aq) + $Cl^-$ (aq) → AgCl (s)

This is the same equation as the last example. How come? Let's look at the steps:

Conventional: $Ag_2SO_4$ (aq) + $AlCl_3$ (aq) →
$$AgCl\ (s) + Al_2(SO_4)_3\ (aq)\ (Unbalanced)$$

Ionic: $Ag^+$ (aq) + $\cancel{SO_4^{2-}}$ (aq) + $\cancel{Al^{3+}}$ (aq) + $Cl^-$ (aq) →
$$AgCl\ (s) + \cancel{Al^{3+}}\ (aq) + \cancel{SO_4^{2-}}\ (aq)$$

Net ionic: $Ag^+$ (aq) + $Cl^-$ (aq) → AgCl (s)

$Al^{3+}$ (aq) and $SO_4^{2-}$ (aq) are spectators.

---

From these two examples we see an important generalization: AgCl (s) will precipitate any time two solutions, one containing $Ag^+$ (aq) and the other containing $Cl^-$ (aq), are mixed. The same net ionic equation will describe the reaction regardless of other ions that may be present. (It is conceivable that a second reaction could occur between the other pair of ions, yielding two net ionic equations; you should be alert to this possibility.) Accordingly, if asked to write the net ionic equation for the reaction between solutions of silver nitrate and sodium chloride, you could skip the conventional and ionic equations, and go directly to $Ag^+$ (aq) + $Cl^-$ (aq) → AgCl (s).

This simple and direct procedure for writing the net ionic equation for a precipitation reaction may be applied to any insoluble ionic compound . . . .

---

**Example 14.24** Write the net ionic equations for the precipitation of the following from aqueous solutions: CuS; $Mg(OH)_2$; $Al_2(CO_3)_3$.

---

**14.24a**
$$Cu^{2+}\ (aq) + S^{2-}\ (aq) → CuS\ (s)$$
$$Mg^{2+}\ (aq) + 2\ OH^-\ (aq) → Mg(OH)_2\ (s)$$
$$2\ Al^{3+}\ (aq) + 3\ CO_3^{2-}\ (aq) → Al_2(CO_3)_3\ (s)$$

In all cases the reaction is the direct combination of the ions in the compound.

TABLE 14.1 Solubilities of Ionic Compounds*

| Ions | Acetate | Bromide | Carbonate | Chlorate | Chloride | Fluoride | Hydrogen Carbonate | Hydroxide | Iodide | Nitrate | Nitrite | Phosphate | Sulfate | Sulfide | Sulfite |
|---|---|---|---|---|---|---|---|---|---|---|---|---|---|---|---|
| Aluminum | I | S | | S | S | I | | I | – | S | | I | S | – | |
| Ammonium | S | S | S | S | S | S | S | – | S | S | S | S | S | S | S |
| Barium | – | S | I | S | S | I | | S | S | S | S | I | I | – | I |
| Calcium | S | S | I | S | S | I | | I | S | S | S | I | I | I | I |
| Cobalt(II) | S | S | I | S | S | – | | I | S | S | | I | S | I | I |
| Copper(II) | S | – | | | S | S | | I | | S | | I | S | I | |
| Iron(II) | S | S | I | | S | I | | I | S | S | | I | S | I | I |
| Iron(III) | – | S | | | S | I | | I | – | S | | I | S | – | |
| Lead(II) | S | I | I | S | I | I | | I | I | S | S | I | I | I | I |
| Lithium | S | S | S | S | S | S | S | S | S | S | S | I | S | S | |
| Magnesium | S | S | I | S | S | I | | I | S | S | S | I | S | – | S |
| Nickel | | S | I | S | S | S | | I | S | S | | I | S | I | I |
| Potassium | S | S | S | S | S | S | S | S | S | S | S | S | S | S | S |
| Silver | I | I | I | S | I | S | | – | I | S | I | I | I | I | I |
| Sodium | S | S | S | S | S | S | S | S | S | S | S | S | S | S | S |
| Zinc | S | S | I | S | S | S | | I | S | S | | I | S | I | I |

*Compounds having solubilities of 0.1 M or more at 20°C are listed as soluble (S); if the solubility is less than 0.1 M, the compound is listed as insoluble (I). A dash (–) identifies an unstable species in aqueous solution, and a blank space indicates lack of data.

If we knew in advance those combinations of ions that yield insoluble compounds we could predict precipitation reactions. Actually these compounds have been identified in the laboratory. Table 14.1 classifies as soluble (S) or insoluble (I) a large number of ionic compounds. For example, the third ion from the bottom of the left column is the silver ion, and the fifth ion across the top row is the chloride ion. The intersection of the third-from-bottom row and the fifth column identifies silver chloride as the compound, and the I in that intersection tells us the compound is insoluble.

It is significant that no substance is completely insoluble, but rather that its solution becomes saturated at an extremely small concentration. When, for example, solid silver chloride reaches equilibrium with its saturated solution, $AgCl (s) \rightleftharpoons Ag^+ (aq) + Cl^- (aq)$, the ion concentrations are very low, roughly 0.00001 molar.

Let's try an example to see how to use this table in predicting precipitation reactions . . . .

---

**Example 14.25**  Solutions of lead(II) nitrate and sodium fluoride are combined. Write the net ionic equation for any precipitation reaction that may occur.

Let us begin by writing the conventional equation . . . .

---

**14.25a**  $Pb(NO_3)_2 (aq) + NaF (aq) \rightarrow PbF_2(?) + NaNO_3(?)$ (Unbalanced)

You will notice that we have placed question marks in the state designation spaces for the products of the reaction. The question is whether each new combination of ions will form a precipitate or remain in solution as dissolved ions. Consulting Table 14.1 we find the symbol I in the intersection of the lead(II) ion row and the fluoride ion column, indicating that $PbF_2$ is insoluble and will therefore precipitate. In the sodium and nitrate ion intersection we find the symbol S, showing $NaNO_3$ to be soluble. We therefore expect the sodium and nitrate ions to remain in solution. Our complete conventional equation is therefore

$$Pb(NO_3)_2 (aq) + NaF (aq) \rightarrow PbF_2 (s) + NaNO_3 (aq) \text{ (Unbalanced)}$$

We can now write the ionic equation by showing the solution inventory of each species, and then the net ionic equation by eliminating the spectators from the ionic equation and balancing. Complete the example.

---

**14.25b**  $Pb^{2+} (aq) + \cancel{NO_3^- (aq)} + \cancel{Na^+ (aq)} + F^- (aq) \rightarrow$
$$PbF_2 (s) + \cancel{Na^+ (aq)} + \cancel{NO_3^- (aq)} \text{ (Unbalanced)}$$
$$Pb^{2+} (aq) + 2 F^- (aq) \rightarrow PbF_2 (s)$$

---

**Example 14.26**  Write the net ionic equation for any reaction that will occur when solutions of nickel chloride and ammonium carbonate are combined.

---

**14.26a** $$Ni^{2+} (aq) + CO_3^{2-} (aq) \rightarrow NiCO_3 (s)$$

The conventional equation would show $NH_4Cl$ and $NiCO_3$ as the two new combinations of ions in this reaction. According to Table 14.1, $NH_4Cl$ is soluble and $NiCO_3$ is insoluble. The precipitating reaction will therefore be between nickel and carbonate ions.

---

**Example 14.27** Write the net ionic equation for any reaction that occurs on combining solutions of $BaCl_2$ and $Ag_2SO_4$.

Caution . . . .

-----------------------------------------------------------------------

**14.27a** $$Ba^{2+} (aq) + SO_4^{2-} (aq) \rightarrow BaSO_4 (s)$$

$$Ag^+ (aq) + Cl^- (aq) \rightarrow AgCl (s)$$

This time *both* new combinations of ions form insoluble compounds, so there will be a double precipitation. Two net ionic equations describe the two individual reactions.

## MOLECULAR PRODUCT REACTIONS

PG 14 R Given reactants that yield a molecular product, write the net ionic equation.

In Chapter 7, page 158, a neutralization reaction was described as a reaction between an acid and a metal hydroxide, yielding water as one product, plus an aqueous solution of the remaining ionic compound. When both acid and hydroxide are ionized in aqueous solutions the net ionic equation is particularly significant.

---

**Example 14.28** Write the net ionic equation for the reaction between hydrochloric acid and an aqueous solution of sodium hydroxide.

First, the conventional equation . . . .

-----------------------------------------------------------------------

**14.28a** $HCl (aq) + NaOH (aq) \rightarrow HOH (l) + NaCl (aq)$ (Unbalanced)

The ionic and net ionic equations follow:

-----------------------------------------------------------------------

**14.28b** $H^+ (aq) + \cancel{Cl^- (aq)} + \cancel{Na^+ (aq)} + OH^- (aq) \rightarrow$
$$HOH (l) + \cancel{Na^+ (aq)} + \cancel{Cl^- (aq)} \text{ (Unbalanced)}$$

$$H^+ (aq) + OH^- (aq) \rightarrow HOH (l)$$

Like a precipitation reaction, neutralization is an ion combination reaction. Any time $H^+$ (aq) ions from one source are added to $OH^-$ (aq) ions from another source, the ions will combine, yielding molecular water as a product. Water is but one of many molecular products. Any time the salt of a weak acid—an acid that ionizes only slightly in aqueous solution—is treated with a solution containing $H^+$ (aq) ions, there will be an ion combination between the hydrogen ion and the anion of the weak acid, forming that acid as a molecular product.

To illustrate, if hydrochloric acid is added to a solution of sodium acetate, $NaC_2H_3O_2$, the conventional equation is

$$\text{HCl (aq)} + NaC_2H_3O_2 \text{ (aq)} \rightarrow HC_2H_3O_2 \text{ (aq)} + NaCl \text{ (aq)} \quad (14.21)$$

To write the ionic equation you list the solution inventory of all species identified by (aq) in the conventional equation, as usual. But in the case of acetic acid, $HC_2H_3O_2$, the molecular product, what is the solution inventory? Acetic acid is not one of the six strong acids. Therefore it must be a weak acid. The solution inventory of a weak acid is the un-ionized acid molecule. As a consequence a weak acid is carried into the ionic equation in exactly the same molecular form it has in the conventional equation. The ionic equation, and ultimately the net ionic equation, are

$$H^+ \text{ (aq)} + \cancel{Cl^- \text{ (aq)}} + \cancel{Na^+ \text{ (aq)}} + C_2H_3O_2^- \text{ (aq)} \rightarrow$$
$$HC_2H_3O_2 \text{ (aq)} + \cancel{Na^+ \text{ (aq)}} + \cancel{Cl^- \text{ (aq)}} \quad (14.22)$$

$$H^+ \text{ (aq)} + C_2H_3O_2^- \text{ (aq)} \rightarrow HC_2H_3O_2 \text{ (aq)} \quad (14.23)$$

If you will compare Equations 14.21–14.23 with the three equations of Example 14.28 you will find them identical except for the substitution of the acetate ion, $C_2H_3O_2^-$, in place of the hydroxide ion, $OH^-$. Both combine with hydrogen ions to form molecular products, $HC_2H_3O_2$ and HOH.

---

**Example 14.29**  What is the net ionic equation for the reaction between sulfuric acid and ammonium nitrite solution?

This reaction is quite similar to the previous example. Take it all the way.

-------------------------------------------------------------------------------------

**14.29a**  $H_2SO_4$ (aq) $+ NH_4NO_2$ (aq) $\rightarrow$
$$HNO_2 \text{ (aq)} + (NH_4)_2SO_4 \text{ (aq) (Unbalanced)}$$

$H^+$ (aq) $+ \cancel{SO_4^{2-} \text{ (aq)}} + \cancel{NH_4^+ \text{ (aq)}} + NO_2^-$ (aq) $\rightarrow$
$$HNO_2 \text{ (aq)} + \cancel{NH_4^+ \text{ (aq)}} + \cancel{SO_4^{2-} \text{ (aq)}}$$

$$H^+ \text{ (aq)} + NO_2^- \text{ (aq)} \rightarrow HNO_2 \text{ (aq)}$$

Because nitrous acid is *not* one of the strong acids, you classify it as weak, and therefore a molecular product. The solution inventory is $HNO_2$ molecules—and so it appears in the net ionic equation.

Three ion combination products yield molecular products, but they are not the products you might expect. Two of the expected products are carbonic and sulfurous acids. If hydrogen ions from one source reach carbonate ions from another, $H_2CO_3$ should form:

$$2\ H^+\ (aq) + CO_3^{2-}\ (aq) \rightarrow H_2CO_3\ (aq)$$

But carbonic acid is unstable, and decomposes to carbon dioxide gas and water. The correct net ionic equation is therefore

$$2\ H^+\ (aq) + CO_3^{2-}\ (aq) \rightarrow CO_2\ (g) + H_2O\ (l)$$

Sulfurous acid, $H_2SO_3$ decomposes similarly to sulfur dioxide and water, but the sulfur dioxide remains in solution rather than bubbling off:

$$2\ H^+\ (aq) + SO_3^{2-}\ (aq) \rightarrow SO_2\ (aq) + H_2O\ (l)$$

The third ion combination product that behaves in an unexpected manner comes from the reaction between solutions of an ammonium salt and a hydroxide. The expected product would be "ammonium hydroxide":

$$NH_4^+\ (aq) + OH^-\ (aq) \rightarrow \text{"NH}_4\text{OH"}$$

Apparently no distinct molecules having the formula $NH_4OH$ ever form. The product instead is simply an aqueous solution of ammonia molecules, $NH_3$ (aq). The proper equation is therefore

$$NH_4^+\ (aq) + OH^-\ (aq) \rightarrow NH_3\ (aq) + H_2O\ (l)$$

This reaction is reversible, and actually comes to an equilibrium in which the solution inventory is overwhelmingly dissolved ammonia molecules.

There is no system by which these three "different" molecular product reactions can be recognized. You simply must be alert to them and recognize them when they appear. Once again, the predicted but unstable formulas are $H_2CO_3$, $H_2SO_3$ and $NH_4OH$.

---

**Example 14.30** Write the net ionic equation for the reaction between solutions of sodium carbonate and hydrochloric acid.

*Solution.* The conventional, ionic and net ionic equations are

$$HCl\ (aq) + Na_2CO_3\ (aq) \rightarrow NaCl\ (aq) + CO_2\ (g) + H_2O\ (l)\ \text{(Unbalanced)}$$

$$H^+\ (aq) + \cancel{Cl^-\ (aq)} + \cancel{Na^+\ (aq)} + CO_3^{2-}\ (aq) \rightarrow$$
$$\cancel{Na^+\ (aq)} + \cancel{Cl^-\ (aq)} + CO_2\ (g) + H_2O\ (l)\ \text{(Unbalanced)}$$

$$2\ H^+\ (aq) + CO_3^{2-}\ (aq) \rightarrow CO_2\ (g) + H_2O\ (l)$$

---

Many compounds are described in chemical handbooks as insoluble in water, but soluble in acids. Aluminum hydroxide is one such compound. The net ionic equation shows why . . . .

**Example 14.31** Write a net ionic equation to describe the reaction that takes place when solid aluminum hydroxide reacts with hydrochloric acid, or nitric acid, or sulfuric acid.

In this case we must begin with *solid* aluminum hydroxide; there are no hydroxide ions in solution from this metal hydroxide. Write the unbalanced conventional equation with hydrochloric acid as a starter . . . .

-------------------------------------------------------------------------

**14.31a** $HCl$ (aq) + $Al(OH)_3$ (s) → HOH (l) + $AlCl_3$ (aq) (Unbalanced)

Complete the example, writing both ionic and net ionic equations.

-------------------------------------------------------------------------

**14.31b** Ionic: $H^+$ (aq) + ~~$Cl^-$ (aq)~~ + $Al(OH)_3$ (s) →
HOH (l) + $Al^{3+}$ (aq) + ~~$Cl^-$ (aq)~~ (Unbalanced)

Net ionic: 3 $H^+$ (aq) + $Al(OH)_3$ (s) → 3 HOH (l) + $Al^{3+}$ (aq)

Now think a bit before writing the net ionic equation for the reaction between solid aluminum hydroxide and nitric acid . . . then go ahead.

-------------------------------------------------------------------------

**14.31c** 3 $H^+$ (aq) + $Al(OH)_3$ (s) → 3 HOH (l) + $Al^{3+}$ (aq)

This is the same as for hydrochloric acid—and sulfuric acid will also yield the same net ionic equation. The chloride, nitrate and sulfate ions are spectators in the three reactions.

*ONE MORE EXAMPLE . . .*

The last example is quite special, but also quite common. Try it, and you will see why.

**Example 14.32** Write the net ionic equation for any reaction that occurs when solutions of ammonium sulfate and potassium nitrate are combined.

-------------------------------------------------------------------------

**14.32a** No reaction occurs.

$(NH_4)_2SO_4$ (aq) + $KNO_3$ (aq) → $NH_4NO_3$ (aq) + $K_2SO_4$ (aq) (Unbalanced)

~~$NH_4^+$ (aq)~~ + ~~$SO_4^{2-}$ (aq)~~ + ~~$K^+$ (aq)~~ + ~~$NO_3^-$ (aq)~~ →
~~$NH_4^+$ (aq)~~ + ~~$NO_3^-$ (aq)~~ + ~~$K^+$ (aq)~~ + ~~$SO_4^{2-}$ (aq)~~ (Unbalanced)

In the ionic equation, all species appear on both sides of the equation; they are all spectators, and no chemical change occurs. There is no reaction. The only "product" of the combination of the two solutions is a single solution having an inventory of all four ions.

## USES OF NET IONIC EQUATIONS

If we consider the balanced conventional equation and net ionic equations for the reaction of Example 14.23,

$$3 \text{ Ag}_2\text{SO}_4 \text{ (aq)} + 2 \text{ AlCl}_3 \text{ (aq)} \rightarrow 6 \text{ AgCl (s)} + \text{Al}_2(\text{SO}_4)_3 \text{ (aq)}$$

$$\text{Ag}^+ \text{ (aq)} + \text{Cl}^- \text{ (aq)} \rightarrow \text{AgCl (s)}$$

we can see the relative usefulness of the two equations. Net ionic equations have the inherent advantage of conveying a better picture of what is actually occurring at the ionic and molecular level than does a conventional equation. Net ionic equations not only are more accurate, but also are simpler equations. Only net ionic equations are suitable for the study of equilibrium reactions (Chapters 15–17). On the other hand, the net ionic equation would be of little use in determining the number of grams of silver chloride that might precipitate from the addition of excess aluminum chloride solution to a solution containing 1.50 grams of dissolved silver sulfate. Only the conventional equation furnishes the molar relationships that are necessary in the solution of most stoichiometry problems.

# CHAPTER 14 IN REVIEW

## 14.1 THE CHARACTERISTICS OF A SOLUTION

14 A   Distinguish between a pure substance, a solution, a suspension and a colloid. (Page 316)

## 14.2 SOLUTION TERMINOLOGY

14 B   Distinguish between terms in the following groups: solute and solvent; concentrated and dilute; solubility, saturated, unsaturated and supersaturated; miscible and immiscible. (Page 317)

## 14.3 THE FORMATION OF A SOLUTION

14 C   Describe the formation of a saturated solution from the time excess solid solute is first placed into a liquid solvent. (Page 318)

14 D   Identify and explain the factors that determine the time required to dissolve a given amount of solute, or to reach equilibrium. (Page 318)

## 14.4 FACTORS THAT DETERMINE SOLUBILITY

14 E   Given the structural formulas of two molecular substances, or other information from which the strength of their intermolecular forces may be estimated, predict if they will dissolve appreciably in each other, and state the criteria on which your prediction is based. (Page 320)

14 F   Predict how the solubility of a gas in a liquid will be affected by a change in the partial pressure of that gas over the liquid. (Page 321)

## 14.5 SOLUTION CONCENTRATION: PERCENTAGE BY WEIGHT

14 G   Given grams of solute and grams of solvent or solution, calculate percentage concentration. (Page 323)

# TERMS AND CONCEPTS

Note: Terms or concepts in the following list that are written in *italics* are from optional sections in the text.

# QUESTIONS AND PROBLEMS

*An asterisk (\*) identifies a question that is relatively difficult, or that extends beyond the performance goals of the chapter.*

### Section 14.1

14.1)  Identify the most important property that distinguishes a solution from a suspension or colloid.

14.55)  "Mixtures of gases are always true solutions." True or false? Explain why.

### Section 14.2

14.2)  Explain why the distinction between solute and solvent is not clearly defined in many solutions.

14.3)  Solution A contains 10 grams of solute dissolved in 100 grams of solvent, while solution B contains only 5 grams of a different solute per 100 grams of solvent. Under what circumstances could we properly classify solution A as dilute and solution B as concentrated?

14.4)  Suggest simple laboratory tests by which you could determine if a given solution is unsaturated, saturated or supersaturated. Explain why your suggestions would distinguish between the different classifications.

14.5)  Contrast the terms miscibility, miscible and immiscible with their counterparts, solubility, soluble and insoluble.

14.56)  Would it be proper to say that a saturated solution is a concentrated solution? Or that a concentrated solution is a saturated solution? Point out the distinctions between these sometimes confused terms.

14.57)  Suggest units in which solubility might be expressed other than grams of solute per 100 grams of solvent.

14.58)  In stating solubility, an important variable must be identified. What is this variable, and how does solubility of a solid solute *usually* depend upon it?

14.59)  Give an example of two immiscible substances other than oil and water.

### Section 14.3

14.6)  Describe the forces that promote the dissolving of a solid solute in a liquid solvent.

14.7)  "A dynamic equilibrium exists when a saturated solution is in contact with excess solute." Explain the meaning of that statement. Is it possible to have a saturated solution *without* excess solute? Explain.

14.60)  Describe the forces that oppose the dissolving of a solute in a liquid solvent.

14.61)  Bakers use confectioner's sugar because it is more finely powdered than the crystals of table (granulated) sugar. Do you think confectioner's sugar would dissolve more or less quickly than table sugar? Why?

14.8) Why is it customary to stir coffee or tea after putting sugar into it?

14.62) Explain the effect of heating on the rate at which a solid dissolves in a liquid.

## Section 14.4

14.9) Which do you suppose would be the better solvent for benzene, $C_6H_6$—water or carbon tetrachloride? Why? The structural formula of benzene may be represented as

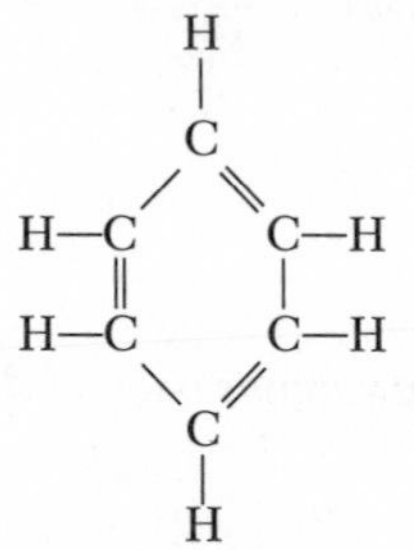

14.63) Suggest why water and liquid HF are good solvents for many ionic salts, but not for waxes and oils having structures such as

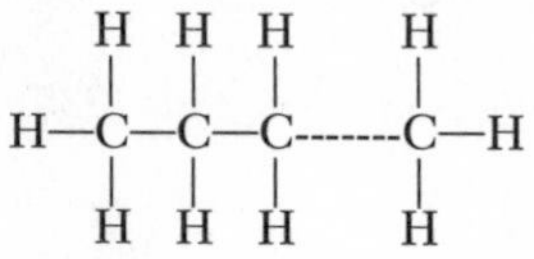

14.10) Suppose you have a spot on some clothing, and water will not take it out. If you have available cyclopentane and methanol (see structures below), which would you choose as the more promising solvent to try? Why?

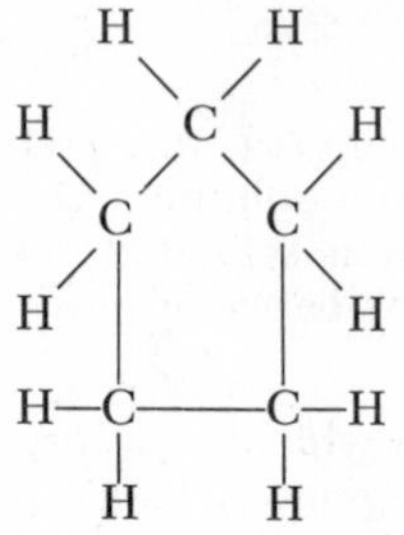

Cyclopentane 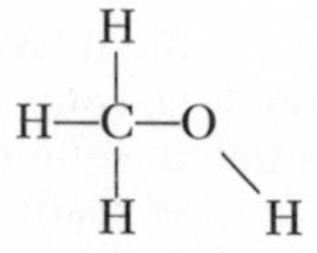          Methanol

14.64) Glycerin and normal hexane (see structures below) are organic compounds of approximately the same molecular weight. Which of these is more apt to be miscible with carbon tetrabromide? Why?

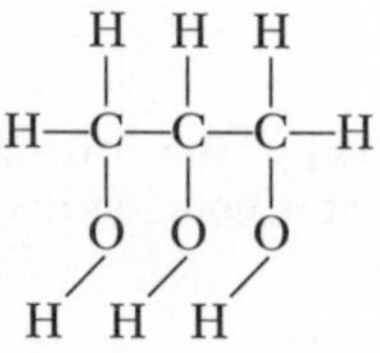

Glycerin

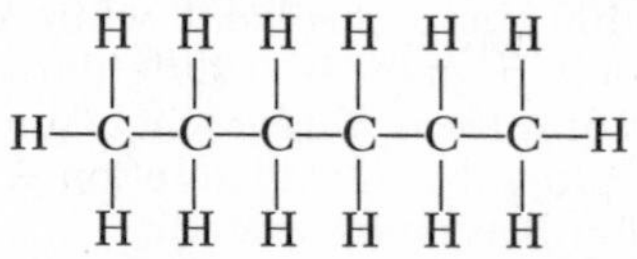

Normal hexane

14.11) "The solubility of carbon dioxide in a carbonated beverage may be increased by raising the air pressure under which it is bottled." Criticize the foregoing statement.

14.65) On opening a bottle of carbonated beverage, many bubbles are released, suggesting the beverage is bottled under high pressure. Yet for safety reasons the pressure cannot be more than slightly greater than atmospheric. How, then, do you account for the substantial reduction in $CO_2$ solubility in an opened bottle?

## Section 14.5

14.12) A sodium chloride solution weighing 32.4 grams was carefully evaporated to dryness. 2.78 grams of NaCl were recovered. Calculate the percentage NaCl in the original solution.

14.13) How many grams of boric acid are required to prepare 75.0 grams of 4.00% solution?

14.66) Calculate the concentration in per cent by weight if a solution is prepared by dissolving 4.23 grams of silver nitrate in 78.4 grams of water.

14.67) Calculate the mass of magnesium sulfate and the milliliters of water required to prepare 245 grams of 10.6% solution.

Section 14.6

14.14)* Calculate the molality of the solution prepared by dissolving 16.9 grams of glucose, $C_6H_{12}O_6$, a type of sugar, in 87.5 grams of water.

14.15)* Find the molality of a 10% NaCl solution. (Hint: Determine first the grams of salt and grams of water in a definite quantity—say 100 grams—of solution.)

14.16)* How many grams of formic acid, HCOOH—a compound first formed by distilling, of all things, red ants, and a source of irritation from their bites—must be dissolved in 50.0 milliliters of water to produce a 1.50 m solution?

Section 14.7

14.17) How many grams of silver nitrate, widely used in the manufacture of photographic chemicals, must be dissolved in the preparation of 150 ml of 0.125 M solution?

14.18) Oxalic acid, a compound used to remove iron stain from fabrics and porcelain, and as a bleach for leather goods, is a dihydrate with the formula $H_2C_2O_4 \cdot 2\ H_2O$. Determine the weight of oxalic acid required for $7.50 \times 10^2$ ml of a 0.480 M solution.

14.19) 16.2 grams of ammonium sulfate, a compound sold at local garden centers for lawn fertilizer, are dissolved in water and diluted to 300.0 ml. Find the molarity of the solution.

14.20) Calculate the molarity of the solution prepared by dissolving 56.2 g $CuSO_4 \cdot 5\ H_2O$, a compound used in electroforming printing plates, in 500.0 ml of solution.

14.21) How many moles of potassium nitrate, a fertilizer and food preservative, are in 45.3 ml of 0.378 M $KNO_3$?

14.22) Concentrated sulfuric acid is 18 molar. What volume of concentrated acid is required to obtain 1.24 moles of $H_2SO_4$?

Section 14.8

14.23)* 65.6 grams of KOH are dissolved in water and diluted to 1.50 liters. Find the normality of the solution.

14.24)* Find the normality of 0.284 M $Na_3PO_4$ in the reaction $H_2SO_4$ (aq) + $Na_3PO_4$ (aq) $\rightarrow$ $Na_2SO_4$ (aq) + $NaH_2PO_4$ (aq).

14.25)* 11.9 g $H_3PO_4$ are dissolved in 100 ml solution and used in the reaction $H_3PO_4$ (aq) + 2 NaOH (aq) $\rightarrow$ 2 HOH (l) + $Na_2HPO_4$ (aq). What is the normality of the phosphoric acid?

14.68)* 24.6 grams of ethanol, $C_2H_5OH$, the alcohol of distilled spirits fame, are dissolved in 55.5 grams of water. Find the molality of the solution.

14.69)* Calculate the molality of the solution described in Problem 14.68.

14.70)* Calculate the weight of malonic acid, $C_3H_4O_4$, that must be dissolved in 0.600 liter of water to produce a 0.75 m solution.

14.71) Calculate the mass of potassium hydroxide required for the preparation of 250 ml of a 2.50 M solution.

14.72) How many grams of $NiCl_2 \cdot 6\ H_2O$, widely used in nickel plating automotive hardware, are dissolved in 0.250 liter of a 1.12 M solution?

14.73) 25.0 ml of a solution of potassium bromide, used in manufacturing photographic paper and film, were carefully evaporated to dryness. 2.35 grams of solute were recovered. What was the molarity of the initial solution?

14.74) A solution is prepared by dissolving 5.11 grams of potassium chromate, $K_2CrO_4$, in 150 ml of water and then diluting to 250.0 ml. Find the molarity.

14.75) Calculate the moles of ammonia present in $8.00 \times 10^2$ ml of 2.25 M $NH_3$.

14.76) How many milliliters of 0.642 M NaOH are necessary to obtain 0.0395 mole of sodium hydroxide?

14.77)* What is the normality of a solution containing 10.2 grams of HCl per liter?

14.78)* What is the normality of 0.350 M $H_2C_2O_4$ used in the reaction NaOH (aq) + $H_2C_2O_4$ (aq) $\rightarrow$ $NaHC_2O_4$ (aq) + HOH (l)?

14.79)* Calculate the normality of a solution of 18.2 grams of $Na_2CO_3$ dissolved in water and diluted to 250.0 milliliters, and then used in the reaction 2 HCl (aq) + $Na_2CO_3$ (aq) $\rightarrow$ $CO_2$ (g) + $H_2O$ (l) + 2 NaCl (aq).

14.26)* Calculate the number of equivalents in 50.0 ml 0.114 N HCl.

14.27)* What volume of 0.200 N $H_3PO_4$ represents 0.00500 equivalents in the reaction $2 \text{ NaOH (aq)} + H_3PO_4 \text{ (aq)} \rightarrow Na_2HPO_4 \text{ (aq)} + 2 \text{ HOH (l)}$?

14.28)* What is the equivalent weight of $H_2C_2O_4$ in $2 \text{ NaOH (aq)} + H_2C_2O_4 \text{ (aq)} \rightarrow Na_2C_2O_4 \text{ (aq)} + 2 \text{ HOH (l)}$?

Section 14.9

14.29) Calculate the number of grams of barium chromate that can be precipitated by adding excess potassium chromate, $K_2CrO_4$, to 50.0 ml of 0.424 M $BaCl_2$.

14.30) Find the number of milliliters of 0.246 M $AgNO_3$ required to precipitate as silver phosphate all the phosphate ion in a solution containing 2.10 grams of sodium phosphate.

14.31) How many milliliters of 6.2 M NaOH must react with aluminum to liberate 2.4 liters of hydrogen measured at STP? The equation is

$2 \text{ Al (s)} + 6 \text{ NaOH (aq)} \rightarrow 2 \text{ Na}_3AlO_3 \text{ (aq)} + 3 H_2 \text{ (g)}$

14.32) Find the molarity of a solution of hydrochloric acid if 16.8 ml are required to react with 10.0 ml 0.862 M NaOH in a titration experiment.

14.33) In a chemical analysis, 14.9 ml of 0.518 M $AgNO_3$ are required to react with all of the nickel chloride in a 10.0 ml sample of a plating solution:

$2 \text{ AgNO}_3 \text{ (aq)} + NiCl_2 \text{ (aq)} \rightarrow$
$\qquad 2 \text{ AgCl (s)} + Ni(NO_3)_2 \text{ (aq)}$

Determine the molarity of the nickel chloride solution.

14.34)* How many grams of "milk of magnesia," $Mg(OH)_2$, will precipitate when 50.0 ml 0.240 M $MgCl_2$ are added to 50.0 ml 0.420 M NaOH?

14.80)* How many equivalents are in 50.0 ml 0.114 N $H_2SO_4$?

14.81)* What volume of 0.200 N $H_3PO_4$ represents 0.00500 equivalents in the reaction $\text{NaOH (aq)} + H_3PO_4 \text{ (aq)} \rightarrow NaH_2PO_4 + HOH \text{ (l)}$?

14.82)* Calculate the equivalent weight of $Na_2CO_3$ in $Na_2CO_3 \text{ (aq)} + HC_2H_3O_2 \text{ (aq)} \rightarrow NaHCO_3 \text{ (aq)} + NaC_2H_3O_2 \text{ (aq)}$.

14.83) Excess sodium hydroxide solution is added to 20.0 ml of 0.184 M $ZnCl_2$. Calculate the number of grams of zinc hydroxide that will precipitate.

14.84) In a reaction that produces hydrogen gas, how many milliliters of 0.569 M HCl are required to react with 0.104 gram of magnesium? How many milliliters of hydrogen, corrected to STP, will be released?

14.85) If you warm the vessel in which solutions of sodium sulfite and hydrochloric acid react in a controlled oxygen-free atmosphere, sulfur dioxide is driven off as a gas:

$Na_2SO_3 \text{ (aq)} + 2 \text{ HCl (aq)} \rightarrow$
$\qquad 2 \text{ NaCl (aq)} + H_2O \text{ (l)} + SO_2 \text{ (g)}$

What volume of $SO_2$, measured after the gas has been adjusted to STP, will be produced by the complete reaction of 35.0 ml 0.924 M $Na_2SO_3$?

14.86) 21.6 ml 0.655 M NaOH are required to titrate a 25.0 ml sample of oxalic acid solution by the following reaction:

$H_2C_2O_4 \text{ (aq)} + 2 \text{ NaOH (aq)} \rightarrow$
$\qquad Na_2C_2O_4 \text{ (aq)} + 2 \text{ HOH (l)}$

Calculate the molarity of the oxalic acid.

14.87) Potassium hydrogen phthalate, $KHC_8H_4O_4$ (MW = 204 g/mole), is used as a primary standard for bases. In one titration 5.34 grams of dissolved $KHC_8H_4O_4$ required 32.5 ml of a sodium carbonate solution:

$2 \text{ KHC}_8H_4O_4 \text{ (aq)} + Na_2CO_3 \text{ (aq)} \rightarrow$
$\qquad 2 \text{ NaKC}_8H_4O_4 \text{ (aq)} + H_2O \text{ (l)} + CO_2 \text{ (g)}$

Calculate the molarity of the sodium carbonate solution.

14.88)* In an effort to recover dissolved silver salts—you may consider them to be $AgNO_3$ (aq) for the purpose of the problem—by the "salting out" process, a workman tosses a cupful of salt into a 25.0 liter recovery tank. If the cupful contained 425 grams of NaCl, and the solution was 0.12 molar in silver ion (or silver nitrate, if you wish), calculate the grams of silver chloride that will precipitate. Was one cupful of salt enough to precipitate all the silver ion, or should he have used more?

14.35)* An industrial waste solution is essentially 1.13 M in $HNO_3$. 61.2 kg of $NaHCO_3$ are used to neutralize 140 gallons of the solution. How many liters of carbon dioxide, measured at STP, will be released, assuming the process proceeds to the complete consumption of the limiting reagent in $NaHCO_3$ (s) + $HNO_3$ (aq) → $H_2O$ (l) + $CO_2$ (g) + $NaNO_3$ (aq)?

14.36)* Calculate the normality of a NaOH solution if 21.9 ml are required to titrate 25.0 ml 0.324 N $H_2SO_4$.

14.37)* 0.512 gram of oxalic acid dihydrate, $H_2C_2O_4 \cdot 2\ H_2O$ is dissolved in water. The solution is titrated with a sodium hydroxide solution of unknown concentration. 16.2 milliliters of the base are required for the reaction $H_2C_2O_4$ (aq) + 2 NaOH (aq) → $Na_2C_2O_4$ (aq) + 2 HOH (l). What is the normality of the base?

14.38)* Find the equivalent weight of an unknown base if 0.452 gram of it requires 31.8 ml 0.169 N HCl for neutralization.

### Section 14.10

14.39)* The specific gravity of any solution of NaCl is greater than 1.00. The specific gravity of any solution of $NH_3$ is less than 1.00. Is specific gravity a colligative property? Why or why not?

14.40)* What will be the boiling temperature of a solution of 96.1 grams of ethylene glycol, $C_2H_6O_2$ (permanent antifreeze), in 100 grams of water? The molal boiling point constant for water is 0.52.

14.41)* The normal freezing point of naphthalene is 80.2°C, and $K_f$ = −6.9. Calculate the freezing point of the solution formed when 4.34 grams of paradichlorobenzene, $C_6H_4Cl_2$, are dissolved in 70.0 grams of naphthalene.

14.42)* Calculate the molal concentration of an aqueous solution that boils at 100.89°C. $K_b$ = 0.52 for water.

### Section 14.11

14.43) If solid solute A is a strong electrolyte and solute B is a nonelectrolyte, state how these solutes differ.

14.44) How do you explain the passage of "electricity" through a solution? The ability of a solution to conduct is considered evidence of what?

14.89)* The phosphate ion is to be precipitated from a 280 liter solution containing sodium and potassium phosphate and ammonium chloride. The phosphate ion concentration, expressed as molarity of $Na_3PO_4$, is 0.704 M. If 55.2 liters of 3.50 M $Ca(NO_3)_2$ are added to the solution, how many kilograms of calcium phosphate will precipitate? Will all of the phosphate ion be precipitated? If not, how many moles of $PO_4^{3-}$ will remain in solution?

14.90)* 25.0 ml of a HCl solution of unknown concentration require 31.4 ml 0.372 N $Na_2CO_3$ in the titration reaction 2 HCl (aq) + $Na_2CO_3$ (aq) → 2 NaCl (aq) + $H_2O$ (l) + $CO_2$ (g). What is the normality of the acid?

14.91)* Sodium hydrogen carbonate is used as a primary standard in finding the normality of $H_2SO_4$ (aq) in the reaction 2 $NaHCO_3$ (aq) + $H_2SO_4$ (aq) → $Na_2SO_4$ (aq) + 2 $H_2O$ (l) + 2 $CO_2$ (g). What is the normality if 0.618 gram of $NaHCO_3$ requires 20.6 milliliters in the reaction?

14.92)* A solution known to contain 0.268 gram of $H_3PO_4$ requires 18.9 ml 0.289 N NaOH in a titration reaction. (a) Calculate the equivalent weight of the $H_3PO_4$. (b) Write the equation for the reaction.

14.93)* Is the partial pressure exerted by one component of a gaseous mixture at a given temperature and volume a colligative property? Justify your answer, pointing out in the process what classifies a property as "colligative."

14.94)* Calculate the freezing point of a solution of 2.16 grams of naphthalene (mothballs), $C_{10}H_8$, in 31.0 grams of benzene. Pure benzene freezes at 5.50°C, and its molal freezing point constant is −5.10.

14.95)* Determine the freezing point of 0.65 m $C_3H_8O_3$, an aqueous solution of glycerol, a substance used in making cosmetics and nitroglycerine. $K_f$ = −1.86 for water.

14.96)* What is the molality of a solution of an unknown solute in acetic acid if the solution freezes at 13.4°C? The normal freezing point of acetic acid is 16.6°C, and $K_f$ = −3.90.

14.97) How does a weak electrolyte differ from a strong electrolyte?

14.98) All soluble ionic compounds are electrolytes. A covalent solute may or may not be an electrolyte. Show how both of these statements are true.

*Write the net ionic equation for each potential reaction described in this section. Use Table 14.1 to determine the possibility of precipitation reactions.*

14.45) Silver nitrate solution is added to a solution of ammonium carbonate.

14.46) Aluminum nitrate solution is added to potassium phosphate solution.

14.47) When combined in an oxygen-free atmosphere, sodium sulfite solution and hydrochloric acid react to form $SO_2$ (aq), water and a solution of sodium chloride.

14.48) Sulfuric acid can neutralize potassium hydroxide.

14.49) If ammonium chloride solution is added to a solution of potassium hydroxide you would expect "$NH_4OH$" to form—but this elusive species is really aqueous ammonia, $NH_3$ (aq), and water.

14.50) Sodium benzoate ($NaC_7H_5O_2$) solution is treated with hydrochloric acid.

14.51) A solution of sodium fluoride is poured into a solution of calcium nitrate.

14.52) Dilute nitric acid is poured over solid barium hydroxide.

14.53) Aluminum reacts with a solution of sodium hydroxide to yield hydrogen gas and a solution of sodium aluminate, $Na_3AlO_3$.

14.54) Write the net ionic equations for Equations 14.8 and 14.9, page 328.

14.99) Barium chloride and sodium sulfite solutions are combined in an oxygen-free atmosphere.

14.100) Copper(II) sulfate and sodium hydroxide solutions are combined.

14.101) Carbon dioxide bubbles appear and water is formed as two of the products of the reaction between solid magnesium carbonate and hydrochloric acid.

14.102) Oxalic acid, $H_2C_2O_4$ (s), is neutralized by reaction with sodium hydroxide solution.

14.103) Hydrogen escapes to the atmosphere when magnesium reacts with hydrochloric acid.

14.104) Solutions of magnesium sulfate and ammonium bromide are combined.

14.105) Solid nickel hydroxide is readily dissolved and neutralized by hydrobromic acid.

14.106) When metallic lithium is added to water, hydrogen is released, leaving behind an aqueous solution of lithium hydroxide.

14.107) Sodium fluoride solution reacts with a solution of nitric acid.

14.108) (a) Hydrochloric acid reacts with a solution of sodium hydrogen carbonate. (b) Hydrochloric acid reacts with a solution of sodium carbonate. (c) A limited quantity of hydrochloric acid reacts with a solution of sodium carbonate, yielding $NaHCO_3$ (aq) as one product.

# CHEMICAL EQUILIBRIUM

## 15.1 PHYSICAL AND CHEMICAL EQUILIBRIA—A REVIEW

The first dynamic equilibrium considered in this text involved liquid benzene, $C_6H_6$ (l), and its saturated vapor, $C_6H_6$ (g), in a closed (sealed) container (page 283). The system was described by a reversible reaction equation:

$$C_6H_6 \text{ (l)} \rightleftarrows C_6H_6 \text{ (g)} \qquad (12.1)$$

Read from left to right, from liquid benzene to gaseous benzene, the equation describes the **forward reaction;** the change from right to left, or vapor to liquid, is the **reverse reaction.** This is a physical equilibrium, as the changes of state between liquid and vapor, evaporation and condensation, are physical changes. Using Figure 12.9, the development of the equilibrium was traced in terms of the rates of the forward and reverse changes. When those rates became equal, the system was said to be at equilibrium. It is equality of rates of reversible changes that is the criterion for an equilibrium.

Another physical equilibrium is the formation of a saturated solution of an ionic solid in a liquid. This was described on page 319, leading ultimately to the equilibrium

$$NaCl \text{ (s)} \rightleftharpoons Na^+ \text{ (aq)} + Cl^- \text{ (aq)} \qquad (14.1)$$

A similar but more significant equilibrium between "insoluble" silver chloride and its saturated solution was mentioned on page 346. The equilibrium equation is

$$AgCl \text{ (s)} \rightleftharpoons Ag^+ \text{ (aq)} + Cl^- \text{ (aq)} \qquad (15.1)$$

In any equilibrium between undissolved solute and its saturated solution, it is the rate of dissolving, or change from solid to solution (forward reaction in Equation 15.1), that is equal to the rate of crystallization, or change from solution to solid (reverse reaction).

In discussing strong and weak acids (page 338), it was pointed out that even weak acids ionize somewhat, reaching equilibrium with a saturated solution of the ions produced. The example was hydrofluoric acid:

$$HF\ (aq) \rightleftharpoons H^+\ (aq) + F^-\ (aq) \tag{14.15}$$

This is a chemical equilibrium, inasmuch as the changes that occur are chemical changes, a fact more apparent if the equation is written to include water and the hydronium ion:

$$HF\ (aq) + H_2O\ (l) \rightleftharpoons H_3O^+\ (aq) + F^-\ (aq) \tag{15.2}$$

At equilibrium, the rate at which hydrofluoric acid molecules ionize to produce hydrogen (or hydronium) and fluoride ions is equal to the rate at which the ions combine to form molecules.

## 15.2 THE COLLISION THEORY OF CHEMICAL REACTIONS

PG   15 A   Explain why, according to the collision theory of chemical reactions, some molecular collisions result in a chemical reaction and others do not.

In the analysis of chemical equilibria, reaction rates are all-important. We therefore begin our study by considering how chemical reactions occur and the factors that influence the rate at which they proceed.

If a chemical reaction is to occur between molecule $A_2$ and molecule $B_2$, it is logical to expect that the two different molecules must come into contact with each other. The view of a chemical reaction as the cumulative effect of a vast number of individual collisions between reacting particles is called the **collision theory of chemical reactions.**

In the hypothetical reaction $A_2 + B_2 \rightarrow 2\ AB$, individual molecules of $A_2$ collide with individual molecules of $B_2$. If all goes well, the four atoms involved rearrange themselves and part as two molecules of AB, as in Figure 15.1A. But not all collisions result in reactions. Quite often—indeed, most of the time—the molecules simply bounce off each other and separate as they approach, as individual molecules of $A_2$ and $B_2$. Whether an individual collision results in a reaction or a "bounce-off" depends upon the "effectiveness" of the collision. In order to produce reaction, existing bonds in the diatomic molecules must be broken, or at least stretched to the breaking point. This requires energy, which must come initially from the kinetic energy of the molecules prior to collision. Colliding molecules frequently do not possess sufficient kinetic energy to bring about reaction (Figure 15.1B). Other collisions may be sufficiently energetic, but lack the proper

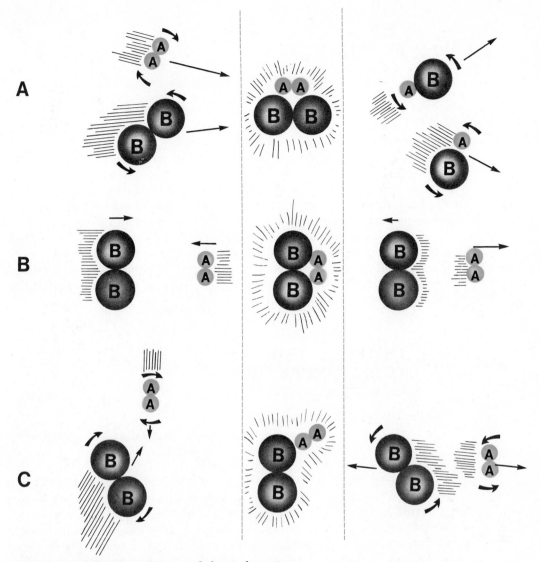

**Figure 15.1**   Molecular collisions and chemical reactions.

A, Two rotating molecules move toward each other on a collision course (left), collide or otherwise interact with sufficient impact and proper orientation to break existing bonds and form new ones (center), and separate as product molecules of the reaction that has occurred (right).

B, Two molecules approach slowly (left), collide with proper orientation but insufficient energy for a reaction (center), and separate as unreacted molecules, just as they were before collision (right).

C, Molecules with sufficient kinetic energy to produce a reaction approach (left), but collide with poor orientation (center), and therefore separate unreacted (right).

orientation for the formation of molecules of AB (Figure 15.1C). *Unless both of these requirements, minimum energy and proper orientation, are met, the collision is ineffective—a bounce-off collision.*

The collision theory of chemical reactions gives us some insight into the factors that determine the rate of a chemical reaction. In essence, *the rate of chemical reaction depends upon the frequency of effective collisions.*

15.3 **THE EFFECT OF CONCENTRATION ON REACTION RATE**

If a reaction rate depends upon frequency of effective collisions, the influence of concentration is readily predictable. The more particles there are in a given space, the more frequently collisions will occur, and the more rapidly the reaction will take place.

The effect of concentration on reaction rate is easily seen in the rate at which objects burn in air compared to the rate of burning in an atmosphere of pure oxygen. If a burning splint is thrust into pure oxygen the burning is brighter, more vigorous, and much faster. In fact, the typical laboratory test for oxygen is to ignite a splint, blow it out, and then, while there is still a faint glow, place it in oxygen. It immediately bursts back into full flame and burns vigorously. Charcoal, phosphorus, and other substances behave similarly.

15.4 **THE EFFECT OF TEMPERATURE ON REACTION RATE**

Chemical reactions occur more rapidly at high temperatures. This is seen experimentally in two simple facts from the kitchen. The changes occurring in cooking are chemical reactions. One way to speed cooking reactions that are performed in boiling water is to use a pressure cooker. Under pressure the boiling point of the water is higher, so the time required for a given cooking operation is reduced. The opposite effect is seen in open cooking at high altitudes, where reduced atmospheric pressure allows water to boil at lower temperatures. Here cooking is significantly slower, the result of reduced reaction rates.

To understand the effect of temperature on reaction rates, we must consider temperature from the standpoint of molecular kinetic energy. In Section 15.2 it was pointed out that not all inter-particle collisions are effective in producing reactions. Only those that have proper orientation and are sufficiently "severe," or energetic, are productive. This suggests that there is a certain minimum kinetic energy that is required to yield a reaction.

If the temperature of reactants is increased, reacting molecules have greater average kinetic energy. Collisions between these molecules will therefore be more energetic and more frequent, both of which predict an increase in reaction rate. This prediction is confirmed in laboratory experiments, but the reaction speed-up is considerably greater than would be expected from the energy and frequency changes alone. Quite often, for example, it is found that an increase of about 10°C is sufficient to double the rate of a reaction. This is because a small temperature increase is often enough to cause a much larger percentage increase in the number of molecules having the minimum energy required to yield reaction-producing collisions. This same idea appeared in the explanation for the large increase in equilibrium vapor pressure with an increase in temperature (page 285). We conclude that *the main reason a chemical change occurs more rapidly at elevated temperatures is because a larger portion of the total number of particles has sufficient kinetic energy to contribute to an effective collision.*

## 15.5 ENERGY CHANGES DURING A PARTICLE COLLISION

PG 15 D Sketch and/or interpret an enthalpy-reaction graph for either an exothermic or an endothermic reaction. Identify (a) the activated complex point; (b) activation energy; and (c) ΔH for the reaction.

The enthalpy-reaction graph, showing ΔH for both an exothermic and an endothermic reaction, was introduced in Chapter 13, page 309. Figure 13.5 is repeated here for your convenience. Enthalpy, H, is the "heat content" of a chemical system. The change in enthalpy, ΔH, corresponds to the heat of reaction, and accounts for nearly all of the energy change accompanying most chemical reactions. It is the only energy change we will consider in this introductory text. For an exothermic reaction the final enthalpy is less than the initial enthalpy, so ΔH is negative; for an endothermic reaction, ΔH is positive. Reactant and product enthalpies are separated on these discontinuous graphs, suggesting that perhaps intermediate energy conditions exist between the initial and final values of enthalpy.

Referring again to the hypothetical reaction in Figure 15.1, it is believed that interaction between the two molecules may commence even before a "collision," in the usual sense of the word. During this interaction bonds between the A atoms and between the B atoms are strained, setting up stresses and higher potential energies within the molecules as they approach. As the new molecules formed by the reaction separate, the interaction between them diminishes and finally disappears. The difference

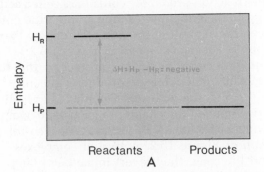
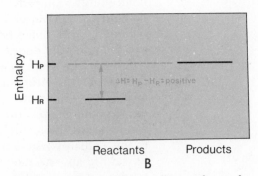

**Figure 13.5** Enthalpy-reaction graph for an exothermic change, in which heat is given off (A), and an endothermic change, in which the reaction system absorbs energy (B).

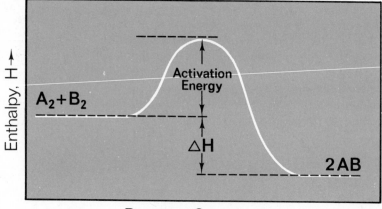

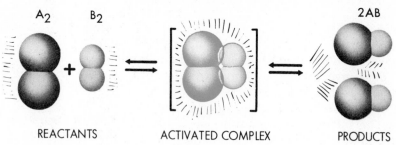

**Figure 15.2** Potential energy diagram for the reaction $A_2 + B_2 \rightarrow 2\,AB$.

between the energy absorbed and the energy released is the enthalpy of reaction, $\Delta H$. This description of *energy absorption* suggests that Figures 13.5A and 13.5B may be completed by drawing in *higher energies during the intereaction,* as in Figure 15.2.

In order for a collision to be effective, the colliding particles must experience a potential energy increase large enough to surpass the summit of the enthalpy-reaction curve, sometimes called the potential energy barrier. The source of this potential energy increase is the kinetic energy of the particles prior to collision. Interactions lacking sufficient kinetic energy to raise the potential energy to the peak will be ineffective, or "bounce-off" collisions. (Figure 15.3 describes a physically analogous system.) This minimum energy required to yield a reaction-producing collision is called **activation energy.** Activation energy on the enthalpy-reaction graph is the potential energy difference between the summit of the potential energy barrier and the potential energy of the reactants, as shown in Figure 15.2.

From the time reactant molecules begin to interact to the time they separate, either as reactant molecules or product molecules, an unstable intermediate species called the **activated complex** is formed. In the hypothetical reaction which opened the chapter the activated complex would be the "molecule" having the formula $A_2B_2$. Its existence is correlated to the enthalpy-reaction graph of Figure 15.2. Activated complexes look forward to an extremely short life span: they revert immediately into the reacting molecules from which they were formed, or separate as product molecules.

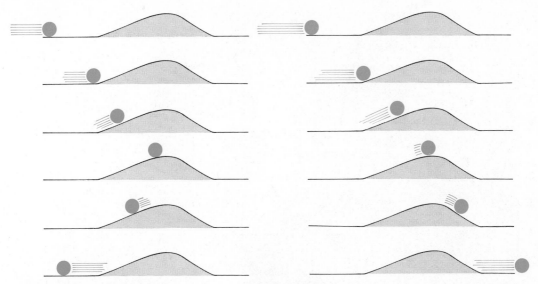

**Figure 15.3** Potential energy barrier. As the ball rolls toward the small hill (potential energy barrier), its speed (kinetic energy factor) determines whether or not it will pass over the hill. In the left series of pictures the slow moving ball slows down even more as it begins to climb the hill in the third picture. In the fourth picture all of its kinetic energy has been converted to potential energy before it reaches the peak. In the fifth picture it is rolling back down the hill, and in the sixth position the ball's potential energy from position four has been restored to kinetic energy as it rolls back to its starting point. In the right series the faster moving ball has more than enough kinetic energy to convert to potential energy at the top of the hill. It therefore passes over the potential energy barrier and rolls down the other side.

## 15.6 THE EFFECT OF A CATALYST ON REACTION RATE

PG   15 E   State qualitatively how a catalyst affects reaction rate.

15 F   Compare a catalyzed and an uncatalyzed reaction on an enthalpy-reaction graph. Explain the catalytic effect.

In driving from one city to another you usually have a choice of two or more alternative routes. One of these would be the quickest, and in all probability your choice. If a superhighway were to be constructed between the cities you would have available another route that would be faster than any of the earlier choices. If your purpose was to make the trip in the minimum time, you would no doubt turn to the superhighway in your future trips.

A **catalyst** for a chemical reaction is somewhat analogous to the superhighway in that it furnishes an alternative "route" for the reactants to change to products. It accomplishes this by changing the reaction path; specifically a lower-energy activated complex is found, requiring a lower activation energy than the uncatalyzed reaction. Figure 15.4 compares the reaction paths for catalyzed and uncatalyzed reactions. $E_a$ indicates the activation energy for the reaction without a catalyst, and $E'_a$ represents the activation energy in the presence of a catalyst. Many molecules having insufficient kinetic energy to overcome the $E_a$ potential energy barrier will be able to cross the lower $E'_a$ hump. With a larger number of molecules engaging in reaction-producing collisions, the reaction is faster.

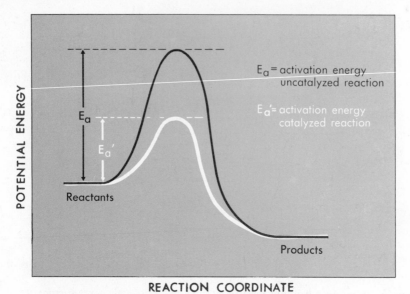

REACTION COORDINATE

**Figure 15.4**  Lowering of activation energy by adding a catalyst.

Catalysts exist in several different forms, and the precise function of many catalysts is not clearly understood. Some catalysts are mixed intimately in the reacting chemicals, while others appear to function simply by providing a surface site on which the reaction may occur. In either case, the catalyst is not permanently affected by the reaction. Many times the catalyst is only a bystander, though an important one from the standpoint of the reaction. In other cases, the catalyst may actually participate in the reaction and undergo chemical change, but eventually it will be regenerated without loss in the overall reaction.

The catalytic reaction most commonly appearing in beginning chemistry laboratories is the making of oxygen by the decomposition of potassium chlorate in the presence of manganese dioxide as a catalyst. A well known industrial process is the catalytic cracking of crude oil, in which large hydrocarbon molecules are broken down into simpler and more useful products in the presence of a catalyst. Biologically, many reactions in living organisms are controlled by catalysts called enzymes.

Some substances interfere with a normal reaction path from reactants to product, forcing the reaction to a higher activation energy route that is slower. Such substances are called **negative catalysts,** or **inhibitors.** Inhibitors are used to control the rates of certain industrial reactions. Sometimes negative catalysts can have disastrous results, as when mercury poisoning prevents the normal biological function of enzymes.

## 15.7  SUMMARY OF FACTORS THAT AFFECT REACTION RATES

We have identified three factors that have a fundamental effect on the rate of a chemical reaction. These are (1) concentration of reactants, with reaction rate being greater at higher concentration; (2) temperature, with reaction rate being greater at higher temperature; and (3) the presence of a catalyst, which increases a reaction rate.

In relating these factors to equilibrium considerations, which will constitute the balance of this chapter, we shall consider only concentration and temperature. These variables affect forward and reverse reaction rates differently. A catalyst, however, has the same effect on both forward and reverse rates, and therefore does not alter a chemical equilibrium, though it does cause the equilibrium to be reached more quickly.

## 15.8 THE DEVELOPMENT OF A CHEMICAL EQUILIBRIUM

PG   15 G   Trace the changes in concentrations of reacting species that lead to a chemical equilibrium.

The role of concentration in chemical equilibrium may be illustrated by tracing the development of an equilibrium. We shall do this in two ways, graphically and by means of the reversible reaction equation. For our purpose the hypothetical reaction $A_2 + B_2 \rightarrow 2\,AB$ will be assumed to take place by the simple collision of $A_2$ and $B_2$ molecules, which separate as two AB molecules, as shown in Figure 15.1. Furthermore, the reverse reaction will be thought of as exactly the reverse process: two AB molecules collide and separate as one $A_2$ molecule and one $B_2$ molecule.

Figure 15.5 is a graph of forward and reverse reaction rates vs. time. Initially at time 0, pure $A_2$ and $B_2$ are introduced to the reaction chamber. At the initial concentrations of $A_2$ and $B_2$ the forward reaction begins at a certain rate, $F_0$. Initially there are no AB molecules present, so the reverse reaction cannot occur. At time 0 the reverse reaction rate, $R_0$, is zero. These points are plotted on the graph.

As soon as the reaction begins, $A_2$ and $B_2$ are consumed, thereby reducing their concentrations in the reaction vessel. As these reactant concentrations decrease, the forward reaction rate declines. Consequently, at time 1, the forward reaction rate will have dropped to $F_1$. During the same interval some AB molecules will have been produced by the forward reaction, and the concentration of AB will be greater than zero. Therefore the reverse reaction will begin with the reverse rate rising to $R_1$ at time 1.

**Figure 15.5** Changes in reaction rates during the development of a chemical equilibrium.

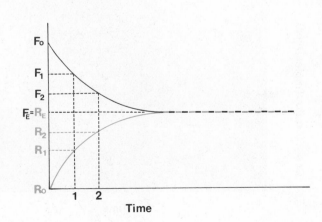

What will happen during the next interval of time, beginning at time 1? At that instant the forward rate is greater than the reverse rate. Therefore $A_2$ and $B_2$ will be consumed by the forward reaction more rapidly than they are produced by the reverse reaction. The net change in the concentrations of $A_2$ and $B_2$ will therefore be downward, causing a further reduction in the forward rate at time 2. Conversely, the forward reaction will produce AB more rapidly than the reverse reaction will consume it. The net change in the concentration of AB is therefore an increase, which in turn, will raise the reverse reaction rate at time 2.

Similar changes will occur over successive intervals until the forward and reverse rates eventually become equal. At this point a dynamic equilibrium will have been established. From this analysis we may state the following generalization: *For any reversible reaction in a closed system, whenever the opposing reactions are occurring at different rates, the faster reaction will gradually become slower, and the slower reaction will become faster. Finally they become equal, and equilibrium is established.*

## 15.9 THE LAW OF CHEMICAL EQUILIBRIUM: THE EQUILIBRIUM CONSTANT

PG 15 H Given any chemical equilibrium equation, write the equilibrium constant expression.

### GAS PHASE EQUILIBRIA

If 1.000 mole of $H_2$ (g) and 1.000 mole of $I_2$ (g) are introduced into a 1.000 liter reaction vessel under pressure, they will react to produce the following equilibrium: $H_2$ (g) $+ I_2$ (g) $\rightleftharpoons 2$ HI (g). At a temperature of 440°C, analysis at equilibrium will show hydrogen and iodine concentrations of 0.218 mole per liter, and a hydrogen iodide concentration of 1.564 moles per liter. Development of these concentrations is shown in black in Figure 15.6. For convenience, chemists enclose the chemical formula of a substance

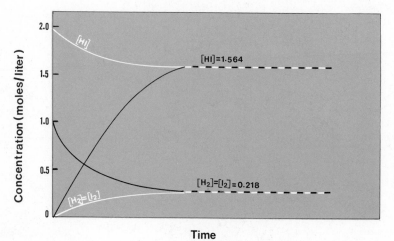

**Figure 15.6** Changes in concentrations during the development of a chemical equilibrium. Black lines represent the development of the equilibrium from hydrogen and iodine each initially at 1.00 mole/liter, and no HI present. White lines show the system reaching the identical equilibrium from hydrogen iodide initially at 2.00 moles/liter, with no hydrogen or iodine present.

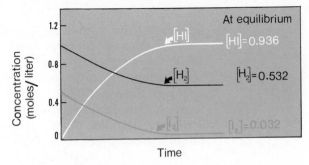

**Figure 15.7** Changes in concentrations during the development of a chemical equilibrium. Hydrogen concentration initially at 1.00 mole/liter; iodine initially at 0.500 mole/liter; no HI present initially.

within brackets to indicate its concentration in moles per liter. Thus concentrations of 0.218 mole per liter of hydrogen and iodine are represented as $[H_2] = [I_2] = 0.218$; and 1.564 moles of hydrogen iodide per liter is shown as $[HI] = 1.564$.

Working in the opposite direction, if gaseous hydrogen iodide is introduced to a reaction chamber at 440°C at an initial concentration of 2.000 moles per liter, it will decompose by the reverse reaction. The system will eventually come to equilibrium with $[H_2] = [I_2] = 0.218$, and $[HI] = 1.564$, exactly the same equilibrium concentrations as in the first example. This illustrates the experimental fact that the position of an equilibrium is the same, regardless of the direction from which it is approached. (See Fig. 15.6, white lines.)

As a third example, if the initial hydrogen iodide concentration is half as much, 1.000 mole per liter, the equilibrium concentration would be half what they are above: $[H_2] = [I_2] = 0.109$ and $[HI] = 0.782$.

If the experiment is repeated at the same temperature, starting this time with hydrogen at 1.000 mole per liter and iodine at 0.500 mole per liter, the reaction again proceeds to equilibrium. The equilibrium concentrations will be $[H_2] = 0.532$, $[I_2] = 0.032$ and $[HI] = 0.936$, as shown in Figure 15.7.

The data of these four equilibria are summarized in Table 15.1. Analysis of these data leads to the observation that, at equilibrium, the ratio of the square of the hydrogen iodide concentration to the product of the hydrogen and iodide concentrations, a ratio that may be expressed mathematically as $\dfrac{[HI]^2}{[H_2][I_2]}$, has a value of 51.5 for all four equilibria. For that matter, this ratio of equilibrium concentrations is always the same regardless of the

TABLE 15.1

| EQUI-LIBRIUM | INITIAL | | | EQUILIBRIUM | | | |
|---|---|---|---|---|---|---|---|
| | $[H_2]$ | $[I_2]$ | $[HI]$ | $[H_2]$ | $[I_2]$ | $[HI]$ | $\dfrac{[HI]^2}{[H_2][I_2]}$ |
| 1 | 1.000 | 1.000 | 0 | 0.218 | 0.218 | 1.564 | 51.5 |
| 2 | 0 | 0 | 2.000 | 0.218 | 0.218 | 1.564 | 51.5 |
| 3 | 0 | 0 | 1.000 | 0.109 | 0.109 | 0.782 | 51.5 |
| 4 | 1.000 | 0.500 | 0 | 0.532 | 0.032 | 0.936 | 51.5 |

initial concentrations of hydrogen, iodine and hydrogen iodide, so long as the temperature is maintained at 440°C.

Analysis of data from countless other equilibrium systems leads to a similar regularity. Expressed in words it is known as the **Law of Chemical Equilibrium: For any equilibrium at a given temperature, the ratio of the product of the concentrations of the species on the right side of the equilibrium equation, each raised to a power equal to its coefficient in the equation, to the corresponding product of the concentrations of the species on the left side of the equation, is a constant.** The constant is called the **equilibrium constant, K.** Thus for the

$$H_2 \text{ (g)} + I_2 \text{ (g)} \rightleftharpoons 2 \text{ HI (g)}$$

equilibrium at 440°C,

$$K = \frac{[HI]^2}{[H_2][I_2]} = 51.5$$

The statement of the equilibrium law sets the procedure by which any equilibrium constant expression may be written. For the general equilibrium $aA + bB \rightleftharpoons cC + dD$, where A, B, C and D are chemical formulas, and a, b, c and d are their coefficients in the equilibrium equation,

1. Write in the *numerator* the concentration of each species on the *right* side of the equation.

$$K = \underline{[C]\ [D]}$$

2. For each species on the right side of the equation, use its coefficient in the equation as an exponent.

$$K = \underline{[C]^c\ [D]^d}$$

3. Write in the *denominator* the concentration of each species on the *left* side of the equation.

$$K = \frac{[C]^c\ [D]^d}{[A]\ [B]}$$

4. For each species on the left side of the equation, use its coefficient in the equation as an exponent.

$$K = \frac{[C]^c\ [D]^d}{[A]^a\ [B]^b} \qquad (15.3)$$

If the equation were written in reverse, $cC + dD \rightleftharpoons aA + bB$, the equilibrium constant would be

$$K' = \frac{[A]^a\ [B]^b}{[C]^c\ [D]^d}, \qquad (15.4)$$

which is obviously the reciprocal of Equation 15.3. In considering any equilibrium constant value it is essential that the constant be associated with a specific equilibrium equation.

**Example 15.1** Write equilibrium constant expressions for each of the following equilibria:

a)  $2 \text{ HI (g)} \rightleftharpoons H_2 \text{ (g)} + I_2 \text{ (g)}$

b)  $HI \text{ (g)} \rightleftharpoons \frac{1}{2} H_2 \text{ (g)} + \frac{1}{2} I_2 \text{ (g)}$

c)  $2 \text{ Cl}_2 \text{ (g)} + 2 H_2O \text{ (g)} \rightleftharpoons 4 \text{ HCl (g)} + O_2 \text{ (g)}$

-------------------------------------------------------------------------

**15.1a**  a) $K = \dfrac{[H_2][I_2]}{[HI]^2}$. Notice that this is the reciprocal of the equilibrium constant developed in the text, when the equation was written $H_2 \text{ (g)} + I_2 \text{ (g)} \rightleftharpoons 2 \text{ HI (g)}$. Its numerical value at 440°C would be 1/51.5, or 0.0194.

b) $K = \dfrac{[H_2]^{1/2}[I_2]^{1/2}}{[HI]}$. The value of this equilibrium constant is *not* the same as above, 0.0194. It is, in fact, the square root of 0.0194, or 0.140. This emphasizes the importance of associating any equilibrium constant expression with a specific chemical equation.

c) $K = \dfrac{[HCl]^4[O_2]}{[Cl_2]^2[H_2O]^2}$. The procedure for writing the equilibrium constant expression is the same regardless of the complexity of the equation.

## SOLIDS AND LIQUID WATER IN EQUILIBRIUM CONSTANT EXPRESSIONS

The equilibrium equation for slightly soluble silver chloride and a saturated aqueous solution of its own ions was given on page 359:

$$AgCl \text{ (s)} \rightleftharpoons Ag^+ \text{ (aq)} + Cl^- \text{ (aq)} \qquad (15.1)$$

The equilibrium constant expression for this equilibrium, written according to the above procedure, is

$$K = \frac{[Ag^+][Cl^-]}{[AgCl]}$$

At this point we might ask, "What is the meaning of the denominator, [AgCl], inasmuch as silver chloride is a solid? How do we express the 'concentration' of a solid in moles/liter?" The answer is that we don't. We could argue that the concentration of a solid is fixed by its density, but that has little if anything to do with reaction rates. Surface area does. The more surface area there is, the faster the reaction is—in *both* directions, forward and reverse. The effect cancels out. The fact is that, so long as there is undissolved solute in contact with the saturated solution, the *amount* of solute does not influence the ion concentrations. It is the indicated product of the ionic concentrations that is constant at equilibrium:

$$K_{sp} = [Ag^+][Cl^-]$$

The subscript in $K_{sp}$ identifies a **solubility product constant,** the special name given to equilibrium constants for low solubility solids.

The same conclusion can be reached by more rigorous reasoning and, more importantly, it is confirmed in a large number of experimental determinations. Accordingly, *when we write a "practical" equilibrium constant expression for an equilibrium that includes a solid as one of the reacting species, the concentration of the solid is omitted.* The resulting equilibrium constant expression is thereby restricted to substances that are variable in concentration.

---

**Example 15.2** Write the equilibrium constant expressions for each of the following equilibria:

a) $CaCO_3$ (s) $\rightleftharpoons$ CaO (s) + $CO_2$ (g)

b) $Li_2CO_3$ (s) $\rightleftharpoons$ 2 $Li^+$ (aq) + $CO_3^{2-}$ (aq)

c) 4 $H_2O$ (g) + 3 Fe (s) $\rightleftharpoons$ 4 $H_2$ (g) + $Fe_3O_4$ (s)

- - - - - - - - - - - - - - - - - - - - - - - - - - - - - - - - - - - - - - - - - - - - - - - - - - - - - - - - - - - - - - -

**15.2a**  a) $K = [CO_2]$

b) $K = [Li^+]^2 [CO_3^{2-}]$

c) $K = \dfrac{[H_2]^4}{[H_2O]^4}$

In all cases the equilibrium constant expression is written in the usual way, except that the solids are omitted.

---

The equilibrium equation for the ionization of acetic acid, a weak acid, by reaction with water may be written

$$HC_2H_3O_2 \text{ (aq)} + H_2O \text{ (l)} \rightleftharpoons H_3O^+ \text{ (aq)} + C_2H_3O_2^- \text{ (aq)} \qquad (15.2)$$

The equilibrium constant expression is

$$K = \frac{[H_3O^+][C_2H_3O_2^-]}{[HC_2H_3O_2][H_2O]}$$

To two significant figures $[H_2O]$ is constant in dilute aqueous solutions. Therefore multiplying both sides of the equation by $[H_2O]$ yields on the left the product of two constants, which is another constant, $K_a$, the so-called **acid constant** for the ionization of a weak acid:

$$K[H_2O] = K_a = \frac{[H_3O^+][C_2H_3O_2^-]}{[HC_2H_3O_2]}$$

The result is similar with all dilute aqueous solution equilibria, and may be confirmed in the laboratory. Therefore, *when we write a "practical" equilibrium constant expression for a dilute aqueous solution equilibrium that includes liquid water as one of the reacting species, the concentration of water, [H₂O] is omitted.* Again the resulting expression includes only those species whose concentrations are variable.

---

**Example 15.3** Write the equilibrium constant expression for each of the following equilibria:

a) $HF\ (aq) + H_2O\ (l) \rightleftharpoons H_3O^+\ (aq) + F^-\ (aq)$

b) $NH_3\ (aq) + H_2O\ (l) \rightleftharpoons NH_4^+\ (aq) + OH^-\ (aq)$

------------------------------------------------------------------------

**15.3a**  a) $K = \dfrac{[H_3O^+][F^-]}{[HF]}$

b) $K = \dfrac{[NH_4^+][OH^-]}{[NH_3]}$

---

## 15.10  THE SIGNIFICANCE OF THE VALUE OF THE EQUILIBRIUM CONSTANT

PG   15 I   Explain what is meant by an equilibrium being "favored" in the forward or reverse direction.

15 J   Given an equilibrium equation and the value of the equilibrium constant, identify which direction will be favored.

By definition, an equilibrium constant is a ratio—a fraction. The numerical value of an equilibrium constant may be very large, very small, or any place in between. The value of 51.5 in our $H_2\ (g) + I_2\ (g) \rightleftharpoons 2\ HI\ (g)$ equilibrium is in the "in between" category. While there is no defined intermediate range, we might think of constants between 0.01 and 100 ($10^{-2}$ to $10^2$) as neither very large nor very small. Equilibria with such constants will have significant quantities of all species present at equilibrium.

To see what is meant by "very large" or "very small" K values, consider an equilibrium similar to the one we studied earlier. Substituting chlorine for iodine we have

$$H_2\ (g) + Cl_2\ (g) \rightleftharpoons 2\ HCl\ (g)$$

At 25°C

$$K = \frac{[HCl]^2}{[H_2][Cl_2]} = 2.4 \times 10^{33},$$

a very large number. If hydrogen and chlorine, each initially at 1.00 mole per liter, were to react to reach equilibrium with HCl, 99.999 999 999 999

999 959% of the hydrogen and chlorine initially present would be converted to HCl in order to satisfy the value of K. Remaining unreacted would be 0.000 000 000 000 000 041% of the elements. This is quite a contrast to the 21.8% of the original hydrogen and iodine that were present at equilibrium with HI when we started with $H_2$ and $I_2$ concentrations of 1.000 mole/liter. For all practical purposes, the hydrogen-chlorine reaction "goes to completion."

This illustration shows that a large equilibrium constant implies that at least one of the reacting species must be almost totally consumed, leaving a denominator concentration close to zero. Such a reaction goes essentially to completion in the forward direction. This is described as an equilibrium in which the forward reaction is favored.

By contrast, if the equilibrium constant is very small, the numerator must include some species at a very low concentration. In such a system the reaction goes nearly to completion in the reverse direction. For example, for the equilibrium $2 N_2 (g) + O_2 (g) \rightleftharpoons 2 N_2O (g)$,

$$K = \frac{[N_2O]^2}{[N_2]^2 [O_2]} = 2.0 \times 10^{-37}$$

This very small value of K indicates that at equilibrium the reverse reaction is strongly favored, with $[N_2O]$ very small.

*To summarize, if an equilibrium constant is very large, the forward reaction is favored; if the constant is very small, the reverse reaction is favored. If the constant is neither large nor small, appreciable quantities of all species will be present at equilibrium.*

## 15.11 EQUILIBRIUM CONSTANT CALCULATIONS (OPTIONAL)

Calculations based on the equilibrium constant cover a wide range of problem types. Thorough understanding of them is essential to the comprehension of many chemical phenomena in the laboratory, in industry, and in living organisms. Only three representative examples will be presented here, leaving the greater number and variety of problems to your more advanced courses.*

If you know the initial concentrations of the reactants in an equilibrium, and the equilibrium concentration of at least one species, you can calculate the value of the equilibrium constant. A gas phase equilibrium may be used to illustrate the approach . . . .

**Example 15.4**  10.0 moles of NO and 6.0 moles of $O_2$ are placed in an empty 1.00 liter reaction vessel at a certain temperature. They react until equilibrium is established according to the equation $2 NO (g) + O_2 (g) \rightleftharpoons 2 NO_2 (g)$. At equilibrium the vessel contains 8.8 moles of $NO_2$. Determine the value of K at that temperature.

*Solution.*  In this example only the equilibrium concentration of $NO_2$ is

*For a thorough consideration of equilibrium problems see Chapters 13–17 of Peters, *Problem Solving for Chemistry*, Second Edition, W. B. Saunders Company, 1976.

given. The other concentrations must be determined by stoichiometry. It is usually helpful to organize your work and your thinking along lines of a table indicating the initial concentration of each species, I; the number of moles per liter of each species reacting, R; and the equilibrium concentration of each species, E. For this problem

| | 2 NO (g) | + | $O_2$ (g) | $\rightleftharpoons$ | 2 $NO_2$ (g) |
|---|---|---|---|---|---|
| I | 10.0 | | 6.0 | | 0 |
| R | | | | | |
| E | | | | | 8.8 |

is the table, filled out as far as possible from the information given in the example. The initial [NO], [$O_2$], and final [$NO_2$] are stated, and the word "empty" indicates zero $NO_2$ present at the beginning.

If there was no $NO_2$ present initially, and 8.8 moles of $NO_2$ are present after equilibrium is reached, it follows that 8.8 moles must have been produced in the reaction. This is the value of R for $NO_2$. Adding that figure to the table, it becomes

| | 2 NO (g) | + | $O_2$ (g) | $\rightleftharpoons$ | 2 $NO_2$ (g) |
|---|---|---|---|---|---|
| I | 10.0 | | 6.0 | | 0 |
| R | | | | | +8.8 |
| E | | | | | 8.8 |

The plus sign indicates that 8.8 moles of $NO_2$ were produced, and that its concentration increased by that amount.

Now that the number of moles of $NO_2$ formed is known, the moles of other species used or produced can be determined from the reaction equation. The equation tells us that 2 moles of NO are consumed while 2 moles of $NO_2$ are formed, or, more generally, the moles of $NO_2$ formed are equal to the moles of NO used. If 8.8 moles of $NO_2$ were formed, 8.8 moles of NO must have been used. Similarly, 1 mole of $O_2$ is used for every 2 moles of $NO_2$ produced. Therefore the production of 8.8 moles of $NO_2$ requires 4.4 moles of $O_2$. The table is now extended to include these reaction values:

| | 2 NO (g) | + | $O_2$ (g) | $\rightleftharpoons$ | 2 $NO_2$ (g) |
|---|---|---|---|---|---|
| I | 10.0 | | 6.0 | | 0.0 |
| R | −8.8 | | −4.4 | | +8.8 |
| E | | | | | 8.8 |

The negative signs indicate that these quantities are consumed, and are to be subtracted from the starting quantities. Beginning with 10.0 moles of NO, 8.8 moles react, leaving 1.2 moles at equilibrium. Similarly 6.0 − 4.4 = 1.6 moles of $O_2$ left at equilibrium. The table may now be completed:

| | 2 NO (g) | + | $O_2$ (g) | $\rightleftharpoons$ | 2 $NO_2$ (g) |
|---|---|---|---|---|---|
| I | 10.0 | | 6.0 | | 0.0 |
| R | −8.8 | | −4.4 | | +8.8 |
| E | 1.2 | | 1.6 | | 8.8 |

The figures in the equilibrium concentration line may now be entered into the equilibrium constant expression and the numerical answer calculated:

$$K = \frac{[NO_2]^2}{[NO]^2 [O_2]} = \frac{(8.8)^2}{(1.2)^2 (1.6)} = 34$$

If the value of K and the starting concentrations of a weak acid are known, the hydrogen ion concentration—and all other concentrations, for that matter—can be calculated. Many chemical processes are very sensitive to hydrogen ion concentrations in aqueous solution. The example that follows is the beginning of the theory that leads to understanding and control of these processes.

---

**Example 15.5**   Find $[H^+]$, $[C_2H_3O_2^-]$ and $[HC_2H_3O_2]$ at equilibrium in 0.10 M $HC_2H_3O_2$. The value of $K_a$ is $1.8 \times 10^{-5}$ for $HC_2H_3O_2$ (aq) $\rightleftharpoons H^+$ (aq) $+ C_2H_3O_2^-$ (aq).

*Solution.*   This time you are given the initial concentration of $HC_2H_3O_2$. Assume the concentrations of $H^+$ and $C_2H_3O_2^-$ to be zero initially. The equilibrium equation shows that each mole of $HC_2H_3O_2$ that ionizes produces one mole of each ion, but there is no indication of the number of moles of acid ionized. Let y represent the moles per liter of acid ionized in reaching equilibrium. This means that y moles of $H^+$ and y moles of $C_2H_3O_2^-$ per liter will be produced; their equilibrium concentrations will be y. The concentration of acetic acid at equilibrium is the acid initially present minus that which ionized, $0.10 - y$. Summarized in a table, this is

| | $HC_2H_3O_2$ (aq) | $\rightleftharpoons$ | $H^+$ (aq) | $+$ | $C_2H_3O_2^-$ (aq) |
|---|---|---|---|---|---|
| I | 0.10 | | 0 | | 0 |
| R | $-y$ | | $+y$ | | $+y$ |
| E | $0.10 - y$ | | y | | y |

On substituting the equilibrium concentrations into the K expression and equating to the known value of K we find ourselves confronted with a not-very-welcome quadratic equation:

$$K_a = \frac{[H^+][C_2H_3O_2^-]}{[HC_2H_3O_2]} = \frac{y^2}{0.10 - y} = 1.8 \times 10^{-5}$$

This equation may be solved by rearranging and using the quadratic formula, but the process is long and usually not necessary. A better way is to take advantage of what we know about significant figures and weak acids. A weak acid ionizes to a very small extent. Is it possible that the amount that ionizes, y, will be negligible compared to the 0.10 mole per liter initially present? If so, application of significant figures rules to $0.10 - y$ will yield 0.10 as the difference. Assuming this to be the case, the denominator becomes simply 0.10, and the problem becomes

$$\frac{y^2}{0.10} = 1.8 \times 10^{-5}$$

$$y = 1.34 \times 10^{-3}$$

This result, rounded off to the two significant figures dictated by both the concentration and equilibrium constant, means that $[H^+] = [C_2H_3O_2^-] = 1.3 \times 10^{-3} = 0.0013$—*if* the assumption that y is negligible compared to 0.10 is valid. The validity can be checked according to the significant figure rule for subtraction (page 48):

$$\frac{\begin{array}{r} 0.10 \\ -0.0013 \end{array}}{0.0987} = 0.10$$

Accordingly, $[HC_2H_3O_2] = 0.10$ at equilibrium, just as it was initially.

The use of the assumption that a term may be negligible in addition or subtraction (*never* in multiplication or division) simplifies many equilibrium calculations. Such an assumption is usually valid if the equilibrium constant is very small, usually $10^{-4}$ or less, and the molecular acid concentration is in the area of 0.1 M or more. Considering measurement uncertainties in K values and solution concentrations, the results are as accurate as those obtained by solving the quadratic formula, rounded off to the proper number of significant figures. Calculated by the quadratic formula, the value of y in Example 15.5 is again $1.3 \times 10^{-3}$.

Solubility equilibria are among the easier applications of equilibrium constants . . . .

---

**Example 15.6** Calculate the solubility of silver chloride in moles per liter if $K_{sp} = 1.7 \times 10^{-10}$ for $AgCl$ (s) $\rightleftharpoons Ag^+$ (aq) + $Cl^-$ (aq).

*Solution.* From the equation, for each mole of AgCl that dissolves, 1 mole of $Ag^+$ and 1 mole of $Cl^-$ appear in the solution. The concentrations of these ions are therefore equal to each other and equal to the solubility of silver chloride in moles per liter. Let s = $[Ag^+]$ = $[Cl^-]$.

$$K_{sp} = [Ag^+][Cl^-] = 1.7 \times 10^{-10}$$
$$s^2 = 1.7 \times 10^{-10}$$
$$s = 1.3 \times 10^{-5}$$

---

The answer to Example 15.6 gives an indication of just how low the solubility of an "insoluble" salt is. The above concentration represents about 0.002 gram per liter of solution, roughly a 0.0002% solution that is saturated with all the solute it can hold. Silver iodide is less soluble by three orders of magnitude, about $10^{-8}$ mole per liter; and silver sulfide has a solubility of about $10^{-13}$ mole per liter, which comes to about 0.00000000001% by weight.

## 15.12   LE CHATELIER'S PRINCIPLE

### *THE CONCENTRATION EFFECT*

PG   15 K   Given the equation for a chemical equilibrium, predict the direction the equilibrium will shift because of a change in the concentration of one species.

In Section 15.8 we traced the development of a chemical equilibrium in terms of the concentration of each species in the equilibrium equation and its effect on the forward and reverse reaction rates. We now return to the equilibrium system

$$H_2 \text{ (g)} + I_2 \text{ (g)} \rightleftharpoons 2 \text{ HI (g)}$$

to consider what might happen if the concentration of one species is changed, either by introducing more from an external source or by withdrawing some.

At equilibrium the forward and reverse reaction rates are equal. This is shown by the equal length arrows in the equation

$$H_2 \text{ (g)} + I_2 \text{ (g)} \rightleftharpoons 2 \text{ HI (g)}$$

The sizes of the formulas represent the relative concentrations of the different species in the reaction. If additional HI is forced into the system, [HI] is increased. This raises the rate of the reverse reaction, in which HI is a reactant:

$$H_2 \text{ (g)} + I_2 \text{ (g)} \rightleftharpoons 2 \text{ HI (g)}$$

The rates no longer being equal, the equilibrium is destroyed.

At this point the unequal reaction rates cause the system to "shift" in the direction of the faster rate—to the left, or in the reverse direction. As a result $H_2$ and $I_2$ are made by the reverse reaction faster than they are used by the forward reaction. Their concentrations increase, and consequently the forward rate increases. Simultaneously HI is used faster than it is produced, reducing both [HI] and the reverse reaction rate. Eventually the rates become equal and a new equilibrium is reached:

$$H_2 \text{ (g)} + I_2 \text{ (g)} \rightleftharpoons 2 \text{ HI (g)}$$

The entire process conforms to the generalization in italics on page 368.

The process just described is an example of **Le Chatelier's Principle.** This principle may be used to predict qualitatively the direction any equilibrium will shift in order to compensate for a disturbance to the equilibrium. Le Chatelier's Principle is stated as follows: **If an equilibrium system is subjected to a change, processes occur that tend to partially counteract the initial change, thereby bringing the system to a new position of equilibrium.** In the hydrogen-iodine–hydrogen iodide example, the system was initially at equilibrium; it was subjected to change by adding HI; processes occurred as the equilibrium shifted in the reverse direction, consuming some, but not all, of the HI added; and it eventually reached a new state of equilibrium.

Changes due to Le Chatelier's Principle can also be analyzed in terms of the equilibrium constant expression. For $H_2 \text{ (g)} + I_2 \text{ (g)} \rightleftharpoons 2 \text{ HI (g)}$, the ratio $\dfrac{[HI]^2}{[H_2][I_2]}$, which we shall call the "concentration quotient," Q, for the moment, has a constant value of 51.5 when the system is at equilibrium at 440°C. When HI is added, as in the example above, [HI] increases. This increases the value of the numerator, making $Q > 51.5$. This temporarily high value of Q can be brought back down to 51.5 by either or both of two numerical changes: the numerator may become smaller and/or the denominator may become larger. In practice both occur as the reaction "shifts in the reverse direction." Some—but not all—of the HI added decomposes to form $H_2$ and $I_2$. As a consequence [HI] is reduced from its high value, while $[H_2]$ and $[I_2]$ simultaneously increase until $Q = 51.5$ once again, at a new position of equilibrium.

A change in the concentration of any species will usually produce a temporary value of Q that is more than or less than the value of the equilibrium constant, K. The system responds by shifting in the direction that restores Q to equality with K, reverse if Q > K, and forward if Q < K, until the new equilibrium is reached.

---

**Example 15.7** The system $N_2$ (g) + 3 $H_2$ (g) $\rightleftharpoons$ 2 $NH_3$ (g) is at equilibrium. Predict qualitatively from Le Chatelier's Principle the direction the equilibrium will shift if ammonia is withdrawn from the reaction chamber. Show that the prediction is confirmed by the value of the concentration quotient, Q, when compared to the value of the equilibrium constant, K.

The equilibrium disturbance is clearly stated: ammonia is withdrawn. In which direction must the reaction shift to *counteract* the removal of ammonia, i.e., to *produce* more ammonia, thereby replacing some of that which has been withdrawn?

-------------------------------------------------------------------------------

**15.7a** The shift is in the *forward* direction. Ammonia is the product of the forward reaction and therefore will be partially restored to its original concentration by a shift in the forward direction.

To confirm that the same conclusion would be reached from the equilibrium constant expression, write that expression.

-------------------------------------------------------------------------------

**15.7b** $\dfrac{[NH_3]^2}{[N_2][H_2]^3}$ = Q = K at equilibrium

If ammonia is removed, will Q be more than K or less than K?

-------------------------------------------------------------------------------

**15.7c** Q < K.  $[NH_3]$ will be reduced by the removal of ammonia. With a smaller numerator, the value of Q becomes less than the value of K.

Now what combination of changes, increase or decrease, in the value of the numerator and denominator will *raise* Q to equivalence with K?

-------------------------------------------------------------------------------

**15.7d** The numerator must increase and the denominator must decrease in order for the value of the fraction to become larger.

Finally, in which direction must the equilibrium shift to bring about these changes? Does that direction correspond with that predicted in 15.7a above?

-------------------------------------------------------------------------------

**15.7e** The reaction must shift in the forward direction, increasing $[NH_3]$ and reducing $[N_2]$ and $[H_2]$. This confirms the prediction in 15.7a above.

**Example 15.8** Predict the direction of a Le Chatelier shift in the equilibrium

$$CH_4 \text{ (g)} + 2 H_2S \text{ (g)} \rightleftharpoons 4 H_2 \text{ (g)} + CS_2 \text{ (g)}$$

caused by each of the following: (a) increase $[H_2S]$; (b) reduce $[CS_2]$; and (c) increase $[H_2]$.

-----------------------------------------------------------------

**15.8a**  a. The shift will be in the **forward** direction, counteracting partially the increase in $[H_2S]$ by consuming some of it.
b. The shift will be in the **forward** direction again, this time to restore some of the $CS_2$ removed.
c. The shift will be in the **reverse** direction, to consume some of the added $H_2$.

## THE VOLUME EFFECT

PG 15 L  Given the equation for a chemical equilibrium involving one or more gases, predict the direction the equilibrium will shift because of a change in the volume of the system.

If the volume of an equilibrium system involving a gaseous component is changed, the concentration of that component will change. *If* this causes the value of Q to differ temporarily from the value of K—it may or may not, as we shall see—the direction of shift required to restore that equivalence may again be predicted in terms of the individual changes in factors that make up Q. The process is complicated, however, by the fact that a *volume* change alters the concentration of *all* gaseous species, not just one. Fortunately, Le Chatelier's Principle furnishes an alternative approach to the problem.

In Chapter 11 we learned that the pressure exerted by a gaseous system is proportional to the total number of gaseous molecules present (see page 250). Reducing the volume of a gaseous system increases the pressure exerted by the system. Le Chatelier's Principle says that a shift will occur that will counteract partially the original change. A pressure increase will be relieved by a reduction in the total number of gaseous molecules, which reduces the pressure exerted by the gas. This leads to the generalization: *If a gaseous equilibrium is compressed, the increased pressure will be partially relieved by a shift in the direction of fewer gaseous molecules; if the system is expanded, the reduced pressure will be partially restored by a shift in the direction of more gaseous molecules.*

Realizing that the coefficients of gases in the equation are in the same proportion as the number of gaseous molecules, let's try some examples:

**Example 15.9** Predict the direction of shift resulting from an expansion in the volume of the following equilibrium system: $2 SO_2 \text{ (g)} + O_2 \text{ (g)} \rightleftharpoons 2 SO_3 \text{ (g)}$.

First, will total pressure increase or decrease as a result of expansion?

-----------------------------------------------------------------

**15.9a** Decrease. Pressure and volume are inversely related.

Next, will the decrease in pressure be counteracted by an increase or decrease in the number of molecules?

-------------------------------------------------------------------------------------

**15.9b** Increase. Pressure is directly related to number of molecules.

Finally, examining the equation, notice that a forward shift finds three reactant molecules, two $SO_2$ and one $O_2$, forming two $SO_3$ product molecules, whereas a reverse shift has two reactant molecules yielding three product molecules. Which direction will the reaction shift, therefore, to increase the total number of molecules?

-------------------------------------------------------------------------------------

**15.9c** Reverse. Three product molecules replace two reactant molecules in a reverse shift. The increase in number of molecules partially restores the total pressure lost in expansion.

---

**Example 15.10** The volume occupied by the equilibrium

$$SiF_4 \text{ (g)} + 2 \text{ } H_2O \text{ (g)} \rightleftharpoons SiO_2 \text{ (s)} + 4 \text{ HF (g)}$$

is reduced. Predict the direction of shift in the position of equilibrium.

Will the shift be in the direction of more gaseous molecules or fewer?

-------------------------------------------------------------------------------------

**15.10a** Fewer. If volume is reduced, pressure increases. Increased pressure will be counteracted by fewer molecules.

Now predict the direction of shift and justify your prediction by stating the numerical loss in molecules predicted by the equation.

-------------------------------------------------------------------------------------

**15.10b** The shift will be in the reverse direction. The reverse reaction has *four gaseous* molecules being reduced to three. Note the number is not five reduced to three, but four. Only the gaseous molecules are involved in pressure adjustments.

---

**Example 15.11** Returning to the familiar $H_2$ (g) + $I_2$ (g) $\rightleftharpoons$ 2 HI (g), predict the direction of the shift that will occur because of a volume increase.

Take it all the way, but be careful.

-------------------------------------------------------------------------------------

**15.11a** There will be no shift. In this instance there are two gaseous molecules on *both* sides of the equation. Therefore there is no gain or loss in number of molecules by a shift in either direction. When the number of gaseous molecules on the two sides of the equilibrium equation is the same, increasing or reducing volume has no effect on the equilibrium.

## THE TEMPERATURE EFFECT

PG 15 M Given a thermochemical equation for a chemical equilibrium, predict the direction the equilibrium will shift because of a change in temperature.

A change in temperature of an equilibrium will change both forward and reverse reaction rates, but the rate changes are not equal. The equilibrium is therefore destroyed temporarily. The events that follow are again predictable by Le Chatelier's Principle . . . .

**Example 15.12** If the temperature of the system

$$PCl_5 \text{ (g)} \rightleftharpoons PCl_3 \text{ (g)} + Cl_2 \text{ (g)} + 22.1 \text{ kcal}$$

is increased, predict the direction of the Le Chatelier shift.

In order to raise the temperature of something, it must be heated. We therefore interpret an increase in temperature as the "addition of heat," and a lowering of temperature as the "removal of heat." In applying Le Chatelier's Principle to a thermochemical equation, heat may be regarded in much the same manner as any chemical species in the equation. Accordingly, if heat is *added* to the equilibrium system shown, in which direction must it shift to *use up*, or *consume*, some of the heat that was added? (If chlorine was *added*, in which direction would the equilibrium shift to use up some of the chlorine?)

-------------------------------------------------------------------------

**15.12a** The equilibrium must shift in the *reverse* direction to use up some of the added heat. An endothermic reaction consumes heat. As the equation is written, the reverse reaction is endothermic.

**Example 15.13** The thermal decomposition of limestone reaches the following equilibrium $CaCO_3 \text{ (s)} + 42 \text{ kcal} \rightleftharpoons CaO \text{ (s)} + CO_2 \text{ (g)}$. Predict the direction this equilibrium will shift if the temperature is reduced.

-------------------------------------------------------------------------

**15.13a** The shift would be in the *reverse* direction. Reduction in temperature is interpreted as the removal of heat. The reaction will respond to replace some of the heat removed—as an exothermic reaction. Heat is produced as the reaction proceeds in the reverse direction.

# CHAPTER 15 IN REVIEW

> 15 L   Given the equation for a chemical equilibrium involving one or more gases, predict the direction the equilibrium will shift because of a change in the volume of the system. (Page 380)
>
> 15 M   Given a thermochemical equation for a chemical equilibrium, predict the direction the equilibrium will shift because of a change in temperature. (Page 382)

# TERMS AND CONCEPTS

Forward and reverse reactions (359)
Collision theory of chemical reactions (360)
Activation energy (364)
Activated complex (364)
Catalyst (365)
Negative catalyst; inhibitor (366)
Law of Chemical Equilibrium (370)

Equilibrium constant (370)
Solubility product constant, $K_{sp}$ (372)
Acid constant, $K_a$ (372)
Shift (as in direction of a reversible reaction) (378)
Le Chatelier's Principle (378)
Concentration quotient, Q (378)

---

# QUESTIONS AND PROBLEMS

*An asterisk (\*) identifies a question that is relatively difficult, or that extends beyond the performance goals of the chapter.*

### Section 15.2

15.1)   According to the collision theory of chemical reactions, what two conditions must be satisfied if a molecular collision is to result in reaction?

15.31)   Explain why a molecular collision can be sufficiently energetic to cause a reaction, yet no reaction occurs as a result of that collision.

### Section 15.3

15.2)   For the hypothetical reaction, A + B → C, what will happen to the rate of reaction if the concentration of A is increased? What will happen if the concentration of B is decreased? Explain why in both cases.

15.32)   For the hypothetical reaction A + B → C, what will happen to the reaction rate if the concentration of A is increased *and* the concentration of B is decreased? Explain.

### Section 15.4

15.3)   "At a given temperature, only a small fraction of the molecules in a sample has sufficient kinetic energy to engage in chemical reaction." What is the meaning of that statement?

15.33)   State the effect of a temperature increase and a temperature decrease on the rate of a chemical reaction. Explain each effect.

### Section 15.5

15.4)   Is ΔH for an exothermic reaction positive or negative? Confirm your answer by reference to an enthalpy-reaction graph for an exothermic reaction.

15.34)   Two reactions, A and B, are the reverse of each other, as M → N and N → M. They therefore have the same activated complex, and their ΔH values are the same in magnitude, but opposite in sign: i.e., $\Delta H_A = +X$, and $\Delta H_B = -X$. Sketch typical enthalpy-reaction graphs for these reactions. Which of your two graphs indicates the greater activation energy?

15.5) Sketch an enthalpy-reaction graph for a reaction in which $\Delta H = -43.6$ kcal. Show how the $\Delta H$ appears on the graph. Indicate also the activation energy, and the position of the activated complex.

15.6) Explain the significance of activation energy. For two reactions that are identical in all respects except activation energy, identify the reaction that would have the higher rate and tell why.

15.35) Sketch an enthalpy-reaction graph for a reaction in which $\Delta H = +9.4$ kcal. Show how $\Delta H$ appears on the graph. Also indicate the activation energy and the activated complex point.

15.36) What is an "activated complex"? Why is it that we cannot list the physical properties of the species represented as an activated complex?

Section 15.6

15.7) Define a catalyst.

15.37) Explain how a catalyst affects reaction rates.

Section 15.8

15.8, 15.9, 15.38 and 15.39: *If nitrogen and hydrogen are brought together at the proper temperature and pressure, they will react until they reach equilibrium:*

$$N_2\ (g) + 3\ H_2\ (g) \rightleftharpoons 2\ NH_3\ (g)$$

*Answer the following questions in regard to the establishment of that equilibrium:*

15.8) When will the forward reaction rate be at a maximum: at the start of the reaction, after equilibrium has been reached, or at some point in between?

15.9) What happens to the concentration of each of the three species between the start of the reaction and the time equilibrium is reached?

15.38) When will the reverse reaction rate be at a maximum: at the start of the reaction, after equilibrium has been reached, or at some point in between?

15.39) On a single set of coordinate axes, sketch graphs of the forward reaction rate vs. time and the reverse reaction rate vs. time from the moment the reactants are mixed to a point beyond the establishment of equilibrium.

Section 15.9

15.10–15.14 and 15.40–15.44: *For each equilibrium equation shown, write the equilibrium constant expression.*

15.10) $2\ SO_2\ (g) + O_2\ (g) \rightleftharpoons 2\ SO_3\ (g)$

15.11) $4\ H_2\ (g) + CS_2\ (g) \rightleftharpoons CH_4\ (g) + 2\ H_2S\ (g)$

15.12) $Cd(OH)_2\ (s) \rightleftharpoons Cd^{2+}\ (aq) + 2\ OH^-\ (aq)$

15.13) $HNO_2\ (aq) \rightleftharpoons H^+\ (aq) + NO_2^-\ (aq)$

15.14) $Ag(CN)_2^-\ (aq) \rightleftharpoons Ag^+\ (aq) + 2\ CN^-\ (aq)$

15.15) "The equilibrium constant expression for a given reaction depends upon how the equilibrium equation is written." Explain the meaning of that statement. You may, if you wish, use the equilibrium equation $N_2\ (g) + 3\ H_2\ (g) \rightleftharpoons 2\ NH_3\ (g)$ to illustrate your explanation.

15.40) $CO\ (g) + H_2O\ (g) \rightleftharpoons CO_2\ (g) + H_2\ (g)$

15.41) $C\ (s) + H_2O\ (g) \rightleftharpoons CO\ (g) + H_2\ (g)$

15.42) $Zn_3(PO_4)_2\ (s) \rightleftharpoons 3\ Zn^{2+}\ (aq) + 2\ PO_4^{3-}\ (aq)$

15.43) $HNO_2\ (aq) + H_2O\ (l) \rightleftharpoons$
$$H_3O^+\ (aq) + NO_2^-\ (aq)$$

15.44) $Cu(NH_3)_4^{2+}\ (aq) \rightleftharpoons Cu^{2+}\ (aq) + 4\ NH_3\ (aq)$

15.45) The equilibrium between nitrogen monoxide, oxygen and nitrogen dioxide may be expressed in the equation

$$2\ NO\ (g) + O_2\ (g) \rightleftharpoons 2\ NO_2\ (g)$$

Write the equilibrium constant expression for this equation. Then express the same equilibrium in at least two other ways, and write the equilibrium constant expression for each. Are the constants numerically equal? Cite some evidence to support your yes or no answer.

## Section 15.10

**15.16)** With regard to the equilibrium

$$F^- (aq) + H_2O (l) \rightleftharpoons HF (aq) + OH^- (aq)$$

what is meant by saying the equilibrium is favored in the reverse direction?

**15.46)** The equilibrium equation for the ionization of formic acid is

$$HCHO_2 (aq) \rightleftharpoons H^+ (aq) + CHO_2^- (aq),$$

and the equilibrium is favored in the reverse direction. If you were to write a net ionic equation in which formic acid is a reactant, would you show formic acid in its ionized form or molecular form? Why?

**15.17)** The equilibrium constant is $2.3 \times 10^{-8}$ for the equilibrium $HCO_3^- (aq) + HOH (l) \rightleftharpoons H_2CO_3 (aq) + OH^- (aq)$. In which direction is the reaction favored at equilibrium? State the basis for your answer.

**15.47)** If sodium cyanide solution is added to silver nitrate solution the following equilibrium will be reached:

$$Ag^+ (aq) + 2\ CN^- (aq) \rightleftharpoons Ag(CN)_2^- (aq)$$

For this equilibrium $K = 5.6 \times 10^{18}$. In which direction is the equilibrium favored? Justify your answer.

## Section 15.11

**15.18)*** At 1350°K the system $2\ SO_2 (g) + O_2 (g) \rightleftharpoons 2\ SO_3 (g)$ reaches equilibrium with $[SO_2] = 0.90$, $[O_2] = 0.45$, and $[SO_3] = 0.60$. Find $K$ at that temperature.

**15.48)*** Calculate the equilibrium constant for the system $Ni(CO)_4 (g) \rightleftharpoons Ni (s) + 4\ CO (g)$ at 400°K if $[Ni(CO)_4] = 1.80$ and $[CO] = 0.800$.

**15.19)*** $[H^+] = 0.0265$ in 1.0 M HF. Calculate $K_a$ for the ionization equilibrium $HF (aq) \rightleftharpoons H^+ (aq) + F^- (aq)$.

**15.49)*** 0.50 M $HCHO_2$ (formic acid) is 2.0% ionized at equilibrium. Calculate the value of the acid constant, $K_a$, for $HCHO_2 (aq) \rightleftharpoons H^+ (aq) + CHO_2^- (aq)$.

**15.20)*** 3.00 moles of HI are introduced into a 1.00 liter reaction chamber at 440°C, the temperature at which $K = 51.5$ for the reaction $H_2 (g) + I_2 (g) \rightleftharpoons 2\ HI (g)$. The compound decomposes until equilibrium is reached. Find the equilibrium concentrations of all species.

**15.50)*** Calculate $[HC_7H_5O_2]$, $[H^+]$ and $[C_7H_5O_2^-]$ in 0.20 M $HC_7H_5O_2$ (benzoic acid) if $K_a = 6.6 \times 10^{-5}$.

**15.21)*** Calculate the solubility of calcium carbonate in moles per liter if $K_{sp} = 8.7 \times 10^{-9}$.

**15.51)*** How many grams of calcium sulfate are dissolved in 10.0 liters of a saturated solution if $K_{sp} = 1.9 \times 10^{-4}$?

**15.22)*** $K_{sp} = 6.5 \times 10^{-9}$ for magnesium fluoride, $MgF_2$. Calculate its solubility in moles per liter.

**15.52)*** What is the solubility (moles per liter) of silver chloride in a salt solution in which $[Cl^-]$ is fixed at 0.010? $K_{sp} = 1.7 \times 10^{-10}$ (Hint: The solubility of AgCl is equal to $[Ag^+]$ at equilibrium.)

## Section 15.12

**15.23)** Predict the direction the equilibrium $SO_2 (g) + NO_2 (g) \rightleftharpoons SO_3 (g) + NO (g)$ will shift if the concentration of NO is increased. Explain or justify your prediction.

**15.53)** If the system $2\ SO_2 (g) + O_2 (g) \rightleftharpoons 2\ SO_3 (g)$ is at equilibrium and the concentration of $O_2$ is reduced, predict the direction the equilibrium will shift. Justify or explain your prediction.

**15.24)** The equilibrium system $COCl_2 (g) \rightleftharpoons CO (g) + Cl_2 (g)$ has some of the chlorine removed. Predict the direction the equilibrium will shift, and explain your prediction.

**15.54)** If additional oxygen is pumped into the equilibrium system

$$4\ NH_3 (g) + 5\ O_2 (g) \rightleftharpoons 4\ NO (g) + 6\ H_2O (g),$$

in which direction will the reaction shift? Justify your answer.

**15.25)** In which direction would the equilibrium $HC_2H_3O_2$ (aq) $\rightleftharpoons$ $H^+$ (aq) + $C_2H_3O_2^-$ (aq) shift if the concentration of the $C_2H_3O_2^-$ ion were increased? Explain your prediction.

**15.26)** Which direction will be favored if the volume of the gaseous equilibrium $N_2O_3$ (g) $\rightleftharpoons$ $N_2O$ (g) + $O_2$ (g) is reduced. Explain.

**15.27)** What shift will occur, according to Le Chatelier's Principle, to the equilibrium

$$C \text{ (s)} + H_2O \text{ (g)} \rightleftharpoons CO \text{ (g)} + H_2 \text{ (g)}$$

if volume is increased. Explain.

**15.28)** The equilibrium CO (g) + $H_2O$ (g) $\rightleftharpoons$ $CO_2$ (g) + $H_2$ (g) + 9.9 kcal is heated. Predict the direction in which the equilibrium will be favored as a result, and justify your prediction in terms of Le Chatelier's Principle.

**15.29)** If you wished to increase the relative amount of HI in the equilibrium

$$H_2 \text{ (g)} + I_2 \text{ (g)} + 6.2 \text{ kcal} \rightleftharpoons 2 \text{ HI (g)}$$

would you heat or cool the system? Explain your decision.

**15.30)** Consider the equilibrium:

$$4 NH_3 \text{ (g)} + 5 O_2 \text{ (g)} \rightleftharpoons$$
$$4 NO \text{ (g)} + 6 H_2O \text{ (g)} + 216.4 \text{ kcal}$$

Determine the direction of the Le Chatelier shift, forward or reverse, for each of the following actions: (a) add ammonia; (b) raise temperature; (c) reduce volume; (d) remove $H_2O$ (g).

**15.55)** Predict the direction of shift for the equilibrium

$$Cu(NH_3)_4^{2+} \text{ (aq)} \rightleftharpoons Cu^{2+} \text{ (aq)} + 4 NH_3 \text{ (aq)}$$

if the concentration of ammonia were reduced. Explain your prediction.

**15.56)** A container holding the equilibrium

$$4 H_2 \text{ (g)} + CS_2 \text{ (g)} \rightleftharpoons CH_4 \text{ (g)} + 2 H_2S \text{ (g)}$$

is enlarged. Predict the direction of the Le Chatelier shift. Explain.

**15.57)** In what direction will

$$CO \text{ (g)} + H_2O \text{ (g)} \rightleftharpoons CO_2 \text{ (g)} + H_2 \text{ (g)}$$

shift as a result of a reduction in volume? Explain.

**15.58)** Which direction of the equilibrium

$$2 NO_2 \text{ (g)} \rightleftharpoons N_2O_4 \text{ (g)} + 14.1 \text{ kcal}$$

will be favored if the system is cooled? Explain.

**15.59)** If your purpose was to increase the yield of $SO_3$ in the equilibrium

$$SO_2 \text{ (g)} + NO_2 \text{ (g)} \rightleftharpoons SO_3 \text{ (g)} + NO \text{ (g)} + 10.0 \text{ kcal}$$

would you use the highest or lowest operating temperature possible? Explain.

**15.60)\*** The solubility of calcium hydroxide is low; $K_{sp} = 5.5 \times 10^{-5}$, which yields a $2.4 \times 10^{-2}$ molar solution at saturation. In acid solutions, with many $H^+$ ions present, calcium hydroxide is quite soluble. Explain this fact in terms of Le Chatelier's Principle. (Hint: Recall what you know of reactions in which molecular products are formed.)

# 16

# ACID-BASE (PROTON TRANSFER) REACTIONS

Most of the chemical reactions that we are familiar with take place in aqueous solution. Some of these we study in the general chemistry laboratory: the formation of a precipitate of silver chloride when water solutions of silver nitrate and sodium chloride are mixed; the neutralization of hydrochloric acid by a water solution of sodium hydroxide; the evolution of hydrogen gas when zinc is added to sulfuric acid. Others we observe in the world around us: the complex series of reactions involved in animal metabolism; the process of photosynthesis in green plants; and the corrosion of metals exposed to air and moisture, all involve species in water solution as reactants.

You will find that a major portion of a later course in chemistry will be devoted to the principles and applications of the many different types of reactions that occur in aqueous solution. In this chapter and the next we shall survey two of the more important of these reaction types: acid-base reactions, and oxidation-reduction reactions.

Acid-base and oxidation-reduction reactions have many similarities, but there is one major difference between them: in acid-base reactions a *proton* is transferred from one species to another; in oxidation-reduction reactions an *electron* is transferred from one substance to the next.

We begin our comparison by examining acid-base reactions.

## 16.1 ACID-BASE CONCEPTS: ARRHENIUS THEORY

PG   16 A   Distinguish between an acid and a base in terms of their classical properties and the ions associated with those properties.

We refer to a water solution as an acid if it shows certain characteristic properties. These include:

1. A sour taste, as shown by vinegar (acetic acid) and lemon juice (citric acid). It is reliably reported that water solutions of HCl, $HNO_3$ and $H_2SO_4$ (hydrochloric, nitric and sulfuric acids) also taste sour.
2. The ability to produce a specific color change in certain substances known as acid-base indicators. Perhaps the most familiar of these is litmus, an organic dye that turns from blue to red in acid solutions. Needless to say, this property affords a safer and a more pleasant test for acidity than taste.
3. The ability to release gaseous carbon dioxide when carbonate ions are added to the solution.
4. The ability to react with and neutralize a base.
5. The ability to react with certain metals, releasing hydrogen gas as one of the products.

A water solution is called a base if it has the following properties:

1. A bitter taste. Sodium bicarbonate, used to relieve stomach acidity, is mildly basic, and better tasting than most other bases.
2. A "soapy" or slippery feeling. This property is shown by such familiar household products as liquid detergents, ammonia and lye. Since feeling a base actually involves the dissolving of a surface layer of skin, it is not recommended as a test for basicity.
3. The ability to change the color of an acid-base indicator. A piece of red litmus paper turns blue when touched by a basic solution.
4. The ability to react with and neutralize an acid.
5. The ability to form a precipitate when added to solutions of certain cations.

Long ago, in 1884, it was proposed by Svante Arrhenius that solutions having the properties of acids have one thing in common: they all release hydrogen ions, $H^+$, in water solutions. Furthermore, all solutions containing hydrogen ions behave like acids. Specifically, in regard to the reaction with carbonate ion, it is the hydrogen ion that combines with the carbonate ion to yield carbonic acid, which is unstable and decomposes to water and carbon dioxide:

$$2 \ H^+ \ (aq) + CO_3{}^{2-} \ (aq) \rightarrow H_2CO_3 \ (aq) \rightarrow H_2O \ (l) + CO_2 \ (g)$$

On evidence such as this, **the Arrhenius concept of an acid is a solution containing hydrogen ions.**

Basic solutions also have their ion that is always present: **the Arrhenius concept of a base is a solution containing hydroxide ion, $OH^-$.** To it may be attributed the properties of bases. The fourth of the properties listed for acids and bases, their ability to neutralize each other, is associated with the net ionic equation for a neutralization developed in Example 14.28, page 347.

$$H^+ \ (aq) + OH^- \ (aq) \rightarrow HOH \ (l)$$

In this reaction the $H^+$ and $OH^-$ ions are literally cancelling each other and all the acid and base properties associated with each other. Hence the name, *neutralization.*

Think back a moment and you will recognize that we have been using the hydrogen-hydroxide ion concept of acids and bases throughout this book. Our approach to nomenclature has been through acids; the formulas of polyatomic anions have been tied to the number of hydrogen ions released by their parent acid (pages 104, 221 and 226). Neutralization reactions were among those used in developing your equation writing skills in Chapter 7 (page 158), at which time the product other than water was defined as a *salt*. The optional topics of equivalents and normality (page 328) and their use in solution stoichiometry (page 334) were related to the hydrogen and hydroxide ions. Finally, in writing net ionic equations, the combination of these ions to form water in neutralization was identified as the most common example of molecule formation reactions (page 347).

The fifth properties of both acids and bases in the above lists are indeed properties of the hydrogen and hydroxide ions, and therefore properties of acids and bases. But the reactions are not acid-base reactions. When zinc is placed in hydrochloric acid, for example, hydrogen is released:

$$\text{Zn (s)} + 2 \text{ HCl (aq)} \rightarrow \text{H}_2 \text{ (g)} + \text{ZnCl}_2 \text{ (aq)}$$

This reaction is typical of the reaction of acids with the so-called *active metals*. As a reaction type, however, it is better classified as an oxidation-reduction reaction that happens to involve an acid, rather than an "acid-base" reaction. The precipitation of metallic hydroxides when a hydroxide solution is added to a solution of cations other than ammonium, alkali metals, and to some extent barium, strontium and calcium, shows only that most hydroxides are insoluble (see page 345). They are simply precipitations that happen to involve a base, rather than acid-base reactions.

## 16.2 ACID-BASE CONCEPTS: BRÖNSTED-LOWRY THEORY

PG 16 B State the Brönsted-Lowry concepts of acids and bases.

16 C Define conjugate acid-base pairs, or identify them in given equations.

### PROTON TRANSFER REACTIONS

The simple hydrogen-hydroxide ion concept of acids and bases is not adequate to describe all of the acid-base reactions we know about, particularly those that occur in the absence of water. In 1923 there was another of those simultaneous events, when Johannes N. Brönsted and Thomas M. Lowry independently proposed a broader acid-base theory that now commemorates both their names as the Brönsted-Lowry theory. By this concept an **acid-base reaction is one in which there is a transfer of a proton from one species to another. The acid is the proton donor; the base is the proton acceptor.** In this sense an acid-base reaction may be thought of as a **proton transfer reaction.**

The proton is the true identity of what we have been calling a hydrogen ion. A hydrogen atom consists of one proton and one electron. Remove the electron, and all that remains is the proton, which is the hydrogen nucleus. Lone protons do not actually exist in aqueous solutions; rather they become hydrated, or attached to a water molecule, forming an ion whose formula could be written $H \cdot H_2O^+$, much as water of hydration formulas are written. More frequently it is written $H_3O^+$, and it is named the **hydronium ion.**

We have seen previously (pages 104 and 372) that certain hydrogen-bearing compounds ionize in water, yielding hydronium ions in solutions we call acids. Hydrogen chloride is typical. The equation is

$$HCl \ (g) + H_2O \ (l) \rightarrow H_3O^+ \ (aq) + Cl^- \ (aq) \qquad (16.1)$$

Among chemists, practice varies between the use of the hydronium ion, $H_3O^+$ and the hydrogen ion, $H^+$. Because it is so much easier to write and balance in equations, the majority probably rests with $H^+$, with the understanding that the species referred to is a hydrated proton. We will side with that majority in this text, except for those occasions where the context requires the hydronium ion, or when the role played by water is significant.

Equation 16.1 for the ionization of HCl is an acid-base, or proton transfer, reaction in the Brönsted-Lowry sense. HCl gives up its hydrogen ion, a proton, and therefore qualifies as the acid; the water molecule receives the proton, thereby satisfying the definition of a base. Lewis diagrams show just how this is done:

$$\mathrm{H{-}\overset{..}{\underset{\underset{H}{|}}{O}}{:} \ + \ H{-}\overset{..}{\underset{..}{C}l}{:} \rightarrow \left[ H{-}\overset{\underset{H}{|}}{O}{-}H \right]^+ + [:\overset{..}{\underset{..}{C}}l:]^-}$$

Base-   Acid-
Proton  Proton
receiver donor

If you were surprised to learn that water is a base, prepare for the second part of that surprise. It is an acid too—in the Brönsted-Lowry sense. Consider what happens when ammonia, $NH_3$, dissolves in water:

$$NH_3 \ (aq) + HOH \ (l) \rightarrow NH_4^+ \ (aq) + OH^- \ (aq) \qquad (16.2)$$

From the insight of Lewis diagrams . . .

$$\mathrm{H{-}\overset{\underset{H}{|}}{\underset{\underset{H}{|}}{N}}{:} \ + \ H{-}\overset{..}{\underset{\underset{H}{|}}{O}}{:} \rightarrow \left[ H{-}\overset{\underset{H}{|}}{\underset{\underset{H}{|}}{N}}{-}H \right]^+ + \left[ :\overset{..}{\underset{\underset{H}{|}}{O}}{:} \right]^-}$$

Base-   Acid-
Proton  Proton
receiver donor

In this reaction it is easily seen that the water molecule is the proton donor, and therefore functions as an acid. Ammonia, the proton receiver, is the base.

This reaction, incidentally, does not take place to a significant extent. Less than 1% of the ammonia in a 1 M solution is converted to ammonium ion, at which point equilibrium is reached. We therefore conclude that the reaction is strongly favored in the reverse direction. The solution inventory is primarily $NH_3$ (aq), not $NH_4^+$ (aq) and $OH^-$ (aq). Of particular interest is the absence of any species whose formula is $NH_4OH$, although this is the formula usually found on laboratory bottles containing this equilibrium.

In Equation 16.1 we saw HCl as an acid, a proton donor. Equation 16.2 had $NH_3$ as a base, a proton acceptor. Do you suppose if we brought HCl and $NH_3$ together, the HCl would give its proton to $NH_3$? The answer, yes, is readily evident if bottles of concentration ammonia and hydrochloric acid are opened next to each other. The solid product is often found on the outside of desk reagent bottles in the laboratory. Conventional and Lewis diagram equations describe the reaction:

$$NH_3 \text{ (g)} + HCl \text{ (g)} \rightarrow NH_4Cl \text{ (s)} \qquad (16.3)$$

| Base- | Acid- |
| Proton | Proton |
| receiver | donor |

$NH_4Cl$ (s) is, of course, an assembly of $NH_4^+$ and $Cl^-$ ions. This last reaction is an acid-base reaction in the absence of both water and the hydroxide ion.

Reviewing Equations 16.1–16.3 we find that all fit into the general equation for a Brönsted-Lowry proton transfer reaction:

$$B \quad + \quad HA \quad \rightarrow \quad HB^+ \quad + \quad A^- \qquad (16.4)$$

| Base- | Acid- |
| Proton | Proton |
| receiver | donor |

In this equation, note that the charges are not "absolute" charges. They indicate, rather, that the acid species, in losing a proton, leaves a species having a charge one less than the acid; and that the base, in gaining a proton, increases by 1 in charge.

## CONJUGATE ACID-BASE PAIRS

It was noted that Equation 16.2 reaches equilibrium. The fact is that most acid-base reactions reach a state of equilibrium. This means the reverse reaction occurs to a measurable extent too. Let's look at the reverse reaction in Equation 16.4. Is not the $HB^+$ donating a proton to $A^-$ in the reverse reaction? In other words, *$HB^+$ is an acid in the reverse reaction, and $A^-$ is a base.*

$$B \quad + \quad HA \quad \leftarrow \quad HB^+ \quad + \quad A^- \qquad (16.5)$$

Acid-
Proton
donor

Base-
Proton
receiver

From this we see that the products of any proton transfer acid-base reaction are *another* acid and base for the reverse reaction.

Combinations such as the acid HA and the base $A^-$ that result from the acid losing its proton are called **conjugate acid-base pairs.** For example, when the ammonium ion functions as an acid and donates a proton,

$$NH_4^+ \text{ (aq)} \rightleftharpoons H^+ \text{ (aq)} + NH_3 \text{ (aq)}$$

$NH_4^+$ and $NH_3$ are a conjugate acid-base pair. In the forward direction, $NH_4^+$ is an acid—a proton donor. What is left after a proton is gone, $NH_3$, is the conjugate base. In the reverse direction ammonia functions as a base by gaining a proton. The result is the conjugate acid of ammonia: $NH_4^+$.

Looking at the general equation for a Brönsted-Lowry proton transfer reaction as an equilibrium equation we can identify two acid-base pairs:

┌──────Conjugate acid-base pair──────┐
$$B \quad + \quad HA \quad \rightleftharpoons \quad HB^+ \quad + \quad A^- \qquad (16.6)$$
└──────Conjugate acid-base pair──────┘

---

**Example 16.1** Nitrous acid engages in a proton transfer reaction with the sulfite ion thus:

$$HNO_2 + SO_3^{2-} \rightleftharpoons NO_2^- + HSO_3^-$$

As the first of several questions regarding this equilibrium, identify the acid and the base for the forward reaction.

-------------------------------------------------------------------

**16.1a** $HNO_2$ is the acid—it donates its proton to $SO_3^{2-}$, which is the base, the proton receiver.

Identify the acid and base for the reverse reaction.

-------------------------------------------------------------------

**16.1b** $HSO_3^-$ is the acid—it donates its proton to $NO_2^-$, which is the base, the proton receiver.

Identify the conjugate of $HNO_2$. Is it a conjugate acid or a conjugate base?

-------------------------------------------------------------------

**16.1c** $NO_2^-$ is the conjugate *base* of $HNO_2$. It is the base that remains after the proton has been donated by the acid.

Identify the other conjugate acid-base pair, and classify each as the acid or the base.

-------------------------------------------------------------------

**16.1d** $SO_3^{2-}$ and $HSO_3^-$ are the other conjugate acid-base pair. $SO_3^{2-}$ is the base, and $HSO_3^-$ is the conjugate acid—the species produced when the base accepts a proton.

**Example 16.2** Identify the conjugate acid-base pairs in

$$HCHO_2 + PO_4^{3-} \rightleftharpoons HPO_4^{2-} + CHO_2^-$$

-----------------------------------------------------------------------------------------------

**16.2a** $HCHO_2$ and $CHO_2^-$ are one conjugate acid-base pair; $PO_4^{3-}$ and $HPO_4^{2-}$ are the second pair.

## 16.3  ACID-BASE CONCEPTS: LEWIS THEORY

PG  16 D  Describe the Lewis theory of acids and bases and the structural features associated with it; identify potential Lewis acids and Lewis bases.

The Arrhenius theory of acids and bases identified acids as substances yielding a hydrogen ion, and bases as sources of hydroxide ions. The Brönsted-Lowry theory expanded the idea of bases to include anything that could receive protons. A third concept, the Lewis theory of acids and bases, expands the idea of acids to include anything that can accept both of the electrons required to form a covalent bond with a Brönsted-Lowry base.

The identifying characteristic of a Lewis acid or a Lewis base rests in the structure of the particle. Lewis diagrams for $OH^-$, $H_2O$ and $NH_3$, all of which have been identified as bases in this chapter, are

$$:\!\overset{..}{O}\!-\!H \qquad \overset{\textstyle :\overset{..}{O}-H}{\underset{\textstyle H}{|}} \qquad \overset{\textstyle H}{\underset{\textstyle H}{\overset{|}{\underset{|}{:\!N\!-\!H}}}}$$

$$OH^- \qquad\qquad H_2O \qquad\qquad NH_3$$

They have in common an unshared electron pair that can serve as a bonding site for anything that can accept an electron pair in forming a covalent bond. The hydrogen ion, $H^+$, with its empty 1s orbital, qualifies:

$$\overset{\textstyle :\overset{..}{O}-H}{\underset{\textstyle H}{|}} \qquad \overset{\textstyle H-\overset{..}{O}-H}{\underset{\textstyle H}{|}} \qquad \overset{\textstyle H}{\underset{\textstyle H}{\overset{|}{\underset{|}{H\!-\!N\!-\!H}}}}$$

$$H_2O \qquad\qquad H_3O^+ \qquad\qquad NH_4^+$$

In each case the base has contributed the electron pair to the bond. **A Lewis base is an electron pair donor.** The hydrogen ion has accepted the electron pair from the base in forming the bond. **A Lewis acid is an electron pair acceptor.**

The classic example of a Lewis acid is boron trifluoride. The boron atom is surrounded by three electron pairs, one pair short of the noble gas octet of electrons. It may acquire these by sharing with a Lewis base, such as ammonia:

$$
\begin{array}{ccccc}
\text{F} & & \text{H} & & \text{F}\ \ \text{H} \\
| & & | & & |\ \ \ | \\
\text{F—B} & + & \text{:N—H} & \rightarrow & \text{F—B—N—H} \\
| & & | & & |\ \ \ | \\
\text{F} & & \text{H} & & \text{F}\ \ \text{H} \\
\text{Acid} & & \text{Base} & &
\end{array}
$$

Other examples of Lewis acid-base reactions appear in the formation of some complex ions and in organic reactions. You will study them in greater detail in more advanced courses.

## 16.4  RELATIVE STRENGTHS OF ACIDS AND BASES

PG   16 E   Distinguish between a strong acid and a weak acid; between a strong base and a weak base.

16 F   Given a table of relative strengths of acids and bases, identify the stronger and weaker of two acids or two bases.

In Section 14.9, page 338, the distinction was made between the relatively few *strong* acids and the many *weak* acids. Strong acids are those that ionize almost completely, whereas weak acids ionize but slightly. Hydrochloric acid is a strong acid; 0.10 M HCl reaches nearly 100% ionization at equilibrium according to the equation $HCl\ (aq) \rightleftharpoons H^+\ (aq) + Cl^-\ (aq)$. Acetic acid is a weak acid; 0.10 M $HC_2H_3O_2$ is only about 1.3% ionized according to the equation $HC_2H_3O_2\ (aq) \rightleftharpoons H^+\ (aq) + C_2H_3O_2^-\ (aq)$. In a Brönsted-Lowry sense, acid strength is a measure of the tendency of an acid to lose protons. *A strong acid loses protons readily—ionizes more completely; a weak acid clings to its protons—does not ionize significantly.*

Table 16.1 is a list of acids arranged in order of decreasing strength: the strongest acids are at the top and the weakest are at the bottom. Each ionization is shown as an equilibrium equation of the general form $HA\ (aq) \rightleftharpoons H^+\ (aq) + A^-\ (aq)$. The ionization of **polyprotic acids,** those capable of donating more than one proton, is particularly notable. Sulfuric acid, for example, loses its first proton almost completely by the reaction $H_2SO_4\ (aq) \rightarrow H^+\ (aq) + HSO_4^-\ (aq)$, entitling sulfuric acid to be included among the strong acids. The hydrogen sulfate ion loses its second proton less readily: $HSO_4^-\ (aq) \rightarrow H^+\ (aq) + SO_4^{2-}\ (aq)$. The second ionization step occurs to only about 11% in a 1.0 M $HSO_4^-$ solution. **Diprotic** (2 protons) and **triprotic** (3 protons) acids always ionize in a stepwise fashion like this, and each succeeding acid is weaker than the one before.

The species on the right side of each equation in Table 16.1 is the conjugate base of the acid on the left. Being a base, it is capable of accepting a proton from a potential donor. *A strong base is one that has a strong attraction for protons; a weak base has a weak attraction for protons.* It follows that the conjugate base of a strong acid is a weak base, and that a weak acid has a strong conjugate base. HCl, for example, a strong acid, loses its proton readily to yield the weak base, Cl⁻. The chloride ion has no tendency to acquire a proton to form HCl by the reverse reaction. Conversely the carbonate ion, $CO_3^{2-}$, has a strong attraction for protons to form weak acid $HCO_3^-$. Thus $CO_3^{2-}$ is a strong base. By this reasoning we conclude that the bases in Table 16.1 are listed in order of *increasing* strength; the weaker bases are at the top, and the stronger bases are at the bottom.

The strongest base in the list is the hydroxide ion. Accordingly, solutions of soluble metal hydroxides are the most common bases used in the laboratory. As shown on page 345, there aren't many soluble hydroxides; only the hydroxides of the alkali metals, plus barium, strontium and calcium, can qualify—and the last three are questionable, having solubility limits of about 0.17 M, 0.04 M and 0.03 M respectively. But these are considerably more soluble than the next in the list; the solubility of $Mg(OH)_2$ is about 0.000003 M—and the solubilities of other hydroxides go down from there.

**TABLE 16.1** Relative Strengths of Acids and Bases

| Acid Name | Acid Formula | Base Formula |
|---|---|---|
| Perchloric | $HClO_4$ | $\rightleftarrows H^+ + ClO_4^-$ |
| Hydroiodic | $HI$ | $\rightleftarrows H^+ + I^-$ |
| Hydrobromic | $HBr$ | $\rightleftarrows H^+ + Br^-$ |
| Hydrochloric | $HCl$ | $\rightleftarrows H^+ + Cl^-$ |
| Nitric | $HNO_3$ | $\rightleftarrows H^+ + NO_3^-$ |
| Sulfuric | $H_2SO_4$ | $\rightleftarrows H^+ + HSO_4^-$ |
| Hydronium ion | $H_3O^+$ | $\rightleftarrows H^+ + H_2O$ |
| Oxalic | $H_2C_2O_4$ | $\rightleftarrows H^+ + HC_2O_4^-$ |
| Sulfurous | $H_2SO_3$ | $\rightleftarrows H^+ + HSO_3^-$ |
| Hydrogen sulfate ion | $HSO_4^-$ | $\rightleftarrows H^+ + SO_4^{2-}$ |
| Phosphoric | $H_3PO_4$ | $\rightleftarrows H^+ + H_2PO_4^-$ |
| Hydrofluoric | $HF$ | $\rightleftarrows H^+ + F^-$ |
| Nitrous | $HNO_2$ | $\rightleftarrows H^+ + NO_2^-$ |
| Formic (methanoic) | $HCHO_2$ | $\rightleftarrows H^+ + CHO_2^-$ |
| Benzoic | $HC_7H_5O_2$ | $\rightleftarrows H^+ + C_7H_5O_2^-$ |
| Hydrogen oxalate ion | $HC_2O_4^-$ | $\rightleftarrows H^+ + C_2O_4^{2-}$ |
| Acetic (ethanoic) | $HC_2H_3O_2$ | $\rightleftarrows H^+ + C_2H_3O_2^-$ |
| Propionic (propanoic) | $HC_3H_5O_2$ | $\rightleftarrows H^+ + C_3H_5O_2^-$ |
| Carbonic | $H_2CO_3$ | $\rightleftarrows H^+ + HCO_3^-$ |
| Hydrosulfuric | $H_2S$ | $\rightleftarrows H^+ + HS^-$ |
| Dihydrogen phosphate ion | $H_2PO_4^-$ | $\rightleftarrows H^+ + HPO_4^{2-}$ |
| Hydrogen sulfite ion | $HSO_3^-$ | $\rightleftarrows H^+ + SO_3^{2-}$ |
| Hypochlorous | $HClO$ | $\rightleftarrows H^+ + ClO^-$ |
| Boric | $H_3BO_3$ | $\rightleftarrows H^+ + H_2BO_3^-$ |
| Ammonium ion | $NH_4^+$ | $\rightleftarrows H^+ + NH_3$ |
| Hydrocyanic | $HCN$ | $\rightleftarrows H^+ + CN^-$ |
| Hydrogen carbonate ion | $HCO_3^-$ | $\rightleftarrows H^+ + CO_3^{2-}$ |
| Monohydrogen phosphate ion | $HPO_4^{2-}$ | $\rightleftarrows H^+ + PO_4^{3-}$ |
| Hydrogen sulfide ion | $HS^-$ | $\rightleftarrows H^+ + S^{2-}$ |
| Water | $HOH$ | $\rightleftarrows H^+ + OH^-$ |

(Left axis: STRENGTH — Increasing / Decreasing; Right axis: STRENGTH — Decreasing / Increasing)

**Example 16.3** Using Table 16.1, list the following acids in order of decreasing strength (strongest first): $HC_2O_4^-$; $NH_4^+$; $H_3PO_4$.

Find the three acids among those listed in the table, and list them from strongest (left) to weakest (right).

-------------------------------------------------------------------------------------

**16.3a** $H_3PO_4$; $HC_2O_4^-$; $NH_4^+$

**Example 16.4** Using Table 16.1, list the following bases in order of decreasing strength (strongest first): $HC_2O_4^-$; $SO_3^{2-}$; $F^-$.

-------------------------------------------------------------------------------------

**16.4a** $SO_3^{2-}$; $F^-$; $HC_2O_4^-$

The ion $HC_2O_4^-$ appears in both Example 16.3 and 16.4, first as an acid and second as a base. It is the intermediate ion in the two-step ionization of oxalic acid, $H_2C_2O_4$. The $HC_2O_4^-$ ion is capable of functioning as an acid by releasing its proton, or as a base by receiving a proton. Substances that are capable of reacting as either acids or bases are called **amphoteric** substances. The intermediate ions resulting from the partial ionization of polyprotic acids are amphoteric.

## 16.5 PREDICTING ACID-BASE REACTIONS

PG 16 G Given a table of the relative strengths of acids and bases and the identity of an acid and a base from the table, (a) write the equation for the single proton transfer reaction between them and (b) predict which direction, forward or reverse, will be favored at equilibrium.

The chemist is interested in predicting whether or not an acid-base reaction will occur if certain reactants are brought together under proper conditions. Obviously there must be a potential proton donor and a potential proton acceptor—there can be no proton transfer reaction without both. From there the decision is based on the relative strengths of the conjugate acid and base pairs. At least to the extent of establishing an equilibrium, the stronger acid and base are the most reactive. They *do* what they must do to qualify as acid and base. The weaker acid and base are more stable—less reactive. It follows that *the stronger acid will always transfer a proton to the stronger base, yielding the weaker acid and base as favored species at equilibrium.* Figure 16.1 summarizes the proton transfer from the stronger acid to the stronger base from the standpoint of positions in Table 16.1.

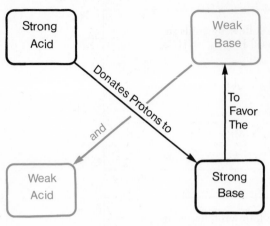

At Equilibrium

**Figure 16.1** Correlation between prediction of an acid-base reaction and position in Table 16.1. Of the two conjugate acid-base pairs in the proton transfer equation, the stronger acid (upper left) will donate protons to the stronger base (lower right), yielding the weaker base (upper right) and weaker acid (lower left). That direction, forward or reverse, will be favored that has the weaker acid and weaker base as products.

If a hydrogen sulfate ion, $HSO_4^-$, a relatively strong acid seeking to dispose of a proton, is exposed to the hydroxide ion, $OH^-$, a strong base seeking a proton, both tendencies can be satisfied by the transfer of the proton from $HSO_4^-$ to $OH^-$:

$$HSO_4^- \text{ (aq)} + OH^- \text{ (aq)} \rightleftharpoons SO_4^{2-} \text{ (aq)} + HOH \text{ (l)}$$

Now identify the conjugate acid-base pairs. In the forward direction the acid is $HSO_4^-$. Its conjugate base for the reverse reaction is $SO_4^{2-}$. Similarly $OH^-$ is the base in the forward reaction, and $HOH$ is the conjugate acid for the reverse reaction. These acid-base roles for the different directions are shown with the letters A for acid and B for base:

$$HSO_4^- \text{ (aq)} + OH^- \text{ (aq)} \rightleftharpoons SO_4^{2-} \text{ (aq)} + HOH \text{ (l)}$$
$$\quad\ \text{A} \qquad\qquad \text{B} \qquad\qquad\quad \text{B} \qquad\qquad \text{A}$$

Now compare the two acids in strength. $HSO_4^-$ is near the top of the list, a much stronger acid than water. We shall therefore label $HSO_4^-$ with SA for strong acid, and water with WA for weak acid. Similarly, compare the bases: $OH^-$ is a stronger base (SB) than $SO_4^{2-}$ (WB).

$$HSO_4^- \text{ (aq)} + OH^- \text{ (aq)} \rightleftharpoons SO_4^{2-} \text{ (aq)} + HOH \text{ (l)}$$
$$\quad\ \text{SA} \qquad\qquad \text{SB} \qquad\qquad \text{WB} \qquad\qquad \text{WA}$$

As you see, the weaker combination is on the right in this equation. This indicates the reaction is favored in the forward direction: the proton transfers spontaneously from the strong proton donor to the strong proton receiver, yielding as products in greater abundance the weaker conjugate base and conjugate acid.

The following procedure is recommended in the prediction of acid-base reactions:

1. For a given pair of reactants, write the equation for the transfer of *one* single proton from one species to the other. (Do not transfer two protons.)

2. Identify the acid and base on each side of the equation.
3. Determine which side of the equation has *both* the weaker acid and the weaker base (they must both be on the same side). That side identifies the products in the favored direction.

---

**Example 16.5** Write the net ionic equation for the reaction between hydrofluoric acid, HF, and the sulfite ion, $SO_3^{2-}$ and predict which side will be favored at equilibrium.

The first step is to write the equation for the single proton transfer reaction between the acid and base. Complete this step.

--------------------------------------------------------------------------------

**16.5a** $\qquad$ $HF\ (aq) + SO_3^{2-}\ (aq) \rightleftharpoons F^-\ (aq) + HSO_3^-\ (aq)$

Next we identify the acid and base on each side of the equation. Do so with letters A and B, as in the foregoing discussion.

--------------------------------------------------------------------------------

**16.5b** $\qquad$ $HF\ (aq) + SO_3^{2-}\ (aq) \rightleftharpoons F^-\ (aq) + HSO_3^-\ (aq)$

$\qquad\qquad$ A $\qquad\quad$ B $\qquad\qquad$ B $\qquad\quad$ A

In each case the acid is the species with a proton to donate. It is transferred from acid HF to base $SO_3^{2-}$ in the forward reaction, and from acid $HSO_3^-$ to base $F^-$ in the reverse reaction.

Finally determine which reaction, forward or reverse, is favored at equilibrium by identifying the side with the weaker acid and base. Identification is by reference to Table 16.1.

--------------------------------------------------------------------------------

**16.5c** The forward reaction is favored at equilibrium. $HSO_3^-$ is a weaker acid than HF, and $F^-$ is a weaker base than $SO_3^{2-}$. These species are the products in the favored direction.

---

**Example 16.6** Write the net ionic equation for the acid-base reaction between $HCO_3^-$ and $ClO^-$ and predict which side will be favored at equilibrium.

Notice that $HCO_3^-$ appears twice in Table 16.1. It can be an acid in one reaction and a base in another. Which will it be in this case? Explain.

--------------------------------------------------------------------------------

**16.6a** $HCO_3^-$ will be an acid in this reaction. The decision must rest on the role of the other reactant. $ClO^-$ has no proton to donate, so it is capable of functioning only as a proton receiver, or as a base. If it is a base to $HCO_3^-$, then $HCO_3^-$ must be the proton donor, or acid.

At this point you should be able to write the proton transfer equation, identify acids and bases, and determine which side will be favored at equilibrium. Go all the way.

--------------------------------------------------------------------------------

**16.6b** $\quad HCO_3^-$ (aq) $+ ClO^-$ (aq) $\rightleftharpoons CO_3^{2-}$ (aq) $+ HClO$ (aq)

The reverse reaction will be favored. This conclusion is based on $HCO_3^-$ being a weaker acid than $HClO$, and $ClO^-$ being a weaker base than $CO_3^{2-}$.

Up to this point we have given primary attention to the direction in which an equilibrium is favored. This does not mean we can ignore the unfavored direction. Consider, for example, the reaction described by Equation 16.2: $NH_3$ (aq) $+ HOH$ (l) $\rightleftharpoons OH^-$ (aq) $+ NH_4^+$ (aq). Though this reaction proceeds only slightly in the forward direction, many of the properties of household ammonia, in particular its cleaning power, depend upon the presence of $OH^-$ ions, few as they are.

## 16.6 THE WATER EQUILIBRIUM: pH

PG   16 H   Write the equation for the ionization of water, the equation for $K_w$ and state the value of $K_w$ at 25°C.

     16 I   Distinguish between an acidic solution, a basic solution and a neutral solution in terms of pH and $[H^+]$.

     16 J   For integer values of pH only, given one of the following calculate any or all of the others: pH; $[H^+]$; $[OH^-]$; pOH.

     16 K   Given the pH, $[H^+]$, $[OH^-]$ or pOH of two or more solutions, rank them in order of acidity, basicity, pH, $[H^+]$, $[OH^-]$ or pOH.

One of the most universally critical equilibria in all of chemistry is represented by the last line in Table 16.1, the ionization of water. Careful control of minute traces of hydrogen and hydroxide ions marks the difference between success and failure in an untold number of industrial chemical processes; and in biochemical systems these concentrations are vital to survival itself.

Though pure water is generally regarded as a nonconductor, a sufficiently sensitive detector discloses that even water contains a tiny concentration of ions. These ions come from the ionization of the water molecule:

$$HOH\ (l) \rightleftharpoons H^+\ (aq) + OH^-\ (aq) \qquad (16.7)$$

From this equation the equilibrium constant expression for water, which is given the special symbol $K_w$, may be written and equated to its value at 25°C:

$$K_w = [H^+]\,[OH^-] = 10^{-14} \qquad (16.8)$$

$[H^+]$ and $[OH^-]$ are the concentrations of the hydrogen and hydroxide ions in moles per liter.

In the event your course syllabus does not include Section 15.9 (page 368) on equilibrium constants, this brief introduction is provided.

In the analysis of chemical equilibria it turns out that certain concentration ratios have constant values, even though the quantities making up those ratios may vary widely. These ratios are called equilibrium constants. In some cases, including the water equilibrium, the equilibrium constant is simply the product of two concentrations, the denominator of the ratio having been absorbed into the numerical value of the constant. In writing equilibrium constant expressions, it is customary to enclose the symbol of a species in square brackets, [ ], to indicate its concentration in moles per liter. Accordingly, Equation 16.8 expresses the experimental fact that at 25°C the product of the hydrogen ion concentration times the hydroxide ion concentration, both measured in moles per liter, is $1.0 \times 10^{-14}$.

A more detailed treatment of equilibrium constants and their applications may be found in Sections 15.9–15.11, pages 368–377.

From the value of $K_w$ we can calculate the theoretical concentrations of hydrogen and hydroxide ions in pure water. The stoichiometry of Equation 16.7 tells us they must be equal. Consequently, if $x = [H^+] = [OH^-]$, then substituting into Equation 16.8 we have

$$x^2 = 10^{-14}$$

$$x = \sqrt{10^{-14}} = 10^{-7} \text{ moles/liter}$$

(If necessary, refer to Appendix I, page 501, on taking square root of exponentials.)

**Water or water solutions in which $[H^+] = [OH^-] = 10^{-7}$ are said to be neutral solutions, neither acidic nor basic. A solution in which $[H^+]$ is greater than $[OH^-]$ is acidic; a solution in which $[OH^-]$ is greater than $[H^+]$ is basic.**

Equation 16.8 indicates an inverse relationship between $[H^+]$ and $[OH^-]$; if one concentration rises, the other must decrease. In fact, if we know either the hydrogen or hydroxide ion concentration, we can, by Equation 16.8, find the other:

---

**Example 16.7**  Find the hydroxide ion concentration in a solution in which $[H^+] = 10^{-5}$.

This problem requires little explanation. Solve Equation 16.8 for $[OH^-]$, substitute and calculate. (You may wish to refresh your memory regarding the division of exponentials on page 501.)

------------------------------------------------------------------------------------------

**16.7a**  $[OH^-] = 10^{-9}$

From Equation 16.8, $[OH^-] = \dfrac{10^{-14}}{[H^+]} = \dfrac{10^{-14}}{10^{-5}} = 10^{-14 \, - \, (-5)} = 10^{-9}$

---

To provide a less complex number by which to measure the small values of hydrogen and hydroxide ions in water and aqueous solutions the chemist uses the "p" concept.* According to this concept, for any number, N, pN is defined by the equation

$$pN = -\log N = \log \frac{1}{N} \qquad (16.9)$$

------------------------

*The need for a "less complex" number is not so apparent when we consider $[H^+]$ as an exponential like $10^{-7}$, but it becomes apparent when we recognize that a more typical value of $[H^+]$ would be $3.2 \times 10^{-7}$.

Applied to $[H^+]$ and $[OH^-]$, it follows that

$$pH = -\log[H^+] = \log\frac{1}{[H^+]}; \quad pOH = -\log[OH^-] = \log\frac{1}{[OH^-]} \quad (16.10)$$

From the theory of logarithms* these equations lead to the relationships

$$[H^+] = 10^{-pH}; \quad [OH^-] = 10^{-pOH} \quad (16.11)$$

The following procedures result:

If hydrogen ion concentration in moles per liter is expressed as 10 raised to some negative exponent, pH is the exponent with its sign changed to positive. Thus if $[H^+] = 10^{-x}$, then $pH = x$.

Conversely, given the pOH of a solution, the hydroxide ion concentration is 10 raised to a power that is the negative of the pOH. Thus if pOH $= y$, then $[OH^-] = 10^{-y}$.**

---

**Example 16.8**   (a) The hydrogen ion concentration of a solution is $10^{-9}$. What is its pH? (b) The pOH of a certain electroplating bath is 4. What is the hydroxide ion concentration?

---------------------------------------------------------------------

**16.8a**   (a) $pH = 9$; (b) $[OH^-] = 10^{-4}$.

(a) $[H^+] = 10^{-pH} = 10^{-9}$; $pH = 9$, the negative of the exponent of 10.

(b) $[OH^-] = 10^{-pOH} = 10^{-4}$. The exponent of 10 is the negative of the pOH.

---

The nature of the equilibrium constant and the logarithmic relationship between pH and pOH yield a simple equation that ties the two together:

$$pH + pOH = 14.0 \quad (16.12)$$

Between Equations 16.8 and 16.12, if you know any one of the group consisting of pH, pOH, $[OH^-]$ or $[H^+]$ you can calculate the others.

---

*The theory of logarithms and its specific application to pH problems are beyond the scope of this book. For a more detailed treatment of this topic you are referred to Peters, *Problem Solving for Chemistry*, Second Edition, W. B. Saunders Company, 1976 (page 279).

**Mathematically, pH and pOH are negative numbers when the corresponding molarities are 1 or more. The "p" concept is not ordinarily used for concentrations greater than 1 molar, so we need not be concerned with negative values of pH or pOH.

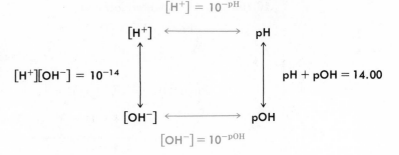

$$[H^+] = 10^{-pH}$$

**Figure 16.2** The "pH loop." Given the value for any corner of the pH loop, all other values may be calculated by progressing around the loop in either direction. Conversion equations are shown for each step around the loop.

$$[H^+][OH^-] = 10^{-14}$$

$$[OH^-] = 10^{-pOH}$$

---

**Example 16.9** Assuming complete ionization of 0.01 M NaOH, find its pH, pOH, $[OH^-]$ and $[H^+]$.

*Solution.* (a) Starting with $[OH^-]$, if 0.01 mole of NaOH is dissolved in one liter of solution, the concentration of the hydroxide ion is 0.01 molar: $[OH^-] = 0.01 = 10^{-2}$.

(b) If $[OH^-] = 10^{-2}$, pOH = 2.

(c) From Equation 16.12, pH = 14 − pOH = 14 − 2 = 12.

(d) If pH = 12, $[H^+] = 10^{-12}$.

---

Two things are worth noting about Example 16.9. First, if we extend the problem by one more step we complete a full "pH loop" (see Fig. 16.2). We began with $[OH^-]$ and went counterclockwise through pOH, pH and $[H^+]$. The $[H^+]$ of $10^{-12}$ can be converted to $[OH^-]$ by Equation 16.8:

$$[OH^-] = \frac{K_w}{[H^+]} = \frac{10^{-14}}{10^{-12}} = 10^{-2}$$

which is the same as the $[OH^-]$ from which we began. The completion of the loop may therefore be used to check the correctness of the other steps in the process.

The second observation from Example 16.9 is that the loop may be circled in either direction. Starting with $[OH^-] = 10^{-2}$ and moving clockwise,

$$[H^+] = \frac{K_w}{[OH^-]} = \frac{10^{-14}}{10^{-2}} = 10^{-12}$$

It follows that pH = 12 and pOH = 14 − 12 = 2, the same results reached by circling the loop in the opposite direction.

You should now be able to make a complete trip around the pH loop....

---

**Example 16.10** The pH of a solution is 3. Calculate the pOH, $[H^+]$ and $[OH^-]$.

You may go either way around the loop, but complete it whichever way you choose, making sure you return to the starting point.

-----------------------------------------------------------------------

**16.10a**

| | Counterclockwise | Clockwise |
|---|---|---|

From pH = 3, [H$^+$] = 10$^{-3}$ $\qquad$ pOH = 14 − 3 = 11

[OH$^-$] = $\dfrac{10^{-14}}{10^{-3}}$ = 10$^{-11}$ $\qquad$ From pOH = 11, [OH$^-$] = 10$^{-11}$

From [OH$^-$] = 10$^{-11}$, pOH = 11 $\qquad$ [H$^+$] = $\dfrac{10^{-14}}{10^{-11}}$ = 10$^{-3}$

pH = 14 − 11 = 3 $\qquad$ From [H$^+$] = 10$^{-3}$, pH = 3

Most of the solutions we work with in the laboratory and essentially all of those involved in biochemical systems have pH values between 1 and 14, corresponding to H$^+$ concentrations between 10$^{-1}$ and 10$^{-14}$, as shown in Table 16.2.

TABLE 16.2 pH Values of Common Liquids

| LIQUID | pH | LIQUID | pH |
|---|---|---|---|
| Human gastric juices | 1.0–3.0 | Cow's milk | 6.3– 6.6 |
| Lemon juice | 2.2–2.4 | Human blood | 7.3– 7.5 |
| Vinegar | 2.4–3.4 | Sea water | 7.8– 8.3 |
| Carbonated drinks | 2.0–4.0 | Saturated Mg(OH)$_2$ | 10.5 |
| Orange juice | 3.0–4.0 | Household ammonia (1–5%) | 10.5–11.5 |
| Tomato juice | 4.0–4.4 | 0.1 M Na$_2$CO$_3$ | 11.7 |

TABLE 16.3 pH and Hydrogen Ion Concentration

| [H$^+$] | [H$^+$] | pH | ACIDITY OR BASICITY* |
|---|---|---|---|
| 1.0 | 10$^0$ | 0 | |
| 0.1 | 10$^{-1}$ | 1 | Strongly Acid |
| 0.01 | 10$^{-2}$ | 2 | |
| 0.001 | 10$^{-3}$ | 3 | |
| 0.0001 | 10$^{-4}$ | 4 | |
| 0.00001 | 10$^{-5}$ | 5 | Weakly Acid |
| 0.000001 | 10$^{-6}$ | 6 | |
| 0.0000001 | 10$^{-7}$ | 7 | Neutral |
| 0.00000001 | 10$^{-8}$ | 8 | |
| 0.000000001 | 10$^{-9}$ | 9 | |
| 0.0000000001 | 10$^{-10}$ | 10 | Weakly Basic |
| 0.00000000001 | 10$^{-11}$ | 11 | |
| 0.000000000001 | 10$^{-12}$ | 12 | |
| 0.0000000000001 | 10$^{-13}$ | 13 | Strongly Basic |
| 0.00000000000001 | 10$^{-14}$ | 14 | |

*Ranges of acidity and basicity are arbitrary.

Let's pause for a moment to develop a "feeling" for pH, what it means. pH is a measure of acidity, at least in the sense that hydrogen ion concentration constitutes acidity. It is an inverse sort of measurement: the higher the pH, the lower the acidity, and vice versa. Table 16.3 brings out this relationship.

On examining Table 16.3, we see that each pH unit represents a factor of 10. Thus a solution of pH 2 is 10 times as acidic as a solution with pH = 3, and 100 times as acidic as the solution of pH 4. In general, the relative acidity in terms of $[H^+]$ is $10^x$, where x is the absolute value difference between the two pH measurements. From this we conclude that a 0.1 M solution of a strong acid, with pH = 1, is one million times as acidic as a neutral solution, with pH = 7. (One million is based on the pH difference, $7 - 1 = 6$. As an exponential, $10^6 = 1,000,000$.)

If you understand the idea behind pH, you should be able to make some comparisons . . . .

---

**Example 16.11**  Arrange the following solutions in order of decreasing acidity (i.e., highest $[H^+]$ first, lowest last): Solution A, pH = 8; Solution B, pOH = 4; Solution C, $[H^+]$ = $10^{-6}$; Solution D, $[OH^-]$ = $10^{-5}$.

To make comparisons, all values should be converted to the same basis, pH, pOH, $[H^+]$ or $[OH^-]$. Because the question asks for a list based on acidity, find the $[H^+]$ for each solution.

--------------------------------------------------------------------------------------

**16.11a**  $[H^+]$ = $10^{-8}$ for A; $10^{-10}$ for B; $10^{-6}$ for C; $10^{-9}$ for D.

Now arrange these $[H^+]$ values in decreasing order. Remember that the exponents are negative . . . .

--------------------------------------------------------------------------------------

**16.11b**  C, A, D, B    $10^{-6} > 10^{-8} > 10^{-9} > 10^{-10}$

**Figure 16.3**  pH papers are impregnated with dyes that, when dipped into solutions, turn certain colors that depend on the pH of the solution. By comparing the color with a standard, the pH of the solution may be estimated. (Photograph by Gregory A. Peters.)

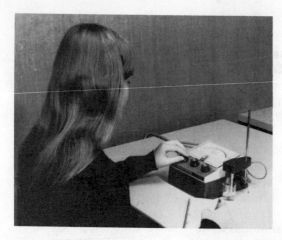

**Figure 16.4** A pH meter for student use. (Photograph by Gregory A. Peters.)

Various methods are used to measure pH in the laboratory. Acid-base indicators have been mentioned already. Each indicator has a specific pH range over which it changes color. Various pH papers (Fig. 16.3), strips impregnated with an indicator dye that functions over a narrow or broad range depending on the pH being measured, are widely used in industry and in laboratories for approximate determinations. More accurate measurements are possible with pH meters (Fig. 16.4), which by proper selection of electrodes may be used to determine minute concentrations of other ions as well.

## 16.7 NONINTEGER pH–[H$^+$] AND pOH–[OH$^-$] CONVERSIONS

It is sad, but true, that in the real world solutions do not come neatly packaged in concentrations that can be expressed as whole-number powers of ten, as we have been considering them. [H$^+$] is more apt to have a value such as $2.5 \times 10^{-4}$, or the pH of a solution is more likely to be 8.32. The chemist must be able to convert each form of expression to the other. Some of the more sophisticated calculators are able to convert directly between a number expressed in exponential notation and its negative logarithm. If you don't happen to have such an instrument available, you must use logarithms. "Two-place" logarithms are sufficient for concentrations expressed in two significant figures, which is typical of pH measurements. Table 16.4 is a table of two-place logarithms for this purpose.

The exponential notation expression for the concentration of the hydrogen ion consists of a *decimal factor* and an *exponential factor*. If [H$^+$] = $4.8 \times 10^{-9}$, the decimal factor is 4.8 and the exponential factor is $10^{-9}$. It can be shown that the pH of a solution is the opposite of the exponent in the exponential factor minus the logarithm of the decimal factor:

$$\text{If } [\text{H}^+] = \text{M} \times 10^{-\text{N}}, \text{ then pH} = \text{N} - \log \text{M} \qquad (16.13)$$

For example, if [H$^+$] = $4.8 \times 10^{-9}$, then pH = $9 - 0.68 = 8.32$, the 0.68 being the logarithm of 4.8 from Table 16.4.

TABLE 16.4   Two-Place Logarithms

|   | 0.0 | 0.1 | 0.2 | 0.3 | 0.4 | 0.5 | 0.6 | 0.7 | 0.8 | 0.9 |
|---|-----|-----|-----|-----|-----|-----|-----|-----|-----|-----|
| 1 | .00 | .04 | .08 | .11 | .15 | .18 | .20 | .23 | .26 | .28 |
| 2 | .30 | .32 | .34 | .36 | .38 | .40 | .41 | .43 | .45 | .46 |
| 3 | .48 | .49 | .51 | .52 | .53 | .54 | .56 | .57 | .58 | .59 |
| 4 | .60 | .61 | .62 | .63 | .64 | .65 | .66 | .67 | .68 | .69 |
| 5 | .70 | .71 | .72 | .72 | .73 | .74 | .75 | .76 | .76 | .77 |
| 6 | .78 | .79 | .79 | .80 | .81 | .81 | .82 | .83 | .83 | .84 |
| 7 | .85 | .85 | .86 | .86 | .87 | .88 | .88 | .89 | .89 | .90 |
| 8 | .90 | .91 | .91 | .92 | .92 | .93 | .93 | .94 | .94 | .95 |
| 9 | .95 | .96 | .96 | .97 | .97 | .98 | .98 | .99 | .99 | 1.00 |

**Example 16.12**   Find the pH of a solution if its hydrogen ion concentration is $7.9 \times 10^{-5}$.

--------------------------------------------------------------------------------

**16.12a**   pH = 4.10

From Equation 16.13, pH = $5 - \log 7.9 = 5 - 0.90 = 4.10$

The procedure for converting a "p" value to exponential notation is not readily reduced to an equation, but is best illustrated by an example. If the pOH of a solution is 6.46, Equation 16.11 indicates that $[OH^-] = 10^{-6.46}$. The exponent is then expressed as the difference between two numbers, a *positive decimal fraction* and the *closest integer more negative than the exponent*. Working in reverse, the closest integer more negative than $-6.46$ is $-7$. Now, what positive decimal fraction must be added to $-7$ to yield $-6.46$?

$$x + [-7] = x - 7 = -6.46$$

$$x = +0.54$$

$$+0.54 - 7 = -6.46$$

(Notice that 0.46, the decimal part of the "p" value, and 0.54 add up to 1.00. A quick way to find the positive decimal fraction is to subtract the decimal part of the p value from 1.00: $1.00 - 0.46 = 0.54$.) We now have

$$[OH^-] = 10^{-6.46} = 10^{0.54-7}$$

The integer part of the difference, $-7$, becomes the exponent of the exponential factor in the concentration. The antilogarithm of the decimal term is the decimal factor of the concentration. From Table 16.4, the antilogarithm of 0.54 is 3.5. Thus if pOH = 6.46, $[OH^-] = 3.5 \times 10^{-7}$.

To summarize the whole procedure:

STEPS

1. Write the $[H^+]$ or $[OH^-]$ as an exponential of 10, with the pH or pOH as the exponent.

2. Rewrite the exponent as the sum of a positive decimal fraction and the closest integer more negative than the exponent.

3. Write the concentration in exponential notation, using the antilog of the positive decimal fraction as the decimal factor, and the negative integer as the exponent of the exponential factor.

EXAMPLE: pOH = 6.46

$[OH^-] = 10^{-6.46}$

Positive deci-          Integer
mal fraction
$[OH^-] = 10^{0.54-7}$

$[OH^-] = 3.5 \times 10^{-7}$

It is always well to check the final answer to be sure it is in the right range. pH = 6.46 is between 6 and 7. This means the answer is between $10^{-7}$ and $10^{-6}$—greater than $10^{-7}$, but less than $10^{-6}$. $3.5 \times 10^{-7}$ falls in this range.

---

**Example 16.13**  Find the hydrogen ion concentration of a solution if its pH is 11.62.

---

**16.13a**  $[H^+] = 2.4 \times 10^{-12}$

$[H^+] = 10^{-11.62} = 10^{0.38-12}$     Antilog $0.38 = 2.4$     $[H^+] = 2.4 \times 10^{-12}$

---

**Example 16.14**  $[OH^-] = 5.2 \times 10^{-9}$ for a certain solution. Calculate in order pOH, pH and $[H^+]$, and then complete the pH loop by recalculating $[OH^-]$ from $[H^+]$.

---

**16.14a**     $pOH = 9 - \log 5.2 = 9 - 0.72 = 8.28$

$pH = 14.00 - 8.28 = 5.72$

$[H^+] = 10^{-5.72} = 10^{0.28-6} = 1.9 \times 10^{-6}$

$[OH^-] = \dfrac{1.0 \times 10^{-14}}{1.9 \times 10^{-6}} = 5.3 \times 10^{-9}$

The variation between $5.2 \times 10^{-9}$ and $5.3 \times 10^{-9}$ arises from round-offs in both calculations and two-place logarithms.

# CHAPTER 16 IN REVIEW

## 16.1 ACID-BASE CONCEPTS: ARRHENIUS THEORY

16 A   **Distinguish between an acid and a base in terms of their classical properties and the ions associated with those properties. (Page 388)**

## 16.2  ACID-BASE CONCEPTS: BRÖNSTED-LOWRY THEORY

16 B    State the Brönsted-Lowry concepts of acids and bases. (Page 390)

16 C    Define conjugate acid-base pairs, or identify them in given equations. (Page 390)

## 16.3  ACID-BASE CONCEPTS: LEWIS THEORY

16 D    Describe the Lewis theory of acids and bases and the structural features associated with it; identify potential Lewis acids and Lewis bases. (Page 394)

## 16.4  RELATIVE STRENGTHS OF ACIDS AND BASES

16 E    Distinguish between a strong acid and a weak acid; between a strong base and a weak base. (Page 395)

16 F    Given a table of relative strengths of acids and bases, identify the stronger and weaker of two acids or two bases. (Page 395)

## 16.5  PREDICTING ACID-BASE REACTIONS

16 G    Given a table of the relative strengths of acids and bases and the identity of an acid and a base from the table, (a) write the equation for the single proton transfer reaction between them and (b) predict which direction, forward or reverse, will be favored at equilibrium. (Page 397)

## 16.6  THE WATER EQUILIBRIUM: pH

16 H    Write the equation for the ionization of water, the equation for $K_w$ and state the value of $K_w$ at 25°C. (Page 400)

16 I    Distinguish between an acidic solution, a basic solution and a neutral solution in terms of pH and $[H^+]$. (Page 400)

16 J    For integer values of pH only, given one of the following calculate any or all of the others: pH; $[H^+]$; $[OH^-]$; pOH. (Page 400)

16 K    Given the pH, $[H^+]$, $[OH^-]$ or pOH of two or more solutions, rank them in order of acidity, basicity, pH, $[H^+]$, $[OH^-]$ or pOH. (Page 400)

## 16.7  NONINTEGER pH–[H⁺] AND pOH–[OH⁻] CONVERSIONS (OPTIONAL) (Page 406)

# TERMS AND CONCEPTS

Arrhenius acid-base theory (389)
Brönsted-Lowry acid-base theory (390)
Proton donor; proton receiver (390)
Hydronium ion, $H_3O^+$ (391)
Conjugate acid-base pairs (393)
Conjugate acid; conjugate base (393)
Lewis acid-base theory (394)

Lewis acid; Lewis base (395)
Electron pair donor; electron pair acceptor (395)
Polyprotic; diprotic; triprotic (395)
Water constant, $K_w$ (400)
Acidic, basic and neutral solutions (401)
pH; pOH (402)

# QUESTIONS AND PROBLEMS

*An asterisk (\*) identifies a question that is relatively difficult, or that extends beyond the performance goals of the chapter.*

16.1) Identify the ions traditionally present in solutions called acids and bases. List three compounds that are commonly regarded as acids, and three that are regarded as bases which contain these ions in their water solutions.

16.26) Identify at least two of the classical properties of acids and two of bases.

### Section 16.2

16.2) Distinguish between an Arrhenius acid and a Brönsted-Lowry acid. Are the two concepts in agreement? Justify your answer.

16.27) Distinguish between an Arrhenius base and a Brönsted-Lowry base. Are the two concepts in agreement? Justify your answer.

16.3) What are the conjugate acids of $OH^-$ and $HCO_3^-$? Write the formulas of the conjugate bases of $H_3O^+$ and $HCO_3^-$.

16.28) Give the formula of the conjugate base of HF; of $H_2PO_4^-$. Give the formula of the conjugate acid of $NO_2^-$; of $H_2PO_4^-$.

16.4) For the reaction $HNO_2$ (aq) + $CN^-$ (aq) $\rightleftharpoons$ $NO_2^-$ (aq) + HCN (aq), identify the acid and base on each side of the equation, i.e., the acid and base for the forward reaction and the acid and base for the reverse reaction.

16.29) For the reaction $HSO_4^-$ (aq) + $C_2O_4^{2-}$ (aq) $\rightleftharpoons$ $SO_4^{2-}$ (aq) + $HC_2O_4^-$ (aq), identify the acid and the base on each side of the equation, i.e., the acid and base for the forward reaction and the acid and base for the reverse reaction.

16.5) Identify both conjugate acid-base pairs in the reaction $HSO_4^-$ (aq) + $C_2O_4^{2-}$ (aq) $\rightleftharpoons$ $SO_4^{2-}$ (aq) + $HC_2O_4^-$ (aq).

16.30) For the reaction $HNO_2$ (aq) + $CN^-$ (aq) $\rightleftharpoons$ $NO_2^-$ (aq) + HCN (aq), identify both conjugate acid-base pairs.

16.6) Identify the conjugate acid-base pairs in $H_2PO_4^-$ (aq) + $HCO_3^-$ (aq) $\rightleftharpoons$ $HPO_4^{2-}$ (aq) + $H_2CO_3$ (aq).

16.31) Identify the conjugate acid-base pairs in $NH_4^+$ (aq) + $HPO_4^{2-}$ (aq) $\rightleftharpoons$ $NH_3$ (aq) + $H_2PO_4^-$ (aq).

### Section 16.3

16.7) What structural feature must be present in a compound for it to qualify as a Lewis acid? Lewis base?

16.32) Explain and/or illustrate by an example what is meant by identifying a Lewis acid as an electron pair acceptor, and a Lewis base as an electron pair donor.

16.8) Is the Arrhenius theory of acids and bases consistent with the Lewis acid-base theory? Explain.

16.33) Is the Brönsted-Lowry acid-base theory consistent with the Lewis acid-base theory? Explain.

16.9) Aluminum chloride, $AlCl_3$, behaves much as a covalent compound rather than as an ionic one. This is illustrated in its ability to form a fourth covalent bond with a chloride ion: $AlCl_3$ + $Cl^- \rightarrow AlCl_4^-$. From the Lewis diagram of the aluminum chloride molecule and the electron configuration of the chloride ion, show that this is an acid-base reaction in the Lewis sense, and identify the Lewis acid and the Lewis base.

16.34) Silver ions form covalent bonds with two ammonia molecules to form a strong complex ion, $Ag(NH_3)_2^+$ in a Lewis acid-base reaction. Identify the acid and base, and explain why this qualifies as an acid-base reaction in the Lewis sense.

16.10)\* The reaction between calcium oxide, an ionic compound, and covalent sulfur trioxide can be described by the following equation, which shows to some extent the bonding structure. Explain how this reaction is an acid-base reaction in the Lewis sense.

16.35)\* Diethyl ether reacts with boron trifluoride by forming a covalent bond between the molecules. Describe the reaction from the standpoint of the Lewis acid-base theory, based on the following "structural" equation:

$$Ca^{2+} \left[ \ddot{:}\ddot{O}\ddot{:} \right]^{2-} + \underset{\underset{\ddot{:}\ddot{O}:}{|}}{\overset{\overset{:\ddot{O}:}{|}}{S}}=\ddot{O} \rightarrow Ca^{2+} \left[ \ddot{:}\ddot{O}-\underset{\underset{:\ddot{O}:}{|}}{\overset{\overset{:\ddot{O}:}{|}}{S}}-\ddot{O}\ddot{:} \right]^{2-}$$

$$\underset{\underset{:\ddot{F}:}{|}}{\overset{\overset{:\ddot{F}:}{|}}{:\ddot{F}-B}} + \underset{\underset{C_2H_5}{}}{:\ddot{O}-C_2H_5} \rightarrow \underset{\underset{:\ddot{F}:\quad C_2H_5}{|}}{\overset{\overset{:\ddot{F}:}{|}}{:\ddot{F}-B-\ddot{O}-C_2H_5}}$$

## Section 16.4

*16.11, 16.12, 16.36 and 16.37: Refer to Table 16.1 in answering these questions.*

**16.11)** What is the difference between a strong acid and a weak acid, according to the Brönsted-Lowry concept? Identify two examples of strong acids and two examples of weak acids.

**16.12)** List the following bases in order of their decreasing strength (strongest base first): $SO_4^{2-}$; $Br^-$; $H_2PO_4^-$; $CO_3^{2-}$.

**16.36)** What is the difference between a strong base and a weak base, according to the Brönsted-Lowry concept? Identify two examples of strong bases and two examples of weak bases.

**16.37)** List the following acids in order of their increasing strength (weakest acid first): $HC_2O_4^-$, $H_2SO_3$; $HOH$; $HClO$.

## Section 16.5

*16.13–16.15 and 16.38–16.40: For each acid and base given, complete a proton transfer equation for the transfer of a single proton. Using Table 16.1, predict the direction in which the resulting equilibrium will be favored.*

**16.13)** $HC_7H_5O_2$ (aq) + $SO_4^{2-}$ (aq) $\rightleftharpoons$

**16.14)** $H_2C_2O_4$ (aq) + $NH_3$ (aq) $\rightleftharpoons$

**16.15)** $H_3PO_4$ (aq) + $CN^-$ (aq) $\rightleftharpoons$

**16.38)** $HC_3H_5O_2$ (aq) + $PO_4^{3-}$ (aq) $\rightleftharpoons$

**16.39)** $HSO_4^-$ (aq) + $CO_3^{2-}$ (aq) $\rightleftharpoons$

**16.40)** $H_2CO_3$ (aq) + $NO_3^-$ (aq) $\rightleftharpoons$

## Section 16.6

**16.16)** How is it that we can classify water as a nonconductor of electricity and yet talk about the ionization of water? If it ionizes, why does it not conduct?

**16.17)** Identify the ranges of the pH scale that we classify as strongly acidic, weakly acidic, strongly basic, weakly basic and neutral, or close to neutral. Select any integer from 1 to 14 and explain what is meant by saying that this number is the pH of a certain solution.

**16.41)** Of what significance is the very small value of $10^{-14}$ for $K_w$, the ionization equilibrium constant for water?

**16.42)** What do we mean when we say one solution is acidic, another is neutral, and another is basic?

*16.18–16.21 and 16.43–16.46: For each question the pH, pOH, $[OH^-]$ or $[H^+]$ of a solution is given. Find each of the other values. Also classify each solution as strongly acidic, weakly acidic, neutral (or close to neutral), weakly basic or strongly basic, as these terms are used in Table 16.3, page 404.*

**16.18)** $[H^+] = 10^{-6}$

**16.19)** pOH = 3

**16.20)** pH = 1

**16.21)** $[OH^-] = 10^{-9}$

**16.43)** $[OH^-] = 10^{-5}$

**16.44)** pH = 2

**16.45)** pOH = 4

**16.46)** $[H^+] = 10^{-11}$

## Section 16.7

*16.22–16.25 and 16.47–16.50: For each question the pH, pOH, $[OH^-]$ or $[H^+]$ of a solution is given. Find each of the other values.*

**16.22)*** $[H^+] = 3.4 \times 10^{-9}$

**16.23)*** pOH = 11.82

**16.24)*** $[OH^-] = 0.26$

**16.25)*** pH = 12.05

**16.47)*** $[OH^-] = 5.9 \times 10^{-4}$

**16.48)*** pH = 8.28

**16.49)*** $[H^+] = 7.2 \times 10^{-13}$

**16.50)*** pOH = 7.35

# 17

# OXIDATION-REDUCTION
# (ELECTRON TRANSFER)
# REACTIONS

In Chapter 16 you found that acid-base reactions may be regarded as proton transfer reactions, in the Brönsted-Lowry sense. Now we will study oxidation-reduction reactions. We will see that they are electron transfer reactions, with many parallels to the proton transfers of the previous chapter.

## 17.1 ELECTRON TRANSFER REACTIONS

PG **17 A** Distinguish between oxidation and reduction in terms of electrons gained or lost.

   **17 B** Given an oxidation half-reaction equation and a reduction half-reaction equation, add them to obtain a balanced redox reaction equation.

If we immerse a strip of zinc into a solution of copper sulfate, the zinc becomes coated with a reddish deposit which we can identify as copper metal. The blue color of the solution, which is characteristic of the hydrated $Cu^{2+}$ ion, fades as the reaction proceeds (see Fig. 17.1). These two observations suggest that $Cu^{2+}$ ions in the solution have been converted to copper atoms. Such a conversion is possible only if each $Cu^{2+}$ ion acquires two electrons:

$$Cu^{2+} \text{ (aq)} + 2e^- \rightarrow Cu \text{ (s)}$$

The source of these electrons becomes apparent when we examine what has happened to the zinc strip. If we carefully scrape off the copper deposit, we find that the zinc weighs less than it did originally. Moreover, we can

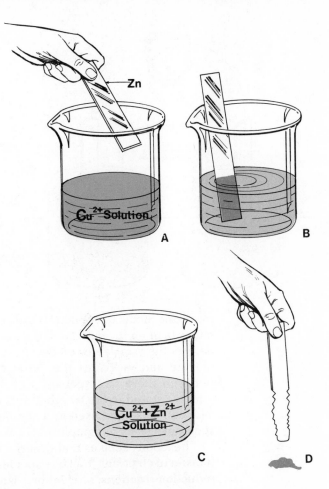

**Figure 17.1** An electron transfer reaction. A, A strip of clean zinc is placed into a solution of $Cu^{2+}$ ion. This ion makes the solution blue. B, The zinc immediately is coated with finely divided copper metal produced when the copper ion gains two electrons to become a copper atom. The blue color becomes lighter as the $Cu^{2+}$ concentration becomes lower. C, After the metal is removed, the solution contains a reduced concentration of blue $Cu^{2+}$ ions plus some colorless $Zn^{2+}$ ions formed as zinc atoms gave up electrons to the copper ions. D, After removing the copper from the zinc, the corroded zinc surface appears.

show by chemical tests that $Zn^{2+}$ ions have entered the solution. In short, zinc atoms have given up electrons, thereby being converted to $Zn^{2+}$ ions:

$$Zn\ (s) \rightarrow Zn^{2+}\ (aq) + 2e^-$$

The net ionic equation for the total reaction that occurs when $Cu^{2+}$ ions come into contact with Zn atoms may be derived by adding the two **half-reaction equations** written above:

$$Cu^{2+}\ (aq) + 2e^- \rightarrow Cu\ (s)$$

$$\underline{Zn\ (s) \rightarrow Zn^{2+}\ (aq) + 2e^-}$$

$$Cu^{2+}\ (aq) + Zn\ (s) \rightarrow Cu\ (s) + Zn^{2+}\ (aq) \qquad (17.1)$$

This reaction could logically be called an **electron transfer reaction.** Electrons have been transferred from zinc atoms to copper(II) ions. Notice that, while no electrons appear in the final equation, the electron transfer character of the reaction is quite evident in the individual half-reactions. Notice also that the number of electrons lost by one species is exactly equal to the number of electrons gained by the other.

There are many different ways in which this reaction can be carried out. One of these is indicated in Figure 17.2. Here the electron transfer

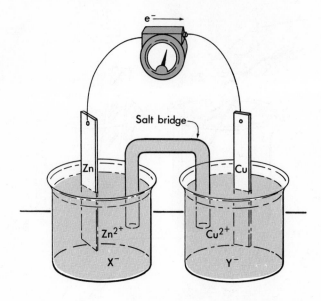

**Figure 17.2** The $Zn$–$Cu^{2+}$ reaction as part of an electrolytic circuit.

from zinc atoms to copper(II) ions takes place through an electrical circuit rather than on the surface of the zinc. The voltmeter in the circuit not only detects the passage of electrons, but also confirms the direction they are moving and measures the "strength" of the electron transfer tendency between these two species. In this so-called *voltaic cell* the "chemical energy" stored in the two reactants is converted to useful electrical energy. The cell was once used as a stationary source of direct current to operate such devices as telegraph relays and doorbells.

Reactions such as that described in Equation 17.1, in which there is a transfer of electrons from one species to another, are referred to as **oxidation-reduction** reactions, often abbreviated to "**redox**" reactions. Any such reaction can always be considered to be the sum of two **half-reactions: a reduction half-reaction in which a species gains electrons** ($Cu^{2+} + 2e^- \rightarrow Cu$), and an **oxidation half-reaction, in which another species loses electrons** ($Zn - 2e^- \rightarrow Zn^{2+}$, or as it is usually written, $Zn \rightarrow Zn^{2+} + 2e^-$). We can apply this analysis to several well known reactions taking place in aqueous solution:

1) The evolution of hydrogen gas on adding zinc to sulfuric acid (the "acid property" that is a redox reaction, mentioned on page 390):

Reduction: $\qquad 2\,H^+\,(aq) + 2e^- \rightarrow H_2\,(g)$

Oxidation: $\qquad \dfrac{Zn\,(s) \rightarrow Zn^{2+}\,(aq) + 2e^-}{}$

Redox: $\qquad 2\,H^+\,(aq) + Zn\,(s) \rightarrow H_2\,(g) + Zn^{2+}\,(aq) \qquad (17.2)$

2) The preparation of bromine by bubbling chlorine gas through a solution of NaBr:

Reduction: $\qquad Cl_2\,(g) + 2e^- \rightarrow 2\,Cl^-\,(aq)$

Oxidation: $\qquad \dfrac{2\,Br^-\,(aq) \rightarrow Br_2\,(l) + 2e^-}{}$

Redox: $\qquad Cl_2\,(g) + 2\,Br^-\,(aq) \rightarrow 2\,Cl^-\,(aq) + Br_2\,(l) \qquad (17.3)$

3) The formation of a "chemical pine tree" (needles of silver—see Fig. 7.6, page 156) by immersing a copper wire into silver nitrate solution:

| | |
|---|---|
| Reduction: | $2\,Ag^+\,(aq) + \cancel{2e^-} \rightarrow 2\,Ag\,(s)$ |
| Oxidation: | $\underline{\hspace{3em} Cu\,(s) \rightarrow Cu^{2+}\,(aq) + \cancel{2e^-} \hspace{3em}}$ |
| Redox: | $2\,Ag^+\,(aq) + Cu\,(s) \rightarrow 2\,Ag\,(s) + Cu^{2+}\,(aq)$     (17.4) |

The development of Equation 17.4 warrants special comment. The usual reduction equation for silver ion is $Ag^+\,(aq) + e^- \rightarrow Ag\,(s)$. Because two moles of electrons are lost in the oxidation reaction, *two moles of electrons must be gained in the reduction reaction.* As has already been noted, the number of electrons lost by one species must equal the number gained by the other species. It is therefore necessary to multiply the usual $Ag^+$ reduction equation by 2 to bring about this equality in electrons gained and lost, and yield their cancellation in the summed-up equations.

---

**Example 17.1** Combine the following half-reactions to produce a balanced redox reaction equation. Indicate which half-reaction is an oxidation reaction, and which is a reduction.

$$Co^{2+}\,(aq) + 2e^- \rightarrow Co\,(s);\ Sn\,(s) \rightarrow Sn^{2+}\,(aq) + 2e^-$$

-----------------------------------------------------------------------------------

**17.1a**

| | |
|---|---|
| Reduction: | $Co^{2+}\,(aq) + \cancel{2e^-} \rightarrow Co\,(s)$ |
| Oxidation: | $\underline{\hspace{3em} Sn\,(s) \rightarrow Sn^{2+}\,(aq) + \cancel{2e^-} \hspace{3em}}$ |
| Redox: | $Co^{2+}\,(aq) + Sn\,(s) \rightarrow Co\,(s) + Sn^{2+}\,(aq)$ |

---

**Example 17.2** Combine the following half-reactions to produce a balanced redox equation. Identify the oxidation half-reaction and reduction half-reaction.

$$Fe^{2+}\,(aq) \rightarrow Fe^{3+}\,(aq) + e^-;\ Al^{3+}\,(aq) + 3e^- \rightarrow Al\,(s)$$

-----------------------------------------------------------------------------------

**17.2a**

| | |
|---|---|
| Reduction: | $Al^{3+}\,(aq) + \cancel{3e^-} \rightarrow Al\,(s)$ |
| Oxidation: | $\underline{\hspace{3em} 3\,Fe^{2+}\,(aq) \rightarrow 3\,Fe^{3+}\,(aq) + \cancel{3e^-} \hspace{3em}}$ |
| Redox: | $Al^{3+}\,(aq) + 3\,Fe^{2+}\,(aq) \rightarrow Al\,(s) + 3\,Fe^{3+}\,(aq)$ |

In this example it is necessary to multiply the oxidation half-reaction equation by 3 in order to balance the electrons gained and lost.

---

Another reaction involving iron and aluminum introduces an additional technique . . . .

**Example 17.3** Arrange and modify as necessary the following half-reactions so they add up to produce a balanced redox equation. Identify the oxidation half-reaction and the reduction half-reaction.

$$Fe^{2+} (aq) + 2e^- \rightarrow Fe (s); \quad Al (s) \rightarrow Al^{3+} (aq) + 3e^-$$

This will extend you a bit when it comes to balancing electrons. If you think about how you would work out the formula for aluminum oxide from $Al^{3+}$ and $O^{2-}$ ions, and apply similar reasoning to this example, you will come up with the answer. Try it.

-----------------------------------------------------------------------

**17.3a** Reduction: $\quad 3 Fe^{2+} (aq) + \cancel{6e^-} \rightarrow 3 Fe (s)$

Oxidation: $\quad \underline{\qquad 2 Al (s) \rightarrow 2 Al^{3+} + \cancel{6e^-}}$

Redox: $\quad 3 Fe^{2+} (aq) + 2 Al (s) \rightarrow 3 Fe (s) + 2 Al^{3+} (aq)$

In this example electrons are transferred two at a time in the iron half-reaction and three at a time in the aluminum half-reaction. The simplest way to equate these is to take the iron half-reaction three times and the aluminum half-reaction twice. This gives six electrons for both half-reactions—just as two $Al^{3+}$ and three $O^{2-}$ balance the positive and negative charges in the ions making up the formula of $Al_2O_3$.

## 17.2 OXIDATION NUMBERS AND REDOX REACTIONS

PG 17 C Given the formula of a chemical species, determine the oxidation number of each element it contains.

17 D Distinguish between oxidation and reduction in terms of oxidation number change.

The redox reactions that we have discussed up to this point have been rather simple ones involving only two reactants. With Equations 17.1 to 17.4 we can see at a glance which species has gained and which has lost electrons. Some relatively common oxidation-reduction reactions are not so readily analyzed. Consider, for example, a reaction that is sometimes used in the general chemistry laboratory to prepare chlorine gas from hydrochloric acid:

$$MnO_2 (s) + 4 H^+ (aq) + 2 Cl^- (aq) \rightarrow Mn^{2+} (aq) + Cl_2 (g) + 2 H_2O (l) \quad (17.5)$$

or the reaction, taking place in a lead storage battery, that produces the electrical spark to start an automobile:

$$Pb (s) + PbO_2 (s) + 4 H^+ (aq) + 2 SO_4^{2-} (aq) \rightarrow 2 PbSO_4 (s) + 2 H_2O (l) \quad (17.6)$$

Looking at these equations, it is by no means obvious which species are gaining and which are losing electrons.

"Electron bookkeeping" in redox reactions like Equations 17.5 and 17.6 is greatly simplified by the concept of **oxidation number,** which was introduced in Section 10.1, page 219. As we saw at that point, using a rather arbitrary set of rules, oxidation numbers may be assigned to each element in a molecule or ion. The rules for assigning oxidation numbers are repeated here for your convenience:

1. **The oxidation number of any elemental substance is 0.** For example, the oxidation number of sodium in Na, of chlorine in $Cl_2$, or of phosphorus in $P_4$, is zero.

2. **The oxidation number of an element as a monatomic ion is equal to the charge of that ion.** The oxidation number of chlorine in $Cl^-$ is $-1$; that of manganese in the $Mn^{2+}$ ion is $+2$.

3. **The oxidation number of combined oxygen is $-2$** except in peroxides $(-1)$ and superoxides $(-\frac{1}{2})$. In water, $H_2O$; sulfuric acid, $H_2SO_4$; sodium hydroxide, NaOH; nitrate ion, $NO_3^-$; hypochlorite ion, $ClO^-$; and nearly all other species, the oxidation state of oxygen is $-2$. We will not encounter peroxides or superoxides in this text.

4. **The oxidation number of combined hydrogen is $+1$,** except in hydrides $(-1)$. In water, sulfuric acid, sodium hydroxide, ammonium ion, $NH_4^+$, and nearly all other species, the oxidation state of hydrogen is $+1$. We will have no further encounter with hydrides in this text.

5. **In any molecular or ionic species, the sum of the oxidation numbers of all elements in the species is equal to the charge on the species.** With this rule we are able to deduce the oxidation numbers of elements not fixed by the first four rules, as the following examples illustrate.

---

**Example 17.4**   What is the oxidation number of manganese in $MnO_2$?

Oxidation rules 1, 2 and 5 do not apply. Rule 5 does not give us a direct answer, but it tells us the *sum* of all oxidation numbers in the species. It is . . . .

---

**17.4a**   Zero. $MnO_2$ is a compound and therefore has zero charge.

We now see that the oxidation number contributed by oxygen plus that from manganese is zero. If we could find the amount from oxygen . . . . Rule three helps. Oxygen contributes how much to the *total?* (Careful!)

---

**17.4b**   $-4$. Each oxygen atom contributes $-2$, so the total from oxygen is $2(-2) = -4$.

The oxidation state of manganese should now be apparent.

---

**17.4c** +4

The total is zero, and oxygen is −4. Therefore −4 + oxidation state of Mn = 0. Consequently the oxidation state of Mn in $MnO_2$ is +4.

---

**Example 17.5**  Find the oxidation state of (a) sulfur in $SO_4^{2-}$; (b) Cr in $HCr_2O_7^-$; (c) Fe in $Fe_2(SO_4)_3$.

---

**17.5a**  (a) +6; (b) +6; (c) +3

(a) $SO_4^{2-}$. Oxygen contributes 4(−2) = −8. The ion has a charge of −2. Therefore −8 + ox. no. of S = −2. Ox. no. of S = +6.

(b) $HCr_2O_7^-$. Hydrogen contributes +1 to the total, by rule 4. Oxygen contributes 7(−2) = −14. The total is the charge on the ion, −1. Therefore +1 + 2 (ox. no. of Cr) − 14 = −1. Solving, the oxidation number of Cr is +6.

(c) $Fe_2(SO_4)_3$. The charge on the sulfate ion is −2, and there are three of them. The sulfate ion contributes 3(−2) = −6 to the total oxidation number of the compound, which, by rule 5, is zero. Two iron ions must contribute a total of +6 to that zero total: 2 (ox. no. of Fe) − 6 = 0, or +3 for each iron(III) ion.

---

Let's glance back to Equations 17.2 and 17.3 and to the redox reaction in Example 17.2 to see what happens to oxidation numbers during a redox reaction. Looking first at the oxidation half-reaction from Equation 17.2, $Zn \rightarrow Zn^{2+} + 2e^-$, the oxidation number of Zn is 0, by rule 1; the oxidation number of $Zn^{2+}$ is +2, by rule 2. The oxidation state of zinc *increased* from 0 to +2 as it was *oxidized*. In Equation 17.3, bromine was oxidized from −1 to 0, again an *increase* during *oxidation*. The pattern is complete in Example 17.2: the *oxidation* of $Fe^{2+}$ to $Fe^{3+}$ represents an *increase* from +2 to +3 in oxidation number. If we examined an endless number of *oxidation* reactions, we would find that in each case there is an *increase* in oxidation number of the element oxidized.

At this point you probably suspect that reduction is accompanied by a reduction in oxidation number. So it is. From the same reactions,

Eq. 17.2: $H^+ \rightarrow H_2$, a reduction from +1 to 0

Eq. 17.3: $Cl_2 \rightarrow Cl^-$, a reduction from 0 to −1

Ex. 17.2: $Al^{3+} \rightarrow Al$, a reduction from +3 to 0

We are now ready to state a second and broader definition of oxidation and reduction: **Oxidation is an increase in oxidation number; reduction is a reduction in oxidation number.** These definitions are more useful in identifying the elements oxidized and reduced in a redox reaction when electron transfer is not apparent. For example, in Equation 17.5,

$$MnO_2 \text{ (s)} + 4\ H^+ \text{ (aq)} + 2\ Cl^- \text{ (aq)} \rightarrow Mn^{2+} \text{ (aq)} + Cl_2 \text{ (g)} + 2\ H_2O \text{ (l)}$$

we can determine the oxidation state of each element in the equation and identify the elements undergoing a change. Oxygen is, by rule 3, in the $-2$ oxidation state both places it appears, and hydrogen is at $+1$, according to rule 4. Chlorine is in elemental form on the right (ox. no. 0 by rule 1), and is a monatomic ion on the left (ox. no. $-1$ by rule 2). The change is from $-1$ to 0, an *increase*. Chlorine therefore has been *oxidized*. We have seen that Mn is at a $+4$ oxidation state in $MnO_2$, as it is on the left. On the right we have a monatomic ion at a $+2$ oxidation state. Mn has *decreased* from $+4$ to $+2$, a *reduction* reaction.

You will note that in this reaction the oxidation numbers of hydrogen and oxygen remain constant at $+1$ and $-2$ respectively. However, $H^+$ ions are by no means spectators; they play a vital role in this reaction by combining with oxygen atoms from $MnO_2$ to form water molecules. Their importance can be demonstrated by attempting to bring about the reaction without them; if we substitute a water solution of sodium chloride for hydrochloric acid, no reaction occurs.

---

**Example 17.6** Determine the element oxidized and the element reduced in a lead storage battery, Equation 17.6:

$$Pb\ (s) + PbO_2\ (s) + 4\ H^+\ (aq) + 2\ SO_4{}^{2-}\ (aq) \rightarrow 2\ PbSO_4\ (s) + 2\ H_2O\ (l)$$

Assign oxidation numbers to as many elements as necessary until you come up with the pair that changed. Then identify the oxidation and reduction changes. (Careful. This one is a bit tricky.)

-----------------------------------------------------------------------------------------

**17.6a** Lead is both oxidized (0 in Pb to $+2$ in $PbSO_4$) and reduced ($+4$ in $PbO_2$ to $+2$ in $PbSO_4$).

---

While oxidation number is a very useful device to keep track of what the electrons are up to in a redox reaction, we should emphasize that it is a man-made concept which has no real physical basis. Unlike the charge of a monatomic ion, the oxidation number of an atom in a molecule or polyatomic ion cannot be measured in the laboratory. It is all very well to talk about "$+4$ manganese" in $MnO_2$ or "$+6$ sulfur" in the $SO_4{}^{2-}$ ion, but we must not fall into the trap of thinking that the elements in these species actually carry positive charges equal to their oxidation numbers.

## 17.3 OXIDIZING AGENTS (OXIDIZERS); REDUCING AGENTS (REDUCERS)

PG 17 E Distinguish between an oxidizing agent and a reducing agent.

The two essential reactants in a redox reaction are given special names to indicate the role they play. The species that brings about oxidation by accepting electrons is referred to as an **oxidizing agent,** or **oxidizer;** the species that donates the electrons so reduction can occur is called a **reducing agent,** or **reducer.** For example, in Equation 17.2,

$$2 \text{ H}^+ \text{ (aq)} + \text{Zn (s)} \rightarrow \text{H}_2 \text{ (g)} + \text{Zn}^{2+} \text{ (aq)}$$

$\text{H}^+$ has accepted electrons from Zn—it has *oxidized* Zn to $\text{Zn}^{2+}$—and is therefore the oxidizing agent. Conversely, Zn has donated electrons to $\text{H}^+$ —it has reduced $\text{H}^+$ to $\text{H}_2$—and is therefore the reducing agent.

The following example summarizes the redox concepts:

---

**Example 17.7**   Consider the redox equation

$$5 \text{ NO}_3^- \text{ (aq)} + 3 \text{ As (s)} + 2 \text{ H}_2\text{O (l)} \rightarrow 5 \text{ NO (g)} + 3 \text{ AsO}_4^{3-} \text{ (aq)} + 4 \text{ H}^+ \text{ (aq)}$$

(a) Determine the oxidation number in each species:
   N:   _____ in $\text{NO}_3^-$, and _____ in NO
   O:   _____ in $\text{NO}_3^-$, _____ in $\text{H}_2\text{O}$, _____ in NO, and _____ in $\text{AsO}_4^{3-}$
   As:  _____ in As, and _____ in $\text{AsO}_4^{3-}$
   H:   _____ in $\text{H}_2\text{O}$, and _____ in $\text{H}^+$.
(b) Identify (1) the element oxidized; (2) the element reduced; (3) the oxidizing agent; (4) the reducing agent.

---

**17.7a**   (a)   N:   +5 in $\text{NO}_3^-$, and +2 in NO
            O:   −2 in all species
            As:  0 in As, and +5 in $\text{AsO}_4^{3-}$
            H:   +1 in both $\text{H}_2\text{O}$ and $\text{H}^+$.
(b) (1) The As is oxidized, increasing in oxidation number from 0 to +5.
    (2) N is reduced, decreasing in oxidation number from +5 to +2.
    (3) $\text{NO}_3^-$ is the oxidizing agent, removing electrons from As.
    (4) The As is the reducing agent, furnishing electrons to $\text{NO}_3^-$.

## 17.4  REDOX REACTIONS AND ACID-BASE REACTIONS COMPARED

PG  17 F   Identify the essential difference between a redox reaction and an acid-base reaction.

    17 G   Distinguish between a strong oxidizing agent and a weak oxidizing agent; between a strong reducing agent and a weak reducing agent.

At this stage in our study of oxidation-reduction reactions, it may be useful to pause briefly and point out how redox reactions resemble acid-base reactions.

(1) Acid-base reactions involve a transfer of protons, redox reactions a transfer of electrons.

(2) In both cases, the reactants are given special names to indicate their role in the transfer process. An acid is a proton donor; a base is a proton

acceptor. A reducing agent is an electron donor; an oxidizing agent is an electron acceptor.

(3) Just as certain species (e.g., $HCO_3^-$, $H_2O$) can either donate or accept protons and thereby behave as an acid in one reaction and a base in another, so we observe that certain species can either accept or donate electrons, acting as an oxidizing agent in one reaction and a reducing agent in another. An example is the $Fe^{2+}$ ion, which can oxidize Zn atoms to $Zn^{2+}$ ions in one reaction:

$$Fe^{2+} \text{ (aq)} + Zn \text{ (s)} \rightarrow Fe \text{ (s)} + Zn^{2+} \text{ (aq)}$$

and reduce $Cl_2$ molecules to $Cl^-$ ions in another:

$$Cl_2 \text{ (g)} + 2 \ Fe^{2+} \text{ (aq)} \rightarrow 2 \ Cl^- \text{ (aq)} + 2 \ Fe^{3+} \text{ (aq)}$$

(4) Just as we can classify acids and bases as "strong" or "weak" depending upon how readily they donate or accept protons, we use the same adjectives to compare the relative strengths of different oxidizing and reducing agents. A "strong" oxidizing agent is one that has a strong attraction for electrons; a "strong" reducing agent gives up electrons readily.

Let's look more closely into the matter of the strength of oxidizing and reducing agents . . . .

## 17.5 STRENGTHS OF OXIDIZING AGENTS AND REDUCING AGENTS

PG 17 H Given a table of relative strengths of oxidizing and reducing agents, identify the stronger and weaker of two oxidizing or two reducing agents.

At the end of Section 16.6, page 406, we pointed out that the pH meter (Fig. 16.4), with proper selection of electrodes, is capable of measurements other than the concentration of hydrogen ions. One of the principal functions of the instrument is to measure electrical potential, or voltage, in a redox reaction. The instrument is, in fact, the voltmeter that appears in Figure 17.2, by means of which the relative strength of the oxidizing and reducing agents may be measured. These data may then be summarized in a table of the relative strengths of oxidizing and reducing agents.

Table 17.1 is a list of oxidizing agents in order of decreasing strength on the left side of the equation, and of reducing agents in order of increasing strength on the right. (Compare to Table 16.1, page 396, which lists acids in order of decreasing strength on the left and bases in order of increasing strength on the right.) The strongest oxidizing agent in the table is the $F_2$ molecule, located at the upper left. Indeed, fluorine is such a powerful oxidizing agent that it is unstable in water solution; it oxidizes water molecules to liberate oxygen.

The two species listed directly below fluorine in the left column of Table 17.1 are quite powerful oxidizing agents. The use of chlorine as a disinfectant in water supplies is due in large measure to its ability to oxidize

TABLE 17.1   Relative Strengths of Oxidizing and Reducing Agents

| | OXIDIZING AGENT | REDUCING AGENT | |
|---|---|---|---|
| ↑ Increasing ——— STRENGTH ——— Decreasing ↓ | $Fe_2 (g) + 2e^-$ | $\rightarrow 2\ F^-$ | ↑ Decreasing ——— STRENGTH ——— Increasing ↓ |
| | $Cl_2 (g) + 2e^-$ | $\rightarrow 2\ Cl^-$ | |
| | $\frac{1}{2}\ O_2 (g) + 2\ H^+ + 2e^-$ | $\rightarrow H_2O$ | |
| | $Br_2 (l) + 2e^-$ | $\rightarrow 2\ Br^-$ | |
| | $NO_3^- + 4\ H^+ + 3e^-$ | $\rightarrow NO\ (g) + 2\ H_2O$ | |
| | $Ag^+ + e^-$ | $\rightarrow Ag\ (s)$ | |
| | $Fe^{3+} + e^-$ | $\rightarrow Fe^{2+}$ | |
| | $I_2 (s) + 2e^-$ | $\rightarrow 2\ I^-$ | |
| | $Cu^{2+} + 2e^-$ | $\rightarrow Cu\ (s)$ | |
| | $2\ H^+ + 2e^-$ | $\rightarrow H_2\ (g)$ | |
| | $Ni^{2+} + 2e^-$ | $\rightarrow Ni\ (s)$ | |
| | $Co^{2+} + 2e^-$ | $\rightarrow Co\ (s)$ | |
| | $Cd^{2+} + 2e^-$ | $\rightarrow Cd\ (s)$ | |
| | $Fe^{2+} + 2e^-$ | $\rightarrow Fe\ (s)$ | |
| | $Zn^{2+} + 2e^-$ | $\rightarrow Zn\ (s)$ | |
| | $Al^{3+} + 3e^-$ | $\rightarrow Al\ (s)$ | |
| | $Na^+ + e^-$ | $\rightarrow Na\ (s)$ | |
| | $Ca^{2+} + 2e^-$ | $\rightarrow Ca\ (s)$ | |
| | $Li^+ + e^-$ | $\rightarrow Li\ (s)$ | |

harmful organic matter. All of us are familiar with one redox reaction in which elementary oxygen participates: the corrosion (rusting) of iron and steel, in which elementary iron is oxidized to the +3 state.

The strongest reducing agents are found at the bottom of the right column. Their reducing strength (ability to donate electrons) decreases steadily as we move up the column. The three metals listed at the bottom (and, indeed, all the Group 1A and 2A elements except magnesium and beryllium) are such powerful reducing agents that they are unstable in water solution. In reaction they reduce $H_2O$ molecules to form elementary hydrogen.

The metals between Na and $H_2$ in the right column of Table 17.1 are stable in pure water; they are, however, strong enough reducing agents to liberate hydrogen gas from solutions of strong acids. The reducing agents above $H_2$ in Table 17.1 are all relatively weak. In particular, the fluoride ion, $F^-$, at the top of the right column holds on to its electrons so tightly that it cannot be oxidized in water solution.

## 17.6   PREDICTING REDOX REACTIONS

PG   17 I   Given a table of relative strengths of oxidizing and reducing agents and the identity of an oxidizer and a reducer from the table, (a) write the net ionic equation for the redox reaction between them, and (b) predict whether the forward or reverse reaction will be favored at equilibrium.

As Table 16.1, page 396, enables us to write acid-base reaction equations and predict the direction that will be favored at equilibrium, Table 17.1

enables us to do the same things for redox reactions. The redox table has a limitation, however. Acid-base reactions are all *single* proton transfer reactions and automatically balanced if taken directly from the table. Redox half-reactions, on the other hand, may, and frequently do, involve unequal numbers of electrons. They must be balanced as in Equation 17.4 and Example 17.2. Examples 17.8 and 17.9 illustrate the process.

---

**Example 17.8**  Write the net ionic equation for the redox reaction between the cobalt(II) ion, $Co^{2+}$, and metallic silver, Ag.

*Solution.*  First, as in a Brönsted-Lowry acid-base reaction there must be a proton giver and a proton taker, so in a redox reaction there must be an electron giver (reducer) and an electron taker (oxidizer). Consulting Table 17.1, we find $Co^{2+}$ among the oxidizers and Ag among the reducers. The reduction half-reaction is

$$\text{Reduction:} \qquad Co^{2+} \text{ (aq)} + 2e^- \rightarrow Co \text{ (s)}$$

To obtain the oxidation half-reaction for silver it is necessary to *reverse* the reduction half-reaction found in the table:

$$\text{Oxidation:} \qquad Ag \text{ (s)} \rightarrow Ag^+ \text{ (aq)} + e^-$$

Multiplying the oxidation equation by 2 to equalize electrons gained and lost, and adding to the reduction equation yields

| Reduction: | $Co^{2+} \text{ (aq)} + 2e^- \rightarrow Co \text{ (s)}$ |
|---|---|
| 2 × Oxidation: | $2\,Ag \text{ (s)} \rightarrow 2\,Ag^+ \text{ (aq)} + 2e^-$ |
| Redox: | $Co^{2+} \text{ (aq)} + 2\,Ag \text{ (s)} \rightarrow Co \text{ (s)} + 2\,Ag^+ \text{ (aq)}$ |

---

The principle underlying the prediction of the favored direction of a redox reaction is the same as for an acid-base reaction. The stronger oxidizing agent—strong by virtue of its attraction for electrons—will take the electrons from a strong reducing agent—strong by virtue of its tendency to donate electrons—to produce the weaker oxidizer and reducer. This is shown diagrammatically in Figure 17.3, which is strikingly similar to Figure

**Figure 17.3**  Correlation between prediction of a redox reaction and position in Table 17.1. The stronger oxidizing agent (upper left) takes electrons from the stronger reducing agent (lower right), yielding the weaker reducing agent (upper right) and weaker oxidizing agent (lower left). That direction, forward or reverse, will be favored that has the weaker reducer and oxidizer as products. The similarity between acid-base and redox reaction predictions may be seen by comparing this illustration with Figure 16.3 (page 405).

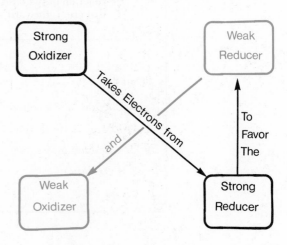

16.1, page 398. In the reaction $2 H^+ (aq) + Zn (s) \rightarrow H_2 (g) + Zn^{2+} (aq)$ (Equation 17.2), the positions in the table establish $H^+$ and $Zn$ as the stronger oxidizer and reducer. The reaction is favored in the forward direction, yielding the weaker oxidizer and reducer, $Zn^{2+}$ and $H_2$. (Note: We will use double arrows when necessary to indicate the equilibrium character of redox reactions.)

---

**Example 17.9**  In which direction, forward or reverse, will the reaction in Example 17.8 be favored?

---

**17.9a**  Reverse

$Ag^+$ is a stronger oxidizer than $Co^{2+}$, and is therefore capable of taking electrons from cobalt atoms. Also, cobalt atoms are a stronger reducer than silver atoms, and therefore readily release electrons to $Ag^+$. The weaker reducer and oxidizer, $Ag$ and $Co^{2+}$, are favored.

---

**Example 17.10**  Write the redox reaction equation between metallic copper and a strong acid, $H^+$, and indicate the direction that is favored.

---

**17.10a**  $2 H^+ (aq) + Cu (s) \rightleftarrows H_2 (g) + Cu^{2+} (aq)$

Reduction:  $2 H^+ (aq) + \cancel{2e^-} \rightleftarrows H_2 (g)$

Oxidation:  $Cu (s) \rightleftarrows Cu^{2+} (aq) + \cancel{2e^-}$

Redox:  $2 H^+ (aq) + Cu (s) \rightleftarrows H_2 (g) + Cu^{2+} (aq)$

The reverse reaction is favored.

---

**Example 17.11**  Write the net ionic equation for the redox reaction between $Al (s)$ and $Ni^{2+} (aq)$, and predict the favored direction, forward or reverse.

---

**17.11a**  $3 Ni^{2+} (aq) + 2 Al (s) \rightleftarrows 3 Ni (s) + 2 Al^{3+} (aq)$. Forward reaction favored.

Reduction half-reaction:  $Ni^{2+} (aq) + 2e^- \rightleftarrows Ni (s)$

Oxidation half-reaction:  $Al (s) \rightleftarrows Al^{3+} (aq) + 3e^-$

$3 \times$ Reduction:  $3 Ni^{2+} (aq) + \cancel{6e^-} \rightleftarrows 3 Ni (s)$

$2 \times$ Oxidation:  $2 Al (s) \rightleftarrows 2 Al^{3+} + \cancel{6e^-}$

$3 Ni^{2+} (aq) + 2 Al (s) \rightleftarrows 3 Ni (s) + 2 Al^{3+}$

One of the properties of acids listed on page 389 is their ability to release hydrogen gas on reaction with "active" metals. Apparently, judging from the result of Example 17.10, copper does not qualify as an active metal. Those metals loosely classified as *active* are the reducers located below hydrogen in Table 17.1. The next example adds further perspective to the reactions between metals and acids . . . .

---

**Example 17.12**   Write the equation for the reaction between copper and nitric acid, and predict which direction is favored.

Copper is in our table of oxidizing and reducing agents, but you will search in vain for $HNO_3$. The solution inventory species of nitric acid ($NO_3^-$ and $H^+$) are present, though. We shall comment on the imbalance between hydrogen ions and nitrate ions shortly. This reaction summarizes our equation writing methods to this point. Take it all the way.

------------------------------------------------------------------------

**17.12a**   $2\,NO_3^-\,(aq) + 8\,H^+\,(aq) + 3\,Cu\,(s) \rightleftharpoons 2\,NO\,(g) + 4\,H_2O\,(l) + 3\,Cu^{2+}\,(aq)$

Forward reaction favored.

$2 \times$ Reduction:       $2\,NO_3^-\,(aq) + 8\,H^+\,(aq) + \cancel{6e^-} \rightleftharpoons 2\,NO\,(g) + 4\,H_2O\,(l)$

$3 \times$ Oxidation:       $\underline{\qquad\qquad\qquad 3\,Cu\,(s) \rightleftharpoons 3\,Cu^{2+}\,(aq) + \cancel{6e^-}\qquad}$

Redox:       $2\,NO_3^-\,(aq) + 8\,H^+\,(aq) + 3\,Cu\,(s) \rightleftharpoons 2\,NO\,(g) + 4\,H_2O\,(l)$
$+ 3\,Cu^{2+}\,(aq)$

Don't worry about those missing nitrate ions, the six unaccounted for from the 8 moles of $HNO_3$ that furnished the 8 $H^+$. They're in there as spectators, just enough to balance the 3 $Cu^{2+}$.

---

Before leaving this series of examples, let's note that any strong acid is theoretically capable of oxidizing an active metal, one below hydrogen in Table 17.1, while it won't touch copper. Copper is attacked by nitric acid, however—not by the hydrogen ion, but by the nitrate ion, which is an even stronger oxidizer. In actual practice we find that some predictions based on Table 17.1 must be modified because of other redox possibilities not included in the table, the concentrations of reaction species and reaction conditions.

## 17.7   WRITING REDOX EQUATIONS

PG   17 J   Given the identity of an oxidizer and reducer, and the identity of their reduced and oxidized products in an acidic solution, write the net ionic equation for the reaction.

Thus far we have considered only redox reactions for which we know the oxidation and reduction half-reactions. We are not always this fortunate. Sometimes we know only the reactants and products. Considering nitric acid, for example, suppose we knew only that the product of the reduction of nitric acid is NO (g). How do we get from this information to the reduction half-reaction given in Table 17.1 so we can write an equation such as the one in the last example?

Let's develop our method using the reduction of $NO_3^-$ to $NO$ in an acidic aqueous solution. We would begin our reduction equation by writing

$$NO_3^- \text{ (aq)} \rightarrow NO \text{ (g)}$$

and then go about balancing the equation. We find nitrogen in balance, but a deficiency of two oxygen atoms on the right. Remembering the reaction takes place in an acidic aqueous solution, we realize there is an abundance of hydrogen ions at our disposal. The $H^+$ ions can form water with the oxygen released by the $NO_3^-$ ion. We therefore **balance oxygen by adding one water molecule for each oxygen atom required,** two in this case.

$$NO_3^- \text{ (aq)} \rightarrow NO \text{ (g)} + 2 \text{ } H_2O \text{ (l)}$$

We turn next to hydrogen: **hydrogen is balanced with $H^+$ ions.** In this case four $H^+$ ions are required to make two water molecules.

$$4 \text{ } H^+ \text{ (aq)} + NO_3^- \text{ (aq)} \rightarrow NO \text{ (g)} + 2 \text{ } H_2O \text{ (l)}$$

Hydrogen, nitrogen and oxygen are now all in balance—but the equation is not. We've encountered quite a few net ionic equations by now, and without any special effort or attention on our part, the total charge of all species on one side of the equation has usually been equal to the total charge of all species on the other side. The charges have not always been zero, but they have always been equal, and therefore balanced. But this is not true in the equation above. Checking charge, we find a net charge of +3 on the left ($+4 - 1$), and zero on the right. **Charge is balanced by adding the required number of electrons to the side that has the more positive charge.** In this case it is the left side:

$$4 \text{ } H^+ \text{ (aq)} + NO_3^- \text{ (aq)} + 3e^- \rightarrow NO \text{ (g)} + 2 \text{ } H_2O \text{ (l)}$$

We have now developed the reduction equation in Table 17.1. Oxidation equations are written in exactly the same way.

When the reactants and products of a redox reaction are known we may write a balanced redox equation by following these steps:

1. Identify the element reduced and the element oxidized. For each, write a partial half-reaction equation, using the element in its original form (element, monatomic ion, or part of a polyatomic ion or compound) on the left, and in its final form on the right.
2. Balance each half-reaction equation separately.
   a. First balance the element oxidized or reduced.
   b. Balance elements other than hydrogen and oxygen, if any.
   c. Balance oxygen by adding water molecules where necessary.
   d. Balance hydrogen by adding hydrogen ions, $H^+$, where necessary.
   e. Balance charges by adding electrons where necessary.

3. Equalize the electrons gained or lost by appropriate multiplication of the half-reaction equations.
4. Add the half-reaction equations to produce the net ionic equation.

Steps 3 and 4 are, of course, what we have done with half-reaction equations in the foregoing sections.

---

**Example 17.13** Elemental iodine and sulfur are the products of a redox reaction between iodide and sulfate ions in an acidic aqueous solution:

$$I^- (aq) + SO_4^{2-} (aq) \rightarrow I_2(s) + S(s)$$

Write the net ionic equation for the reaction.

Identify the element reduced, and the element oxidized.

---

**17.13a** Sulfur is reduced (ox. no. change +6 to 0) and iodine oxidized ($-1$ to 0).

The oxidation reaction is the simplest. Write its partial half-reaction equation.

---

**17.13b** $I^- (aq) \rightarrow I_2 (s)$

Balance atoms first, and then charges in a single step.

---

**17.13c** $2 I^- (aq) \rightarrow I_2 (s) + 2e^-$. (This one happens to be in Table 17.1.)

Now for the reduction reaction. Write the partial half-reaction equation.

---

**17.13d** $SO_4^{2-} (aq) \rightarrow S (s)$

To balance we observe that the element reduced is already in balance, and the only other element is oxygen. According to step 2c, oxygen is balanced by adding the necessary water molecules. Complete that step.

---

**17.13e** $SO_4^{2-} (aq) \rightarrow S (s) + 4 H_2O (l)$. Four oxygen atoms in a sulfate ion require four water molecules.

Next comes the hydrogen balancing, using $H^+$ ions.

---

**17.13f** $8 H^+ (aq) + SO_4^{2-} (aq) \rightarrow S (s) + 4 H_2O (l)$

And last, add to the positive side the electrons that will bring the charges into balance.

---

**17.13g**  $8 H^+ (aq) + SO_4^{2-} (aq) + 6e^- \rightarrow S (s) + 4 H_2O (l)$

On the left in 17.13f there are 8 + charges from hydrogen ion and 2 − charges from sulfate ion, a net of 6 +. On the right the net charge is zero. Charge is balanced by adding 6 electrons to the left (positive) side.

Now that you have the two half-reaction equations, complete writing the net ionic equation (steps 3 and 4) as in prior sections.

-----------------------------------------------------------------------

**17.13h**  $8 H^+ (aq) + SO_4^{2-} (aq) + 6 I^- (aq) \rightarrow S (s) + 4 H_2O (l) + 3 I_2 (s)$

Reduction:  $8 H^+ (aq) + SO_4^{2-} (aq) + 6e^- \rightarrow S (s) + 4 H_2O (l)$

3 × Oxidation:  $6 I^- (aq) \rightarrow 3 I_2 (s) + 6e^-$
_____

Redox:  $8 H^+ (aq) + SO_4^{2-} (aq) + 6 I^- (aq) \rightarrow S (s) + 4 H_2O (l) + 3 I_2 (s)$

Let's check to make sure the equation is, indeed, balanced:
—the atoms balance (six I, one S, four O and eight H atoms on each side);
—the charge balances ($+8 - 2 - 6 = 0 + 0 + 0$).
Moreover, the equation also has the lowest set of whole number coefficients. Sometimes the procedure we have followed results in an equation that can be divided by two or some other integer.

**Example 17.14**  The permanganate ion, $MnO_4^-$, is a strong oxidizing agent, and will oxidize chloride ion to chlorine in an acidic solution. Manganese ends up as a monatomic manganese(II) ion. Write the net ionic equation for the redox reaction.

This is a challenging example, but watch how it falls into place following the procedure we have outlined. Begin by extracting from the verbal description of the reaction the partial oxidation half-reaction and the partial reduction half-reaction equations, and write them separately.

-----------------------------------------------------------------------

**17.14a**  Reduction:  $MnO_4^- (aq) \rightarrow Mn^{2+} (aq)$ (Ox. no. +7 to +2)

Oxidation:  $Cl^- (aq) \rightarrow Cl_2 (g)$ (Ox. no. −1 to 0)

Completing the oxidation half-reaction should be easy—both atoms and charge.

-----------------------------------------------------------------------

**17.14b**  $2 Cl^- (aq) \rightarrow Cl_2 (g) + 2e^-$. With a switch from iodine to chlorine, this is the same oxidation reaction as in the last example.

Can you take the reduction to a completed half-reaction equation? First oxygen, then hydrogen, and finally charge . . . .

-----------------------------------------------------------------------

**17.14c**  $8 H^+ (aq) + MnO_4^- (aq) + 5e^- \rightarrow Mn^{2+} (aq) + 4 H_2O (l)$

By steps:  Oxygen:  $MnO_4^- (aq) \rightarrow Mn^{2+} (aq) + 4 H_2O$ (four waters for four oxygen atoms).

Hydrogen: $8\,H^+\,(aq) + MnO_4^-\,(aq) \rightarrow Mn^{2+}\,(aq) + 4\,H_2O$ (eight $H^+$ for four waters).

Charge: $+8 - 1$ on left is $+7$; $+2 + 0$ on right is $+2$. Charge is balanced by adding 5 negatives on the left, or 5 electrons, as in the final answer.

Now combine the half-reaction equations for the net ionic equation.

---------------------------------------------------------------------------------------

### 17.14d

$2 \times$ Reduction: $16\,H^+\,(aq) + 2\,MnO_4^-\,(aq) + \cancel{10e^-} \rightarrow 2\,Mn^{2+}\,(aq) + 8\,H_2O\,(l)$

$5 \times$ Oxidation: $\qquad\qquad\qquad\qquad 10\,Cl^-\,(aq) \rightarrow 5\,Cl_2\,(g) + \cancel{10e^-}$

Redox: $16\,H^+\,(aq) + 2\,MnO_4^-\,(aq) + 10\,Cl^-\,(aq) \rightarrow 2\,Mn^{2+}\,(aq) + 8\,H_2O\,(l)$
$\qquad\qquad\qquad\qquad\qquad\qquad\qquad\qquad\qquad\qquad\qquad\qquad + 5\,Cl_2\,(g)$

Checking:
—atoms balance (sixteen H, two Mn, eight O and ten Cl)
—charges balance ($+16 - 2 - 10 = +4$)
Can you imagine a trial and error approach to an equation such as this?

# CHAPTER 17 IN REVIEW

## 17.1  ELECTRON TRANSFER REACTIONS

**17 A**  Distinguish between oxidation and reduction in terms of electrons gained or lost. (Page 412)

**17 B**  Given an oxidation half-reaction equation and a reduction half-reaction equation, add them to obtain a balanced redox reaction equation. (Page 412)

## 17.2  OXIDATION NUMBERS AND REDOX REACTIONS

**17 C**  Given the formula of a chemical species, determine the oxidation number of each element it contains. (Page 416)

**17 D**  Distinguish between oxidation and reduction in terms of oxidation number change. (Page 416)

## 17.3  OXIDIZING AGENTS (OXIDIZERS); REDUCING AGENTS (REDUCERS)

**17 E**  Distinguish between an oxidizing agent and a reducing agent. (Page 419)

## 17.4  REDOX REACTIONS AND ACID-BASE REACTIONS COMPARED

**17 F**  Identify the essential difference between a redox reaction and an acid-base reaction. (Page 420)

**17 G**  Distinguish between a strong oxidizing agent and a weak oxidizing agent; between a strong reducing agent and a weak reducing agent. (Page 420)

## 17.5 STRENGTHS OF OXIDIZING AGENTS AND REDUCING AGENTS

**17 H** Given a table of relative strengths of oxidizing and reducing agents, identify the stronger and weaker of two oxidizing or two reducing agents. (Page 421)

## 17.6 PREDICTING REDOX REACTIONS

**17 I** Given a table of relative strengths of oxidizing and reducing agents and the identity of an oxidizer and a reducer from the table, (a) write the net ionic equation for the redox reaction between them, and (b) predict whether the forward or reverse reaction will be favored at equilibrium. (Page 422)

## 17.7 WRITING REDOX EQUATIONS

**17 J** Given the identity of an oxidizer and reducer, and the identity of their reduced and oxidized products in an acidic solution, write the net ionic equation for the reaction. (Page 425)

# TERMS AND CONCEPTS

Half-reaction equation (413)
Electron transfer reaction (413)
Oxidation-reduction (redox) reaction (414)
Half-reaction (414)

Reduction (414)
Oxidation (414)
Oxidation number (417)
Oxidizing agent; oxidizer (419)
Reducing agent; reducer (420)

# QUESTIONS AND PROBLEMS

*To save space, the designations (s), (l), (g) and (aq) are omitted. All ions below are in aqueous solution.*

Section 17.1

17.1) Define oxidation; define reduction. It is sometimes said that oxidation and reduction are simultaneous processes—that you cannot have one without the other. From the standpoint of your definitions, explain why.

17.2) Classify each of the following half-reaction equations as oxidation or reduction half-reactions:
(a) $2 \, Cl^- \rightarrow Cl_2 + 2e^-$
(b) $Na \rightarrow Na^+ + e^-$
(c) $Sn^{2+} \rightarrow Sn^{4+} + 2e^-$
(d) $O_2 + 4 \, H^+ + 4e^- \rightarrow 2 \, H_2O$

17.3) Combine the following half-reaction equations to produce a balanced redox equation: $Ni^{2+} + 2e^- \rightarrow Ni$; $Mg \rightarrow Mg^{2+} + 2e^-$.

17.22) Using any example of a redox reaction, explain why such reactions are described as electron transfer reactions.

17.23) Classify each of the following half-reaction equations as oxidation or reduction half-reactions:
(a) $Zn \rightarrow Zn^{2+} + 2e^-$
(b) $2 \, H^+ + 2e^- \rightarrow H_2$
(c) $Fe^{2+} \rightarrow Fe^{3+} + e^-$
(d) $NO + 2 \, H_2O \rightarrow NO_3^- + 4 \, H^+ + 3e^-$

17.24) Combine the following half-reaction equations to produce a balanced redox equation: $Cr \rightarrow Cr^{3+} + 3e^-$; $Cl_2 + 2e^- \rightarrow 2 \, Cl^-$.

## Section 17.2

*17.4, 17.5, 17.25 and 17.26: Give the oxidation state of the element whose symbol is underlined in each chemical substance.*

17.4)  $\underline{Mg}^{2+}$; $\underline{Cl}^-$; $\underline{Cl}O^-$; $K\underline{Cl}O_3$

17.5)  $\underline{N}_2O_5$; $\underline{N}H_4^+$; $\underline{Mn}O_4^-$; $Na_2H\underline{P}O_3$

17.25)  $\underline{Al}^{3+}$; $\underline{S}^{2-}$; $\underline{S}O_3^{2-}$; $Na_2\underline{S}O_4$

17.26)  $\underline{N}_2O_3$; $\underline{N}O_3^-$; $\underline{Cr}O_4^{2-}$; $NaH_2\underline{P}O_4$

*17.6–17.8 and 17.27–17.29: For each half-reaction (1) identify the element experiencing oxidation or reduction, (2) state "oxidized" or "reduced" and (3) show the change in oxidation number. Example: $2\ Cl^- \rightarrow Cl_2 + 2e^-$. Chlorine oxidized from −1 to 0.*

17.6)  a) $Cu^{2+} + 2e^- \rightarrow Cu$
   b) $Co^{3+} + e^- \rightarrow Co^{2+}$

17.7)  a) $H_2O + SO_3^{2-} \rightarrow SO_4^{2-} + 2\ H^+ + 2e^-$
   b) $PH_3 \rightarrow P + 3\ H^+ + 3e^-$

17.8)  a) $2\ HF \rightarrow F_2 + 2\ H^+ + 2e^-$
   b) $MnO_4^{2-} + 2\ H_2O + 2e^- \rightarrow$
   $MnO_2 + 4\ OH^-$

17.27)  a) $Br_2 + 2e^- \rightarrow 2\ Br^-$
   b) $Pb^{2+} + 2\ H_2O \rightarrow PbO_2 + 4\ H^+ + 2e^-$

17.28)  a) $8\ H^+ + IO_4^- + 8e^- \rightarrow I^- + 4\ H_2O$
   b) $4\ H^+ + O_2 + 4e^- \rightarrow 2\ H_2O$

17.29)  a) $NO_2 + H_2O \rightarrow NO_3^- + 2\ H^+ + e^-$
   b) $2\ Cr^{3+} + 7\ H_2O \rightarrow$
   $Cr_2O_7^{2-} + 14\ H^+ + 6e^-$

## Section 17.3

17.9)  In the reaction between copper(II) oxide and hydrogen, identify the oxidizing agent and the reducing agent:

$$CuO + H_2 \rightarrow Cu + H_2O$$

17.10)  Name the oxidizing and reducing agents in the reaction for preparing chlorine:

$$MnO_2 + 4\ H^+ + 2\ Cl^- \rightarrow Mn^{2+} + Cl_2 + 2\ H_2O$$

17.30)  Identify the oxidizing and reducing agents in $Cl_2 + 2\ Br^- \rightarrow 2\ Cl^- + Br_2$

17.31)  What is the oxidizing agent in the equation for the storage battery, $Pb + PbO_2 + 4\ H^+ + 2\ SO_4^{2-} \rightarrow 2\ PbSO_4 + 2\ H_2O$? What does it oxidize? Also name the reducing agent and the species it reduces.

## Section 17.4

17.11  Show how redox and acid-base reactions parallel each other—how they are similar, but also what makes them different.

17.32)  Explain how a strong acid is similar to a strong reducer. Also, how a strong base compares to a strong oxidizer.

## Section 17.5

17.12)  Which is the stronger oxidizing agent, $Ag^+$ or $H^+$? What is the basis of your selection? What is the meaning of the statement that one substance is a stronger oxidizer than another?

17.13)  Arrange the following reducing agents in order of *decreasing* strength, i.e., strongest reducer first: $H_2$; $Al$; $Cl^-$; $Fe^{2+}$.

17.33)  Identify the stronger reducer between $Zn$ and $Fe^{2+}$. On what basis do you make your decision? What is the significance of one reducer being stronger than another?

17.34)  Arrange the following oxidizers in order of *increasing* strength, i.e., weakest oxidizing agent first: $Na^+$; $Br_2$; $Fe^{2+}$; $Cu^{2+}$.

## Section 17.6

*17.14–17.16 and 17.35–17.37: Write the redox equation for the two redox reactants given, using Table 17.1 as a source of the required half-reactions. Then predict the direction in which the reaction will be favored at equilibrium.*

17.14)  $Ni + Zn^{2+} \rightleftharpoons$

17.15)  $Fe^{3+} + Co \rightleftharpoons$

17.16)  $O_2 + H^+ + Ca \rightleftharpoons$

17.35)  $Br_2 + I^- \rightleftharpoons$

17.36)  $H^+ + Br^- \rightleftharpoons$

17.37)  $NO + H_2O + Fe^{2+} \rightleftharpoons$

Section 17.7

*17.17–17.21 and 17.38–17.42: Each "equation" identifies an oxidizer and a reducer, as well as the oxidized and reduced products of the redox reaction. Write the separate oxidation and reduction half-reaction equations, assuming the reaction takes place in an acidic solution, and add them to produce a balanced redox equation.*

17.17)  $Ag + SO_4^{2-} \rightarrow Ag^+ + SO_2$

17.18)  $NO_3^- + Zn \rightarrow NH_4^+ + Zn^{2+}$

17.19)  $Cr_2O_7^{2-} + Fe^{2+} \rightarrow Cr^{3+} + Fe^{3+}$

17.20)  $I^- + MnO_4^- \rightarrow I_2 + MnO_2$

17.21)  $BrO_3^- + Br^- \rightarrow Br_2$

17.38)  $S_2O_3^{2-} + Cl_2 \rightarrow SO_4^{2-} + Cl^-$

17.39)  $Sn + NO_3^- \rightarrow H_2SnO_3 + NO_2$

17.40)  $C_2O_4^{2-} + MnO_4^- \rightarrow CO_2 + Mn^{2+}$

17.41)  $Cr_2O_7^{2-} + NH_4^+ \rightarrow Cr_2O_3 + N_2$

17.42)  $As_2O_3 + NO_3^- \rightarrow AsO_4^{3-} + NO$

# 18

# NUCLEAR CHEMISTRY

## 18.1 THE DAWN OF NUCLEAR CHEMISTRY

*Serendipity.* This pleasant sounding word refers to the tendency to find valuable things you are not looking for—an accidental discovery, in other words. Serendipity has played an important part in many scientific discoveries, but rarely has an unplanned observation led to a discovery as far reaching as that of Henri Becquerel. What he stumbled across now affects the lives of you and me, and potentially of every creature on this planet. Becquerel discovered nuclear chemistry.

Becquerel became interested in X-rays soon after they were discovered, also accidentally, by Wilhelm Roentgen in 1895. Becquerel was also interested in fluorescence and wondered if the two were related. Fluorescence can be described as the emission of one color of light by an object that is being exposed to another color of light. X-rays are like light, except that they lie outside the visible spectrum. They are high-energy radiations that are capable of penetrating solids such as human tissue and black paper, neither of which is penetrated by ordinary sunlight. Becquerel's plan was to put some fluorescent material, a uranium salt, on top of unexposed photographic film wrapped in black paper, and then place them in the sunlight. The sunlight could not expose the film by itself, but if the film *was* exposed, it would show that the fluorescent emission of the uranium salt had penetrated the paper in a manner similar to that of X-rays.

The day Becquerel selected for his experiment was overcast. After waiting in vain for the sun to come out, he put the assembled uranium salt and wrapped film into a drawer, to be used the next sunny day. The weather continued to be dull for several days. Becquerel finally decided to develop the film to see if any penetration had occurred even on the cloudy day. To his amazement, the film was strongly exposed. The only possible conclusion was that penetrating energy emissions were going on all the time in the darkness of the drawer, and that they had nothing whatever to do with sunlight. It was later shown that, though their penetrating ability is similar to that of X-rays, the emissions were not X-rays but an entirely new form of energy ray.

It is thus that Becquerel discovered **radioactivity,** a natural, spontaneous process that has been going on in the nuclei of atoms since the beginning of time.

Once the door to nuclear chemistry was opened, progress was rapid. While working with a uranium-bearing mineral called *pitchblende*, Marie and Pierre Curie discovered that a second element, thorium, gave off penetrating radiations. They also found two new radioactive elements: polonium, which is about 400 times as radioactive as uranium, and radium, which is again many times more radioactive than polonium.

## 18.2   NATURAL RADIOACTIVITY

PG   18 A   Define radioactivity. Name, identify from a description, or describe the three types of natural radioactive emission.

**Radioactivity is the spontaneous emission of rays resulting from the disintegration, or decay, of an atomic nucleus.** Three kinds of rays can be identified. If a beam consisting of all three rays is directed into an electric field, as in Figure 18.1, the individual rays are separated. One, called an alpha ray, or α-ray,* is attracted toward the negatively charged plate, indicating it has a positive charge. The α-ray has little penetrating power; it can be stopped by a few sheets of paper (see Fig. 18.2). Alpha rays are now known to be nuclei of helium atoms, having the nuclear symbol $^4_2He$.† They are commonly called alpha particles, or α particles.

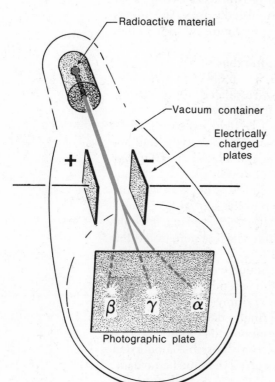

Figure 18.1   Paths of alpha, beta and gamma emissions in an electric field. Alpha rays are attracted toward the negative plate, indicating they have a positive charge. Beta rays are attracted toward the positive plate, indicating a negative charge. Gamma rays are unaffected, showing they have no electrical charge.

---

*α, β and γ are the Greek letters alpha, beta and gamma by which radioactive rays are named.

†Recall from page 119 that the nuclear symbol is $^A_Z Sy$, where Sy is the symbol of the element, Z is the nuclear charge, or atomic number, and A is the mass number, the total number of protons plus neutrons in the nucleus of the isotope. It follows that the number of neutrons is A − Z.

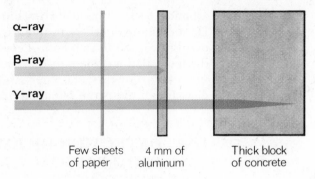

**Figure 18.2** Penetration power of radioactive emissions. A few sheets of paper are unable to stop beta or gamma rays, but they will stop alpha emissions. Gamma rays will pass through an aluminum sheet 4 mm thick, but beta rays will not. A heavy layer of concrete or lead is required to stop gamma radiations.

The second kind of ray also turns out to be a beam of particles, but they are negatively charged and therefore attracted to the positively charged plate (Fig. 18.1). Initially called beta rays, or $\beta$-rays, they have been identified as electrons. The nuclear symbol for a $\beta$ particle is $_{-1}^{0}e$, indicating zero atomic mass and a $-1$ charge emerging from the nucleus. $\beta$ particles have considerably more penetrating power than $\alpha$ particles, but they can be stopped by a sheet of aluminum about 4 millimeters thick (see Fig. 18.2).

The third kind of radiation is the gamma ray, or $\gamma$-ray. Gamma rays are not particles, but very high-energy electromagnetic rays. Because of their high energy, gamma rays have high penetrating power. They can be stopped only by thick layers of lead or heavy concrete walls, as shown in Figure 18.2. Not having an electric charge, gamma rays are not deflected by an electric field (Fig. 18.1).

## 18.3 INTERACTION OF RADIOACTIVE EMISSIONS WITH MATTER

PG  18 B  Describe how emissions from radioactive substances affect gases and living organisms.

When $\alpha$, $\beta$ or $\gamma$ radiation collides with an atom or molecule, some of its energy will be given to the target particle. The collision changes the electron arrangement in the species hit. An electron may be excited to a higher energy level, leading to electromagnetic radiation as the electron drops back to its ground state level. The electron may be knocked all the way out of an atom or molecule, ionizing it. Air, or any gas, can be ionized by a radioactive substance. If the radiation strikes chemically bonded atoms, it often breaks those bonds, thereby causing a chemical reaction.

The ionization of molecules in air by radioactive emission has a present day application in one kind of home smoke detector. These "ionization" detectors generally use a tiny chip of americium-241, a radioactive isotope of an element not found in nature. The ionization of air causes a small current to flow through the air inside the detector. When smoke enters, it disrupts the current and sets off the alarm.

It is the ability of radiation to initiate chemical change that is responsible for its effect on body tissue, leading to illness and even death in extreme cases. These effects have been studied among the survivors of the atomic bombs dropped at Hiroshima and Nagasaki at the close of World War II, and among workers in nuclear plants and laboratories who have been overly exposed to radiation. People working in such plants wear a small device called a dosimeter that monitors the amount of radiation that falls upon them. By this and other safety measures the danger to workers associated with industrial radiation has been minimized and controlled.

## 18.4 DETECTION OF EMISSIONS FROM RADIOACTIVE SUBSTANCES

PG 18 C  Define or describe a cloud chamber. Explain the principle by which it operates.

18 D  Define or describe a Geiger counter. Explain the principle by which it operates.

There are several ways by which radioactivity may be detected. Perhaps the most obvious, but not necessarily the most convenient, is through its effect on photographic film, the property that caused it to be discovered in the first place. Another is based on its ability to produce visible emissions in fluorescent materials. Luminous watch dials, for example, are painted with a mixture including zinc sulfide and a tiny trace of radium sulfate. Radiations from the radium cause fluorescence in the zinc sulfide, making it visible in the dark. A *scintillation counter* may be used to measure radiant energy by fluorescence.

Perhaps the best visible evidence of radiation is found in a **cloud chamber** (see Fig. 18.3). Radioactive emissions are directed through an air space

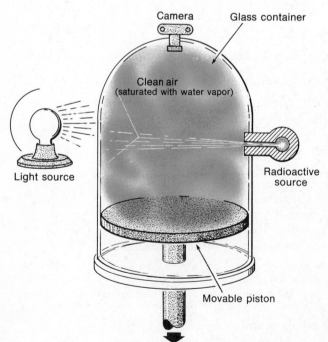

**Figure 18.3** Wilson cloud chamber. Moving the piston down causes the air in the chamber to become supersaturated in water vapor. Radioactive emissions ionize the air through which they pass, causing condensation of visible droplets that leave a trail behind the emitted ray.

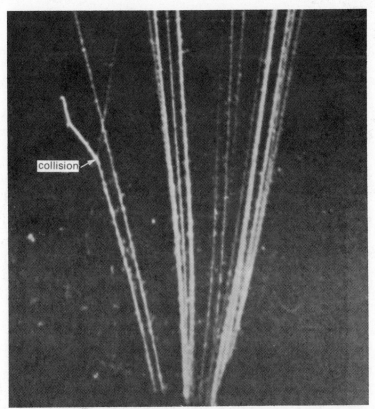

**Figure 18.4** Alpha particle tracks as they appear in a cloud chamber. This particular photograph is historically significant. At the point marked "collision," an alpha particle struck a nitrogen nucleus, producing a proton and an oxygen-17 nucleus. This was the first artificial nuclear transformation (see page 444).

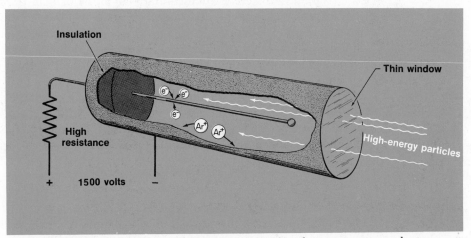

**Figure 18.5** Schematic drawing of a Geiger counter. The tube contains argon at low pressure. A high electrical potential is established between a positively charged wire in the center and the negatively charged case. When radioactive emissions enter the tube window, electrons are knocked out of the argon atoms, ionizing the gas. Electrons are accelerated toward the positively charged wire, and argon ions move toward the negatively charged case, producing more ions en route. The buildup of ions is such that a sharp electrical discharge occurs, producing an audible "click." These discharges may be counted and recorded, indicating the intensity of the radioactivity.

that is supersaturated with some vapor, often water. As the molecules in the air are ionized, the vapor condenses on the ions. The condensate is visible, and leaves a growing track behind the radiation as it travels through the space, much as a vapor trail forms behind a high flying jet aircraft on a clear day. Cloud chamber tracks may be photographed; one historically significant photograph is shown as Figure 18.4.

The **Geiger counter** is probably the best known instrument for detecting and measuring radiation. It consists of a tube filled with a gas, as shown in Figure 18.5. The gas is ionized by radiation entering the tube through a window, permitting a surge of electrical current between two electrodes. The current is related to the quantity of radiation received, and can be read quantitatively on a meter. Many Geiger counters are equipped to produce audible clicks, the frequency of which is an indication of the intensity of radiation.

## 18.5 NATURAL RADIOACTIVE DECAY SERIES—NUCLEAR EQUATIONS

PG 18 E   Describe a natural radioactive decay series.

18 F   Given the identity of a radioactive isotope and the particle it emits, write a nuclear equation for the emission.

While gamma rays have the greatest penetrating power and highest energy of the three forms of nuclear radiation, and while they usually accompany both alpha and beta emissions, it is true that these particular rays do not affect nuclear composition. For the balance of the chapter, therefore, we will consider only alpha and beta particles in nuclear reactions.

When a radioactive nucleus emits an alpha or beta particle, there is a **transmutation** of an element, or a change from one element to another. Recall that the elemental identity of an atom depends upon the number of protons in the nucleus. If an atom changes from one element to another, there must be a change in the number of protons. A change of this kind is represented by a nuclear equation showing the nuclear symbols of the reactants and products. The emission observed by Becquerel was an alpha particle emission. The nuclear product remaining after a $^{238}_{92}U$ nucleus emits an $\alpha$ particle is a thorium nucleus, $^{234}_{90}Th$. In other words, a $^{238}_{92}U$ nucleus has disintegrated into a $^{4}_{2}He$ nucleus and a $^{234}_{90}Th$ nucleus. This may be shown in a nuclear equation:

$$^{238}_{92}U \rightarrow {}^{4}_{2}He + {}^{234}_{90}Th \qquad (18.1)$$

Notice that the above equation is "balanced" in both neutrons and protons. The total number of neutrons and protons is 238, the mass number of the uranium isotope. The total mass number of the two products is 234 + 4, again 238. In terms of protons, the 92 in a uranium nucleus are accounted for by 90 in the thorium nucleus plus 2 in the helium nucleus. A nuclear equation is balanced if the sums of the mass numbers on the two sides of the equation are equal, and if the sums of the atomic numbers are equal.

The $^{234}_{90}Th$ nucleus resulting from the disintegration of uranium-238 is

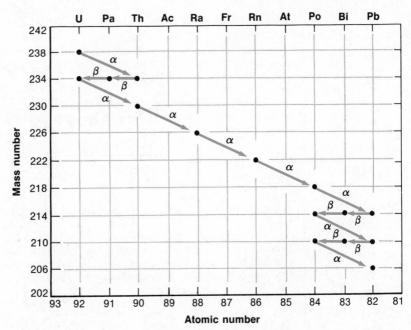

**Figure 18.6** Radioactive decay series. This series begins with $^{238}_{92}$U, and after 8 alpha emissions and 6 beta emissions, produces $^{206}_{82}$Pb as a stable end product.

also radioactive. It emits a beta particle, $_{-1}^{0}$e, and produces an isotope of protactinium, $^{234}_{91}$Pa:

$$^{234}_{90}\text{Th} \rightarrow \,^{234}_{91}\text{Pa} + \,^{0}_{-1}\text{e} \qquad (18.2)$$

In a $\beta$ particle emission the mass numbers of the reactant and product isotopes are the same, while the atomic number increases by 1. In effect, a neutron in the nucleus has divided into a proton and an electron, the electron being ejected.

The two radioactive disintegration steps described by Equations 18.1 and 18.2 are but the first two of fourteen steps that begin with $^{238}_{92}$U. There are eight $\alpha$ particle emissions and six $\beta$ particle emissions, leading ultimately to a stable isotope of lead, $^{206}_{82}$Pb. This entire **natural radioactive decay series** is described in Figure 18.6. There are two other natural disintegration series, as they are also called. One begins with $^{232}_{90}$Th and ends with $^{208}_{82}$Pb, and the other passes from $^{235}_{92}$U to $^{207}_{82}$Pb.

---

**Example 18.1** Write the nuclear equation for the changes that occur in the U-238 disintegration series when $^{226}_{88}$Ra ejects an $\alpha$ particle. Ra is the symbol for radium, one of the elements discovered by Pierre and Marie Curie in their study of radioactivity.

In writing a nuclear equation, one product will be the particle ejected. The mass number of the other product will be such that, when added to the mass number of the ejected particle, they will total the mass number of the original isotope. What is the mass number of the second product of the emission of an alpha particle from a $^{226}_{88}$Ra nucleus?

---

**18.1a** 222

The reactant isotope has a mass number of 226. It emits a particle having a mass number of 4. This leaves $226 - 4 = 222$ as the mass number of the remaining particle.

Now find the atomic number of the second product particle. The atomic number of the starting isotope is 88. It emitted a particle having 2 protons. How many protons are left in the nucleus of the other product?

------------------------------------------------------------------------------

**18.1b** 86

If two protons are emitted from a nucleus having 88 protons, 86 will remain.

You now know the mass number and the atomic number of the second product of an alpha particle emission from $^{226}_{88}Ra$. Using a periodic table you can find the elemental symbol of this product, and assemble all three symbols into the required nuclear equation.

------------------------------------------------------------------------------

**18.1c** $^{226}_{88}Ra \rightarrow {}^{4}_{2}He + {}^{222}_{86}Rn$

The second product is a radioactive isotope of the noble gas radon, whose atomic number is 86. This isotope continues the natural emission series by ejecting another $\alpha$ particle.

---

**Example 18.2** Write the nuclear equation for the emission of a $\beta$ particle from bismuth-210, $^{210}_{83}Bi$.

The method is the same. Remember the beta particle, $_{-1}^{0}e$, has zero mass number, and an effective atomic number of $-1$. Both mass number and atomic number must be conserved in the equation.

------------------------------------------------------------------------------

**18.2a** $^{210}_{83}Bi \rightarrow {}^{0}_{-1}e + {}^{210}_{84}Po$

In the emission of a $\beta$ particle, which has effectively no mass, the mass number of the radioactive isotope and the product isotope are the same. The product isotope has an atomic number greater by one than the radioactive isotope, an increase of one proton. Po is the symbol that corresponds to atomic number 84. The element is polonium, the other element discovered by the Curies in their investigation of radioactivity. The name of the element was selected to honor Mme. Curie's native Poland.

## 18.6 HALF-LIFE

PG 18 G Describe or illustrate what is meant by the half-life of a radioactive isotope.

18 H Given the original quantity of a radioactive isotope and its half-life, and either the final quantity or elapsed time, find the time or quantity not given. (Whole-number half-lives only.)

The rate at which a single step in nuclear disintegration occurs is measured by its **half-life, the time required for the disintegration of one half of the radioactive atoms in a sample.** Each radioactive isotope has its own unique half-life, commonly designated $t_{\frac{1}{2}}$. The half-lives of some isotopes are very long; the half-life of $^{238}_{92}U$ is 4.5 billion years. Other half-lives are very short. The half-life of $^{234}_{90}Th$ (see Equation 18.1, page 438) is a comfortable 24.1 days; but the half-life of the particle that appears after $^{234}_{90}Th$ emits a $\beta$ particle, $^{234}_{91}Pa$, is only 1.18 minutes. The half-life of one species in the same radioactive disintegration series, $^{214}_{84}Po$, is only 0.00016 second. Needless to say, people who work with radioactive substances with such short half-lives are rather pressed for time.

Figure 18.7 illustrates the meaning of half-life. Shown in black along the vertical axis is the fraction of the original sample that is left at the end of any number of half-lives, plotted horizontally in black. One half of the material present at the beginning of each half-life cycle will remain at the end of that cycle. At the end of the first cycle, ½ of the original sample remains. At the end of the second half-life, ½ of ½, or ¼ remains. This is the same as saying that the fraction of the original sample remaining after *two* half-lives is $(\frac{1}{2})^2$. At the end of *three* half-lives, the amount remaining will be ½ of ¼, or ⅛, which is the same as $(\frac{1}{2})^3$. At the end of $n$ half-lives the fraction of the original sample that remains is $(\frac{1}{2})^n$. Expressed as an equation,

$$\text{Quantity remaining after n half-lives} = \text{initial quantity} \times (\tfrac{1}{2})^n \quad (18.3)$$

To further illustrate the significance of the half-life cycle, the decay of a sample of $^{210}_{83}Bi$, the radioactive isotope used in Example 18.2, is shown in color. The half-life of that isotope is 5.0 days. If you begin with 16 grams on the

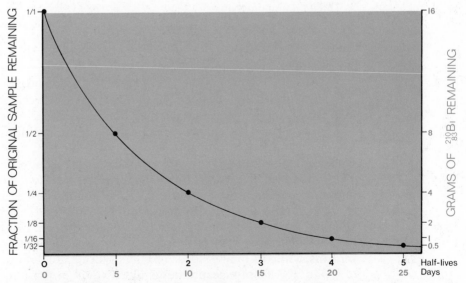

**Figure 18.7**  Half-life decay curve for a radioactive substance. During each half-life period, half of the material present at the beginning of the period undergoes radioactive disintegration. This is shown by the black numbers. Colored numbers trace the decay of 16 grams of $^{210}_{83}Bi$ through five half-lives of five days each. The fraction of a sample remaining after n half-lives is $(\frac{1}{2})^n$.

first day of April, only 8 grams will remain on April 5. By April 10 your sample will be reduced to 4 grams, and on April 15, the middle of the month, it will be down to 2 grams. One gram will still be there five days later, on April 20, dropping to 0.5 gram on April 25. At the end of April you will have only 0.25 gram of your original 16 to carry into the month of May. Using Equation 18.3 for the six half-life cycles,

$$16 \text{ g} \times \left(\frac{1}{2}\right)^6 = \frac{16 \text{ g}}{2^6} = \frac{16 \text{ g}}{2 \times 2 \times 2 \times 2 \times 2 \times 2} = 0.25 \text{ g}$$

---

**Example 18.3**   The half-life of $^{45}_{19}K$ is 20 minutes. If you have a sample containing 2050 micrograms of this isotope at noon, how many micrograms will remain at 3 o'clock in the afternoon?

First of all, through how many half-lives will the decay process pass between noon and 3 o'clock, at 20 minutes for each half-life?

-------------------------------------------------------------------------------

**18.3a**   9 half-lives

At 1 half-life every 20 minutes, that's 3 half-lives per hour. In 3 hours that will be $3 \times 3 = 9$ half lives. Or, by dimensional analysis,

$$3 \text{ hours} \times \frac{60 \text{ minutes}}{1 \text{ hour}} \times \frac{1 \text{ half-life}}{20 \text{ minutes}} = 9 \text{ half-lives}$$

From here the solution comes by direct substitution into Equation 18.3:

-------------------------------------------------------------------------------

**18.3b**   4.0 micrograms

$$2050 \text{ } \mu\text{g} \times \left(\frac{1}{2}\right)^9 = \frac{2050 \text{ } \mu\text{g}}{2^9} = \frac{2050 \text{ } \mu\text{g}}{2 \times 2 \times 2 \times 2 \times 2 \times 2 \times 2 \times 2 \times 2} = 4.0 \text{ } \mu\text{g}$$

---

   This half-life rate of decay of radioactive substances is the basis of an interesting application, **radiocarbon dating,** or simply "carbon dating." Living organisms, both plant and animal, have carbon as the principal chemical element. Most carbon atoms are of the isotope $^{12}_{6}C$; but a small portion of the carbon in atmospheric carbon dioxide is $^{14}_{6}C$, a radioactive isotope with a half-life of 5720 years. While a plant or animal is alive, it takes in this isotope from the atmosphere, while the same isotope in the organism itself is disappearing by nuclear disintegration. A steady-state equilibrium type of situation exists while the system lives, maintaining a constant ratio of $^{14}_{6}C$ to $^{12}_{6}C$. When the organism dies, the disintegration of $^{14}_{6}C$ continues in the remains, while its intake stops. This leads to a gradual reduction in the ratio of $^{14}_{6}C$ to $^{12}_{6}C$. From measurements of this ratio in the remains of very old organic matter that was once a part of a living species it is possible to estimate its age. Carbon dating has produced evidence of man's presence on earth as long ago as 14 to 15 thousand years, and recent findings indicate he may date back as many as 100,000 years.

Similar dating techniques are also applied to mineral deposits. Analysis of geological deposits have yielded rocks with an estimated age of 3.0 to 4.5 billion years, the latter figure being an approximation of the age of the earth. The oldest moon rocks analyzed to date indicate an age of about 3.5 billion years.

---

**Example 18.4** 1.12 grams (1120 milligrams) of a certain radioisotope are produced in one laboratory on a university campus, and then rushed to a radioisotope laboratory for tests. By the time it arrives, only 70 milligrams remain, the rest having been lost by radioactive decay during transport. If the half-life of the isotope is 2.1 minutes, how many minutes elapsed between the time the isotope was prepared and the time it arrived in the radiochemistry laboratory?

*Solution.* For a whole number of half-lives, which we are considering under Performance Goal 18 H, the most direct approach to a dating question is to find the quantity that was present one, two, three or more half-lives ago until reaching the starting quantity. If a radioactive isotope loses half its mass during one half-life, the quantity present at the beginning of that half-life must have been twice the mass at the end of the half-life. Therefore one half-life (2.1 minutes) before the mass was 70 milligrams, it was 2 × 70, or 140 milligrams. One half-life before that (2 half-lives before 70 milligrams) the mass was 2 × 140 = 280 milligrams. Three half-lives before 70 it was 2 × 280, or 560 milligrams. And four half-lives before 70 there were 2 × 560, or 1120 milligrams, the starting quantity. It therefore took four half-lives to transport the material from one laboratory to the other, or 4 × 2.1 = 8.4 minutes. In summary,

| Half-lives before 70 mg | 0 | 1 | 2 | 3 | 4 |
|---|---|---|---|---|---|
| Minutes before 70 mg | 0 | 2.1 | 4.2 | 6.3 | 8.4 |
| Milligrams remaining | 70 | 140 | 280 | 560 | 1120 |

---

## 18.7 NUCLEAR REACTIONS AND ORDINARY CHEMICAL REACTIONS COMPARED

PG 18 I List or identify four ways in which nuclear reactions differ from ordinary chemical reactions.

Now that you have seen the nature of a nuclear change and the type of equation by which it is described, we will pause to compare nuclear reactions with the others you have been studying throughout this book. There are four areas of comparison:

1. In ordinary chemical reactions, the chemical properties of an element depend only on the electrons outside the nucleus, and the properties are essentially the same for all isotopes of the element. The nuclear properties of the various isotopes of an element are quite different, however. In

the radioactive decay series beginning with U-238, $^{234}_{90}$Th emits a $\beta$ particle, whereas a bit farther down the line $^{230}_{90}$Th ejects an $\alpha$ particle. Both $^{214}_{82}$Pb and $^{210}_{82}$Pb are $\beta$ particle emitters toward the end of the series, while the final product, $^{206}_{82}$Pb, has a stable nucleus, emitting neither alpha nor beta particles, nor gamma rays.

2. Radioactivity is independent of the state of chemical combination of the radioactive isotope. The reaction of $^{210}_{83}$Bi occurs for atoms of that particular isotope whether they are in pure elemental bismuth, combined in bismuth chloride, $BiCl_3$, bismuth sulfate, $Bi_2(SO_4)_3$, or any other bismuth compound, or if they happen to be present in the low melting bismuth alloy used for fire protection in sprinkler systems in large buildings.

3. Nuclear reactions usually result in the formation of different elements because of changes in the number of protons in the nucleus of an atom. In ordinary chemical reactions the atoms keep their identity while changing from one compound as a reactant to another as a product.

4. Both nuclear and ordinary chemical changes involve energy, but the amount of energy for a given amount of reactant in a nuclear change is enormous—greater by several orders of magnitude*—compared to the energies of ordinary chemical reactions. Further comment on this appears in Sections 18.11 and 18.13.

## 18.8 NUCLEAR BOMBARDMENT AND ARTIFICIAL RADIOACTIVITY

PG  18 J  Define nuclear bombardment.

18 K  Distinguish between natural radioactivity and artificial radioactivity.

18 L  Define or identify transuranium elements.

18 M  Given the identity of all except one of the isotopes and particles in a nuclear bombardment, write the nuclear equation for the reaction.

In natural radioactive decay, we find an example of the alchemist's get-rich-quick dream of converting one element to another. But the natural process did not yield the gold coveted by the alchemist; rather it produced the element lead, with which the dreamer wanted to begin his transmutation. The question was still present after the discovery of radioactivity: can man cause the transmutation of one element into another?

In 1919, Rutherford produced a "Yes" answer to that question. He found he could "bombard" the nucleus of a nitrogen atom with a beam of alpha particles from a radioactive source, jarring a proton out of the nucleus and producing an atom of oxygen-17:

$$^{14}_{7}N + ^{4}_{2}He \rightarrow ^{17}_{8}O + ^{1}_{1}H$$

The oxygen isotope produced is stable; the experiment did not yield any

---

*"Orders of magnitude" refers to multiples of ten. For example, things that differ by three orders of magnitude differ by a ratio of one thousand to one—$10 \times 10 \times 10 = 10^3 = 1000$.

man-made radioactive isotopes. Similar experiments were conducted with other elements, using high-speed alpha particles as atomic bullets. It was found that most of the elements up to potassium can be changed to other elements by nuclear bombardment. None of the isotopes produced were radioactive.

One experiment during this period was first thought to yield a nuclear particle that emitted some sort of high-energy radiation, perhaps a gamma ray. In 1932 James Chadwick correctly interpreted the experiment, however, and in doing so he became the first person to identify the neutron. The reaction comes from bombarding a beryllium atom with a high-energy $\alpha$ particle:

$$\ce{^{9}_{4}Be} + \ce{^{4}_{2}He} \rightarrow \ce{^{12}_{6}C} + \ce{^{1}_{0}n}$$

where $^{1}_{0}n$ is the nuclear symbol for the neutron, with a mass number of 1 and zero charge.

Two years later, in 1934, Irene Curie, the daughter of Pierre and Marie Curie, and her husband, Frederic Joliot, used high-energy $\alpha$ particles to produce the first man-made radioactive isotope. Their target was boron-5; the product was a radioactive nitrogen nucleus not found in nature:

$$\ce{^{10}_{5}B} + \ce{^{4}_{2}He} \rightarrow \ce{^{13}_{7}N} + \ce{^{1}_{0}n}$$

When $^{13}_{7}N$ decays it emits a particle having the mass of an electron and a charge equal to that of an electron, except that it is positive. This "positive electron" is called a **positron,** and it is represented by the symbol $^{0}_{1}e$. The decay equation is

$$\ce{^{13}_{7}N} \rightarrow \ce{^{13}_{6}C} + \ce{^{0}_{1}e}$$

Today hundreds of *radioisotopes* have been produced in laboratories all over the world, and they find broad use in medicine, industry and research. Many have been made possible by the use of various kinds of *particle accelerators*, which use electrical fields to increase the kinetic energy of charged particles used to bombard nuclei. Among the earliest and best known accelerators is the cyclotron (see Fig. 18.8), designed by E. O. Lawrence at the University of California, Berkeley. Other more powerful accelerators are approximately circular in shape (see Fig. 18.9), and running almost unnoticed beneath a busy freeway in the foothills just west of the campus of Stanford University is a two-mile long linear accelerator.

One of the more exciting areas of research with bombardment reactions has been the production of elements that do not exist in nature. Except for trace quantities, no natural elements having atomic numbers greater than 92 have ever been discovered. In 1940, however, it was found that $^{238}_{92}U$ is capable of capturing a neutron:

$$\ce{^{238}_{92}U} + \ce{^{1}_{0}n} \rightarrow \ce{^{239}_{92}U} \qquad (18.4)$$

The newly formed isotope is unstable, progressing through two successive

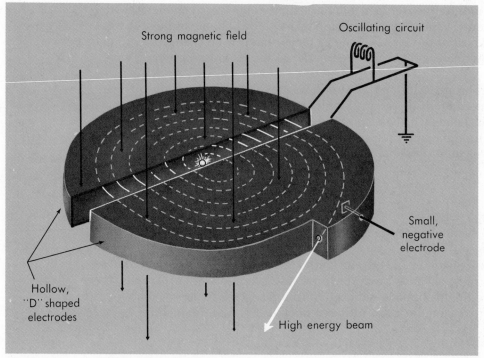

Strong magnetic field

Oscillating circuit

Small, negative electrode

Hollow, "D" shaped electrodes

High energy beam

**Figure 18.8** The cyclotron consists of two oppositely charged, evacuated "dees" placed between the poles of a powerful electromagnet. Positive ions, originating at the center, enter the upper dee, which is originally at a negative potential. They pass through this dee in a curved path. At the instant they reenter the central corridor, the polarity of the dees is reversed, and the particles enter the lower dee at an increased velocity. This procedure is repeated over and over; the particles move at higher and higher velocities in paths of greater and greater radius. Eventually they are deflected from the periphery of one of the dees to strike the target.

$\beta$ particle emissions, yielding isotopes of the elements having atomic numbers 93 and 94:

$$^{239}_{92}U \rightarrow {}^{0}_{-1}e + {}^{239}_{93}Np \text{ (neptunium)} \qquad (18.5)$$

$$^{239}_{93}Np \rightarrow {}^{0}_{-1}e + {}^{239}_{94}Pu \text{ (plutonium)} \qquad (18.6)$$

Neptunium, plutonium and all of the other man-made elements having atomic numbers greater than 92 are called the **transuranium elements.** To the time of this writing, all elements up to atomic number 106 have been identified. All of the transuranium isotopes are radioactive, and a few have been isolated only in isotopes with very short half-lives and in extremely small quantities. Some of the bombardments yielding transuranium products use relatively heavy isotopes as bullets. For example, einsteinium-247 is produced by bombarding uranium-238 with ordinary nitrogen nuclei:

$$^{238}_{92}U + {}^{14}_{7}N \rightarrow {}^{247}_{99}Es + 5 {}^{1}_{0}n$$

The principles for writing equations for nuclear bombardment reactions are the same as for writing equations for radioactive decay. The sums of the mass numbers on the two sides of the equation must be equal, and

**Figure 18.9**  Particle accelerator at the Fermi National Laboratory, near Batavia, Illinois.
A. Aerial view of main accelerator. B. Magnets in interior of main accelerator, which is
4 miles in circumference and 1.27 miles in diameter. (Fermi Lab Photo.)

the sums of the atomic numbers must be equal. The following examples
illustrate the procedures:

**Example 18.5**  (a) When $^{25}_{12}Mg$ is bombarded with an alpha
particle, $^{4}_{2}He$, the aluminum isotope $^{28}_{13}Al$ and another particle are the products. Write the equation.

(b) $^{10}_{4}Be$ and another particle are the prod-
ucts of the bombardment of $^{13}_{6}C$ with a neutron, $^{1}_{0}n$. Write the
nuclear equation.

In both examples you are given three of the four species appearing in the equation. By equating mass numbers and protons you can figure out the identity of the missing particle. Complete both equations.

---

**18.5a**  (a) $^{25}_{12}Mg + ^{4}_{2}He \rightarrow ^{28}_{13}Al + ^{1}_{1}H$

(b) $^{13}_{6}C + ^{1}_{0}n \rightarrow ^{10}_{4}Be + ^{4}_{2}He$

## 18.9 USES OF RADIOISOTOPES

Shortly after radioactivity was discovered, it was thought that the radiations had certain curative powers. Radium compounds were made, and radium solutions were bottled and sold for drinking and bathing, before the harmful effects of radiation exposure were well understood. Today's medical practitioners are much wiser, and they have devised sophisticated ways to examine their patients, diagnose their illnesses and treat their disorders, all using man-made radioisotopes.

One means of radioisotope examination involves injecting a radioactive sodium isotope into the bloodstream, and then tracing its progress through the body with a suitable detector such as a Geiger counter. If some portion of the body shows a low radiation count, it is an indication of a circulatory problem in that area. The normal absorption of iodine by the thyroid glands is checked by adding a radioactive iodine isotope to drinking water, followed by monitoring the radiation of that isotope. The same isotope is used to treat cancer of the thyroid glands. Radioactive isotopes of cobalt, phosphorus, radium, yttrium, strontium and cesium are used to destroy cancerous tissue in different forms of radiation therapy.

Industrial applications of radioisotopes include studies of piston wear and corrosion resistance. Petroleum companies use radioisotopes to monitor the progress of certain oils through pipelines. The thickness of thin sheets of metal, plastic and paper is subject to continuous production control by using a geiger counter to measure the amount of radiation that passes through the sheet; the thinner the sheet, the more radiation that will be detected by the counter. Quality control laboratories can detect small traces of radioactive elements in a metal part.

Scientific research is another major application of radioisotopes. Chemists use "tagged" atoms as *radioactive tracers* to study the mechanism, or series of individual steps, in complicated reactions. For example, by using water containing radioactive oxygen it has been determined that the oxygen in the $C_6H_{12}O_6$ formed in the photosynthesis reaction.

$$6\ CO_2\ (g) + 6\ H_2O\ (l) \rightarrow C_6H_{12}O_6\ (s) + 6\ O_2\ (g)$$

comes entirely from the carbon dioxide, and all oxygen from water is released as oxygen gas. Archaeologists use neutron bombardment to produce radioactive isotopes in an artifact, which makes it possible to analyze the item without destroying it. Biologists employ radioactive tracers in the water absorbed by the roots of plants to study the rate at which the water is distributed throughout the plant system. These are but a few of the many ingenious applications that have been devised for this useful tool of science.

## 18.10  NUCLEAR FISSION

PG  18 N  Define or identify a nuclear fission reaction.

18 O  Define or identify a chain reaction.

In 1938, during the period when Nazi Germany was moving steadily toward war, dramatic and far-reaching events were taking place in her laboratories. A team made up of Otto Hahn, Fritz Strassman and Lise Meitner was working with neutron bombardment of uranium. In the products of the reaction they were finding, of all things, atoms of barium and krypton, and other elements far removed in both atomic mass and atomic number from the uranium atoms and neutrons used to produce them. The only explanation was, at that time, unbelievable: the nucleus must be splitting into two nuclei of smaller mass. This kind of reaction is called **nuclear fission.**

Only 0.7% of all naturally occurring uranium is $^{235}_{92}U$, the isotope that is capable of the **chain reaction** required to keep a fission reaction going. In the fission of U-235 there are many products; it is not possible to write a single equation to show what happens. A representative equation is

$$^{235}_{92}U + ^{1}_{0}n \rightarrow ^{94}_{38}Sr + ^{139}_{54}Xe + 3\,^{1}_{0}n$$

Notice it takes a neutron to initiate the reaction. Notice also that the reaction produces *three* neutrons. If one or two of these collide with other fissionable uranium nuclei, there is the possibility of another fission or two. And the neutrons from those reactions can trigger others, repeatedly, as long as the supply of nuclei lasts. This is what is meant by a *chain reaction* (see Fig. 18.10), in which a nuclear product of the reaction becomes a nuclear reactant in the next step, thereby continuing the process.

The number of neutrons produced in the fission of $^{235}_{92}U$ varies with each reaction. Some reactions yield two neutrons per uranium atom; others, like the above, yield three; and still others produce four or more. The average is about 2.5. If the quantity of uranium, or any other fissionable isotope, is large enough that most of the neutrons produced are captured within the sample, rather than escaping to the surroundings, the chain reaction will be sustained. The minimum quantity required for this purpose is called the **critical mass.**

Because uranium-235 is less than 1% of all naturally occurring uranium, it is not a very satisfactory source of nuclear fuel. Fortunately the plutonium isotope, $^{239}_{94}Pu$, produced from $^{238}_{92}U$, the more abundant uranium isotope (Equations 18.4–18.6, page 445), has a long half-life (24,360 years) and is fissionable. $^{239}_{94}Pu$ has been used in the production of atomic bombs and is also used in some nuclear power plants to generate electrical energy. It is made in a **breeder reactor,** the name given to a device whose purpose is to produce fissionable fuel from nonfissionable isotopes.

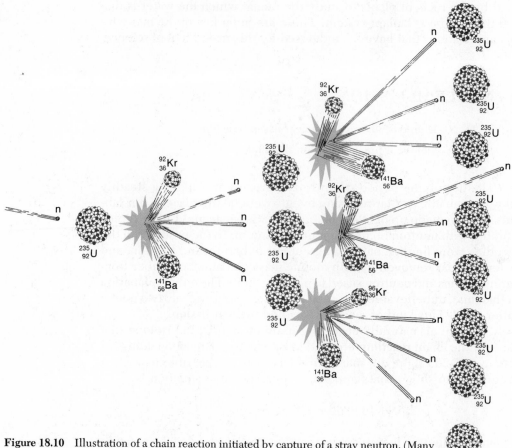

**Figure 18.10** Illustration of a chain reaction initiated by capture of a stray neutron. (Many pairs of different isotopes are produced but only one kind of pair is shown.)

## 18.11 NUCLEAR ENERGY

PG 18 P Explain the source of energy in a nuclear reaction.

The term *atomic energy* is widely used; a better term is *nuclear energy*, for it is the nucleus that is the source of the enormous energy associated with a nuclear reaction. In Chapter 2 (page 20) it was pointed out that the conservation of mass and energy laws must be combined to account for nuclear energy, in which matter is converted to energy. The conversion is expressed by Einstein's equation, $\Delta E = \Delta mc^2$, in which $\Delta E$ is the change in energy, $\Delta m$ is the change in mass and c is the speed of light. The amount of energy for a given quantity of matter is huge. For example, it can be shown that the energy released by changing a given mass of matter to energy is about 2.5 *billion* times greater than the energy released by burning the same mass of coal.

An atom of carbon-12 can be used to illustrate the mass-energy relationship. By the definition of an atomic mass unit, a carbon-12 atom has a mass of exactly 12 atomic mass units. The atom consists of a nucleus and 6 electrons. If the mass of the electrons (0.000549 amu each) is subtracted from the mass of the atom, the difference is the mass of the nucleus:

| | |
|---|---|
| Mass of atom: | 12.000000 amu |
| Mass of electrons ($6 \times 0.000549$): | $-0.003294$ amu |
| Mass of nucleus: | 11.996706 amu |

The nucleus is made up of 6 protons (1.00727 amu each) and 6 neutrons (1.00867 amu each). The sum of the masses of these component parts is

| | |
|---|---|
| Mass of protons ($6 \times 1.00728$): | 6.04368 amu |
| Mass of neutrons ($6 \times 1.00867$): | $+6.05202$ amu |
| Total | 12.09570 amu |

The difference between the total mass of the *parts* of the nucleus and the actual mass of the *whole* nucleus is $12.09570 - 11.996706 = 0.09899$ amu. This 0.09899 amu of mass is lost—converted to energy—when 6 protons and 6 neutrons combine to form a nucleus of $^{12}_{6}C$. The energy represented by this difference is called the **binding energy;** it is the energy that holds the nucleus together against the tremendous repulsion forces between the tightly packed positively charged protons.

In nuclear changes some of this binding energy is lost as one nucleus is destroyed, and another quantity of energy is absorbed as new nuclei form. The differences between these energies for all atoms involved represents the energy of a nuclear reaction.

## 18.12 ELECTRICAL ENERGY FROM NUCLEAR FISSION

Aside from hydroelectric plants located on major rivers, most of the electrical energy consumed in the world comes from generators driven by steam. Traditionally the steam comes from boilers fueled by oil, gas or coal. The fast dwindling supplies of these fossil fuels, and the uncertainties surrounding the availability and cost of petroleum from the countries where it is so abundant, have turned attention to nuclear fission as an alternative energy source.

A schematic diagram of a nuclear power plant is shown in Figure 18.11. The turbine, generator and condenser are similar to those found in any fuel-burning power plant. The nuclear fission reactor has three main components: the fuel elements, the control rods and the moderator. The fuel elements are simply long trays that hold fissionable material in the reactor. As the fission reaction proceeds, fast moving neutrons are released. These neutrons are slowed down by a moderator, which is water in the reactor illustrated. When the slower neutrons collide with more fissionable material the reaction is continued. The reaction rate is governed by cadmium control rods, which absorb excess neutrons. At times of peak power demand the control rods are largely withdrawn from the reactor, permitting as many neutrons as necessary to find fissionable nuclei. When demand drops, the control rods are pushed in, absorbing neutrons and limiting the reaction.

The energy released in a nuclear reactor appears as heat. This heat is

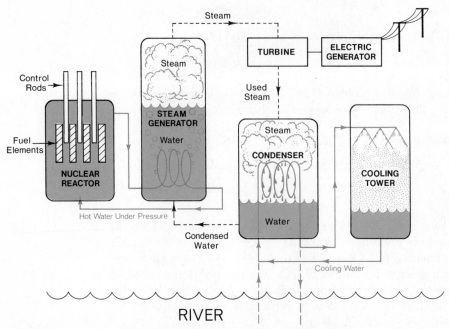

**Figure 18.11** Schematic illustration of a nuclear power plant. Nuclear fission occurs in the reactor. Control rods are largely withdrawn during period of high power demand, permitting much energy to be released in the fission process. Inserting control rods during low demand periods limits nuclear reaction and energy release. Fission energy is used to convert water to steam in the steam generator. High pressure steam is delivered to the turbine, where it drives the generator that produces electrical energy, which is distributed to the user. Spent steam from the turbine is converted to liquid water in the condenser, and then recycled back to the steam generator. Cooling water for the condenser comes from a cooling tower, to which it is recycled. Makeup cooling water, and sometimes the cooling water itself, is drawn from a river, lake or ocean.

transferred to the water, which is in the liquid state under a pressure of 50 atmospheres or more. The water from the reactor enters a heat exchanger at about 300°C. After giving up some of its heat in the heat exchanger it is returned to the reactor, as shown.

Low pressure water that runs through a coil in the heat exchanger absorbs the heat coming from the high temperature reactor water. The low pressure water forms steam, which is delivered to the turbines that drive the generators. Spent steam from the turbines is sent through a condenser where residual heat is removed and the steam condenses to water. The water is then cycled through the reactor heat exchanger again.

Nuclear power plants are confronted with serious problems. One is heat pollution released into a nearby body of water. This sometimes has a harmful effect on plant and animal life, destroying the ecological balance of the area.

Two other major problems must be resolved before nuclear power plants will be wholly accepted. First is the fact that many of the waste products of the fission reaction are radioactive. Their disposal must be accomplished in a way that will prevent the escape of radiation into the area surrounding the reactor. One method is to accumulate the materials in large containers that may be buried in the earth, as in Figure 18.12. The second major problem is the threat of an accident that might release large quantities of radiation on all the people in the area.

**Figure 18.12**  Burial of nuclear waste. Land burial trench at the Oak Ridge National Laboratory reservation. Each day's accumulation of waste containers is buried by 3 or more feet of earth. This is one of several partial solutions to the problem of disposal of radioactive waste.

The possible consequences of a nuclear accident underlie genuine concern over the growing number of nuclear power plants being used today, under construction for tomorrow, and in the design stage for the day after tomorrow. To the date of this writing, the safety record of nuclear power facilities is unblemished; but there have been a few "close calls" that are the cause of great alarm. The need for absolutely foolproof safeguards is clearly evident.

## 18.13  NUCLEAR FUSION

PG  18 Q  Define or identify a nuclear fusion reaction.

There is nothing new about nuclear energy. Man didn't invent it. In fact, without knowing it, man has been enjoying its benefits since the beginning of recorded time, and beyond. In its common form, though, we do not call it nuclear energy. We call it solar energy. It is the energy that comes from the sun.

The energy the earth derives from the sun comes from another type of nuclear reaction called **nuclear fusion, in which two small nuclei combine to form a larger nucleus.** The smaller nuclei are "fused" together, you might say. The typical fusion reaction believed to be responsible for the heat energy radiated by the sun is represented by the equation

$$^{2}_{1}H + ^{3}_{1}H \rightarrow ^{4}_{2}He + ^{1}_{0}n$$

Fusion processes are, in general, more energetic than fission reactions. The fusion of one gram of hydrogen in the above reaction yields about four times as much energy as the fission of an equal mass of uranium-235. So far man has been able to produce only one kind of fusion reaction, and that has been the explosion of a hydrogen bomb.

Much research effort is being made to develop nuclear fusion as a source of useful energy. It has several advantages over a fission reactor. It presents more energy per given quantity of fuel. The isotopes required for fusion are far more abundant than those needed for fission. Perhaps the biggest advantage is that fusion yields no radioactive waste, removing both the need for extensive disposal systems and the danger of accidental release of radiation to the atmosphere.

The major obstacle to be overcome in obtaining energy from a fusion reaction is the extremely high temperatures required to start and sustain the reaction. This is no problem in the sun, where temperatures are more than one million degrees. On earth, no substance now known can hold the reactants at the required temperature. The only presently known way to reach the necessary temperature in a hydrogen bomb is to explode a small fission bomb first. The problem is not considered insurmountable, however, and fusion could well be the ultimate answer to the world's energy needs in the 21st century.

# CHAPTER 18 IN REVIEW

**18.1 THE DAWN OF NUCLEAR CHEMISTRY (Page 433)**

**18.2 NATURAL RADIOACTIVITY**

**18 A** **Define radioactivity. Name, identify from a description or describe the three types of natural radioactive emission. (Page 434)**

**18.3 INTERACTION OF RADIOACTIVE EMISSIONS WITH MATTER**

**18 B** **Describe how emissions from radioactive substances affect gases and living organisms. (Page 435)**

**18.4 DETECTION OF EMISSIONS FROM RADIOACTIVE SUBSTANCES**

**18 C** **Define or describe a cloud chamber. Explain the principle by which it operates. (Page 436)**

**18 D** **Define or describe a Geiger counter. Explain the principle by which it operates. (Page 436)**

**18.5 NATURAL RADIOACTIVE DECAY SERIES—NUCLEAR EQUATIONS**

**18 E** **Describe a natural radioactive decay series. (Page 438)**

18 F   Given the identity of a radioactive isotope and the particle it emits, write a nuclear equation for the emission. (Page 438)

### 18.6   HALF-LIFE

18 G   Describe or illustrate what is meant by the half-life of a radioactive isotope. (Page 440)

18 H   Given the original quantity of a radioactive isotope and its half-life, and either the final quantity or elapsed time, find the time or quantity not given. (Whole-number of half-lives only.) (Page 440)

### 18.7   NUCLEAR REACTIONS AND ORDINARY CHEMICAL REACTIONS COMPARED

18 I   List or identify four ways in which nuclear reactions differ from ordinary chemical reactions. (Page 443)

### 18.8   NUCLEAR BOMBARDMENT AND ARTIFICIAL RADIOACTIVITY

18 J   Define nuclear bombardment. (Page 444)

18 K   Distinguish between natural radioactivity and artificial radioactivity. (Page 444)

18 L   Define or identify transuranium elements. (Page 444)

18 M   Given the identity of all except one of the isotopes and particles in a nuclear bombardment, write the nuclear equation for the reaction. (Page 444)

### 18.9   USES OF RADIOISOTOPES (Page 448)

### 18.10   NUCLEAR FISSION

18 N   Define or identify a nuclear fission reaction. (Page 449)

18 O   Define or identify a chain reaction. (Page 449)

### 18.11   NUCLEAR ENERGY

18 P   Explain the source of energy in a nuclear reaction. (Page 450)

### 18.12   ELECTRICAL ENERGY FROM NUCLEAR FISSION (Page 451)

### 18.13   NUCLEAR FUSION

18 Q   Define or identify a nuclear fusion reaction. (Page 453)

## TERMS AND CONCEPTS

# QUESTIONS AND PROBLEMS

*An asterisk (\*) identifies a question that is relatively difficult, or that extends beyond the performance goals of the chapter.*

## Section 18.2

18.1) What is meant by radioactivity?

18.2) Identify the three types of emission normally associated with radioactive decay. What is each type of emission made of; that is, what is its "structure?" Write the nuclear symbol, if any, for each kind of emission.

## Section 18.3

18.3) When an emission from a radioactive substance passes through air or any other gas, what effect does it have?

## Section 18.4

18.4) What is a cloud chamber? By what process do "tracks" of radioactive emission particles appear?

## Section 18.5

18.5) What is meant by the expression *transmutation of an element?*

18.6) Describe the change in mass number and atomic number that accompanies an alpha particle emission from a radioactive nucleus.

18.7) Write nuclear equations for beta emissions from $^{212}_{82}Pb$ and from $^{231}_{90}Th$.

18.8) Write nuclear equations for the ejection of an alpha particle from $^{228}_{90}Th$ and from $^{222}_{86}Rn$.

## Section 18.6

18.9) What is meant by the half-life of a radioactive substance? What fraction of an original sample remains after the passage of six half-lives?

18.10) The half-life of $^{210}_{82}Pb$ is 22 years. How many grams of that isotope will remain after 66 years if the original sample had a mass of 100 grams?

18.11) One of the more hazardous radioactive isotopes in the fallout of atomic bombs is strontium-90, $^{90}_{38}Sr$, for which $t_{\frac{1}{2}} = 28$ years. If 600 grams of Sr-90 descended on a family farm on the day a child was born in 1950, how many grams will still be on that farm when his granddaughter is born in 2006?

18.27) What is the meaning of the terms *disintegration* and *decay* in relation to radioactivity?

18.28) Compare the three principal forms of radioactive emission in terms of mass, electrical charge and penetrating power.

18.29) Explain how rays from radioactive substances can cause injury or damage to internal tissue in a living organism.

18.30) What is a Geiger counter? How does it work?

18.31) What is a *natural radioactive decay series?* How many such series have been found?

18.32) If a radioactive nucleus emits a beta particle, what change will occur in the atomic number and mass number of the nucleus that remains?

18.33) Write nuclear equations for beta emissions from $^{228}_{89}Ac$ and from $^{214}_{83}Bi$.

18.34) Write nuclear equations for alpha decay of $^{216}_{84}Po$; of $^{234}_{92}U$.

18.35) Suggest some of the difficulties that might surround the determination of physical properties of the radioactive isotope of an element that has a short half-life. What fraction of the sample would remain after the passage of four half-lives?

18.36) $^{222}_{86}Rn$ has a half-life of roughly half a week. If you had 50 milligrams of that isotope, how many milligrams would be left after you returned from a three week vacation?

18.37) A radiochemistry research laboratory ends its workday at 5 PM, and resumes at 8 AM the next morning. If they had in storage 4.8 centigrams of $^{24}_{11}Na$ ($t_{\frac{1}{2}} = 15$ hours) at the close of a work day, how much would they have at the beginning of the next work day? What would be the beginning supply on the Tuesday morning after a three day weekend that began at 2 PM the previous Friday, if they left 24.0 centigrams in their vault at the start of the holiday?

18.12) The half-life of $^{13}_{7}N$ is 10.0 minutes. How long will it take for the mass of that particular isotope in a sample of ammonium chloride to be reduced to 0.120 gram if the sample originally contained 0.960 gram?

18.13) Analysis of charcoal taken from an ancient campfire yielded a $^{14}_{6}C/^{12}_{6}C$ ratio that is 0.125 times the ratio in trees living today. This is interpreted as meaning that of the $^{14}_{6}C$ originally present, only $\frac{1}{8}$ remains today. If $t_{\frac{1}{2}} = 5720$ years for carbon-14, estimate the age of the sample.

**Section 18.7**

18.14) Explain why the chemical properties of an element are the same for all isotopes in an ordinary chemical change, but depend on the particular isotope for a nuclear change.

18.15)* If the uranium in pure $UCl_4$ and $UBr_4$ has all isotopes in their normal percentage distribution in nature, which will exhibit the greatest amount of radioactivity, 100 grams of $UCl_4$ or 100 grams of $UBr_4$? How about 0.10 mole of $UCl_4$ or 0.10 mole of $UBr_4$? Explain both answers.

18.16) A fundamental idea in Dalton's atomic theory is that atoms of an element can be neither created nor destroyed. How, then, can you account for the fact that the number of lead atoms in the world is constantly increasing?

**Section 18.8**

18.17) What is the meaning of the expression *nuclear bombardment?*

18.18)* Particles used for nuclear bombardment reactions frequently do not have sufficient kinetic energy when obtained from their natural sources. Identify some of the "particle accelerators" that have been developed to increase their kinetic energy.

18.19) What are transuranium elements? Where on the periodic table are they located?

18.38) The half-life of iodine-131, an isotope used in radiotherapy, is eight days. If a hospital wishes to have available at all times material containing no less than 10 micrograms of the isotope, how soon after the inventory shows 80 micrograms will they have to restock, assuming none of the supply is used during the period?

18.39) The full decay process for $^{238}_{92}U$ may be expressed by the overall equation

$$^{238}_{92}U \rightarrow {}^{206}_{82}Pb + 8\,^{4}_{2}He + 6\,_{-1}^{0}e$$

The half-life for the process is $4.5 \times 10^9$ years. Assuming all the lead is from the uranium originally present, how old is a rock sample that contains equal quantities of $^{238}_{92}U$ and $^{206}_{82}Pb$?

18.40) Two bottles of the same lead compound are carelessly left exposed. Explain the circumstances under which one of these bottles might be hazardous but not the other.

18.41) An ore sample containing a certain quantity of radioactive uranium disintegrates at 7000 counts per minute, a way of expressing rate of radioactive decay when measured with a Geiger counter. If all the uranium in the sample is extracted and isolated as a pure element, would you expect the rate of decay to remain at 7000 counts per minute, would it be less than 7000, or would it be more? (Disregard any loss of the radioactive isotope because of disintegration during the extraction process.) Explain your answer.

18.42) A radiochemical laboratory prepares a sample of pure KCl containing a measurable amount of $^{43}_{19}K$, a radioactive isotope that emits a beta particle and has a half-life of 22.4 hours. The compound is securely stored overnight in an inert atmosphere. The next day the compound will no longer be pure. Why? With what element would you expect it to be contaminated?

18.43) How does artificial radioactivity differ from natural radioactivity?

18.44)* Describe the principle by which a particle accelerator increases the kinetic energy of a particle used in nuclear bombardment. What major nuclear particle cannot be accelerated? Why?

18.45)* The first eleven transuranium elements were discovered at the Lawrence Radiation Laboratory, named after the inventor of the cyclotron, at the University of California, Berkeley. (Other laboratories participated with the Berkeley group in the discovery of some of the elements.) The names of the elements represent the continuation of a series already stated with uranium, some nationalism, some state and school loyalties, and the desire to honor men who have made major contributions to chemistry, mostly in the area of nuclear chemistry. You will find it interesting to review those names to see if you can connect them with their sources, most of which are identified in this chapter.

*Questions 18.19–18.22 and 18.46–18.48: Complete each of the following nuclear bombardment equations by supplying the nuclear symbol for the missing species:*

18.20) $^{98}_{42}Mo + ^{2}_{1}H \rightarrow ? + ^{1}_{0}n$

18.21) $^{238}_{92}U + ^{4}_{2}He \rightarrow 3\,^{1}_{0}n + ?$

18.22) $? + ^{2}_{1}H \rightarrow ^{60}_{27}Co + ^{1}_{1}H$

18.46) $^{44}_{20}Ca + ^{1}_{1}H \rightarrow ? + ^{1}_{0}n$

18.47) $^{252}_{98}Cf + ^{10}_{5}B \rightarrow 5\,^{1}_{0}n + ?$

18.48) $^{106}_{46}Pd + ^{4}_{2}He \rightarrow ^{109}_{47}Ag + ?$

## Section 18.10

18.23) Explain what is meant by a *fission* reaction. You will find the significance of this term easier to remember if you associate it with another science course you probably have taken. Even without the course, looking the word up in the dictionary should fix its significance in your thought.

18.49) What is a *chain reaction*? What essential feature must be present in a reaction before it can become a chain reaction?

## Section 18.11

18.24) What is the source of the enormous energy released in a nuclear reaction?

18.50) For all practical purposes, the Law of Conservation of Mass is obeyed in ordinary chemical reactions, but not in nuclear reactions. Would the mass of the products of an atomic bomb explosion be more than or less than the mass of the reactants? Explain your answer.

## Section 18.12

18.25) List some of the principal advantages that are associated with nuclear power plants as a source of electrical energy.

18.51)* In January, 1977, when this question was written, there were serious concerns about using nuclear reactors as a source of electrical energy. List those concerns. In the period since 1977, have there been changes that have removed or reduced the earlier objections to nuclear energy? If so, identify them. Has anything happened over the same period to show that the worries of 1977 were justified, and that continuing to build nuclear power plants was an unwise procedure? If so, identify the events. How do you feel about nuclear power sources today?

## Section 18.13

18.26) What is a *fusion* reaction? How does it differ from a fission reaction?

18.52) Why is nuclear fusion more promising as a source of electrical energy than nuclear fission? What major obstacle prevents us from building nuclear fusion power plants?

# INTRODUCTION TO
# ORGANIC CHEMISTRY

Chapter 19 is a brief survey of organic chemistry. It is unlike the first eighteen chapters, in which we could expect to find mastery of specific chemical concepts. While it is reasonable to expect comparable achievement in selected areas of this chapter, most topics are presented so briefly it is unrealistic to establish for them the kind of performance goals that have accompanied the other chapters. In their place the following chapter-wide performance goals are offered for your guidance:

1. Distinguish between organic and inorganic chemistry.
2. Define a hydrocarbon.
3. Distinguish between saturated and unsaturated hydrocarbons.
4. Write, recognize or otherwise identify (1) the structural unit, or functional group, (2) the general formula, and (3) molecular or structural formulas and/or names of specific examples, of the following classes of organic compounds: alkanes, alkenes, alkynes, aromatic hydrocarbons, alcohols, ethers, aldehydes, ketones, carboxylic acids and esters.
5. Define and give examples of isomerism.

## 19.1  THE NATURE OF ORGANIC CHEMISTRY

In the early development of the science of chemistry the logical starting point was a study of substances that occur in nature. As in the organization of any body of accumulated knowledge, substances were classified into groups having certain common characteristics. One classification system for natural substances assigns them to groups labeled animal, vegetable or mineral. So far in this text attention has been directed almost entirely to the last of these three, minerals and the compounds that may be derived from them. These substances constitute that area of chemistry commonly known as *inorganic chemistry.*

Animal and vegetable substances are, or at one time were, composed of living organisms, a distinction that sets them apart from inorganic substances. **Organic chemistry** was therefore originally defined as the chemistry of living organisms, extended to include those compounds directly derived from living organisms by natural processes of decay. It was learned, however, that compounds called "organic" can be produced from inorganic chemicals. The scope of organic chemistry was therefore expanded to include these substances as well. With the further passage of time it was recognized that all compounds called "organic" contain the element *carbon.* Consequently, the modern day definition of organic chemistry is **the chemistry of carbon compounds.** This definition acknowledges the exception of certain carbon-bearing mineral substances such as carbonates, cyanides, oxides of carbon, and a few other compounds.

Today it is generally acknowledged that the only unique features of organic chemistry are the aforementioned presence of carbon in all organic compounds, and the vast and rapidly growing number of organic compounds —many times more than the total number of known compounds that do not contain carbon. All of the chemical principles we have studied in the context of inorganic chemicals, such as bonding, reaction rates, equilibrium and others, apply equally to organic compounds. In particular, a clear picture of bonding and the structure of molecules is the cornerstone of all that we understand about organic chemistry, as the remaining pages will show.

Some science-fiction writers have speculated about the possibility that "life" can be based on some element other than carbon. Silicon is a possibility. The particular value of carbon to life on earth is due to its ability to form long chains of carbon atoms. Without these chains, none of the complex proteins that comprise living organisms could be formed. No other element can form such chains on earth. However, perhaps somewhere in the universe, a form of life based on completely different chemical structures has found an environment suitable for it.

## 19.2 THE MOLECULAR STRUCTURE OF ORGANIC COMPOUNDS

Table 19.1 summarizes the covalent bonding properties of carbon, hydrogen, oxygen, nitrogen and the halogens, the elements most frequently found in organic compounds. All the bond geometries are indicated—the linear, planar or three dimensional shapes, and, where constant, the actual bond angles. Of particular significance is the number of covalent bonds atoms of the different elements can form, when each atom contributes one electron to the bond. This is determined by the electron configuration of the atom. The bonding relationships of these elements are fundamental to your understanding of the structure of organic compounds.

When they are all single bonds, the geometry of the four bonds of carbon imposes a limitation on the accuracy by which molecular shape may be described in a book. The directional orientation of the bonds around the

TABLE 19.1   Bonding in Organic Compounds

| ELEMENT | NUMBER OF BONDS* | BOND GEOMETRY | | | |
|---|---|---|---|---|---|
| | | *Single Bond* | *Double Bond* | *Double Bonds* | *Triple Bond* |
| Carbon | 4 | C (Tetrahedral— 109.5° angles) | C= (Planar— 120° angles) | =C= (Linear— 180° angle) | —C≡ (Linear— 180° angle) |
| Hydrogen | 1 | H— | | | |
| Halogens | 1 | :Ẍ— | | | |
| Oxygen | 2 | Ö (Bent Structure) | Ö= | | |
| Nitrogen | 3 | N̈ (Pyramidal Structure) | N̈ (Bent Structure) | | :N≡ |

*Number of bonds to which an atom of the element shown can contribute *one* electron.

carbon atom is tetrahedral (see Fig. 19.1), a shape not accurately shown on a two dimensional page. If, in the figure, bond axis number 1 rising from the top of the atom is *in* the plane of the paper, then bond axis number 2 is coming *out* of the plane of the paper toward you, while bond axes 3 and 4 are extending *behind* the plane of the paper.

**Figure 19.1** Tetrahedral bond orientation about a carbon atom.

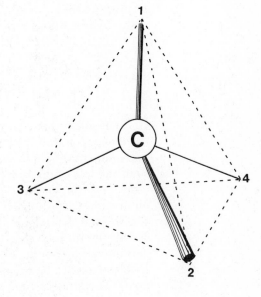

Rather than wrestle with these difficulties in accurate illustrations, the chemist usually represents the four bonds radiating from a carbon atom thus:

$$-\overset{\displaystyle |}{\underset{\displaystyle |}{C}}-$$

These bonds are identical from the standpoint of directional orientation. Thus in a three-carbon chain,

$$x-\overset{\displaystyle x}{\underset{\displaystyle x}{C}}-\overset{\displaystyle z}{\underset{\displaystyle z}{C}}-\overset{\displaystyle y}{\underset{\displaystyle y}{C}}-y$$

all three x positions are geometrically equal, three y positions are equal, and two z positions are equal. It follows that

$$H-\overset{\displaystyle H}{\underset{\displaystyle H}{C}}-\overset{\displaystyle Br}{\underset{\displaystyle H}{C}}-\overset{\displaystyle H}{\underset{\displaystyle H}{C}}-Cl \quad \text{and} \quad H-\overset{\displaystyle H}{\underset{\displaystyle H}{C}}-\overset{\displaystyle H}{\underset{\displaystyle Br}{C}}-\overset{\displaystyle Cl}{\underset{\displaystyle H}{C}}-H$$

are identical compounds.

## HYDROCARBONS

The simplest type of organic compound is the hydrocarbon. As the name suggests, **hydrocarbons** consist of two elements, hydrogen and carbon. A hydrocarbon may be classified into one of several categories based upon its structure: (a) alkanes; (b) alkenes; (c) alkynes; and (d) aromatic hydrocarbons. The first three of these are sometimes grouped together as the *aliphatic* hydrocarbons, in which the carbon atoms are arranged in chains. Representative aliphatic hydrocarbons are shown in Table 19.2; the aromatic hydrocarbons will be considered separately.

## 19.3   SATURATED HYDROCARBONS: THE ALKANES

The **alkanes** are known as **saturated hydrocarbons.** This means that each carbon atom is bonded by four single covalent bonds to four other atoms, the maximum number possible according to the octet rule (see page 193). Structural models of the first three alkane molecules, methane, ethane and propane, are shown in Figure 19.2. Notice the tetrahedral orientation of atoms bonded to carbon in all three molecules. Notice also with propane that this tetrahedral arrangement yields a zig-zag pattern to the carbon chain. Furthermore, each side of the molecule is free to rotate around the single carbon-carbon bonds, so that the "end" carbon atoms may, at one moment, be close to each other, and at a different moment they may be relatively far apart. The straight lines we draw on paper are not true representations of carbon chains.

TABLE 19.2 Aliphatic Hydrocarbons

| CARBON ATOMS | ALKANE | ALKENE | ALKYNE |
|---|---|---|---|
| 1 | H—C—H (with H above and below) $CH_4$ Methane | | |
| 2 | H—C—C—H (with H's) $C_2H_6$ Ethane | H—C=C—H (with H's) $C_2H_4$ Ethylene Ethene | H—C≡C—H $C_2H_2$ Acetylene Ethyne |
| 3 | H—C—C—C—H (with H's) $C_3H_8$ Propane | H—C=C—C—H (with H's) $C_3H_6$ Propylene Propene | H—C≡C—C—H (with H's) $C_3H_4$ Propyne |
| Structural Unit or Functional Group | —C— | C=C | —C≡C— |

## THE GENERAL FORMULA FOR THE ALKANES—A HOMOLOGOUS SERIES

If you examine the molecular and structural formulas of methane, ethane and propane, you will find a pattern developing. Each additional carbon atom is accompanied by two more hydrogen atoms. The alkane with four carbon atoms is butane, $C_4H_{10}$; five carbon atoms yield pentane, $C_5H_{12}$; and so forth, with each additional step extending the chain by a —$CH_2$ structural unit.

A series of compounds in which each member differs from the members before and after it by the same structural unit is called a **homologous series.** The alkane series may be represented by the general formula, $C_nH_{2n+2}$, where n is the number of carbon atoms in the molecule. With this general formula you can produce the molecular formula for any member of the series. For octane, the alkane with 8 carbon atoms, n = 8. The number of hydrogen atoms is 2(8) + 2 = 18. The formula of octane is therefore $C_8H_{18}$.

**Figure 19.2** Ball-and-stick and space-filling models of the first three members of the alkane hydrocarbon series. (Photograph by Janice Peters.)

The formulas and names of the first ten alkanes are shown in Table 19.3. Also shown are the melting and boiling points of the so-called **normal** alkanes, those in which the carbon atoms form a continuous chain. You will recall from Chapter 12 (page 278) that intermolecular forces between nonpolar molecules increase with increasing molecular size, and that stronger intermolecular attractions yield higher boiling points. Accordingly, alkanes having fewer than 5 carbons have the weakest intermolecular attractions, low boiling points, and are gases at normal temperatures. All are used as

TABLE 19.3  The Alkane Series

| MOLECULAR FORMULA | NAME | NUMBER OF CARBON ATOMS | PREFIX | MELTING POINT (°C) | BOILING POINT (°C) |
|---|---|---|---|---|---|
| $CH_4$ | Methane | 1 | Meth- | −183 | −162 |
| $C_2H_6$ | Ethane | 2 | Eth- | −172 | −89 |
| $C_3H_8$ | Propane | 3 | Prop- | −187 | −42 |
| $C_4H_{10}$ | Butane | 4 | But- | −138 | 0 |
| $C_5H_{12}$ | Pentane | 5 | Pent- | −130 | 36 |
| $C_6H_{14}$ | Hexane | 6 | Hex- | −95 | 69 |
| $C_7H_{16}$ | Heptane | 7 | Hept- | −91 | 98 |
| $C_8H_{18}$ | Octane | 8 | Oct- | −57 | 126 |
| $C_9H_{20}$ | Nonane | 9 | Non- | −54 | 151 |
| $C_{10}H_{22}$ | Decane | 10 | Dec- | −30 | 174 |

fuels for stoves. Methane is the main constituent of "natural gas." It is also found in large quantities in the forbidding atmosphere of planets like Jupiter and Saturn.

Intermolecular forces are stronger between the larger alkane molecules from $C_5H_{12}$ to $C_{17}H_{35}$. These higher boiling compounds are liquids at room temperature. Several of the lower molecular weight liquid alkanes, notably octane, are present in gasoline. Diesel fuel and lubricating oils are made up largely of higher molecular weight liquid alkanes. Alkanes with molecular weights greater than 250 are normally solids at room temperature.

Because of isomerism (see next subsection), a molecular formula does not adequately identify a compound. Structural formulas, on the other hand, are complex, though frequently the only satisfactory representation of a compound. A compromise sometimes used is the *condensed formula,* or *line formula,* in which the structure is indicated by repeating the groups it contains. Line formulas for a few sample alkanes are shown below:

| | |
|---|---|
| Ethane, $C_2H_6$ | $CH_3CH_3$ |
| Propane, $C_3H_8$ | $CH_3CH_2CH_3$ |
| Butane, $C_4H_{10}$ | $CH_3CH_2CH_2CH_3$ |
| Octane, $C_8H_{18}$ | $CH_3CH_2CH_2CH_2CH_2CH_2CH_2CH_3$ |

When line formulas become long, as in the case of octane, they are sometimes shortened by grouping the $CH_2$ units: $CH_3(CH_2)_6CH_3$.

The alkane hydrocarbons also serve to introduce organic nomenclature. Table 19.3 illustrates the system for the first ten alkanes. Each alkane is named by combining a prefix and a suffix. The prefix indicates the number of carbons in the chain. The first ten prefixes used in this nomenclature system appear in the fourth column of Table 19.3. The suffix identifying an alk*ane* is -*ane*. Thus the name of methane comes from combining the prefix *meth-*, indicating one carbon atom, with the suffix -*ane* indicating an alkane. We shall see shortly that these prefixes are used in naming other organic compounds and groups as well.

## ISOMERISM IN THE ALKANE SERIES

Not all alkanes have their carbons bonded in a continuous chain; some have branches. The smallest alkane in which this is possible is butane, which has two possible structures:*

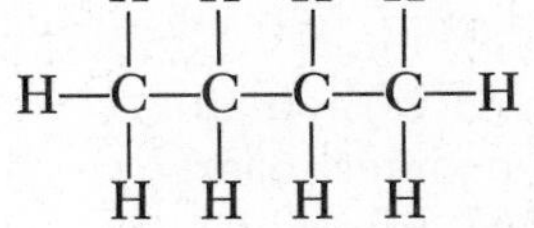

n-butane
("normal butane")

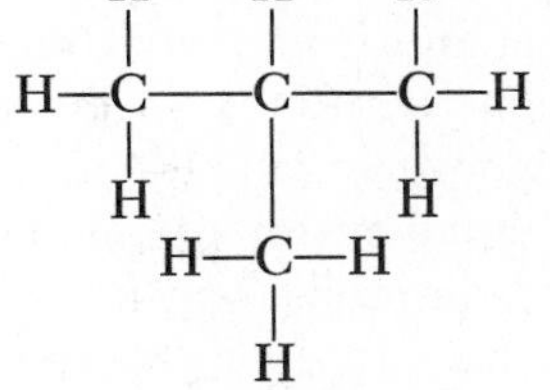

Isobutane

*The length of lines representing bonds has no significance. Different lengths are used only to separate on paper those parts of the molecule that are not bonded to each other.

You will observe that in the compound at the left, called normal butane, or $n$-butane, the four carbons are in a single chain. In isobutane, the structure at the right, there are but three carbons in the chain, with the fourth carbon branching off the middle carbon of the three. Both compounds have the same molecular formula, $C_4H_{10}$. These compounds are **isomers: two compounds having the same molecular formula but different structures are called isomers.** It should be realized that isomers are distinctly different compounds, having different physical and chemical properties.

The number of isomers that are possible increases rapidly with the number of carbon atoms in the compound. There are three different pentanes; their structures, showing only the carbon skeletons to make the diagrams less "cluttered," are

| $n$-pentane | Isopentane | Neopentane |

There are five isomeric hexanes, nine heptanes, and seventy-five possible decanes. It is possible to draw over 300,000 isomeric structures for $C_{20}H_{42}$, and more than one hundred million for $C_{30}H_{62}$. Obviously not all of them have been prepared and identified! This does give us some idea, though, why there are so many organic compounds.

(You might like to try your hand at writing isomers of the alkanes. See if you can sketch the five isomers of hexane. In doing so, be sure they are all different. You would be wise to start with the longest chain possible, then shorten it by one, and again shorten it by one, drawing all possible structures each time until all combinations are exhausted. The correct diagrams are shown on page 469.)

Distinguishing between isomeric structures requires some extension of nomenclature rules. As noted earlier, a continuous straight chain alkane is called a normal alkane; hence the name "normal butane," written $n$-butane (page 465). In a normal alkane each carbon atom is bonded to no more than two other carbon atoms. The branched chain isomer of butane is called isobutane. In it, one carbon atom is bonded to three other carbon atoms. The *normal* and *iso-* terminology is carried forth to the isomers of pentane, and is expanded to *neo*pentane to provide for the third isomer in which one carbon atom is bonded to four other carbon atoms.

Beyond pentane the number of isomers becomes so large it is necessary to develop a systematic nomenclature. Several different systems exist and are in current use, but the one most widely adopted is the IUPAC* system. Before we can look upon this system, however, we must establish the concept of *alkyl groups*, which may be derived directly from the alkane hydrocarbons.

---

*International Union of Pure and Applied Chemistry—see page 218.

## ALKYL GROUPS

As in inorganic chemistry we have found it convenient to assign names to certain groups of atoms that function as units in forming chemical compounds (e.g., sulfate, nitrate, ammonium), so in organic chemistry it is convenient to identify **alkyl groups** that may be derived from an alkane. If, on paper, we isolate a hydrogen atom from methane, $CH_4$, we get $—CH_3$. This $—CH_3$ group, appearing in the structural formula of a compound, is called a **methyl** group, the term being made up of the prefix *meth-* for one carbon (see Table 19.3) and the suffix *-yl* applied to all alkyl groups. If we compare two compounds,

and

we see that the colored H in the first compound has been replaced by a $—CH_3$ group, or methyl group, in the second. If the replacement group has two carbon atoms,

it is an ethyl group, $—C_2H_5$, one hydrogen short of the alkane ethane, $C_2H_6$. All the alkyl groups are similarly named.

Frequently we wish to show a bonding situation in which *any* alkyl group may appear. The letter R is used for this purpose. Thus R—OH could be $CH_3OH$, $C_2H_5OH$, $C_3H_7OH$, or any other alkyl group attached to an —OH group.

Some chemists consider alkyl groups as **functional groups** also. A functional group is **an atom or group of atoms that both establishes the identity of a class of compounds and determines its chemical properties.** The structural units in the bottom row of Table 19.2 (page 463) may be considered as the functional groups of the aliphatic hydrocarbons. Other functional groups will appear later in the chapter.

## NAMING THE ALKANES BY THE IUPAC SYSTEM

We are now ready to describe the IUPAC system of naming isomers of the alkanes, a system which, incidentally, applies equally well to other compounds we shall encounter shortly. The system follows a set of rules:

1. **Identify as the parent alkane the longest continuous chain.** For example, in the compound having the structure

$$
\begin{array}{ccccc}
| & | & | & | & | \\
-\text{C}- & \text{C}- & \text{C}- & \text{C}- & \text{C}- \\
| & | & | & | & | \\
 & & & \text{C}- & \\
 & & & | & \\
 & & & \text{C}- & \\
 & & & | &
\end{array}
$$

the longest chain is six carbons long, not five as you might first expect. This is readily apparent if we number the carbon atoms in the original representation of the structure and an equivalent layout:

$$
\begin{array}{ccccc}
|_6 & |_5 & |_4 & |_3 & | \\
-\text{C}- & \text{C}- & \text{C}- & \text{C}- & \text{C}- \\
| & | & | & |_2 & | \\
 & & & \text{C}- & \\
 & & & |_1 & \\
 & & & \text{C}- & \\
 & & & | &
\end{array}
\qquad
\begin{array}{cccccc}
 & & & & & | \\
 & & & & & \text{C}- \\
|_6 & |_5 & |_4 & |_3 & |_2 & |_1 \\
-\text{C}- & \text{C}- & \text{C}- & \text{C}- & \text{C}- & \text{C}- \\
| & | & | & | & | & |
\end{array}
$$

2. **Identify by number the carbon atom to which the alkyl group (or other species) is bonded to the chain.** In the example compound this is the *third* carbon, as shown. Notice that counting always begins at that end of the chain that places the branch on the *lowest* number carbon atom possible.

3. **Identify the branched group (or other species).** In this example the branch is a methyl group, —CH$_3$, shown in color.

$$
\begin{array}{cccccc}
| & | & | & | & & \text{H} \\
 & & & & & | \\
-\text{C}- & \text{C}- & \text{C}- & \text{C}- & & \text{C}-\text{H} \\
| & | & | & | & & | \\
 & & & \text{C}- & & \text{H} \\
 & & & | & & \\
 & & & \text{C}- & & \\
 & & & | & &
\end{array}
$$

These three items of information are combined to produce the name of the compound, *3-methylhexane*. The 3 comes from the third carbon (step 2); methyl comes from the branch group (step 3); and hexane is the parent alkane (step 1).

Sometimes the same branch appears more than once in a single compound. This situation is governed by the following rule:

4. **If the same alkyl group, or other species, appears more than once, indicate the number of appearances by di-, tri-, tetra-, etc., and show the location of each branch by number.**

For example,

$$
\begin{array}{ccccc}
 & & \text{Cl} & \text{Cl} & \\
| & | & | & | & | \\
-\text{C}- & \text{C}- & \text{C}- & \text{C}- & \text{C}- \\
| & | & | & | & |
\end{array}
$$

would be 2,3-dichloropentane. In the other direction, to write the structural formula for 1,1,5-tribromohexane, we would establish a six-carbon skeleton and attach bromines as required, two to the first carbon and one to the fifth:

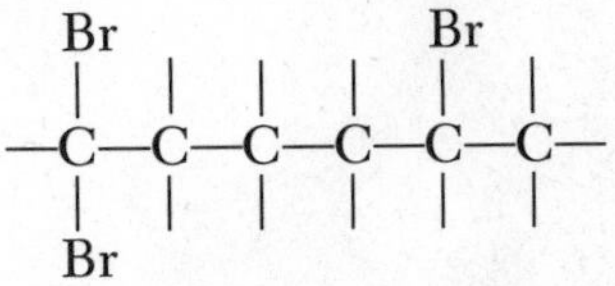

$$
\begin{array}{c}
\text{Br} \qquad\qquad\qquad \text{Br} \\
| \quad | \quad | \quad | \quad | \quad | \\
-\text{C}-\text{C}-\text{C}-\text{C}-\text{C}-\text{C}- \\
| \quad | \quad | \quad | \quad | \quad | \\
\text{Br}
\end{array}
$$

5. **If two or more different alkyl groups, or other species, are attached to the parent chain, they are named in order of increasing group size or in alphabetical order.** By this rule the compound

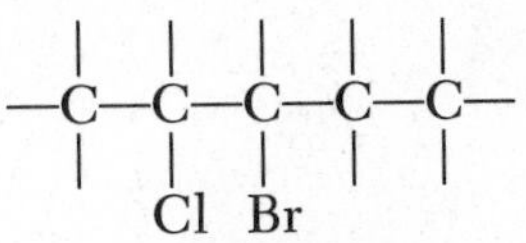

$$
\begin{array}{c}
| \quad | \quad | \quad | \quad | \\
-\text{C}-\text{C}-\text{C}-\text{C}-\text{C}- \\
| \quad | \quad | \quad | \quad | \\
\quad\; \text{Cl} \;\; \text{Br}
\end{array}
$$

would be named 3-bromo-2-chloropentane. The formula for 2,2-dibromo-4-chloroheptane would be

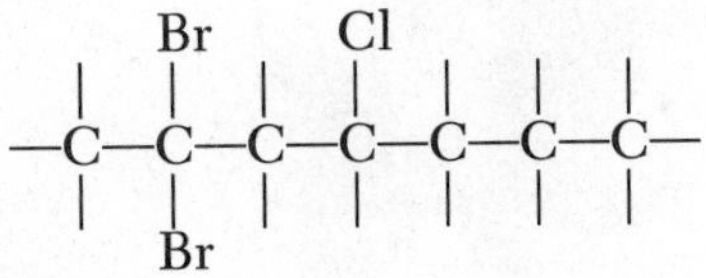

$$
\begin{array}{c}
\text{Br} \qquad \text{Cl} \\
| \quad | \quad | \quad | \quad | \quad | \quad | \\
-\text{C}-\text{C}-\text{C}-\text{C}-\text{C}-\text{C}-\text{C}- \\
| \quad | \quad | \quad | \quad | \quad | \quad | \\
\quad\;\; \text{Br}
\end{array}
$$

The five isomers of hexane, models of which are pictured in Figure 19.3, further illustrate the application of these rules:

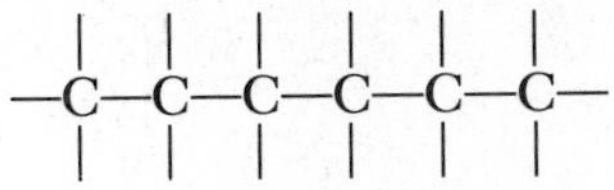

$$
\begin{array}{c}
| \quad | \quad | \quad | \quad | \quad | \\
-\text{C}-\text{C}-\text{C}-\text{C}-\text{C}-\text{C}- \\
| \quad | \quad | \quad | \quad | \quad | \\
\end{array}
$$

*n*-hexane

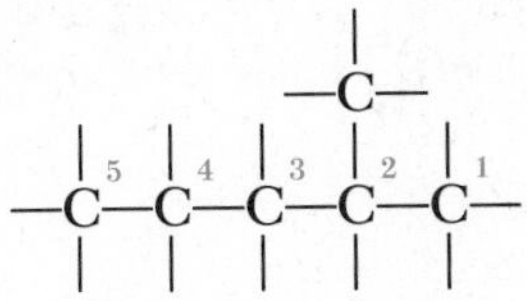

2-methylpentane

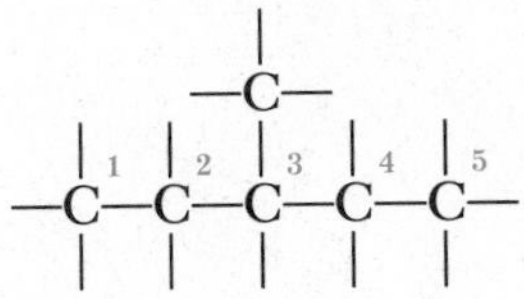

3-methylpentane

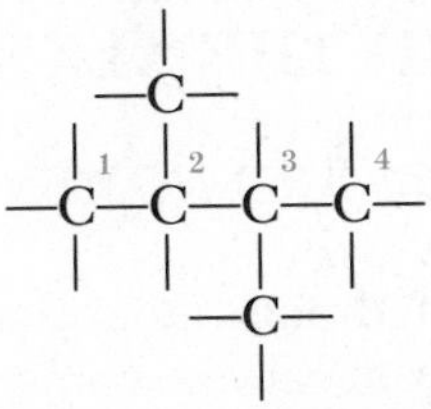

2,3-dimethylbutane

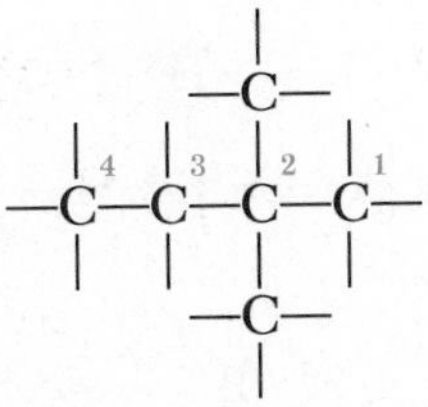

2,2-dimethylbutane

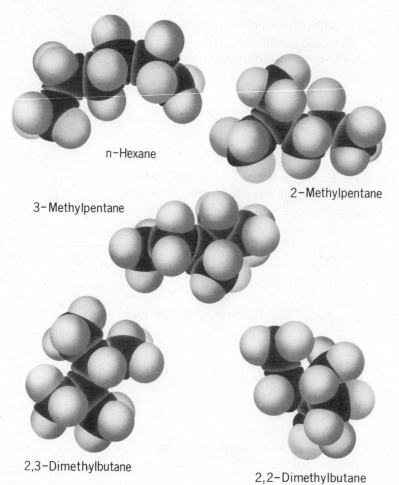

n–Hexane

3–Methylpentane

2–Methylpentane

2,3–Dimethylbutane

2,2–Dimethylbutane

**Figure 19.3**   Space-filling models of the isomeric hexanes.

19.4 ## UNSATURATED HYDROCARBONS: THE ALKENES AND THE ALKYNES

### STRUCTURE AND NOMENCLATURE

If one hydrogen atom, complete with its electron, is removed from each of two adjacent carbon atoms in an alkane (A below), each carbon is left with a single unpaired electron (B). These electrons may then form a second bond between the two carbon atoms (C):

Each carbon atom now is bonded to three other atoms, one less than the maximum permitted by the octet rule. The hydrocarbon is therefore said to be **unsaturated.** An aliphatic hydrocarbon with a double bond is called an **alkene.** Figure 19.4 shows two models of the simplest alkene.

Removal of another hydrogen atom from each of the doubly bonded carbon atoms in an alkene yields a triple bond:

$$H-\overset{\overset{\displaystyle H}{|}}{C}=\overset{\overset{\displaystyle H}{|}}{C}-H \xrightarrow{-2\ H\cdot} H-\dot{C}=\dot{C}-H \longrightarrow H-C\equiv C-H$$

Each of the carbon atoms is now bonded to two other atoms. This compound is also unsaturated. An aliphatic hydrocarbon containing a triple bond is called an **alkyne.** Models of acetylene, the most common alkyne, are shown in Figure 19.4.

Both the alkenes and the alkynes make up a new homologous series. Just as with the alkanes, each series may be extended by adding —$CH_2$ units. Longer chains may have more than one multiple bond, but we will not consider such compounds in this text. The general formula for an alkene is $C_nH_{2n}$, and for an alkyne, $C_nH_{2n-2}$.

Table 19.4 gives the names and formulas of some of the simpler unsaturated hydrocarbons. The IUPAC nomenclature system for the alkenes matches that of the alkanes. The suffix designating the alkene hydrocarbon series is -*ene*, just as -*ane* identifies an alkane. The same prefixes are used to show the total number of carbon atoms in the molecule. For example, pentene is $C_5H_{10}$, hexene is $C_6H_{12}$, and octene is $C_8H_{16}$. The common names for the alkenes are produced similarly, except that the suffix is -*ylene*. These names are firmly entrenched in reference to the lower alkenes: $C_2H_4$ is almost always called ethylene, $C_3H_6$ is propylene and $C_4H_8$ is butylene.

Acetylene, $C_2H_2$, the first member of the alkyne series, is always called by its common name. The next alkynes are sometimes named as derivatives

**Figure 19.4** Ball-and-stick and space-filling models of the first members of the alkene and alkyne hydrocarbon series. (Photograph by Janice Peters.)

TABLE 19.4   Unsaturated Hydrocarbons

| HYDROCARBON SERIES | N | FORMULAS | | NAMES | |
|---|---|---|---|---|---|
| | | Molecular | Structural | IUPAC | Common |
| Alkene, $C_nH_{2n}$ | 2 | $C_2H_4$ | $\begin{array}{c} H \quad\quad H \\ \searrow \quad\quad \swarrow \\ C{=}C \\ \nearrow \quad\quad \nwarrow \\ H \quad\quad H \end{array}$ | Ethene | Ethylene |
| | 3 | $C_3H_6$ | $\begin{array}{c} H \quad\quad H \\ \searrow \quad\quad \mid \\ C{=}C{-}C{-}H \\ \nearrow \quad\quad \mid \; \mid \\ H \quad\quad H\;H \end{array}$ | Propene | Propylene |
| | 4 | $C_4H_8$ | $\begin{array}{c} H \quad\quad H\;H \\ \searrow \quad\quad \mid \; \mid \\ C{=}C{-}C{-}C{-}H \\ \nearrow \quad\quad \mid \; \mid \; \mid \\ H \quad\quad H\;H\;H \end{array}$ | Butene | Butylene |
| Alkynes, $C_nH_{2n-2}$ | 2 | $C_2H_2$ | $H{-}C{\equiv}C{-}H$ | Ethyne | Acetylene |
| | 3 | $C_3H_4$ | $\begin{array}{c} H \\ \mid \\ H{-}C{\equiv}C{-}C{-}H \\ \mid \\ H \end{array}$ | Propyne | — |

of acetylene, as methyl acetylene, $C_3H_4$, and ethyl acetylene, $C_4H_6$. The IUPAC system is more often employed for all alkynes except acetylene. The same prefixes are used, and the alkyne suffix is *-yne*. Thus, for the alkynes with 2, 3 and 4 carbon atoms, the formal names are ethyne, propyne and butyne, respectively.

## ISOMERISM AMONG THE UNSATURATED HYDROCARBONS

All of the possibilities for isomerism among the alkanes are duplicated in the alkenes and alkynes. Moreover, the unsaturated hydrocarbons introduce a second form of isomerism, and the alkenes even a third. The unique isomerism they both have concerns the location of the multiple bond, which can be anywhere in the chain. Double and triple bonds are handled similarly. In the simplest example, butene may have either of these two structures:

$$\begin{array}{c} \mid_1 \quad \mid_2 \quad \mid_3 \quad \mid_4 \\ {-}C{=}C{-}C{-}C{-} \\ \mid \quad \mid \end{array} \qquad\qquad \begin{array}{c} \mid_1 \quad \mid_2 \quad \mid_3 \quad \mid_4 \\ {-}C{-}C{=}C{-}C{-} \\ \mid \quad\quad \mid \; \mid \end{array}$$

1-butene                                           2-butene

The number appearing before the name is the lowest number possible to identify the carbon atom to which the double bond is attached. The compound

$$-\overset{|}{\underset{|}{C}}-\overset{|}{\underset{|}{C}}-\overset{|}{C}=\overset{|}{C}-\overset{|}{\underset{|}{C}}-$$

is 2-pentene, because the double bond is attached to the *second* carbon atom counting from the right.

A form of isomerism shown only by alkenes arises because a double bond allows no rotation about its axis. This leads to two possible arrangements around the double bond. The first alkene in which these options appear is butene:

$$\underset{\text{cis-2-butene}}{\overset{\displaystyle H_3C \diagdown \diagup CH_3}{\underset{\displaystyle H \diagup \diagdown H}{C=C}}} \qquad \underset{\text{trans-2-butene}}{\overset{\displaystyle H_3C \diagdown \diagup H}{\underset{\displaystyle H \diagup \diagdown CH_3}{C=C}}}$$

The two methyl groups attached to the double-bonded carbons can be on the *same* side of the double bond, as in *cis*-2-butene, or on *opposite* sides, as in *trans*-2-butene. *Cis*- and *trans*- are prefixes meaning, respectively, *on this side* and *across*. The latter is perhaps most easily remembered by association with such words as transcontinental, suggesting across a continent.

## 19.5 SOURCES AND PREPARATION OF ALIPHATIC HYDROCARBONS

### *PETROLEUM PRODUCTS*

Alkanes and alkenes are natural products which have resulted from the decay over millions of years of organic compounds that once constituted living plants and animals. They are found today as petroleum, mixtures of hydrocarbons containing up to 30–40 carbon atoms in the molecule. Different components of petroleum may be separated by fractional distillation, illustrated in Figure 19.5. This process separates "fractions" that boil at different temperatures. The lower alkanes and alkenes, up to four or five carbons per molecule, may be obtained in pure form by this method. The boiling points of larger compounds are too close for their complete separation, so chemical methods must be employed to obtain pure samples.

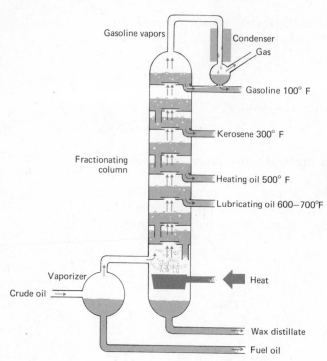

Gasoline vapors

Condenser

Gas

Gasoline 100° F

Kerosene 300° F

Fractionating column

**Figure 19.5** Diagram of a fractional distillation column, showing the various levels from which the petroleum fractions are removed.

Heating oil 500° F

Lubricating oil 600–700°F

Vaporizer

Crude oil

Heat

Wax distillate

Fuel oil

## PREPARATION OF ALKENES

Two ways alkenes are produced are the *dehydration* of alcohols and the *dehydrohalogenation* of an alkyl halide. These two impressive terms describe very similar processes which are quite simple, at least in principle if not in practice. Dehydration is removal of water; dehydrohalogenation is removal of a hydrogen and a halogen. As an example, propanol is an alcohol (see page 480). Its formula is $C_3H_7OH$. Under certain conditions a molecule of water may be separated from a molecule of the alcohol, producing propene:

$$
\underset{\text{Propanol}}{
\begin{array}{c}
\text{H } \text{ H } \text{ H}\\
| \quad | \quad |\\
\text{H}-\text{C}-\text{C}-\text{C}-\text{H}\\
| \quad | \quad |\\
\text{H } \text{ H } \text{OH}
\end{array}}
\xrightarrow{\text{H}_2\text{SO}_4}
\underset{\text{Propene}}{
\begin{array}{c}
\text{H } \text{ H } \text{ H}\\
| \quad | \quad |\\
\text{H}-\text{C}-\text{C}=\text{C}-\text{H}\\
|\\
\text{H}
\end{array}}
+ \underset{\text{Water}}{\text{HOH}}
$$

An alkyl halide is an alkane in which a halogen atom has been substituted for a hydrogen atom; or viewed in another way an alkyl halide is an alkyl group bonded to a halogen. The molecule is attacked with a base in the presence of an alcohol.

$$
\underset{\text{Propyl halide}}{
\begin{array}{c}
\text{H } \text{ H } \text{ H}\\
| \quad | \quad |\\
\text{H}-\text{C}-\text{C}-\text{C}-\text{H}\\
| \quad | \quad |\\
\text{H } \text{ H } \text{ X}
\end{array}}
+ \underset{\text{Base}}{\text{KOH}}
\xrightarrow{\text{alcohol}}
\underset{\text{Propene}}{
\begin{array}{c}
\text{H } \text{ H } \text{ H}\\
| \quad | \quad |\\
\text{H}-\text{C}-\text{C}=\text{C}-\text{H}\\
|\\
\text{H}
\end{array}}
+ \underset{\text{Salt}}{\text{KX}} + \underset{\text{Water}}{\text{HOH}}
$$

## PREPARATION OF ALKANES

There are several industrial and laboratory methods by which alkanes may be prepared. One of the more important is the catalytic **hydrogenation** of an alkene. Hydrogenation is the reaction of a substance with hydrogen. The general reaction for the hydrogenation of an alkene is

$$C_nH_{2n} + H_2 \xrightarrow{\text{cat.}} C_nH_{2n+2}$$

## PREPARATION OF ACETYLENE

One alkyne is of major importance—acetylene. It is produced commercially in a two-step process in which calcium oxide reacts with coke (carbon) at high temperatures to produce calcium carbide and carbon monoxide:

$$\text{CaO (s)} + 3 \text{ C (s)} \rightarrow \text{CaC}_2 \text{ (s)} + \text{CO (g)}$$

Calcium carbide then reacts with water to produce acetylene:

$$\text{CaC}_2 \text{ (s)} + 2 \text{ H}_2\text{O (l)} \rightarrow \text{C}_2\text{H}_2 \text{ (g)} + \text{Ca(OH)}_2 \text{ (s)}$$

## 19.6 CHEMICAL PROPERTIES OF ALIPHATIC HYDROCARBONS

The combustibility—ability to burn in air—of the hydrocarbons is probably one of the most important of all chemical reactions to modern man. As components of liquid and gaseous fuels, hydrocarbons are among the most heavily processed and distributed chemical products in the world. When burned in an excess of air, the end products are water and carbon dioxide (see page 152).

One major distinction separates the chemical properties of saturated hydrocarbons from the unsaturated hydrocarbons. By opening a multiple bond in an alkene or alkyne, the compound is capable of reacting by **addition,** simply by adding atoms of some element to the molecule. By contrast, an alkane molecule is literally saturated; there is no room for an atom to join the molecule without first removing a hydrogen atom. A reaction in which a hydrogen atom in an alkane is replaced by an atom of another element is called a **substitution** reaction.

Both alkanes and alkenes undergo *halogenation* reactions—reaction with a halogen. These reactions serve to show the difference between addition and substitution reactions:

**ADDITION REACTION:**

| Propene | Chlorine | 1,2-dichloropropane |

**SUBSTITUTION REACTION:**

$$\underset{\text{Propane}}{H-\overset{\overset{\displaystyle H}{|}}{\underset{\underset{\displaystyle H}{|}}{C}}-\overset{\overset{\displaystyle H}{|}}{\underset{\underset{\displaystyle H}{|}}{C}}-\overset{\overset{\displaystyle H}{|}}{\underset{\underset{\displaystyle H}{|}}{C}}-H} + \underset{\text{Chlorine}}{Cl_2} \xrightarrow{\text{Heat or light}} \underset{\text{1-chloropropane}}{H-\overset{\overset{\displaystyle H}{|}}{\underset{\underset{\displaystyle H}{|}}{C}}-\overset{\overset{\displaystyle H}{|}}{\underset{\underset{\displaystyle H}{|}}{C}}-\overset{\overset{\displaystyle H}{|}}{\underset{\underset{\displaystyle Cl}{|}}{C}}-H} + \underset{\substack{\text{Hydrogen}\\\text{chloride}}}{HCl}$$

The substituted chlorine atom may appear on either an end carbon atom or the middle carbon; the actual product is usually a mixture of 1-chloropropane and 2-chloropropane.

Normally, addition reactions are more readily accomplished than substitution reactions. This is hinted in the reaction conditions specified above. The addition of a halogen to an alkene will occur easily at room temperature, whereas the substitution of a halogen for a hydrogen in an alkane requires either high temperature or ultraviolet light. This shows that unsaturated hydrocarbons are more reactive than saturated hydrocarbons.

Hydrogenation is also an addition reaction. We have already indicated that the hydrogenation of an alkene may be used to produce an alkane. Hydrogenation of an alkyne is a stepwise process, which may often be controlled to give the intermediate alkene as a product:

$$\underset{\text{Alkyne}}{R-C\equiv C-R'} \xrightarrow[\text{cat.}]{H_2} \underset{\text{Alkene}}{R-\overset{\overset{\displaystyle H}{|}}{C}=\overset{\overset{\displaystyle H}{|}}{C}-R'} \xrightarrow[\text{cat.}]{H_2} \underset{\text{Alkane}}{R-\overset{\overset{\displaystyle H}{|}}{\underset{\underset{\displaystyle H}{|}}{C}}-\overset{\overset{\displaystyle H}{|}}{\underset{\underset{\displaystyle H}{|}}{C}}-R'}$$

A particularly interesting addition reaction is the addition of ethylene to itself. It is an example of **polymerization**. Polymerization is the process whereby small molecular units called **monomers** join together to form giant molecules called **polymers**. Ethylene polymerizes as follows:

Ethylene monomers          Segment of polyethylene

In this reaction the double bond of each ethylene molecule opens, and each carbon atom joins to a carbon atom of another molecule, producing the chain shown. The chains formed yield molecules of molecular weights in the area of 20,000. Most plastics are polymers: the example above is the familiar polyethylene used for squeeze bottles, toys, packaging, etc.

*Pyrolysis* is a decomposition by heat. The pyrolysis of petroleum, consisting primarily of high molecular weight alkanes is called *cracking*, and is usually done in the presence of a catalyst. Pyrolysis is conducted at temperatures in the range of 400–600°C. The usual products are alkanes of fewer carbon atoms, alkenes and hydrogen.

Cracking is one of the major operations in extracting gasoline from raw petroleum. It increases both the yield and quality of gasoline. Yield is increased because some of the longer chain hydrocarbons are reduced to acceptable length for use as gasoline (5–10 carbon atoms per molecule). Quality is improved because the alkenes resulting from the reaction have good antiknock properties.*

## 19.7 THE AROMATIC HYDROCARBONS

Historically the term **aromatic** was associated with a series of compounds found in such pleasant smelling substances as oil of cloves, vanilla, wintergreen, cinnamon and others. Ultimately it was found that the key structure in these compounds is the benzene ring.

Benzene has been studied thoroughly in an attempt to determine its structure. Its molecular formula is $C_6H_6$. It is also known to be a ring compound. How this structure is to be represented in print is a problem, and a universally agreed upon answer has yet to be found. Two common forms are:

I                                        II

Structure I is perhaps the most complete representation in that it shows all the carbon and hydrogen atoms, as well as alternating single and double bonds that would satisfy the requirements of the octet rule. This structure is not in agreement with experimental fact, however; among other things, all carbon-carbon bonds are known to be alike, rather than some being single, some double. Structure II is a compromise that suggests equality among these bonds. It is also understood that each "corner" of the hexagon represents a carbon atom that forms the equivalent of four covalent bonds, three within the benzene ring, and one without. If the fourth bonded atom is not shown, it is understood to be hydrogen.

---

*"Knocking" in an internal combustion engine is a sharp detonation of the fuel-air mixture, rather than a smooth explosion, producing the familiar knocking sound heard in automobiles when accelerating too quickly, or when climbing a hill. Knocking is rated on an arbitrary scale in which n-heptane is given an **octane number** of zero, and 2,2,4-trimethyl pentane ("isooctane") is rated at 100 in octane number.

An alkyl group, halogen, or other species may replace a hydrogen in the benzene ring.

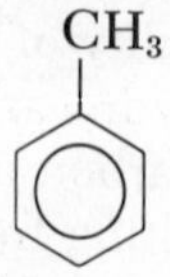

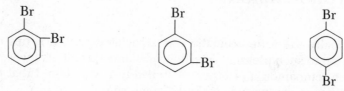

Toluene 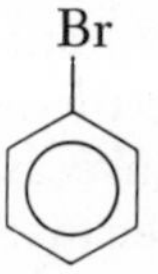                    Bromobenzene

If two bromines substitute for hydrogens on the same ring we must consider three possible isomers:

1,2-dibromobenzene      1,3-dibromobenzene      1,4-dibromobenzene
o-dibromobenzene        m-dibromobenzene        p-dibromobenzene

Two names are given for each isomer. The number system, which counts locations around the ring beginning at the substituted position that yields the lowest numbers, is more formal and serves any number of substituents. The other names are pronounced *ortho*-dibromobenzene, *meta*-dibromobenzene, and *para*-dibromobenzene. *Ortho-*, *meta-* and *para-* are prefixes commonly used when two hydrogens have been replaced from the benzene ring. Relative to position X, the other positions are shown:

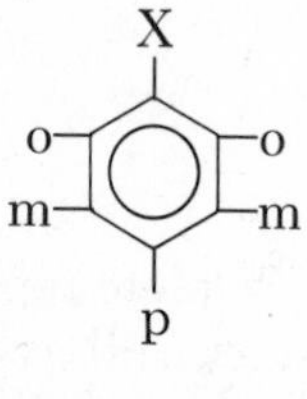

The physical properties of benzene and its derivatives are quite similar to those of other hydrocarbons. The compounds are nonpolar, insoluble in polar solvents such as water, but generally soluble in nonpolar solvents. In fact, benzene is widely used as the solvent for many nonpolar organic compounds. Like other hydrocarbons of comparable molecular weight, benzene is a liquid at room temperature. Members of the homologous series increase in boiling point in the usual manner as the number of carbon atoms increases.

The principal industrial source for benzene has been as a byproduct of the preparation of coke from coal. More recently commercial methods have been developed by which certain petroleum products are converted to aromatic hydrocarbons. For example, toluene may be prepared from *n*-heptane:

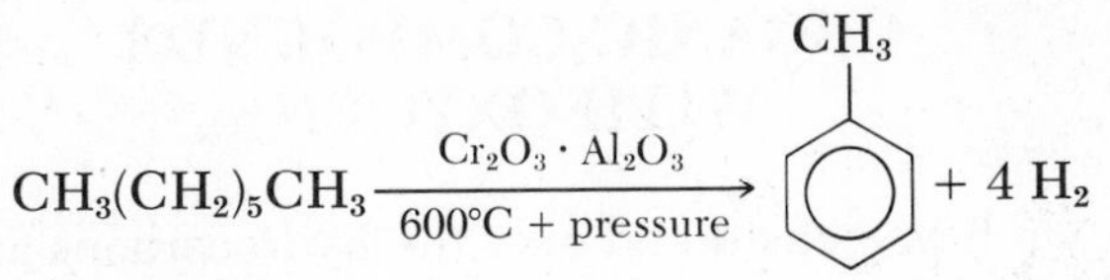

$$CH_3(CH_2)_5CH_3 \xrightarrow[\text{600°C + pressure}]{Cr_2O_3 \cdot Al_2O_3}$$ $$+ \ 4 \ H_2$$

*n*-heptane  Toluene

Perhaps the most significant—and surprising—chemical property of benzene is that, despite its high degree of unsaturation, it does not normally engage in addition reactions. The most important reaction of benzene itself is the substitution reaction in which one hydrogen is displaced from the benzene ring. Numerous substituents are possible, including the halogens:

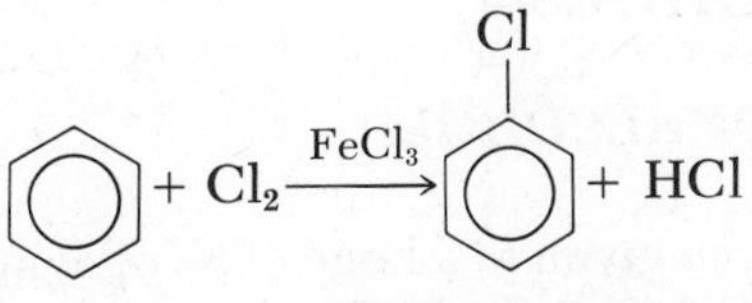

Benzene  Chlorobenzene

Substitutions with nitric and sulfuric acids yield, respectively, nitrobenzene and benzenesulfonic acid. Second substitutions on the same ring are possible, though more difficult to bring about. Substitution reactions may also be performed on benzene derivatives, such as toluene, yielding isomers of nitrotoluene, for example. A triple nitro substitution produces 2,4,6-trinitrotoluene, better known simply as TNT.

## 19.8 SUMMARY OF THE HYDROCARBONS

Four types of hydrocarbons we have considered are summarized in Table 19.5:

TABLE 19.5  Hydrocarbons

| TYPE | NAME | FORMULA | SATURATION | STRUCTURE |
|---|---|---|---|---|
| Aliphatic Open Chain | Alkane | $C_nH_{2n+2}$ | Saturated | $-\overset{|}{\underset{|}{C}}-$ |
| | Alkene | $C_nH_{2n}$ | Unsaturated | $C=C$ |
| | Alkyne | $C_nH_{2n-2}$ | Unsaturated | $-C\equiv C-$ |
| Aromatic | —— | —— | Unsaturated | 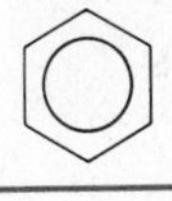 |

# ORGANIC COMPOUNDS WITH OXYGEN

Thus far we have considered only the hydrocarbons and their derivatives. The third element most commonly found in organic chemicals is oxygen. Capable of forming two bonds (see Table 19.1), oxygen serves as a connecting link between two other elements; or, double bonded, usually to carbon, it is a terminal point in a functional group. We shall now examine functional groups that contain oxygen: the alcohols; ethers; aldehydes and ketones; acids and esters.

## 19.9  THE ALCOHOLS

### THE STRUCTURE OF ALCOHOLS

**Alcohol** is the name given to a large class of compounds containing the **hydroxyl** group, —OH.* This functional group is not to be confused with the hydroxide ion of inorganic chemistry, which exists as an entity in ionic compounds and solutions. In alcohols the hydroxyl group is covalently bonded to an alkyl or other hydrocarbon group. Thus the general formula for an alcohol is R-OH, where R represents the alkyl group.

As shown in Table 19.1, and also in the two models of ethyl alcohol, $C_2H_5OH$ in Figure 19.6, the bond angle around an oxygen atom is close to the tetrahedral angle—about 105°. Thus the alcohol molecule is a water molecule in which one hydrogen has been replaced by an alkyl group:

$$
\begin{array}{ccc}
\overset{\displaystyle O}{\diagup\ \diagdown} & \overset{\displaystyle O}{\diagup\ \diagdown} & \overset{\displaystyle O}{\diagup\ \diagdown} \\
\text{H}\qquad\text{H} & \text{H} & \text{R}\qquad\text{H} \\
\text{Water} & \begin{array}{c}\text{Functional}\\\text{Group}\end{array} & \text{Alcohol}
\end{array}
$$

This structural similarity correctly suggests similar intermolecular forces and therefore similar physical properties. The lower alcohols (1-3 carbon atoms) are liquids with boiling points ranging from 65°C to 97°C, comparable with water but well above the boiling points of alkanes of about the same

---

*Some chemists refer to the hydroxyl group as the *hydroxy group.*

**Figure 19.6** Ball-and-stick and space-filling models of ethanol (ethyl alcohol), $C_2H_5OH$. (Photograph by Janice Peters.)

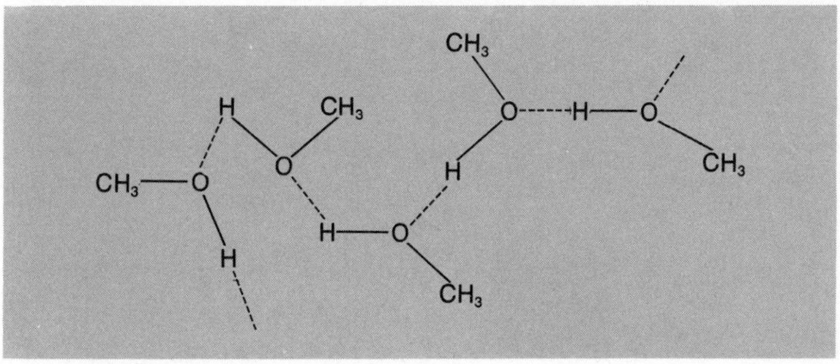

**Figure 19.7** Hydrogen bonding in methanol.

molecular weight. This is attributable largely to hydrogen bonding, very much in evidence in the lower alcohols (see Fig. 19.7). Hydrogen bonding also accounts for the complete miscibility (solubility) between lower alcohols and water. As usual, boiling points rise with increasing molecular weight. Solubility drops off sharply as the alkyl chain lengthens and the molecule assumes more the character of the parent alkane.

## NAMES OF THE ALCOHOLS

Alcohols are best known by their common names, which originate in the name of the alkyl group to which the hydroxyl group is bonded. This system names the alkyl group, followed by "alcohol." Thus $CH_3OH$ is *methyl alcohol* and $C_2H_5OH$ is *ethyl alcohol*. When we reach propyl alcohol we are confronted with two isomers;

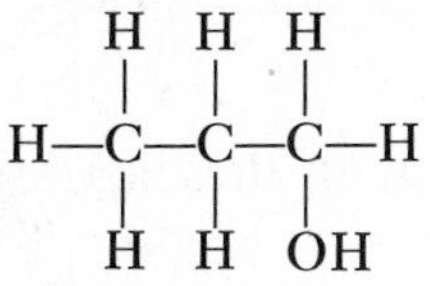

*n*-propyl alcohol 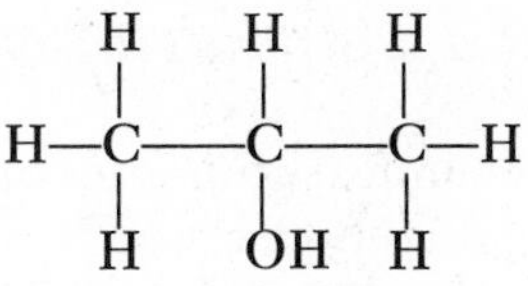                    isopropyl alcohol

As usual, the IUPAC nomenclature is more precise in identifying the higher alcohols. The *e* at the end of the corresponding alkane is replaced with the suffix -*ol* and the result is the name of the alcohol. Thus methyl alcohol becomes *methanol*, and ethyl alcohol is formally *ethanol*. The propane isomers are distinguished by stating the number of the carbon atom to which the hydroxyl group is bonded. Accordingly n-propyl alcohol becomes *1-propanol* and isopropyl alcohol is designated *2-propanol*.

## SOURCES AND PREPARATION OF ALCOHOLS

**HYDRATION OF ALKENES.** The major industrial source of several of our most important alcohols is the hydration of alkenes obtained from the cracking of petroleum. Beginning with ethylene, for example, the reaction may be summarized

$$\underset{\text{Ethylene}}{H-\overset{\overset{\displaystyle H}{|}}{C}=\overset{\overset{\displaystyle H}{|}}{C}-H} + HOH \xrightarrow{\text{H}_2\text{SO}_4} \underset{\text{Ethyl alcohol}}{H-\overset{\overset{\displaystyle H}{|}}{\underset{\underset{\displaystyle H}{|}}{C}}-\overset{\overset{\displaystyle H}{|}}{\underset{\underset{\displaystyle OH}{|}}{C}}-H}$$

**FERMENTATION OF CARBOHYDRATES.** Making ethyl alcohol by the fermentation of sugars in the presence of yeast is probably the oldest synthetic chemical process known:

$$\underset{\text{Glucose (sugar)}}{C_6H_{12}O_6} \xrightarrow{\text{yeast}} 2\ CO_2 + \underset{\text{Ethyl alcohol}}{2\ C_2H_5OH}$$

A solution that is 95% ethyl alcohol (190 proof) may be obtained from the final mixture by fractional distillation. The mixture also yields two other products that are of commercial importance today: *n*-butyl alcohol and acetone (see page 485).

**SYNTHESIS OF METHYL ALCOHOL.** Methyl alcohol is sometimes called wood alcohol because it was once made by the destructive distillation (heating in absence of air) of wood. It is now produced by the catalytic hydrogenation of carbon monoxide at high pressure and temperature:

$$CO + 2\ H_2 \xrightarrow[\text{250 atm} + \text{300°C}]{\text{ZnO} + \text{Cr}_2\text{O}_3} CH_3OH$$

## CHEMICAL PROPERTIES OF ALCOHOLS

The chemical properties of alcohols are essentially the chemical properties of the functional group, —OH. In some reactions the C—OH bond is broken, separating the entire hydroxyl group. This is true in the dehydration of alcohols to form alkenes (page 474). In other reactions of alcohols the O—H bond within the hydroxyl group is broken. We will postpone discussion of these reactions until the structures of the various products— aldehydes, ketones, carboxylic acids and esters—have been examined.

## SOME COMMON ALCOHOLS

Methyl alcohol is an important industrial chemical with production measured in the billions of pounds annually. It is a raw material for the production of many chemicals, particularly formaldehyde, which is widely used in the plastics industry. It is also used in antifreezes, commercial solvents, and as a denaturant, or additive to ethyl alcohol to make it unfit for human consumption. Taken internally, methyl alcohol is a deadly poison, frequently causing blindness in less than lethal doses.

In addition to its uses in beverages, ethyl alcohol is used in organic solvents and in the preparation of various organic compounds such as chloroform and ether. Its production is also measured in the billions of pounds annually.

Other widely used alcohols include isopropyl alcohol, which is sold as rubbing alcohol, and *n*-butyl alcohol, used in lacquers in the automobile industry. Alcohols containing more than one hydroxyl group are also common. Permanent antifreeze in automobiles is ethylene glycol, which has two hydroxyl groups in the molecule. Glycerine, or glycerol, a tri-hydroxyl alcohol, has many uses in the manufacture of drugs, cosmetics, explosives and other chemicals.

## 19.10  THE ETHERS

On page 480 we pointed out that structurally an alcohol might be considered as a water molecule in which one hydrogen has been replaced by an alkyl group. An **ether** may be similarly considered, except that *both* hydrogens are replaced by an alkyl group. The functional group that identifies an ether is simply the oxygen atom bonded to two alkyl groups:

$$
\begin{array}{ccc}
\overset{\displaystyle O}{\diagup\ \diagdown} & \overset{\displaystyle O}{\diagup\ \diagdown} & \overset{\displaystyle O}{\diagup\ \diagdown} \\
H \qquad H & & R \qquad R'
\end{array}
$$

|   Water   |  Functional group  |   Ether   |

The R′ indicates that the functional groups may or may not be identical. For example, methyl ethyl ether and ethyl ether have the structures:

$$
\begin{array}{cc}
\overset{\displaystyle O}{\diagup\ \diagdown} & \overset{\displaystyle O}{\diagup\ \diagdown} \\
CH_3 \qquad C_2H_5 & C_2H_5 \qquad C_2H_5
\end{array}
$$

|   Methyl ethyl ether   |   Ethyl ether   |

Figure 19.8 shows models of methyl ether.

Ether molecules are less polar than alcohol molecules, and there is no opportunity for hydrogen bonding between them. Intermolecular attractions are therefore lower, as are the dependent boiling points. Up to three carbons, ethers are gases at room conditions, and the familiar ethyl ether is a volatile liquid that boils at 35°C. The solubility of an ether in water is about

**Figure 19.8** Ball-and-stick and space-filling models of methyl ether, $CH_3OCH_3$. (Photograph by Janice Peters.)

the same as the solubility of its isomeric alcohol, primarily because of hydrogen bonding between the ether molecule and water molecule.

All ethers are called "ether," and identified specifically by naming first the two alkyl groups that are bonded to the functional group. If the groups are identical, the prefix *di-* may be used, as in diethyl ether.

Under properly controlled conditions, ethers can be prepared by dehydrating alcohols. At 140°C, and with constant alcohol addition to replace the ether as it distills from the mixture, ethyl ether is formed from two molecules of ethanol:

Ethyl alcohol          Ethyl alcohol          Ethyl ether

Aside from combustion, ethers are relatively unreactive compounds, being quite resistant to attack by active metals, strong bases and oxidizing agents. They are, however, highly flammable and must be handled cautiously in the laboratory.

The isolated word "ether" generally prompts one to think of the anesthetic that is so identified. This compound is ethyl ether; its line formula is $C_2H_5$—O—$C_2H_5$. Recently its isomer, methyl propyl ether, $CH_3$—O—$C_3H_7$, has been gaining popularity as a substitute in this use; it has fewer objectionable after-effects than ethyl ether. Ethyl ether is also used as a solvent for fats from foods and animal tissue in the laboratory.

## 19.11 THE ALDEHYDES AND KETONES

**Aldehydes** and **ketones** are characterized by the **carbonyl functional group,**

If at least one hydrogen atom is bonded to the carbonyl carbon, the compound is an aldehyde, RCHO; if two alkyl groups are attached, the compound is a ketone, R—CO—R′.

| Aldehyde | Ketone |

The simplest carbonyl compound is formaldehyde, HCHO, which has two hydrogen atoms bonded to the carbonyl carbon. If a methyl group replaces one of the hydrogens of formaldehyde, the result is acetaldehyde, $CH_3CHO$ (Fig. 19.9). Replacement of both formaldehyde hydrogens with methyl groups yields acetone.

| Formaldehyde | Acetaldehyde | Acetone |

The carbonyl group is polar, thereby making ketone and aldehyde molecules polar, though not as polar as alcohols. Only formaldehyde is definitely a gas at room temperature (boiling point −21°C); acetaldehyde is at its boiling point at 20°C. Aldehydes and ketones of up to about five carbons enjoy some solubility in water, no doubt attributable to the polarity of both the solvent and solute molecules and hydrogen bonding. The liquid "formaldehyde" we encounter in the laboratory is actually a water solution, sold under the trade name "Formalin."

**Figure 19.9** Ball-and-stick and space-filling models of acetaldehyde and acetone. (Photograph by Janice Peters.)

The lower aldehydes are best known by their common names. The IUPAC nomenclature system for aldehydes employs the name of the parent hydrocarbon, substituting the suffix -*al* for the final *e* to identify the compound as an aldehyde. Thus the IUPAC name for formaldehyde is methanal; for acetaldehyde, ethanal; and so forth.

Ketones are named by one of two systems. The first duplicates the method of naming ethers: identify each alkyl group attached to the carbonyl group, followed by the class name, ketone. Accordingly, methyl ethyl ketone has the structure

$$\underset{\underset{CH_3 \qquad C_2H_5}{}}{\overset{\overset{O}{\parallel}}{C}}$$

Under the IUPAC system the number of carbons in the longest chain carrying the carbonyl carbon establishes the hydrocarbon base, which is followed by -*one* to identify the ketone as the class of compound. Methyl ethyl ketone, having four carbons, would be called *butanone*. Two isomers of pentanone would be 2-*pentanone* and 3-*pentanone*, the number being used to designate the carbonyl carbon.

$$\underset{\underset{\text{2-pentanone}}{CH_3 \qquad CH_2CH_2CH_3}}{\overset{\overset{O}{\parallel}}{C}} \qquad\qquad \underset{\underset{\text{3-pentanone}}{CH_3CH_2 \qquad CH_2CH_3}}{\overset{\overset{O}{\parallel}}{C}}$$

Aldehydes and ketones may be prepared by oxidation of alcohols. If the product is to be a ketone, the alcohol must be a *secondary* alcohol, in which the hydroxyl group is bonded to an interior carbon:

$$\underset{\underset{R'}{|}}{\overset{\overset{H}{|}}{R-C-OH}} + \tfrac{1}{2}\,O_2 \rightarrow \underset{\underset{R'}{|}}{R-C=O} + H_2O$$

$$\quad\text{Secondary alcohol} \qquad\qquad\qquad \text{Ketone}$$

Care must be taken not to over-oxidize aldehyde preparations, as aldehydes are easily oxidized to carboxylic acids (see next section).

Aldehydes and ketones may also be produced by the hydration of alkynes. If the triple bond is on an end carbon, an aldehyde is produced; if on an internal carbon, the result is a ketone. A typical reaction is the commercial preparation of acetaldehyde:

$$H-C\equiv C-H + HOH \rightarrow \underset{\underset{H}{|}}{\overset{\overset{H}{|}}{H-C}}-\overset{\overset{O}{\diagup\!\!\diagup}}{C}{\diagdown}_{H}$$

$$\text{Acetaldehyde}$$

The double bond of the carbonyl group is susceptible to addition reactions, just as the double bond in the alkenes. One such reaction is the catalytic hydrogenation of ketones to secondary alcohols, in which the carboxyl group is bonded to a carbon atom *within* the chain.

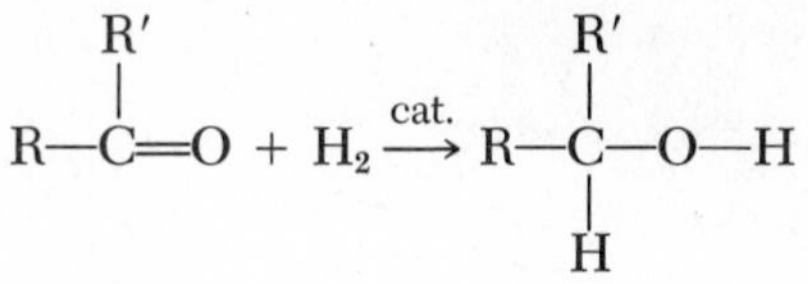

$$
\begin{array}{ccc}
\underset{\displaystyle R}{\overset{\displaystyle R'}{\text{R—C=O}}} + H_2 & \xrightarrow{\text{cat.}} & \underset{\displaystyle H}{\overset{\displaystyle R'}{\text{R—C—O—H}}}
\end{array}
$$

Ketone            Secondary alcohol

Oxidation reactions occur quite readily with aldehydes, but are resisted by ketones. The product when an aldehyde is oxidized is a carboxylic acid:

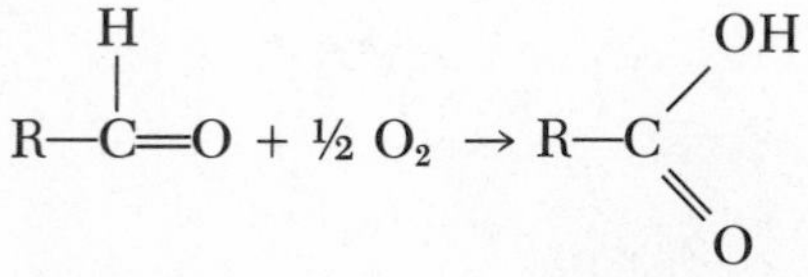

$$
\underset{\displaystyle}{\overset{\displaystyle H}{\text{R—C=O}}} + \tfrac{1}{2}\,O_2 \ \rightarrow \ \text{R—C}\begin{smallmatrix}\text{OH}\\ \\ \text{O}\end{smallmatrix}
$$

Aldehyde            Carboxylic acid

Formaldehyde is probably the best known carbonyl compound. It is widely used as a preservative. Large quantities are consumed in the manufacture of resins and in the preparation of numerous organic compounds. Acetaldehyde finds use in the manufacture of acetic acid, ethyl acetate and other organic products.

Acetone is the most important ketone. It is a solvent for many organic chemicals, including cellulose derivatives, varnish, lacquer, plastics and resins. Methyl ethyl ketone finds application in the petroleum industry, although it is most familiar as an ingredient of fingernail polish remover.

## 19.12 CARBOXYLIC ACIDS AND ESTERS

In the last section we saw that oxidation of an aldehyde produces a **carboxylic acid,** the general formula of which is frequently represented as RCOOH. The functional group, —COOH, shown at the left below, is a combination of a carbonyl group and a hydroxyl group, appropriately called the **carboxyl group.** You can probably pick out the carboxyl group in the formic and acetic acid models in Figure 19.10. In an **ester,** the carboxyl carbon may be bonded to a hydrogen atom or an alkyl group, and the carboxyl hydrogen is replaced by another alkyl group, as shown:

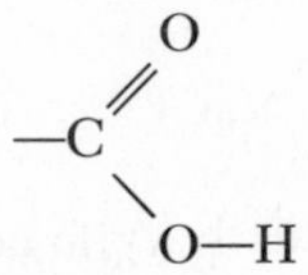

$$
\text{—C}\begin{smallmatrix}\text{O}\\ \\ \text{O—H}\end{smallmatrix}
\qquad\qquad
\text{R—C}\begin{smallmatrix}\text{O}\\ \\ \text{O—R}'\end{smallmatrix}
$$

Carboxyl Group 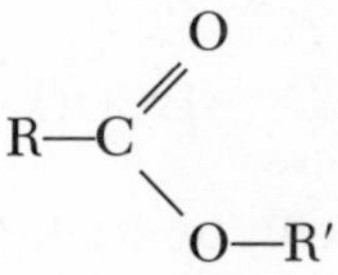                 Ester

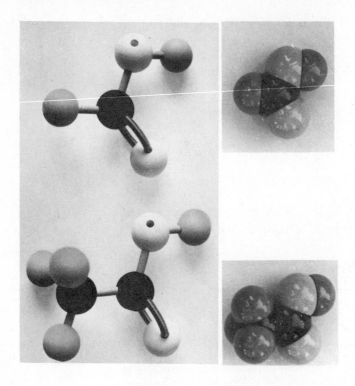

**Figure 19.10** Ball-and-stick and space-filling models of formic and acetic acids. (Photograph by Janice Peters.)

The geometry of the carboxyl group results in strong dipole attractions and hydrogen bonding between molecules. As a consequence, boiling points tend to be high compared to compounds of comparable molecular weight. Formic acid, for example, boils at 100.5°C. Lower acids are completely miscible in water, but solubility drops off as the aliphatic chain lengthens and the molecule behaves more like a hydrocarbon.

Common names continue to be used for most acids. Formic acid, HCOOH, with only a hydrogen attached to the carboxyl group, is the simplest of the carboxylic acids. Next in the series is acetic acid, in which the methyl group is bonded to the carboxyl group: $CH_3COOH$. The names of many acids are associated with their sources or some physical property associated with them. Butyric acid, $C_3H_7COOH$, for example, is responsible for the odor of rancid butter, for which the Latin word is *butyrum*. The IUPAC system for naming carboxylic acids drops the *e* from the alkane of the same number of carbon atoms, and replaces it with *-oic*. Thus HCOOH is *methanoic acid;* $CH_3COOH$ is *ethanoic acid;* $C_2H_5COOH$ is *propanoic acid*, and so forth.

Formic acid is prepared commercially from sodium formate, produced by the reaction of carbon monoxide and sodium hydroxide. In a typical molecular product reaction (page 347), sodium formate reacts with hydrochloric acid to yield formic acid and sodium chloride:

$$HCOONa + HCl \rightarrow HCOOH + NaCl$$

Acetic acid, by far the most important of the carboxylic acids, is produced by stepwise oxidation of ethanol, first to acetaldehyde and then to acetic acid:

$$H-\overset{\overset{\displaystyle H}{|}}{\underset{\underset{\displaystyle H}{|}}{C}}-\overset{\overset{\displaystyle H}{|}}{\underset{\underset{\displaystyle H}{|}}{C}}-OH \xrightarrow{KMnO_4} H-\overset{\overset{\displaystyle H}{|}}{\underset{\underset{\displaystyle H}{|}}{C}}-\overset{\overset{\displaystyle H}{|}}{C}=O \xrightarrow{KMnO_4} H-\overset{\overset{\displaystyle H}{|}}{\underset{\underset{\displaystyle H}{|}}{C}}-C\overset{OH}{\underset{O}{}}$$

$$\text{Ethyl alcohol} \qquad \text{Acetaldehyde} \qquad \text{Acetic acid}$$

Carboxylic acids are weak acids that release a proton from the carboxyl group on ionization.* Acetic acid, for example, ionizes in water as follows:

$$CH_3COOH\ (aq) \rightleftharpoons CH_3COO^-\ (aq) + H^+\ (aq)$$

The ionization takes place but slightly; only about 1% of the acetic acid molecules ionize. The solution consists primarily of molecular $CH_3COOH$. This notwithstanding, acetic acid participates in typical acid reactions such as neutralization,

$$CH_3COOH\ (aq) + OH^-\ (aq) \rightarrow HOH\ (l) + CH_3COO^-\ (aq)$$

and the release of hydrogen on reaction with a metal,

$$2\ CH_3COOH\ (aq) + Zn\ (s) \rightarrow 2\ CH_3COO^-\ (aq) + Zn^{2+}\ (aq) + H_2\ (g)$$

Metal acetate salts may be obtained by evaporating the resulting solutions to dryness.

The reaction between an acid and an alcohol is called **esterification.** The products of the reaction are an ester and water. A typical esterification reaction is

$$H-\overset{\overset{\displaystyle H}{|}}{\underset{\underset{\displaystyle H}{|}}{C}}-\overset{\overset{\displaystyle H}{\|}}{C}-O-H + H-O-\overset{\overset{\displaystyle H}{|}}{\underset{\underset{\displaystyle H}{|}}{C}}-H \rightleftharpoons H-\overset{\overset{\displaystyle H}{|}}{\underset{\underset{\displaystyle H}{|}}{C}}-\overset{\overset{\displaystyle O}{\|}}{C}-O-\overset{\overset{\displaystyle H}{|}}{\underset{\underset{\displaystyle H}{|}}{C}}-H + HOH$$

$$\text{Acetic acid} \qquad \text{Methanol} \qquad \text{Methyl acetate}$$
$$\text{Acid} \qquad\qquad \text{Alcohol} \qquad\qquad \text{Ester}$$

Notice how the water molecule is formed: the *acid contributes the entire hydroxyl group,* while the *alcohol furnishes only the hydrogen.*

The names of esters are derived from the parent alcohol and acid. The first term is the alkyl group associated with the alcohol; the second term is the name of the anion derived from the acid. In the example above, methyl alcohol (methanol) yields methyl as the first term, and acetic acid yields acetate as the second term.

Carboxylic acids engage in typical proton transfer acid-base type reac-

---

*In more advanced consideration of organic reactions, the term *acid* is also used in reference to Lewis acids (page 394). This is why the adjective *carboxylic* is used to identify an organic acid containing the carboxyl group.

tions with ammonia to produce salts. The ammonium salt so produced may then be heated, which causes it to lose a water molecule. The resulting product is called an amide. Compared to the original acid, an amide substitutes an —$NH_2$ group for the —OH group of the acid (see Section 19.14).

$$\underset{\text{Acid}}{R-\overset{\overset{\textstyle O}{\|}}{C}-O-H} + NH_3 \longrightarrow \underset{\text{Salt}}{\left[R-\overset{\overset{\textstyle O}{\|}}{C}-O\right]^- NH_4^+} \overset{\Delta}{\longrightarrow} \underset{\text{Amide}}{R-\overset{\overset{\textstyle O}{\|}}{C}-NH_2} + H_2O$$

Formic acid and acetic acid are the two most important carboxylic acids. Formic acid is the source of irritation in the bite of ants and other insects, or the scratch of nettles. A liquid with a sharp, irritating odor, it is used in manufacturing esters, salts, plastics and other chemicals. Acetic acid is present to about 4–5% in vinegar, and is responsible for its odor and taste. Acetic acid is among the least expensive organic acids, and is therefore a raw material in many commercial processes that require a carboxylic acid. Sodium acetate is one of several important salts of carboxylic acids. It is used to control the acidity of chemical processes and in the preparation of soaps and pharmaceutical agents.

Ethyl acetate and butyl acetate are two of the relatively few esters produced in large quantity. Both are used as solvents, particularly in the manufacture of lacquers. Other esters are involved in the plastics industry, and some find application in the medicinal fields. Esters are responsible for the odor of most fruit and flowers, leading to their use in the food and perfume industries.

# ORGANIC COMPOUNDS WITH NITROGEN

Nitrogen is one of the important elements found in living organisms, so we might expect to find it in many organic compounds. We will mention but briefly two classes of nitrogen-bearing organic compounds, the amines and the amides.

## 19.13 AMINES

**Amines** are organic derivatives of ammonia. In much the same way that an alcohol can be viewed structurally as a water molecule in which a hydrogen atom has been replaced by an alkyl group, and ether molecules are water molecules with both hydrogen atoms replaced by alkyl groups, so an amine is an ammonia molecule with one, two or three hydrogen atoms replaced by alkyl groups. The number of hydrogens so replaced distinguishes between a primary, secondary and tertiary amine. Amines are named by identifying the alkyl groups that are bonded to the nitrogen atom, using appropriate prefixes if two or three identical groups are present, followed by the suffix -*amine*. Illustrative examples follow:

$$H-\overset{\cdot\cdot}{\underset{\underset{H}{|}}{N}}-H \qquad CH_3-\overset{\cdot\cdot}{\underset{\underset{H}{|}}{N}}-H \qquad CH_3-\overset{\cdot\cdot}{\underset{\underset{H}{|}}{N}}-C_2H_5 \qquad CH_3-\overset{\cdot\cdot}{\underset{\underset{CH_3}{|}}{N}}-C_2H_5$$

| Ammonia | Methylamine | Methylethylamine | Dimethylethylamine |
|---------|-------------|------------------|--------------------|
|         | Primary amine | Secondary amine | Tertiary amine |

As ammonia is polar and capable of forming hydrogen bonds, so are the primary and secondary amines, but to a lesser extent. Tertiary amines are essentially nonpolar. These structural features contribute in the usual way to the physical properties of the amines. All three methylamines and ethylamines are gases, with boiling points in the range of −6°C to 11°C. Other amines are liquids with boiling points that increase with molecular weight and more complex structure. Amines can form hydrogen bonds with water molecules; therefore lower amines—particularly primary and secondary amines—are very soluble in water, and less soluble in nonpolar solvents.

Because of the unshared electron pair in the nitrogen atom, amines behave as Brönsted-Lowry or Lewis bases. A typical reaction is

$$\underset{\text{Dimethylamine}}{H-\overset{\overset{\displaystyle CH_3}{|}}{\underset{\underset{\displaystyle CH_3}{|}}{N}}:} \;+\; HCl \;\rightarrow\; \underset{\substack{\text{Dimethylammonium}\\\text{chloride}}}{\left[ H-\overset{\overset{\displaystyle CH_3}{|}}{\underset{\underset{\displaystyle CH_3}{|}}{N}}-H \right]^{+}\; Cl^{-}}$$

By reaction first with nitrous acid and then hydrogen, dimethylamine is made into dimethyl hydrazine, $(CH_3)_2NNH_2$, which is used as a rocket propellant. Dimethylamine and trimethylamine are both used in making anion exchange resins. Dyes, drugs, herbicides, fungicides, soaps, insecticides and photographic developers are among the chemical products made from amines. Aniline, or phenylamine, an aromatic amine, is among the more important materials used in dye making.

## 19.14  AMIDES

On page 490, an **amide** was shown to be a derivative of a carboxylic acid in which the hydroxyl part of the carboxyl group is replaced by an $NH_2$ group. For example,

$$\underset{\text{Acetic acid}}{CH_3-C\overset{\displaystyle O}{\underset{\displaystyle OH}{\big<}}} \qquad \text{becomes} \qquad \underset{\text{Acetamide}}{CH_3-C\overset{\displaystyle O}{\underset{\displaystyle NH_2}{\big<}}}$$

by substitution of the $-NH_2$ in place of the $-OH$, as shown. An amide is named by replacing the *-ic acid* name of the acid with *amide*.

Amides are polar compounds that are capable of strong hydrogen bonding between the electronegative oxygen of the carboxyl group of one molecule and the electropositive amide hydrogen of the next. As a consequence the amides as a group have higher melting and boiling points than otherwise similar compounds. Only formamide, $HCONH_2$, is a liquid at room temperature; all higher amides are solids. Polarity and hydrogen bonding predict accurately the solubility of the lower amides in water.

The amide structure appears in an important biochemical system, protein, as a connecting link between amino acids. The linkage is commonly called a *peptide linkage*. This linkage has the form

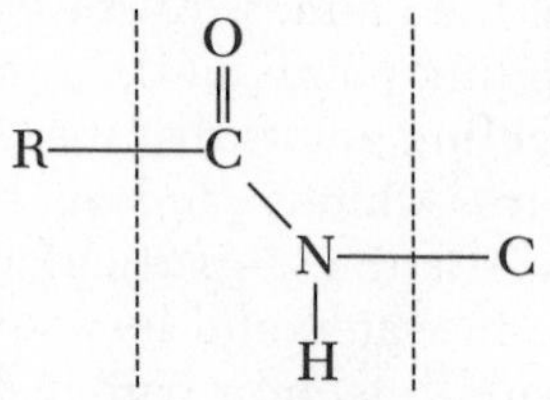

Peptide linkage

An *amino acid* is an acid in which an amine group is substituted for a hydrogen atom in the molecule. The amino acids involved in the protein structure have the general formula

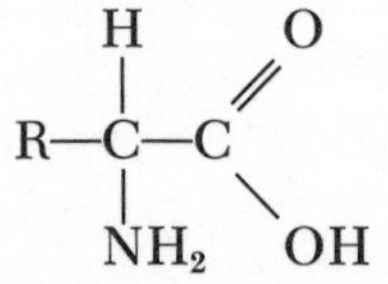

in which the amine and carboxyl groups are bonded to the same carbon atom. The peptide linkage is formed when the carboxyl group of one amino acid and the amine group of another combine by removing a water molecule

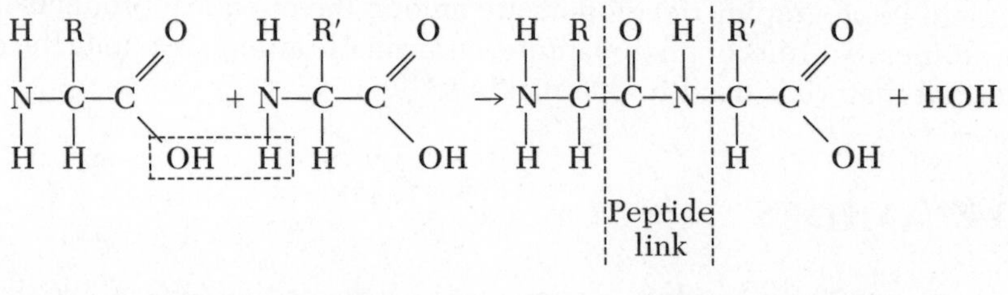

Proteins are chains of such links between as many as 18 different amino acids producing huge molecules with molar weights ranging from about 34,500 to 50,000,000.

\* \* \* \* \*

## 19.15 SUMMARY OF THE ORGANIC COMPOUNDS OF CARBON, HYDROGEN, OXYGEN AND NITROGEN

The eight types of organic compounds of carbon, hydrogen, oxygen and nitrogen we have considered are summarized in Table 19.6.

TABLE 19.6   Classes of Organic Compounds

| COMPOUND CLASS | GENERAL FORMULA | FUNCTIONAL GROUP | NAMES* |
|---|---|---|---|
| Alcohol | R—OH | —OH | Alkyl group + *alcohol;* methyl alcohol<br>Alkane prefix + *-ol:* methanol |
| Ether | R—O—R' |  O (with bonds) | Name both alkyl groups + *ether:* methyl ethyl ether<br>Alkyl group + *-oxy-* + alkane: methoxyethane |
| Aldehyde | R—CHO | 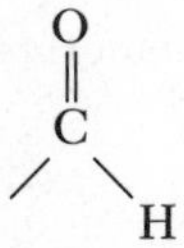 O=C—H | Common prefix + *-aldehyde:* formaldehyde<br>Alkane prefix + *-al:* methanal |
| Ketone | R—CO—R' | 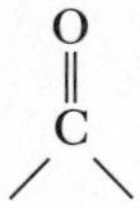 O=C | Name both alkyl groups + *ketone:* methyl ethyl ketone; methyl n-propyl ketone<br>(Number carbonyl carbon) + alkane prefix + *-one:* butanone; 2-pentanone |
| Acid | R—COOH | 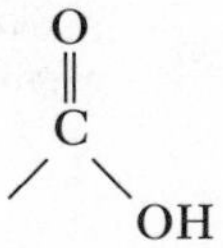 O=C—OH | Common name + acid: formic acid<br>Alkane prefix + *-oic* + *acid:* methanoic acid |
| Ester | R—CO—OR' | 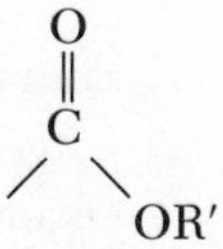 O=C—OR' | Alcohol alkyl group + acid anion: methyl acetate<br>Alcohol alkyl group + acid alkane prefix + *-oate;* methyl ethanoate |
| Amine | $RNH_2$<br>$R_2NH$<br>$R_3N$ | 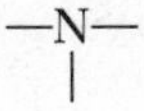 —N— | Name alkyl group(s) + *-amine:* methylamine<br>*Amino-* + alkane: aminomethane |
| Amide | $R—CONH_2$ | 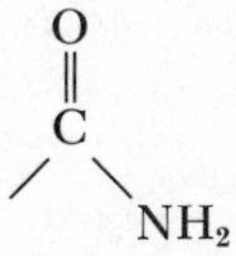 O=C—$NH_2$ | Common acid prefix + *-amide:* formamide<br>Alkane prefix + *-amide:* methanamide |

*Common name followed by IUPAC name.

# TERMS AND CONCEPTS

Organic chemistry (460)
Hydrocarbon (462)
Aliphatic hydrocarbon (462)
Saturated hydrocarbon (462)

Alkane (462)
Homologous series (463)
Normal alkane (464)
Condensed (line) formula (465)

# QUESTIONS AND PROBLEMS

## Section 19.1

19.1) Distinguish between organic chemistry and inorganic chemistry. Define organic chemistry.

19.41) Name some of the classes of carbon-bearing compounds that are not included in organic chemistry.

## Section 19.3

19.2) Define hydrocarbon.

19.42) Distinguish between saturated and unsaturated hydrocarbons.

19.3) Define an alkane, and explain how it is an example of a homologous series.

19.43) Write the molecular formula for the alkane having 19 carbon atoms in its molecules. Explain how you determined this formula.

19.4) What is a condensed formula, or line formula? What advantage does it have over a molecular formula?

19.44) Write a structural formula for $CH_3(CH_2)_4CH_3$.

19.5) Write both the structural formula and condensed (line) formula for the normal alkane having 9 carbon atoms in its molecules.

19.45) Write the name of

$$
\begin{array}{c}
\text{H} \quad \text{H} \quad \text{H} \quad \text{H} \\
| \quad\; | \quad\; | \quad\; | \\
\text{H—C—C—C—C—H} \\
| \quad\; | \quad\; | \quad\; | \\
\text{H} \quad \text{H} \quad \text{H} \quad \text{H}
\end{array}
$$

19.6) Define isomerism. From the following list of molecular and condensed formulas, select two that are isomers and explain why they may be so classified:
  (a) $CH_3CH(CH_3)CH_2CH_2CH(CH_3)CH_2CH_3$;
  (b) $CH_3CH_2C(CH_3)_2CH_2CH_3$;
  (c) $CH_3(CH_2)_8CH_3$;
  (d) $C_6H_{14}$;
  (e) $C_9H_{20}$.

19.46) Draw structural diagrams for all the isomers of heptane.

19.7) What is meant by an alkyl group? How are alkyl groups named? Give three examples of alkyl groups, both name and formula.

19.47) What is the significance of the letter R in a formula such as R—Cl? Write two structural formulas that might be described by R—Cl.

19.8)

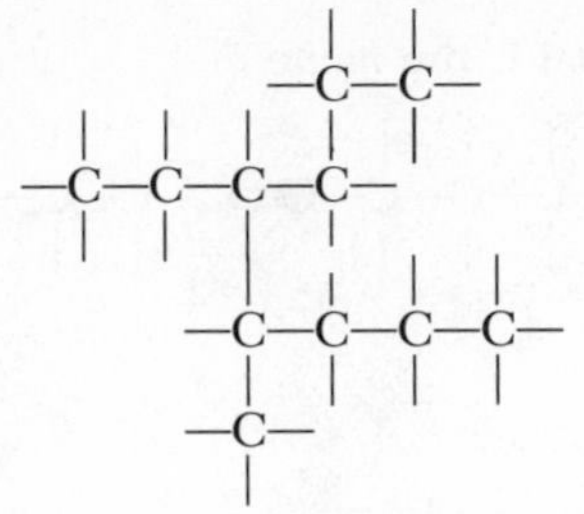

is the skeleton structure for an alkane. Identify the alkane on which its IUPAC name would be based (e.g., propane, butane, etc.). Justify your choice.

19.48)  Re-draw the skeleton structure of the alkane shown in question 19.8 so its identification would be more evident.

*19.9–19.16 and 19.49–19.56: Write the names of those compounds whose structural formulas are given, and the structural formulas for the compounds whose names are given.*

19.9)  3-methylheptane

19.10)  2,4-dimethyloctane

19.11)

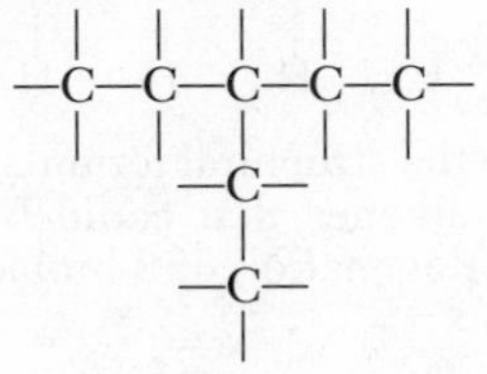

19.12)

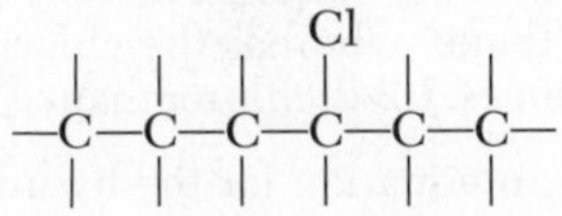

19.13)

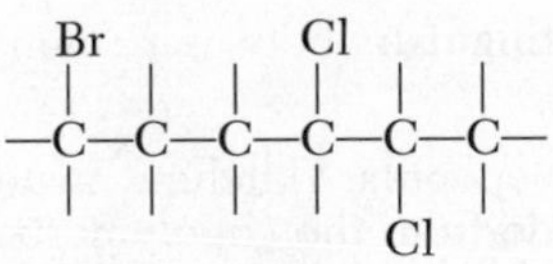

19.14)  1,1-dibromoethane

19.15)  1,1,4,5-tetrabromopentane

19.16)  1,2,2-tribromo-1-chlorobutane

19.49)  3-ethylhexane

19.50)  2,2,4-trimethyl-3,6-diethyloctane.

19.51)

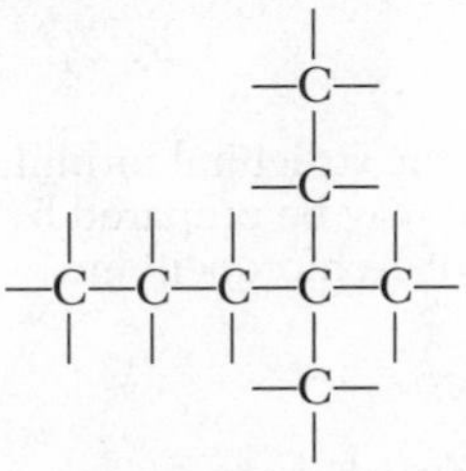

19.52)

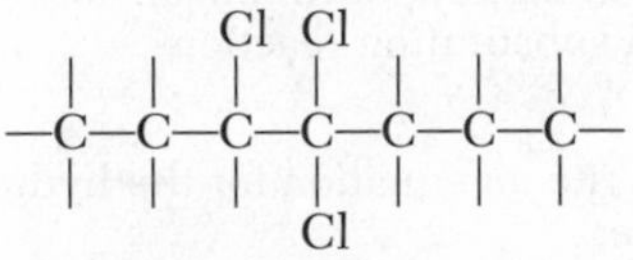

19.53)

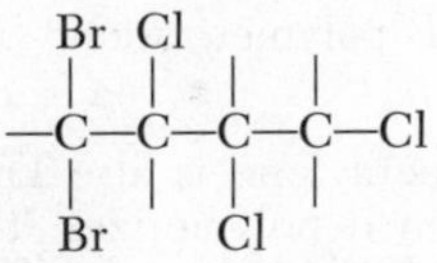

19.54)  1,2,3-trichloropropane

19.55)  2,3-dibromo-4-chlorohexane

19.56)  1,1,1-tribromo-3-chloropentane

## Section 19.4

19.17)  What structural feature identifies an alkene? Write the general formula for an alkene.

19.18)  Write structural formulas and names for the first three alkynes.

19.19)  Explain in words the difference between 1-hexene, 2-hexene and 3-hexene.

19.20)  What are *cis-*, *trans-* isomers?

19.57)  What structural feature identifies an alkyne? Write the general formula for an alkyne.

19.58)  Write structural formulas and names for the first three alkenes.

19.59)  Write the structural formula for 3-hexyne.

19.60)  Write structural formulas for the *cis-* and *trans-* isomers of 3-hexene, and state which is which.

19.21)   What is the name of

$$CH_3 \qquad C_2H_5$$
$$\diagdown \qquad \diagup$$
$$C=C$$
$$\diagup \qquad \diagdown$$
$$H \qquad \qquad H$$

19.61)   What is the name of

$$-C-C-C-C\equiv C-C-C-$$

### Section 19.5

19.22)   Write the formula and name of the alkene that can be prepared by the dehydration of the following alcohol:

$$H-\overset{\overset{\displaystyle H}{|}}{\underset{\underset{\displaystyle H}{|}}{C}}-\overset{\overset{\displaystyle H}{|}}{\underset{\underset{\displaystyle H}{|}}{C}}-\overset{\overset{\displaystyle H}{|}}{\underset{\underset{\displaystyle H}{|}}{C}}-\overset{\overset{\displaystyle H}{|}}{\underset{\underset{\displaystyle H}{|}}{C}}-\overset{\overset{\displaystyle H}{|}}{\underset{\underset{\displaystyle H}{|}}{C}}-\overset{\overset{\displaystyle H}{|}}{\underset{\underset{\displaystyle H}{|}}{C}}-OH$$

19.62)   Which of the two alcohol structures shown could be used to produce 2-butene? Explain.

(a)
$$H-\overset{\overset{\displaystyle OH}{|}}{\underset{\underset{\displaystyle H}{|}}{C}}-\overset{\overset{\displaystyle H}{|}}{\underset{\underset{\displaystyle H}{|}}{C}}-\overset{\overset{\displaystyle H}{|}}{\underset{\underset{\displaystyle H}{|}}{C}}-\overset{\overset{\displaystyle H}{|}}{\underset{\underset{\displaystyle H}{|}}{C}}-H$$

(b)
$$H-\overset{\overset{\displaystyle H}{|}}{\underset{\underset{\displaystyle H}{|}}{C}}-\overset{\overset{\displaystyle H}{|}}{\underset{\underset{\displaystyle H}{|}}{C}}-\overset{\overset{\displaystyle OH}{|}}{\underset{\underset{\displaystyle H}{|}}{C}}-\overset{\overset{\displaystyle H}{|}}{\underset{\underset{\displaystyle H}{|}}{C}}-H$$

19.23)   Write the structural formula and name of the alkene that may be prepared by the dehydrohalogenation of 1-chloropentane.

19.63)   Write the structural formula and name of two isomeric alkenes that could be formed by the dehydrohalogenation of 2-bromobutane.

### Section 19.6

19.24)   Distinguish between an addition reaction and a substitution reaction.

19.64)   Explain why the chlorination of ethane yields chloroethane, whereas the chlorination of ethylene produces 1,2-dichloroethane.

19.25)   Write an equation for the hydrogenation of propene.

19.65)   Write an equation for the hydrogenation of propyne.

19.26)   What is polymerization?

19.66)   Distinguish between monomers and polymers.

19.27)   Chloroethylene is also known as vinyl chloride. When it polymerizes it becomes the well-known polyvinyl chloride of which shower curtains, phonograph records, raincoats and plastic pipe are made. Show how three vinyl chloride monomers would polymerize to form "PVC."

19.67)   The popular kitchen material *Saran Wrap* is made from the copolymerization (polymerization of two monomers) of chloroethylene and vinylidene chloride, $CH_2=CCl_2$. Show how this copolymerization occurs, and write the structure through at least eight carbon atoms.

### Section 19.7

19.28)   How do aromatic hydrocarbons differ from aliphatic hydrocarbons?

19.68)   Write the symbol used to represent the benzene ring and describe what is signified and/or understood about what it represents.

19.29)   Name the compounds represented by these three formulas:

19.69)   Draw a structural formula for 1,3,5-trichlorobenzene.

(a)              (b)              (c)

## Section 19.9

19.30) What is the functional group of an alcohol? Write both its name and formula. Write the general formula of an alcohol.

19.31) Write the structural formula for 1-pentanol.

## Section 19.10

19.32) Show how water, alcohols and ethers are related structurally.

19.33) Write the structural formula for methyl propyl ether.

## Section 19.11

19.34) How are aldehydes and ketones alike, and how do they differ?

## Section 19.12

19.35) Write the general formula for a carboxylic acid, as well as the formula of the functional group.

19.36) Account for the high boiling points of carboxylic acids compared to the boiling points of other compounds of comparable molecular weight.

19.37) Demonstrate by equation the acid character of the carboxylic group.

## Section 19.13

19.38) Explain the relationship between ammonia and the amines.

19.39) What is the name of

$$CH_3-\underset{\underset{H}{|}}{N}-C_3H_7$$

## Section 19.14

19.40) Compare the functional groups of carboxylic acids and amides.

19.70) Account for the high boiling points and water solubility of alcohols compared to the same properties of alkanes of comparable molecular weight.

19.71) Give the common and IUPAC name of

$$H-\overset{\overset{\displaystyle H}{|}}{\underset{\underset{\displaystyle H}{|}}{C}}-\overset{\overset{\displaystyle H}{|}}{\underset{\underset{\displaystyle H}{|}}{C}}-\overset{\overset{\displaystyle H}{|}}{\underset{\underset{\displaystyle H}{|}}{C}}-\overset{\overset{\displaystyle H}{|}}{\underset{\underset{\displaystyle H}{|}}{C}}-OH$$

19.72) Write a structural equation that shows how methyl ether might be prepared from an alcohol.

19.73) "Each ether with two or more carbon atoms has an alcohol with which it is isomeric." Show that this statement is true.

19.74) Write structural formulas for butanal and butanone.

19.75) Write the structural formula for hexanoic acid. What is the name of $C_4H_9COOH$?

19.76) Account for the high solubility of the lower carboxylic acids.

19.77) Write the equation for the reaction between formic acid, $HCOOH$, and propanol, and name the ester formed.

19.78) "Amines are bases in many chemical reactions." Explain why this is so.

19.79) Write the structural formula of dibutylamine.

19.80) Explain by structural formula how a peptide link is formed.

# REVIEW OF MATHEMATICS

## PART A:  ALGEBRA AND ARITHMETIC

We here present a brief review of arithmetic and algebra to the point that it is used or assumed in this text. Formal mathematical statement and development will be avoided. The sole purpose of this section is to refresh your memory in those areas where it may be needed, in order that you may perform and understand the calculations in this book.

**1. ADDITION:**   $a + b$. Example: $2 + 3 = 5$.

**2. SUBTRACTION:**   $a - b$. Example: $5 - 3 = 2$. Subtraction may be thought of as the addition of the opposite, or negative, of a number. In that sense, $a - b = a + (-b)$. Example: $5 - 3 = 5 + (-3) = 2$.

**3. MULTIPLICATION:**   $a \times b = ab = a \cdot b = a(b) = (a)(b) = b \times a = ba = b \cdot a = b(a)$. The foregoing are equivalent expressions that factor a is to be multiplied by factor b. Reversing the sequence of the factors, i.e., $a \times b = b \times a$, indicates that factors may be taken in any order when two or more are multiplied together. Examples: $2 \times 3 = 2 \cdot 3 = 2(3) = (2)(3) = 3 \times 2 = 3 \cdot 2 = 3(2) = 6$.

**Grouping of factors:**   $(a)(b)(c) = (ab)(c) = (a)(bc)$. Factors may be grouped in any way in multiplication. Example: $(2)(3)(4) = (2 \times 3)(4) = (2)(3 \times 4) = 24$.

**Multiplication by 1:**   $n \times 1 = n$. If any number is multiplied by 1, the product is the original number. Examples: $6 \times 1 = 6$; $3.25 \times 1 = 3.25$.

**Multiplication of fractions:**   $\dfrac{a}{b} \times \dfrac{c}{d} \times \dfrac{e}{f} = \dfrac{ace}{bdf}$

If two or more fractions are to be multiplied, the product is equal to the product of the numerators divided by the product of the denominators. Example:

$$4 \times \frac{9}{2} \times \frac{1}{6} = \frac{4}{1} \times \frac{9}{2} \times \frac{1}{6} = \frac{4 \times 9 \times 1}{1 \times 2 \times 6} = \frac{36}{12} = 3$$

**4. DIVISION:** $a \div b = a/b = \frac{a}{b}$. The foregoing are equivalent expressions that a is to be divided by b. Example:

$$12 \div 4 = 12/4 = \frac{12}{4} = 3$$

**Special case:** If any number is divided by itself the quotient (result) is equal to 1. Examples: $\frac{4}{4} = 1$; $\frac{n}{n} = 1$.

**Division by 1:** $\frac{n}{1} = n$. If any number is divided by 1, the quotient is the original number. Examples: $\frac{6}{1} = 6$; $\frac{3.25}{1} = 3.25$. From this it follows that any number may be expressed as a fraction having 1 as the denominator. Examples: $4 = \frac{4}{1}$; $9.12 = \frac{9.12}{1}$; $m = \frac{m}{1}$.

**5. RECIPROCALS:** If n is any number, the reciprocal of n is $\frac{1}{n}$; if $\frac{a}{b}$ is any fraction, the reciprocal of $\frac{a}{b}$ is $\frac{b}{a}$. The first part of the foregoing sentence is actually a special case of the second part: If n is any number, it is equal to $\frac{n}{1}$. Its reciprocal is therefore $\frac{1}{n}$.

A reciprocal is sometimes referred to as the inverse (more specifically, the multiplicative inverse) of a number. This is because **the product of any number multiplied by its reciprocal equals 1.** Examples:

$$2 \times \frac{1}{2} = \frac{2}{2} = 1 \qquad n \times \frac{1}{n} = \frac{n}{n} = 1$$

$$\frac{4}{3} \times \frac{3}{4} = \frac{12}{12} = 1 \qquad \frac{m}{n} \times \frac{n}{m} = \frac{mn}{mn} = 1$$

Division may be regarded as multiplication by a reciprocal:

$$a \div b = \frac{a}{b} = a \times \frac{1}{b} \qquad \text{Example: } 6 \div 2 = \frac{6}{2} = 6 \times \frac{1}{2} = 3$$

$$a \div b/c = \frac{a}{b/c} = a \times \frac{c}{b} \qquad \text{Example: } 6 \div \frac{2}{3} = \frac{6}{2/3} = 6 \times \frac{3}{2} = 9$$

**6. SUBSTITUTION.** If $d = b + c$, then $a(b + c) = ad$. Any number or expression may be substituted for its equal in any other expression. Example: $7 = 3 + 4$. Therefore $2(3 + 4) = 2 \times 7$.

**7. "CANCELLATION":** $\frac{ab}{ca} = \frac{ab}{ca} = \frac{b}{c}$. The process commonly called cancellation is actually a combination of grouping of factors (see 3), substitution (see 6) of 1 for a number divided by itself (see 4), and multiplication by 1 (see 3). Note the steps in the following examples:

$$\frac{xy}{yz} = \frac{yx}{yz} = \left(\frac{y}{y}\right)\left(\frac{x}{z}\right) = 1 \cdot \frac{x}{z} = \frac{x}{z}$$

$$\frac{24}{18} = \frac{6 \times 4}{6 \times 3} = \frac{6}{6} \times \frac{4}{3} = 1 \times \frac{4}{3} = \frac{4}{3}$$

Note that only *factors* and *multipliers* can be cancelled. There is no cancellation in $\frac{a+b}{a+c}$.

**8. EXPONENTIALS:** An exponential indicates the number of times the same factor is to be used in multiplication. For example, $10^3$ means 10 is to be used as a factor 3 times: $10^3 = 10 \times 10 \times 10 = 1000$.

**Raising a product to a power:** $(ab)^n = a^n \times b^n$. When the product of two or more factors is raised to some power, each factor is raised to that power. Example:

$$(2 \times 5y)^3 = 2^3 \times 5^3 \times y^3 = 8 \times 125 \times y^3 = 1000\, y^3$$

**Raising a fraction to a power:** $\left(\frac{a}{b}\right)^n = \frac{a^n}{b^n}$. When a fraction is raised to some power, the numerator and denominator are both raised to that power. Example:

$$\left(\frac{2x}{5}\right)^3 = \frac{2^3 x^3}{5^3} = \frac{8x^3}{125} = 0.064\, x^3$$

**Multiplication of exponentials:** $a^m \times a^n = a^{m+n}$. To multiply exponentials, add the exponents. Example: $10^3 \times 10^4 = 10^7$.

**Division of exponentials:** $a^m \div a^n = \frac{a^m}{a^n} = a^{m-n}$. To divide exponentials, subtract the denominator exponent from the numerator exponent. Example: $10^7 \div 10^4 = \frac{10^7}{10^4} = 10^{7-4} = 10^3$.

**Square root of exponentials:** $\sqrt{a^{2n}} = a^n$. To find the square root of an exponential, divide the exponent by 2. Example: $\sqrt{10^6} = 10^3$. If the exponent is odd, see below.

**Square root of a product:** $\sqrt{ab} = \sqrt{a} \times \sqrt{b}$. The square root of the product of two numbers equals the product of the square roots of the numbers. Example: $\sqrt{9 \times 10^{-6}} = \sqrt{9} \times \sqrt{10^{-6}} = 3 \times 10^{-3}$.

Using this principle, by adjusting a decimal point you may take the square root of an exponential having an odd exponent. Example:

$$\sqrt{10^5} = \sqrt{10 \times 10^4} = \sqrt{10} \times \sqrt{10^4} = 3.16 \times 10^2$$

The same technique may be used in taking the square root of a number expressed in exponential notation. Example:

$$\sqrt{1.8 \times 10^{-5}} = \sqrt{18 \times 10^{-6}} = \sqrt{18} \times \sqrt{10^{-6}} = 4.2 \times 10^{-3}$$

**9. SOLVING AN EQUATION FOR AN UNKNOWN QUANTITY:** Most problems in this course can be solved by dimensional analysis methods.

There are occasions, however, when algebra must be used, particularly in relation to the gas laws. Solving an equation for an unknown involves manipulating the equation in such a manner that the unknown is left as the only item on either side of the equation, all known items being on the other side. "Manipulating" an equation takes on several forms, but the essential feature is that **whatever is done to one side of the equation must also be done to the other.** The resulting relationship then remains an equality, a valid equation. Among the operations that may be performed on both sides of an equation are addition, subtraction, multiplication, division and taking square root.

The objective in each example that follows is to solve the equation for the unknown, x. The steps of the algebraic solutions are shown, as well as the operation performed on both sides of the equation. Each example is accompanied by a practice problem that is solved by the same method. You should be able to solve the problem, even if you have not yet reached that point in the book where such a problem is likely to appear. Answers to these practice problems may be found on page 513.

$$(1) \quad x + a = b$$

$$x + a - a = b - a \qquad \text{Subtract a.}$$

$$x = b - a \qquad \text{Simplify.}$$

*PRACTICE: 1) If* $P = p_{O_2} + p_{H_2O}$, *find* $p_{O_2}$ *when* $P = 748$ *torr and* $p_{H_2O} = 24$ *torr.*

$$(2) \quad ax = b$$

$$\frac{ax}{a} = \frac{b}{a} \qquad \text{Divide by a.}$$

$$x = \frac{b}{a} \qquad \text{Simplify.}$$

*PRACTICE: 2) At a certain temperature,* $PV = k$. *If* $P = 1.23$ *atm and* $k = 1.62$ $(\ell)(atm)$, *find V.*

$$(3) \quad \frac{x}{a} = b$$

$$\frac{ax}{a} = ba \qquad \text{Multiply by a.}$$

$$x = ba \qquad \text{Simplify.}$$

*PRACTICE: 3) In a fixed volume,* $\frac{P}{T} = k$. *Find P if* $k = \frac{2.4 \ torr}{°K}$ *and* $T = 300°K$.

(4)  $\dfrac{a}{x} = b$

$\dfrac{\cancel{x}a}{\cancel{x}} = bx$  Multiply by x.

$a = bx$  Simplify.

$x = \dfrac{a}{b}$  See (2).

PRACTICE: 4) At what value of T will P = 760 torr when $k = \dfrac{2.4\ torr}{°K}$ and $\dfrac{P}{T} = k$?

(5)  $\dfrac{a}{x} = \dfrac{b}{c}$

$\dfrac{c\cancel{x}}{b} \cdot \dfrac{a}{\cancel{x}} = \dfrac{\cancel{b}}{\cancel{c}} \cdot \dfrac{\cancel{c}x}{\cancel{b}}$  Multiply by $\dfrac{cx}{b}$.

$\dfrac{ac}{b} = x$  Simplify.

PRACTICE: 5) For gases at constant volume, $\dfrac{P_1}{T_1} = \dfrac{P_2}{T_2}$. If $P_1 = 0.80$ atm at $T_1 = 320°K$, at what value of $T_2$ will $P_2 = 1.00$ atm?

(6)  $\dfrac{a}{b+x} = c$

$(b+x)\dfrac{a}{(b+x)} = c(b+x)$  Multiply by (b+x).

$a = c(b+x)$  Simplify.

$\dfrac{a}{c} = \dfrac{\cancel{c}(b+x)}{\cancel{c}}$  Divide by c.

$\dfrac{a}{c} = b+x$  Simplify.

$\dfrac{a}{c} - b = x$  See (1).

PRACTICE: 6) In how many grams of water must you dissolve 20.0 g of salt to make a 25% solution? The formula is

$$\dfrac{g\ salt}{g\ salt + g\ water} \times 100 = \%;\ or$$

$$\dfrac{g\ salt}{g\ salt + g\ water} = \dfrac{\%}{100}$$

# PART B:  EXPONENTIAL NOTATION

## EXPRESSING NUMBERS IN EXPONENTIAL NOTATION

Problems in chemistry frequently involve extremely large or very small numbers. For example, the mass of a single atom of hydrogen is 0.00000000000000000000000166 gram. In one liter of hydrogen measured at one atmosphere pressure and 0°C there are 53,770,000,000,000,000,000,000 hydrogen atoms.

Numbers such as these can be represented more conveniently in a form called *exponential notation* (also known as *scientific notation*). To illustrate,

$$3425 = 3.425 \times 10^3$$

$$0.000000794 = 7.94 \times 10^{-7}$$

In each case the number, expressed in normal decimal form, has been rewritten as the product of two factors. The first factor is an ordinary decimal number, usually between 1 and 10, which we will call the *decimal factor*. The second factor is a power of 10 we will call the *exponential factor*.

> *PRACTICE: 7) Identify the decimal factor and the exponential factor in each of the following numbers, correctly written in exponential notation:*
>
> *7a)  $3.91 \times 10^5$        7b)  $2.0090 \times 10^{-18}$        7c)  $1.00 \times 10^1$*
>
> *(Answers to practice problems are on page 509.)*

*     *     *     *     *

To convert an ordinary number to exponential notation, you first count the number of places, n, the decimal point must be moved so that it will appear immediately after the first nonzero digit. You then multiply the original number by 1 in the form $10^n \times 10^{-n}$.* Then group the factors in such a way that the product of the first two yields the required decimal factor, and simplify. For example, to express 3425 in exponential notation, the decimal must be moved three places to the left to locate it between the 3 and the 4. Then

$$3425 = 3425 \times 10^3 \times 10^{-3} = (3425 \times 10^{-3}) \times 10^3 = 3.425 \times 10^3$$

Applying the same procedure to 0.000000794, the decimal point must be moved 7 places to the right. Therefore

$$0.000000794 = 0.000000794 \times 10^7 \times 10^{-7} = (0.000000794 \times 10^7) \times 10^{-7} =$$

$$7.94 \times 10^{-7}$$

In essence, if you move the decimal to the left (decimal factor is SMALLER than number), the exponent is positive (exponential factor is

---

*$10^n \times 10^{-n} = 10^0 = 1$. Any number raised to the zeroth power equals 1.

LARGE); if you move the decimal to the right (decimal factor LARGER), the exponent is negative (exponential factor SMALL).

*PRACTICE:* 8) *Express each of the following numbers in exponential notation:*

*8a)* 3,762,199    *8b)* 198.75    *8c)* 0.000098
*8d)* 10.080      *8e)* 0.00460

\*        \*        \*        \*        \*

In converting a number expressed in exponential notation to an ordinary decimal number, the *absolute value* of the exponent (its *numerical* value, regardless of sign) tells you how many places the decimal point must be moved. If the exponent is positive, the number is large, so the decimal is moved to the right; if the exponent is negative, the number is small, so the exponent goes to the left. For example, in $7.89 \times 10^5$ the exponent is positive. This indicates a large number, so the decimal is moved 5 places to the right, the same number of places as the value of the exponent: 789,000. In $5.37 \times 10^{-4}$ the negative exponent identifies a small number, so the decimal point is moved 4 places to the left: 0.000537.

*PRACTICE:* 9) *Express each of the following numbers in ordinary decimal form:*

*9a)* $3.49 \times 10^{-11}$    *9b)* $5.16 \times 10^4$    *9c)* $1.0 \times 10^1$    *9d)* $3.75 \times 10^0$

## *MULTIPLICATION WITH EXPONENTIAL NOTATION*

In multiplying numbers expressed in exponential notation, the factors are rearranged, placing together all the decimal factors in one group and all the exponential factors in a second group. The two groups are multiplied separately. The decimal result and the experimental result are then combined as the exponential notation expression of the product. For example:

$$
\begin{aligned}
(3.96 \times 10^4)(5.19 \times 10^{-7}) &= 3.96 \times 10^4 \times 5.19 \times 10^{-7} \\
&= 3.96 \times 5.19 \times 10^4 \times 10^{-7} \\
&= (3.96 \times 5.19)(10^4 \times 10^{-7}) \\
&= 20.6 \times 10^{4-7} \\
&= 20.6 \times 10^{-3}
\end{aligned}
$$

Adjusting the decimal so the decimal factor is between 1 and 10 requires that it be moved 1 place to the left. The decimal factor is thus made smaller. This is offset by making the exponential factor larger by *increasing* the exponent by 1, the same number as the number of places the decimal was moved in the decimal factor. To *increase* an exponent, you make it *more positive*. To increase $-3$ by 1, it becomes $-2$: $-3 + 1 = -2$. Thus

$$20.6 \times 10^{-3} = 2.06 \times 10^{-2}$$

*PRACTICE:* 10) *Perform each of the following calculations and express the result in exponential notation.*

10a) $(3.26 \times 10^4)(1.54 \times 10^6) =$  10e) $(35,780)(761.9) =$

10b) $(8.39 \times 10^{-7})(4.53 \times 10^9) =$  10f) $(0.00091)(235.6) =$

10c) $(6.73 \times 10^{-3})(9.11 \times 10^{-3}) =$  10g) $(0.0890)(0.000000726) =$

10d) $(2.93 \times 10^5)(4.85 \times 10^6)(5.58 \times 10^{-3}) =$

## DIVISION WITH EXPONENTIAL NOTATION

Division of numbers expressed in exponential notation is performed in essentially the same manner as multiplication. The decimal factors and exponential factors are separated into two groups; the value of each group is calculated separately; and finally the decimal result is combined with the exponential result as the exponential notation form of the quotient. To illustrate,

$$3.96 \times 10^4 \div 5.19 \times 10^{-7} = \frac{3.96 \times 10^4}{5.19 \times 10^{-7}}$$

$$= \left(\frac{3.96}{5.19}\right)\left(\frac{10^4}{10^{-7}}\right) = 0.763 \times 10^{11} = 7.63 \times 10^{10}$$

PRACTICE: 11) *Perform each of the following calculations and express the result in exponential notation.*

11a) $\frac{8.94 \times 10^6}{4.35 \times 10^4} =$  11b) $\frac{5.08 \times 10^{-3}}{7.23 \times 10^{-5}} =$

11c) $\frac{(3.05 \times 10^{-6})(2.19 \times 10^{-3})}{5.48 \times 10^{-5}} =$  11d) $\frac{(867)(7.2 \times 10^{-7})}{0.000000000927} =$

11e) $\frac{(7180)(9,420,000)}{(258,000)(0.000618)} =$

## ADDITION AND SUBTRACTION IN EXPONENTIAL NOTATION

As in arithmetic, addition and subtraction of exponentials requires that digit values (hundreds, units, tenths, etc.) be aligned vertically. This may be achieved by adjusting decimals and exponents so the exponents of all numbers are the same, and then proceeding as in ordinary arithmetic. Shown below is the addition of $6.44 \times 10^{-7} + 1.3900 \times 10^{-5}$ in three ways: with exponents of $10^{-5}$, with exponents of $10^{-7}$, and as decimal numerals.

$$
\begin{array}{llll}
1.3900 \times 10^{-5} = & 1.3900 \times 10^{-5} = & 139.00 \times 10^{-7} = & 0.000013900 \\
6.44 \times 10^{-7} \phantom{x} = & 0.0644 \times 10^{-5} = & \phantom{1}6.44 \times 10^{-7} = & 0.000000644 \\
\hline
& 1.4544 \times 10^{-5} = & 145.44 \times 10^{-7} = & 0.000014544
\end{array}
$$

PRACTICE: 12) *Add or subtract the following numbers:*

12a) $5.2 \times 10^5 + 3.98 \times 10^4 =$

12b) $3.971 \times 10^2 + 1.98 \times 10^{-1} =$

12c) $5.63 \times 10^4 + 2.93 \times 10^3 + 5.42 \times 10^5 =$

12d) $1.05 \times 10^{-4} - 9.7 \times 10^{-5} =$

12e) $\frac{9.02 \times 10^7}{265.9} - \frac{7.89 \times 10^3}{2.37 \times 10^{-2}} =$

# PART C:   THE MEANING OF PER: PERCENT

## *THE MEANING OF "PER"*

If you drive 100 miles in 4 hours, what is your average speed? Twenty five miles per hour, you say. Fine. But how did you get it? Was it not by *dividing* 100 by 4? In other words, 100 miles *per* 4 hours is the same as 25 miles *per* 1 hour; or $100 \div 4 = 25 \div 1$; or $\frac{100}{4} = \frac{25}{1}$.

Four concepts converge here, and you should recognize them as all meaning the same thing. *Division problems* (1) may be expressed as *fractions* (2) in which the numerator is divided by the denominator. Furthermore, the result is frequently expressed as some quantity *per unit* of something else. The term *"per"* means *arithmetic division* (3). Finally, "per" signifies an *equivalence* (4), as 25 miles $\simeq$ 1 hour.

The most familiar use of "per" is in percent. *Cent-* is an expression referring to 100, as there are 100 *cent*imeters in a meter, or 100 *cents* in a dollar. *Percent,* then, means *parts per 100 parts.* This establishes percentage as an equivalency relationship that may be used in dimensional analysis. For example, if 40% of the 1200 people attending a concert are men, how many men are in the audience? The 40% may be interpreted as an equivalence between the number of men and the number of people in the audience: 40 men $\simeq$ 100 people. By dimensional analysis,

$$1200 \text{ people} \times \frac{40 \text{ men}}{100 \text{ people}} = 480 \text{ men}$$

The more familiar way of solving this problem is to convert the percentage figure to a decimal fraction by moving the decimal two places left (dividing by 100), and then multiplying by the total figure. In other words, the fraction of the audience that are men is 0.40. The total number of men is $0.40 \times 1200 = 480$.

*PRACTICE: 13) 82% of the registered voters in a certain precinct cast ballots in a recent election. If there are 1482 registered voters in the district, how many ballots were cast?*

\*     \*     \*     \*     \*

The equivalency approach to percentage problems is even more useful in the reverse direction. For example: a state-supported university establishes a quota such that 35% of the freshmen admitted in a certain year will be from out of state. If 2891 incoming freshmen were from other states, what was the total number of freshmen enrolled that term? The dimensional analysis setup uses the percentage equivalency 35 out-of-state students $\simeq$ 100 total students. Applying to the given quantity,

$$2891 \text{ out-of-state students} \times \frac{100 \text{ total students}}{35 \text{ out-of-state students}} = 8260 \text{ total students}$$

The conventional approach to this example is algebraic:

$$35\% \text{ of the total students} = \text{out-of-state students}$$

$$0.35 \times \text{total students} = 2891$$

$$\text{total students} = \frac{2891}{0.35} = 8260$$

*PRACTICE: 14) The total number of pages in a monthly magazine is planned on the basis of the number of advertising pages sold. If 58 pages of advertising sold for a forthcoming issue is to represent 46% of the pages in the magazine, what will be the total number of pages?*

\*     \*     \*     \*     \*

A mathematical expression that defines percentage is

$$\% = \frac{\text{part quantity}}{\text{total quantity}} \times 100$$

If you have a count of the number of each kind of part making up a total quantity, you can calculate the percentage of any part, using the above equation. For example, the weather bureau of a certain city recorded measurable precipitation on 104 days in 1977. What percentage of the days of that year had recorded rainfall or snow?

$$\frac{\text{Part quantity}}{\text{Total quantity}} \times 100 = \frac{104}{365} \times 100 = 28.5\%$$

*PRACTICE: 15) Over a given test period at the counting station of a fish ladder bypass around a hydroelectric dam, 89 Chinook salmon, 14 steelhead trout and 21 fish of other kinds were tallied. Calculate the percentage of Chinook salmon.*

# APPENDIX II:

## ANSWERS TO PRACTICE PROBLEMS

1) $p_{O_2} = P - p_{H_2O} = 748 - 24 = 724$ torr

2) $V = \dfrac{k}{P} = \dfrac{1.62\ (\ell)(\text{atm})}{1.23\ \text{atm}} = 1.32\ \ell$

3) $P = kT = \dfrac{2.4\ \text{torr}}{°K} \times 300°K = 720$ torr

4) $T = \dfrac{P}{k} = 760\ \text{torr} \times \dfrac{°K}{2.4\ \text{torr}} = 317°K$

5) $T_2 = \dfrac{T_1 P_2}{P_1} = \dfrac{(320°K)(1.00\ \text{atm})}{0.80\ \text{atm}} = 400°K$

6) Because of the complexity of this problem, it is easier to substitute the given values into the original equation and then solve for the unknown. The steps in the solution correspond to those in the appendix.

$$\dfrac{\text{g salt}}{\text{g salt} + \text{g water}} = \dfrac{\%}{100}$$

$$\dfrac{20.0}{20.0 + \text{g water}} = \dfrac{25}{100}$$

$$20.0 = 0.25(20.0 + \text{g water}) = 5.0 + 0.25(\text{g water})$$

$$20.0 - 5.0 = 0.25(\text{g water})$$

$$\text{g water} = \dfrac{15.0}{0.25} = 60\ \text{g water}$$

7)

| | DECIMAL FACTOR | EXPONENTIAL FACTOR |
|---|---|---|
| 7a) | 3.91 | $10^5$ |
| 7b) | 2.0090 | $10^{-18}$ |
| 7c) | 1.00 | $10^1$ |

8) a) $3.762199 \times 10^6$;  b) $1.9875 \times 10^2$;  c) $9.8 \times 10^{-5}$
   d) $1.0080 \times 10^1$;  e) $4.60 \times 10^{-3}$.

9) a) 0.0000000000349;  b) 51,600;  c) 10;  d) 3.75

10) a) $5.02 \times 10^{10}$
    b) $38.0 \times 10^2 = 3.80 \times 10^3$
    c) $61.3 \times 10^{-6} = 6.13 \times 10^{-5}$
    d) $79.3 \times 10^8 = 7.93 \times 10^9$

    e) $27.26 \times 10^6 = 2.726 \times 10^7$
    f) $21 \times 10^{-2} = 2.1 \times 10^{-1}$
    g) $64.6 \times 10^{-9} = 6.46 \times 10^{-8}$

11) a) $2.06 \times 10^2$;     b) $0.703 \times 10^2 = 7.03 \times 10^1$
    c) $1.22 \times 10^{-4}$;     d) $6.7 \times 10^5$;     e) $4.24 \times 10^8$

12) a) $\begin{array}{r} 5.2 \ \ \times 10^5 \\ 0.398 \times 10^5 \\ \hline 5.6 \ \ \times 10^5 \end{array}$
    b) $\begin{array}{r} 3.971 \ \ \ \times 10^2 \\ 0.00198 \times 10^2 \\ \hline 3.973 \ \ \ \times 10^2 \end{array}$
    c) $\begin{array}{r} 0.563 \ \ \times 10^5 \\ 0.0293 \times 10^5 \\ 5.42 \ \ \ \times 10^5 \\ \hline 6.01 \ \ \ \times 10^5 \end{array}$

    d) $\begin{array}{r} 10.5 \times 10^{-5} \\ - \ 9.7 \times 10^{-5} \\ \hline 0.8 \times 10^{-5} = 8 \times 10^{-6} \end{array}$

    e) $\dfrac{9.02 \times 10^7}{265.9} = 3.39 \times 10^5$

    $\dfrac{7.89 \times 10^3}{2.37 \times 10^{-2}} = 3.33 \times 10^5$

    $\begin{array}{r} 3.39 \times 10^5 \\ -3.33 \times 10^5 \\ \hline 0.06 \times 10^5 = 6 \times 10^3 \end{array}$

13) $1482 \ \text{registered} \times \dfrac{82 \ \text{voters}}{100 \ \text{registered}} = 1215 \ \text{voters}$; or $1482 \times 0.82 = 1215$

14) $58 \ \text{pages sold} \times \dfrac{100 \ \text{pages total}}{46 \ \text{pages sold}} = 126 \ \text{pages total}$; or $0.46 \times \text{total} = 58$
    $$\text{total} = 126$$

15) Total fish $= 89$ Chinook $+ 14$ steelhead $+ 21$ others $= 124$ fish
    $\dfrac{89 \ \text{Chinook}}{124 \ \text{total}} \times 100 = 71.8\%$ Chinook (arbitrarily rounded to 3 significant figures)

# CONVERSION TABLE—
# METRIC AND ENGLISH UNITS

*MASS*

454 grams = 1 pound
28.3 grams = 1 ounce
1 kilogram = 2.2 pounds

*LENGTH*

2.54 centimeters = 1 inch
30.5 centimeters = 1 foot
1 meter = 39.4 inches
1 meter = 1.09 yards
1.61 kilometers = 1 mile

*CAPACITY*

1 liter = 1.06 quarts
3.785 liters = 1 gallon
4.546 liters = 1 imperial gallon

*PRESSURE*

$$\left.\begin{array}{l} 760.0 \text{ torr} \\ 760.0 \text{ mm mercury} \\ 76.00 \text{ cm mercury} \\ 101.3 \text{ kilopascals} \end{array}\right\} = 1 \text{ atm} = \left\{\begin{array}{l} 14.69 \text{ lb/in}^2 \\ 29.92 \text{ in mercury} \end{array}\right.$$

*ENERGY*

4.184 joules = 1 calorie = 0.00397 British thermal unit

# APPENDIX IV:

## COMMON NAMES OF CHEMICALS

| COMMON NAME | CHEMICAL NAME | FORMULA |
|---|---|---|
| Alumina | Aluminum oxide | $Al_2O_3$ |
| Baking soda | Sodium hydrogen carbonate | $NaHCO_3$ |
| Bluestone | Copper(II) sulfate pentahydrate | $CuSO_4 \cdot 5\ H_2O$ |
| Borax | Sodium tetraborate decahydrate | $Na_2B_4O_7 \cdot 10\ H_2O$ |
| Brimstone | Sulfur | $S$ |
| Carbon tetrachloride | Tetrachloromethane | $CCl_4$ |
| Chloroform | Trichloromethane | $CHCl_3$ |
| Cream of tartar | Potassium hydrogen tartrate | $KHC_4H_4O_6$ |
| Diamond | Carbon | $C$ |
| Dolomite | Calcium magnesium carbonate | $CaCO_3 \cdot MgCO_3$ |
| Epsom salts | Magnesium sulfate heptahydrate | $MgSO_4 \cdot 7\ H_2O$ |
| Freon (refrigerant) | Dichlorodifluoromethane | $CCl_2F_2$ |
| Galena | Lead(II) sulfide | $PbS$ |
| Grain alcohol | Ethyl alcohol; ethanol | $C_2H_5OH$ |
| Graphite | Carbon | $C$ |
| Gypsum | Calcium sulfate dihydrate | $CaSO_4 \cdot 2\ H_2O$ |
| Hypo | Sodium thiosulfate | $Na_2S_2O_3$ |
| Laughing gas | Dinitrogen oxide | $N_2O$ |
| Lime | Calcium oxide | $CaO$ |
| Limestone | Calcium carbonate | $CaCO_3$ |
| Lye | Sodium hydroxide | $NaOH$ |
| Marble | Calcium carbonate | $CaCO_3$ |
| MEK | Methyl ethyl ketone | $CH_3COC_2H_5$ |
| Milk of magnesia | Magnesium hydroxide | $Mg(OH)_2$ |
| Muriatic acid | Hydrochloric acid | $HCl$ |
| Oil of vitriol | Sulfuric acid (conc.) | $H_2SO_4$ |
| Plaster of Paris | Calcium sulfate hemihydrate | $CaSO_4 \cdot \frac{1}{2}\ H_2O$ |
| Potash | Potassium carbonate | $K_2CO_3$ |
| Pyrites (fool's gold) | Iron disulfide | $FeS_2$ |
| Quartz | Silicon dioxide | $SiO_2$ |
| Quicksilver | Mercury | $Hg$ |
| Rubbing alcohol | Isopropyl alcohol | $(CH_3)_2CHOH$ |
| Sal ammoniac | Ammonium chloride | $NH_4Cl$ |
| Salt | Sodium chloride | $NaCl$ |
| Saltpeter | Sodium nitrate | $NaNO_3$ |
| Slaked lime | Calcium hydroxide | $Ca(OH)_2$ |
| Sugar | Sucrose | $C_{12}H_{22}O_{11}$ |
| Washing soda | Sodium carbonate decahydrate | $Na_2CO_3 \cdot 10\ H_2O$ |
| Wood alcohol | Methyl alcohol; methanol | $CH_3OH$ |

# GREEK PREFIXES USED
# IN CHEMICAL NOMENCLATURE

| NUMBER | PREFIX |
|--------|--------|
| 1 | Mono- |
| 2 | Di- |
| 3 | Tri- |
| 4 | Tetra- |
| 5 | Penta- |
| 6 | Hexa- |
| 7 | Hepta- |
| 8 | Octa- |
| 9 | Nona- |
| 10 | Deca- |
| 12 | Dodeca- |

# APPENDIX VI:

## WATER VAPOR PRESSURE

| Temperature (°C) | Vapor Pressure (torr) | Temperature (°C) | Vapor Pressure (torr) |
|---|---|---|---|
| 0 | 4.6 | 28 | 28.3 |
| 5 | 6.5 | 29 | 30.0 |
| 10 | 9.2 | 30 | 31.8 |
| 15 | 12.8 | 31 | 33.7 |
| 16 | 13.6 | 32 | 35.7 |
| 17 | 14.5 | 33 | 37.7 |
| 18 | 15.5 | 34 | 40.0 |
| 19 | 16.5 | 35 | 42.2 |
| 20 | 17.5 | 40 | 55.3 |
| 21 | 18.6 | 45 | 71.9 |
| 22 | 19.8 | 50 | 92.5 |
| 23 | 21.1 | 60 | 149.4 |
| 24 | 22.4 | 70 | 233.7 |
| 25 | 23.8 | 80 | 355.1 |
| 26 | 25.2 | 90 | 525.8 |
| 27 | 26.7 | 100 | 760.0 |

## A

**absolute zero**—the temperature predicted by extrapolation of experimental data where translational kinetic energy theoretically becomes zero; the zero of the absolute temperature scale, which is equivalent to −273.15°C.

**acid**—a substance that yields hydrogen (hydronium) ions in aqueous solution (Arrhenius definition); a substance that donates protons in chemical reaction (Brönsted-Lowry definition); a substance that forms covalent bonds by accepting a pair of electrons (Lewis definition).

**acidic solution**—an aqueous solution in which the hydrogen ion concentration is greater than the hydroxide ion concentration; a solution in which the pH is less than 7.

**activated complex**—an intermediate molecular species presumed to be formed during the interaction (collision) of reacting molecules in a chemical change.

**activation energy**—the energy barrier that must be overcome to start a chemical reaction.

**alcohol**—an organic compound consisting of an alkyl group and at least one hydroxyl group, having the general formula ROH.

**aldehyde**—a compound consisting of a carbonyl group bonded to a hydrogen on one side, and a hydrogen, alkyl or aryl group on the other, having the general formula RCHO.

**aliphatic hydrocarbon**—an alkane, alkene or alkyne.

**alkaline earth metal**—a metal from Group IIA of the periodic table.

**alkali metal**—a metal from Group IA of the periodic table.

**alkane**—a saturated hydrocarbon containing only single bonds, in which each carbon atom is bonded to four other atoms.

**alkene**—an unsaturated hydrocarbon containing a double bond, and each carbon atom that is double bonded is bonded to a total of three atoms.

**alkyl group**—an alkane hydrocarbon group lacking one hydrogen atom, having the general formula $C_nH_{2n+1}$, and frequently symbolized by the letter R.

**alkyne**—an unsaturated hydrocarbon containing a triple bond, and each carbon atom that is triple bonded is bonded to a total of two atoms.

**alpha ($\alpha$) particle**—the nucleus of a helium atom, often emitted in nuclear disintegration.

**amide**—a derivative of a carboxylic acid in which the hydroxyl group is replaced by a —NH₂ group, and having the general formula RCONH₂.

**amine**—an ammonia derivative in which one or more hydrogens are replaced by an alkyl or aryl group.

**amorphous**—a substance that is without definite structure or shape.

**anhydride (anhydrous)**—a substance that is without water, or from which water has been removed.

**anion**—a negatively charged ion.

**aqueous**—pertaining to water.

**aromatic hydrocarbon**—a hydrocarbon containing a benzene ring.

**atmosphere** (pressure unit)—a unit of pressure based on atmospheric pressure at sea level, and capable of supporting a mercury column 760 mm high.

**atom**—the smallest particle of an element that can combine with atoms of other elements in forming chemical compounds.

**atomic mass unit**—a unit of mass that is exactly 1/12 of the mass of an atom of carbon-12.

**atomic number**—the number of protons in an atom of an element.

**atomic weight**—the number that expresses the average mass of the atoms of an element compared to the mass of an atom of carbon-12 at a value of exactly 12; the average mass of the atoms of an element expressed in atomic mass units.

**Avogadro's number**—the number of carbon atoms in exactly 12 grams of carbon-12; the number of units in 1 mole ($6.02 \times 10^{23}$).

## B

**barometer**—laboratory device for measuring atmospheric pressure.

**base**—a substance that yields hydroxide ions in aqueous solution (Arrhenius definition); a substance that accepts protons in chemical reaction (Brönsted-Lowry definition); a substance that forms covalent bonds by donating a pair of electrons (Lewis definition).

**basic solution**—an aqueous solution in which the hydroxide ion concentration is greater than the hydrogen ion concentration; a solution in which the pH is greater than 7.

**beta (β) particle**—a high-energy electron, often emitted in nuclear disintegration.

**binary compound**—a compound consisting of two elements.

**boiling point**—the temperature at which vapor pressure becomes equal to the pressure above a liquid; the temperature at which vapor bubbles form spontaneously anyplace within a liquid.

**boiling point elevation**—the difference between the boiling point of a solution and the boiling point of the pure solvent.

**bombardment (nuclear)**—the striking of a target nucleus by an atomic particle, causing a nuclear change.

**bond**—*see chemical bond.*

**bond angle**—the angle formed by the bonds between two separate atoms and a common central atom.

**bonding electrons**—the electrons transferred or shared in forming chemical bonds; valence electrons.

*C*

**calorie**—a unit of heat; the quantity of heat required to raise the temperature of one gram of water from 14.5°C to 15.5°C; 1 calorie = 4.184 joules.

**calorimeter**—laboratory device for measuring heat flow.

**carbonyl group**—an organic functional group, $\diagdown C{=}O$, characteristic of aldehydes and ketones.

**carboxyl group**—an organic functional group, $-C\diagup^{O}\diagdown_{OH}$ , characteristic of carboxylic acids.

**carboxylic acid**—an organic acid containing the carboxyl group, having the general formula RCOOH.

**catalyst**—a substance that increases the rate of a chemical reaction by lowering activation energy; the catalyst is either a non-participant in the reaction, or it is regenerated. *See inhibitor.*

**cathode**—the negative electrode in a cathode ray tube; the electrode at which reduction occurs in an electrochemical cell.

**cation**—a positively charged ion.

**chain reaction**—a reaction that has, as a product, one of its own reactants; that product becomes a reactant, thereby perpetuating the original reaction.

**charge cloud**—*see electron cloud.*

**chemical bond**—a general term that sometimes includes all of the electrostatic attractions, between atoms, molecules and ions, but more often refers to covalent and ionic bonds; *see covalent bond, ionic bond.*

**chemical change**—a change in which one or more substances disappear and one or more new substances are formed.

**chemical family**—a group of elements having similar chemical properties because of similar valence electron configuration, appearing in the same column of the periodic table.

**chemical properties**—the types of chemical change in which a substance is capable of reacting.

**cloud chamber**—a device in which condensation tracks form behind certain radioactive emissions as they travel through a supersaturated vapor.

**colligative properties**—physical properties of mixtures that depend upon concentration of particles irrespective of their identity.

**colloid**—a non-settling dispersion of aggregated ions or molecules intermediate in size between the particles in a true solution and those in a suspension.

**compound**—a pure substance that can be broken down into two or more other pure substances by a chemical change.

**conjugate acid-base pair**—a Brönsted-Lowry acid and the base derived from it when it loses a proton; or a Brönsted-Lowry base and the acid developed from it when it accepts a proton.

**coulomb**—a unit of electrical charge.

**covalent bond**—the chemical bond between two atoms that share a pair of electrons.

**crystalline solid**—a solid in which the ions and/or molecules are arranged in a definite geometric pattern.

*D*

**density**—the mass of a substance per unit volume.

**diatomic**—that which has two atoms.

**dilute**—that which is relatively weak in concentration.

**dipole**—a polar molecule.

**diprotic acid**—an acid capable of yielding two protons per molecule in complete ionization.

**dispersion forces**—weak electrical forces of attraction between molecules, temporarily produced by the shifting of electrons within molecules.

**distillation**—the process of separating components of a mixture by boiling off and condensing the more volatile component.

**dynamic equilibrium**—a state in which opposing changes occur at equal rates, resulting in zero net change over a period of time.

*E*

**electrode**—conductor by which electric charge enters or leaves an electrolyte.

**electrolysis**—passage of electric charge through an electrolyte.

**electrolyte**—a substance which, when dissolved, yields a solution that conducts electricity; a solution or other medium that conducts electricity by ionic movement.

**electron**—subatomic particle carrying a unit negative charge and having a mass of $9.1 \times 10^{-28}$ gram, or 1/1837 of the mass of a hydrogen nucleus, found outside the nucleus of the atom.

**electron (charge) cloud**—region of space around or between atoms that is occupied by electrons.

**electron configuration**—the orbital arrangement of electrons in ions or atoms.

**electron-dot diagram (structure)**—*see Lewis diagram.*

**electronegativity**—a scale of relative electron attracting ability of atoms of one element when bonded to atoms of another element by a single covalent bond.

**electron orbit**—the circular or elliptical path supposedly followed by an electron around an atomic nucleus, according to the Bohr theory of the atom.

**electron orbital**—a mathematically described region in space within an atom in which there is a high probability that an electron will be found.

**electron pair geometry**—a description of the distribution of bonding and unshared electron pairs around a bonded atom.

**electron pair repulsion**—the principle that electron pair geometry is the result of repulsion between electron pairs around a bonded atom, causing them to be as far apart as possible.

**electrostatic force**—force of attraction or repulsion between electrically charged objects.

**element**—a pure substance that cannot be decomposed into other pure substances by ordinary chemical means.

**empirical formula**—a formula that represents the lowest integral ratio of atoms of the elements in a compound.

**endothermic reaction**—a reaction that absorbs energy from the surroundings, having a positive $\Delta H$.

**energy**—the ability to do work.

**enthalpy**—the heat content of a chemical system.

**enthalpy of reaction**—*see heat of reaction.*

**equilibrium**—*see dynamic equilibrium.*

**equilibrium constant**—with reference to an equilibrium equation, the ratio in which the numerator is the product of the concentrations of the species on the right side of the equation, each raised to a power corresponding to its coefficient in the equation, and the denominator is the corresponding product of the species on the left side of the equation; symbol: $K$, $K_c$ or $K_{eq}$.

**equilibrium vapor pressure**—*see vapor pressure.*

**equivalent**—that quantity of an acid (or base) that yields or reacts with one mole of $H^+$ (or $OH^-$) in a chemical reaction; that quantity of a substance that gains or loses one mole of electrons in a redox reaction.

**ester**—an organic compound formed by the reaction between a carboxylic acid and an alcohol, having the general formula R—CO—OR′.

**ether**—an organic compound in which two alkyl and/or aryl groups are bonded to the same oxygen, having the general formula R—O—R′.

**excited state**—the state of an atom in which one or more electrons have absorbed energy—become "excited"—to raise them to energy levels above ground state.

**exothermic reaction**—a reaction that gives off energy to its surroundings.

### F

**family**—*see chemical family.*

**fission**—a nuclear reaction in which a large nucleus splits into two smaller nuclei.

**formula, chemical**—a combination of chemical symbols and subscript numbers that represent the elements in a pure substance and the ratio in which the atoms of the different elements appear.

**formula unit**—a real (molecular) or hypothetical (ionic) unit particle represented by a chemical formula.

**formula weight**—the weight of one formula unit of a substance in amu; the molar weight of formula units of a substance.

**fractional distillation**—separation of a mixture into fractions whose components boil over a given temperature range.

**freezing point depression**—the difference between the freezing point of a solution and the freezing point of the pure solvent.

**fusion**—the process of melting; also, a nuclear reaction in which two small nuclei combine to form a larger nucleus.

### G

**gamma (γ) ray**—a high-energy photon (light) emission in radioactive disintegration.

**Geiger counter**—an electrical device for detecting and measuring the intensity of radioactive emission.

**gram atomic weight**—the number of grams of an element that is numerically equal to its atomic weight.

**gram molecular weight**—the number of grams of a molecular compound that is numerically equal to its molecular weight.

**gram formula weight**—the number of grams of an ionic compound that is numerically equal to its formula weight.

**ground state**—the state of an atom in which all electrons occupy the lowest possible energy levels.

**group**—the elements comprising a vertical column in the periodic table.

## H

**half-life**—the time required for the disintegration of one-half of the radioactive atoms in a sample.

**half reaction**—the oxidation or reduction half of an oxidation-reduction reaction.

**halogen**—the name of the chemical family consisting of fluorine, chlorine, bromine and iodine; any member of the halogen family.

**heat of fusion (solidification)**—the heat flow when one gram of a substance changes between a solid and a liquid at its normal melting point.

**heat of reaction**—change of enthalpy in a chemical reaction.

**heat of vaporization (condensation)**—the heat flow when one gram of a substance changes between a liquid and a vapor at its normal boiling point.

**heterogeneous matter**—matter having a non-uniform composition, usually with visibly different parts or phases.

**homologous series**—a series of compounds in which each member differs from the one next to it by the same structural unit.

**homogeneous matter**—matter having a uniform appearance and uniform properties throughout.

**hydrate**—a crystalline solid that contains water of hydration.

**hydrocarbon**—an organic compound consisting of carbon and hydrogen.

**hydrogen bond**—an intermolecular bond (attraction) between a hydrogen atom in one molecule and a highly electronegative atom (fluorine, oxygen or nitrogen) of another polar molecule; the polar molecule may be of the same substance containing the hydrogen, or a different substance.

**hydronium ion**—a hydrated hydrogen ion, $H_3O^+$.

**hydroxyl group**—an organic functional group, —OH, characteristic of alcohols.

## I

**ideal gas**—a hypothetical gas that behaves according to the ideal gas model over all ranges of temperature and pressure.

**ideal gas equation**—the equation $PV = nRT$ that relates quantitatively the pressure, volume, quantity and temperature of an ideal gas.

**immiscible**—insoluble (usually used only in reference to liquids).

**indicator**—an organic dye that changes from one color to another over a range in pH, used to signal the end of a titration reaction.

**inhibitor**—a substance added to a chemical reaction to retard its rate; sometimes called a negative catalyst.

**ion**—an atom or group of covalently bonded atoms that is electrically charged by virtue of an excess or deficiency of electrons.

**ion combination reaction**—when two solutions are combined, the formation of a precipitate or molecular compound by a cation from one solution and an anion from the second solution.

**ionic bond**—the chemical bond arising from the attraction forces between oppositely charged ions in an ionic compound.

**ionic compound**—a compound in which ions are held by ionic bonds.

**ionic equation**—a chemical equation in which dissociated ionic compounds are shown in ionic form.

**ionization**—the formation of an ion from a molecule or atom.

**ionization energy**—the energy required to remove an electron from an atom (or ion); usually expressed in kilocalories per mole.

**isoelectronic**—having the same electron configuration.

**isomers**—two compounds having the same molecular formulas but different structural formulas and different physical and chemical properties.

**isotopes**—two or more atoms of the same element that have different atomic masses, because of different numbers of neutrons.

**IUPAC**—International Union of Pure and Applied Chemistry.

## J

**joule**—the SI energy unit, defined as a force of one newton applied over a distance of one meter; 1 joule = 0.239 calorie.

## K

**K**—the symbol for an equilibrium constant; $K_a$ is the constant for the ionization of a weak acid; $K_{sp}$ is the constant for the equilibrium between a slightly soluble ionic compound and a saturated solution of its ions; $K_w$ is the constant for the ionization of water.

**Kelvin temperature scale**—an absolute temperature scale on which the degrees are the same size as Celsius degrees, with $0°K$ at absolute zero, or $-273.15°C$.

**ketone**—a compound consisting of a carbonyl group bonded on each side to an alkyl or aryl group, having the general formula R—CO—R′.

**kinetic energy**—energy of motion; translational kinetic energy is equal to ½ mass × (velocity)².

**kinetic molecular theory**—the general theory that all matter consists of minute particles in constant motion, with different degrees of freedom distinguishing between solids, liquids and gases.

**kinetic theory of gases**—that portion of the kinetic molecular theory that describes gases and from which the model of an ideal gas is developed.

### L

**latent heat**—the heat flow when one gram (or one mole) of a substance changes from one state to another.

**Le Chatelier's Principle**—if an equilibrium system is subjected to a change, processes occur that tend to counteract partially the initial change, thereby bringing the system to a new position of equilibrium.

**Lewis diagram, structure or symbol**—a diagram representing the valence electrons and covalent bonds in an atomic or molecular species.

**limiting reagent**—the reactant first totally consumed in a reaction, thereby determining the maximum yield possible.

**line spectrum**—the spectral lines that appear when light emitted from a sample is analyzed in a spectroscope.

### M

**macromolecular crystal**—a crystal made up of a large but indefinite number of atoms covalently bonded to each other to form a huge molecule.

**manometer**—a laboratory device for measuring gas pressure.

**mass**—quantity of matter in a particular sample.

**mass number**—the total number of protons plus neutrons in the nucleus of an atom.

**mass spectroscope**—a laboratory device whereby a flow of gaseous ions may be analyzed in regard to their charge and/or mass.

**matter**—that which occupies space and has mass.

**metal**—an element that possesses metallic properties, such as luster, ductility, malleability, good conductivity of heat and electricity, tendency to form monatomic cations.

**miscible**—soluble (usually used only in reference to liquids).

**mixture**—a sample of matter containing two or more pure substances.

**molality**—solution concentration expressed in moles of solute per kilogram of solvent.

**molar heat of fusion (solidification)**—the heat flow when one mole of a substance changes between a solid and a liquid at its normal melting point.

**molar heat of vaporization (condensation)**—the heat flow when one mole of a substance changes between a liquid and a vapor at its normal boiling point.

**molarity**—solution concentration expressed in moles of solute per liter of solution.

**molar volume**—the volume occupied by one mole, usually of a gas.

**molar weight**—the mass of one mole of any substance.

**mole**—that quantity of any species that contains the same number of units as the number of atoms in exactly 12 grams of carbon-12.

**molecular compound**—a compound whose fundamental particles are molecules rather than ions.

**molecular crystal**—a molecular solid in which the molecules are arranged according to a definite geometric pattern.

**molecular geometry**—a description of the shape of a molecule.

**molecular weight**—the number that expresses the average mass of the molecules of a compound compared to the mass of an atom of carbon-12 at a value of exactly 12; the average mass of the molecules of a compound expressed in atomic mass units.

**molecule**—the smallest unit particle of a pure substance that can exist independently and possess the identity of the substance.

**monatomic**—that which has only one atom.

**monomer**—the individual chemical structural unit from which a polymer may be developed.

**monoprotic acid**—an acid capable of yielding one proton per molecule in complete ionization.

### N

**negative catalyst**—*see inhibitor.*

**net ionic equation**—an ionic equation from which all spectators have been removed.

**neutralization**—the reaction between an acid and a base to form a salt and water.

**neutron**—an electrically neutral subatomic particle having a mass of $1.7 \times 10^{-24}$ gram, approximately equal to the mass of a proton, or 1 atomic mass unit, found in the nucleus of the atom.

**noble gas**—the name of the chemical family of relatively unreactive elemental gases appearing in Group O of the periodic table.

**nonelectrolyte**—a substance which, when dissolved, yields a solution that is a nonconductor of electricity; a solution or other fluid that does not conduct electricity by ionic movement.

**nonpolar**—pertaining to a bond or molecule having a symmetrical distribution of electric charge.

**normal boiling point**—the temperature at which a substance boils in an open vessel at one atmosphere pressure.

**normality**—solution concentration in equivalents per liter.

**nucleus**—the extremely dense central portion of the atom that contains the neutrons and protons which constitute nearly all the mass of the atom and all of the positive charge.

### O

**octet rule**—the general rule that atoms tend to form stable bonds by sharing or transferring electrons until the atom is surrounded by a total of eight electrons.

**orbit**—*see electron orbit.*

**orbital**—*see electron orbital.*

**organic chemistry**—the chemistry of carbon compounds other than carbonates, cyanides, carbon monoxide and carbon dioxide.

**oxidation**—chemical reaction with oxygen; a chemical change in which the oxidation number (state) of an element is increased; also, the loss of electrons in a redox reaction.

**oxidation number**—a number assigned to each element in a compound, ion or elemental species by an arbitrary set of rules. Its two main functions are to organize and simplify the study of oxidation-reduction reactions and to serve as a base for one branch of chemical nomenclature.

**oxidation state**—*see oxidation number.*

**oxidizer, oxidizing agent**—the substance that takes electrons from another species, thereby oxidizing it.

**oxyacid**—an acid that contains oxygen.

### P

**partial pressure**—the pressure one component of a mixture of gases would exert if it alone occupied the same volume as the mixture at the same temperature.

**Pauli Exclusion Principle**—the principle that says, in effect, that no more than two electrons can occupy the same orbital.

**period (periodic table)**—a horizontal row of the periodic table.

**pH**—a way of expressing hydrogen ion concentration; the negative of the logarithm of the hydrogen ion concentration.

**phase**—a visibly distinct part of a heterogeneous sample of matter.

**physical change**—a change in the physical form of a substance without changing its chemical identity.

**physical properties**—properties of a substance that can be observed and measured without subjecting the substance to a chemical change.

**pOH**—a way of expressing hydroxide ion concentration; the negative logarithm of the hydroxide ion concentration.

**polar**—pertaining to a bond or molecule having an unsymmetrical distribution of electric charge.

**polyatomic**—pertaining to a species consisting of two or more atoms; usually said of polyatomic ions.

**polymer**—a chemical compound formed by bonding two or more monomers; frequently in plastics, a huge macromolecule.

**polymerization**—the reaction in which monomers combine to form polymers.

**polyprotic acid**—an acid capable of yielding more than one proton per molecule on complete ionization.

**potential energy**—energy possessed by a body by virtue of its position in an attractive or repulsive force field.

**precipitate**—an insoluble ionic compound that forms when a solution containing its cation is added to a solution containing its anion; any solid that is formed in and settles from a solution; also, to cause a solid to form in a solution by any means.

**pressure**—force per unit area.

**principal energy level**—the main energy levels within the electron arrangement in an atom. They are quantized by a set of integers beginning at $n = 1$ for the lowest level, $n = 2$ for the next, and so forth; also called the principal quantum number.

**proton**—a subatomic particle carrying a unit positive charge and having a mass of $1.7 \times 10^{-24}$ gram, almost the same as the mass of a neutron, found in the nucleus of the atom.

**pure substance**—a sample consisting of only one kind of matter, either compound or element.

### Q

**quantization of energy**—the existence of certain discrete electron energy levels within an atom such that electrons may have any one of these energies, but no energy between two such levels.

**quantum mechanical model of the atom**—an atomic concept that recognizes four quantum numbers by which electron energy levels may be described.

### R

**R**—a symbol used to designate any alkyl group; the ideal gas constant, 0.0821 $(\ell)(\text{atm})/(\text{mole})(°K)$.

**radioactivity**—spontaneous emission of rays and/or particles from an atomic nucleus.

**redox**—a term coined from REDuction–OXidation to refer to oxidation–reduction reactions.

**reducer, reducing agent**—the substance that loses electrons to another species, thereby reducing it.

**reduction**—a chemical change in which the oxidation number (state) of an element is reduced.

**reversible reaction**—a chemical reaction in which the products may react to re-form the original reactants.

### S

**salt**—the product of a neutralization reaction other than water; an ionic compound containing neither the hydrogen ion, $H^+$, nor the hydroxide ion, $OH^-$.

**saturated hydrocarbon**—a hydrocarbon that contains only single bonds, in which each carbon atom is bonded to four other atoms.

**saturated solution**—a solution of such concentration that it is or would be in a state of equilibrium with excess solute present.

**significant figures**—the digits in a measurement that are known to be accurate plus one doubtful digit.

**solubility**—the quantity of solute that will dissolve in a given quantity of solvent, or in a given quantity of solution, at a specified temperature, to establish an equilibrium between the solution and excess solute; frequently expressed in grams of solute per 100 grams of solvent.

**solubility product constant**—*under K, see $K_{sp}$*.

**solute**—the substance dissolved in the solvent; sometimes not clearly distinguishable from the solvent (see below), but usually the lesser of the two.

**solution**—a homogeneous mixture of two or more substances of molecular or ionic particle size, the concentration of which may be varied, usually within certain limits.

**solution inventory**—a precise identification of the chemical species present in a solution, in contrast with the solute from which they may have come; i.e., sodium ions and chloride ions, rather than sodium chloride.

**solvent**—the medium in which the solute is dissolved; *see solute.*

**specific gravity**—the ratio of the density of a substance to the density of some standard, usually water at 4°C.

**specific heat**—the quantity of heat required to raise the temperature of one gram of a substance one degree Celsius.

**spectator (ion)**—a species present at the scene of a reaction but not a participant in it.

**spectroscope**—a laboratory instrument used to analyze spectra.

**spectrum (plural: spectra)**—the result of a dispersion of a beam of light into its component colors; also the result of a dispersion of a beam of gaseous ions into its component particles, distinguished by mass and electric charge.

**standard temperature and pressure (STP)**—arbitrarily defined conditions of temperature (0°C) and pressure (1 atmosphere) at which gas volumes and quantities are frequently measured and/or compared.

**stoichiometry**—the quantitative relationships between the substances involved in a chemical reaction, established by the equation for the reaction.

**STP**—abbreviation for standard temperature and pressure (see above).

**strong acid**—an acid that ionizes almost completely in aqueous solution; an acid that loses its protons readily.

**strong base**—a base that dissociates almost completely in aqueous solution; a base that has a strong attraction for protons.

**strong electrolyte**—a substance which, when dissolved, yields a solution that is a good conductor of electricity because of nearly complete ionization or dissociation.

**strong oxidizer (oxidizing agent)**—an oxidizer that has a strong attraction for electrons.

**strong reducer (reducing agent)**—a reducer that releases electrons readily.

**sublevel**—the levels into which the principal energy levels are divided according to the quantum mechanical model of the atom; usually specified s, p, d and f.

**supersaturated**—a state of solution concentration which is greater than the equilibrium concentration (solubility) at a given temperature and/or pressure.

*T*

**tetrahedral**—related to a tetrahedron; usually used in reference to the orientation of four covalent bonds radiating from a central atom toward the vertices of a tetrahedron, or to the 109°28' angle formed by any two corners of the tetrahedron and the central atom as its vertex.

**tetrahedron**—a regular four-sided solid, having congruent equilateral triangles as its four faces.

**thermochemical equation**—a chemical equation which includes an energy term, or for which $\Delta H$ is indicated.

**thermochemical stoichiometry**—stoichiometry expanded to include the energy involved in a chemical reaction, as defined by the thermochemical equation.

**titration**—the controlled and measured addition of one solution to another.

**torr**—a unit of pressure equal to the pressure unit millimeter of mercury.

**transition element; transition metal**—an element from one of the B groups or Group VIII of the periodic table.

**transmutation**—conversion of an atom from one element to another by means of a nuclear change.

**transuranium elements**—man-made elements whose atomic numbers are greater than 92.

**triprotic acid**—an acid capable of yielding three protons in complete ionization.

*U*

**unsaturated hydrocarbon**—a hydrocarbon that contains one or more multiple bonds, such that two or more carbons are each bonded to a total of three atoms (double bond) or two atoms (triple bond).

## V

**valence electrons**—the highest energy s and p electrons in an atom that determine the bonding characteristics of an element.

**van der Waals forces**—a general term for all kinds of weak intermolecular attractions.

**vapor pressure**—the partial pressure exerted by a specific vapor component in a mixture of gases. Frequently refers to the partial pressure of a vapor that is in equilibrium with its liquid state at a given temperature.

## W

**water of crystallization, water of hydration**—water molecules that are included as structural parts of crystals formed from aqueous solutions.

**weak acid**—an acid that ionizes only slightly in aqueous solution; an acid that does not donate protons readily.

**weak base**—a base that dissociates only slightly in aqueous solution; a base that has a weak attraction for protons.

**weak electrolyte**—a substance which, when dissolved, yields a solution that is a poor conductor of electricity because of limited ionization or dissociation.

**weak oxidizer (oxidizing agent)**—an oxidizer that has a weak attraction for electrons.

**weak reducer (reducing agent)**—a reducer that does not release electrons readily.

**weight**—a measure of the force of gravitational attraction of a body to the earth.

## Y

**yield**—the amount of product from a chemical reaction.

# ANSWERS TO QUESTIONS AND PROBLEMS

## CHAPTER 2

2.1) a – P; b – C; c – C; d – P; e – P

2.2) a – C; b – P; c – P; d – C; e – P

2.3) Gas: volume and shape variable; particle movement totally random within entire volume of closed container. Liquid: volume fixed, shape variable; particle movement random within fluid volume at bottom of container. Solid: volume and shape fixed; particle movement limited to vibration in fixed position.

2.4) Particles of solid are rigidly fixed in position, giving definite shape that turns corner as a unit. Liquid particles, not fixed in position, flow ahead while bowl turns corner.

2.5) "Fluid" refers to substance in which particles are not in fixed position, and therefore flow. Applies to liquids and gases only.

2.6) Homogeneous refers to sameness in appearance and behavior. The composition of a homogeneous substance is the same throughout.

2.7) a and c heterogeneous; b, d and e homogeneous.

2.8) Pure substance: single kind of matter, unlike any other kind; mixture: two or more pure substances. Properties of pure substances are fixed; properties of mixtures vary with composition. Liquid is a mixture because it has variable boiling point, a physical property.

2.9) Pure substance. Density constant in situation that would have resulted in varying concentration and density if liquid was a mixture.

2.10) Element cannot be decomposed into simpler pure substances, while a compound can.

2.11) Elements: tin, nitrogen, iron; compounds: carbon dioxide, baking soda.

2.12) Mass is conserved in a chemical change. The combined mass of all reactants (carbon and oxygen) is equal to the combined mass of all products (carbon dioxide).

2.13) Electrostatic force is force of attraction or repulsion between charged bodies. Gravitational force: attraction only.

2.14) Exothermic, heat evolved, or given off; endothermic, heat absorbed. Boiling water is endothermic, as you must add heat to the water to boil it.

2.15) a – K; b – P; c – K; d – K; e – P; f – P; g – P

2.16) "Chemical (potential) energy" is released as atoms in the fuel are rearranged in electrical force fields.

## CHAPTER 3

3.1) $\dfrac{122\ \cancel{qts}}{}\ \bigg|\ \dfrac{1\ \text{gal}}{4\ \cancel{qts}} = 30.5\ \text{gal}$

3.2) $\dfrac{0.35\ \cancel{ton}}{}\ \bigg|\ \dfrac{2000\ \cancel{lbs}}{1\ \cancel{ton}}\ \bigg|\ \dfrac{16\ \text{oz}}{1\ \cancel{lb}} = 11{,}200\ \text{oz}$

3.3) $\dfrac{5\ \cancel{month}}{}\ \bigg|\ \dfrac{1500\ \cancel{bikes}}{1\ \cancel{month}}\ \bigg|\ \dfrac{2\ \text{wheels}}{1\ \cancel{bike}} = 15{,}000\ \text{wheels}$

3.4) $\dfrac{2500\ \cancel{ft}}{}\ \bigg|\ \dfrac{1\ \cancel{mile}}{5280\ \cancel{ft}}\ \bigg|\ \dfrac{1\ \cancel{hour}}{35\ \cancel{miles}}\ \bigg|\ \dfrac{60\ \text{min}}{1\ \cancel{hour}} = 0.81\ \text{minute}$

3.5) $\dfrac{2.0\ \cancel{lbs}}{}\ \bigg|\ \dfrac{1\ \cancel{batch}}{75\ \cancel{lbs}}\ \bigg|\ \dfrac{12\ \cancel{min}}{1\ \cancel{batch}}\ \bigg|\ \dfrac{1\ \cancel{hr}}{60\ \cancel{min}}\ \bigg|\ \dfrac{\cancel{\$}18}{1\ \cancel{hr}}\ \bigg|\ \dfrac{100\cancel{\text{¢}}}{\cancel{\$}} = 9.6\text{¢}$

3.6) $\dfrac{3.6\ \cancel{yd^3}}{}\ \bigg|\ \dfrac{3^3\ \text{ft}^3}{1\ \cancel{yd^3}} = 97.2\ \text{ft}^3$

3.7) $\dfrac{53.3\ \cancel{yd}}{}\ \bigg|\ \dfrac{100\ \cancel{yd}}{}\ \bigg|\ \dfrac{3^2\ \text{ft}^2}{1\ \cancel{yd^2}} = 48{,}000\ \text{ft}^2$ (3 significant figures)

3.8) $\dfrac{28.0\ \cancel{in}}{}\ \bigg|\ \dfrac{2.54\ \text{cm}}{1\ \cancel{in}} = 71.1\ \text{cm}$

3.9) $\dfrac{148\ \cancel{km}}{}\ \bigg|\ \dfrac{1\ \text{mile}}{1.61\ \cancel{km}} = 91.9\ \text{miles}$

3.10) $\dfrac{40.8\ \cancel{yds}}{}\ \bigg|\ \dfrac{1\ \text{meter}}{1.09\ \cancel{yds}} = 37.4\ \text{meters}$

3.11) $5 \times 10$. As delivered, a "two-by-four" does not measure $2'' \times 4''$, but is closer to $1.5'' \times 3.5''$. The more accurate metric dimensions are 4 cm $\times$ 9 cm.

3.12) $\dfrac{1\ \cancel{in}}{8}\ \bigg|\ \dfrac{2.54\ \cancel{cm}}{1\ \cancel{in}}\ \bigg|\ \dfrac{10\ \text{mm}}{1\ \cancel{cm}} = 3.175\ \text{mm}$

3.13) $\dfrac{5''}{8} = 0.625'';\ \dfrac{3.625\ \cancel{in}}{}\ \bigg|\ \dfrac{2.54\ \cancel{cm}}{1\ \cancel{in}}\ \bigg|\ \dfrac{10\ \text{mm}}{1\ \cancel{cm}} = 92.075\ \text{mm}$

3.14) $\dfrac{0.786\ \cancel{m}}{}\ \bigg|\ \dfrac{1000\ \text{mm}}{1\ \cancel{m}} = 786\ \text{mm}$

3.15) $\dfrac{5.9 \times 10^3\ \cancel{Å}}{}\ \bigg|\ \dfrac{10^{-8}\ \text{cm}}{1\ \cancel{Å}} = 5.9 \times 10^{-5}\ \text{cm}$

3.16) $\dfrac{2.50\ \cancel{yd^3}}{}\ \bigg|\ \dfrac{1^3\ \text{m}^3}{1.09^3\ \cancel{yd^3}} = 1.93\ \text{m}^3$

3.17) $\dfrac{5.00\ \cancel{gal}}{}\ \bigg|\ \dfrac{3.785\ \ell}{1\ \cancel{gal}} = 18.9\ \ell$

3.18) "Weightlessness" refers to absence of a gravitational field. Mass refers to quantity of matter, regardless of gravitational field. An astronaut is therefore never "massless."

3.19) $\dfrac{4.80\ \cancel{oz}}{}\ \bigg|\ \dfrac{28.3\ \text{g}}{1\ \cancel{oz}} = 136\ \text{g}$

3.20) $\dfrac{0.85\ \cancel{carat}}{}\ \bigg|\ \dfrac{200\ \cancel{mg}}{1\ \cancel{carat}}\ \bigg|\ \dfrac{1\ \text{g}}{10^3\ \cancel{mg}} = 0.17\ \text{g}$

3.21) $\dfrac{79\text{¢}}{1\ \cancel{lb}}\ \bigg|\ \dfrac{2.20\ \cancel{lbs}}{1\ \text{kg}}\ \bigg|\ \dfrac{1\ \text{dollar}}{100\text{¢}} = \$1.74/\text{kg}$

3.22) 0.89 g/cm$^3$. The density of the substance is 0.89 as great as the density of water.

3.23) $\dfrac{50.0\ \cancel{ml}}{}\ \bigg|\ \dfrac{1.60\ \text{g}}{1\ \cancel{ml}} = 80.0\ \text{g}$

3.24) $\dfrac{68.3\ g\ \big|\ 1\ cm^3}{\big|\ 19.3\ g} = 3.54\ cm^3$

3.25) $\dfrac{2.04 \times 10^3\ g}{150\ ml} = 13.6\ g/ml;\ Sp.\ g. = 13.6$

3.26) $\dfrac{1\ qt\ \big|\ 1\ \ell\ \big|\ 13.6\ g\ \big|\ 1\ lb}{\big|\ 1.06\ qts\ \big|\ 0.001\ \ell\ \big|\ 454\ g} = 28.3\ lbs$

3.27) $32 + 1.8(805) = 1481°F$      3.28) $120 - 32 = 1.8(°C)$      $°C = 48.9$

3.29) $-196°C;\ 96.1°C;\ 1349°C$      3.30) $-407°F;\ 115°F;\ 2768°F$

3.31) 4      3.32) 2      3.33) 3

3.34) $8.36\ g/cm^3$      3.35) 125 g      3.36) $5.06 \times 10^{-4}\ \ell$

3.37)
$\begin{aligned}
0.475\ &g\\
3.40\ &g\\
1.8\ \ \ &g\\
\underline{12.92\ \ \ }&\underline{g}\\
18.595\ g &= 18.6\ g
\end{aligned}$

3.38) $\dfrac{85.0\ g\ \big|\ 1\ cm^3}{\big|\ 1.74\ g} = 48.9\ cm^3$

3.39) $\dfrac{12\ hr\ \big|\ 2\ uses\ \big|\ 10.36\ ml}{\big|\ 1\ hr\ \big|\ 1\ use} = 248.6\ ml$

3.40) $\dfrac{519\ lbs\ \big|\ 454\ g\ \big|\ 1^3\ ft^3\ \big|\ 1^3\ in^3}{1\ ft^3\ \big|\ 1\ lb\ \big|\ 12^3\ in^3\ \big|\ 2.54^3\ cm^3} = 8.32\ g/cm^3$

3.41) $\dfrac{3.0 \times 10^3\ gal\ \big|\ 3785\ ml\ \big|\ 1.29\ g\ \big|\ 1\ lb\ \big|\ 1\ ton}{\big|\ 1\ gal\ \big|\ 1\ ml\ \big|\ 454\ g\ \big|\ 2000\ lbs} = 16\ tons$

3.42) $\dfrac{(113 - 62.0)\ g}{47.2\ ml} = 1.1\ g/ml$

# CHAPTER 4

4.1) See page 59 of text.

4.2) 46 grams is fixed quantity of sodium. Oxygen quantities combining with 46 grams of sodium are 16 and 32 grams. $^{16}/_{32} = \frac{1}{2}$, a ratio of small, whole numbers.

4.3) Atoms are divisible; all atoms of a given element are not alike.

4.4) Proton and neutron are approximately equal in mass, and each is about 1837 times as massive as electron.

4.5) In neutral atom, number of protons = number of electrons, which may or may not be equal to the number of neutrons.

4.6) Be, Mg, Ca, Sr, Ba, Ra. 11 – 18.

4.7) (a) period; (b) family; (c) family; (d) period.

4.8) Alpha particles passed through the open space of the atom, between the atomic nuclei.

4.9) Atoms consist of extremely tiny and extremely dense positively charged nuclei, surrounded by "open space" thinly populated with electrons of very small mass.

4.10) Rutherford experiment established that massive atomic particles were in nucleus. When found, proton and neutron proved to be heavy particles.

4.11) In a continuous light spectrum one color blends into another, as in a rainbow. In a discrete line spectrum there is a series of sharp lines, separated by black areas.

4.12) Each electron is permitted to have certain discrete energies and no others.

4.13) An atom in the ground state has all electrons in the lowest possible energy states. An atom in an excited state has one or more electrons at energy levels above the ground state.

4.14) Atom described as dense nucleus with electrons moving around it in circular orbits. Add a nucleus to the center of Figure 4.10, page 69, for the sketch.

4.15) Hydrogen atoms are the only neutral atoms that fit the Bohr model. The model also fails to account for the energy radiations that would accompany an electron moving in a circular orbit.

4.16) "Principal energy levels" are those recognized by Bohr and on which the quantum concept applied to atoms is based. Their energies increase as their number increases; $n = 1$ is lowest in energy, $n = 2$ is next, $n = 3$ is higher, and so forth.

4.17) "Orbits" are circular or elliptical paths electrons travel around nucleus in Bohr model. "Orbitals" are mathematically defined regions of space about the nucleus where there is a high probability of locating an electron. See Figure 4.11, page 73, for shapes.

4.18) The possible electron populations of an orbital are 0, 1 and 2.

4.19) N: $1s^2 2s^2 2p^3$. Ti: $1s^2 2s^2 2p^6 3s^2 3p^6 4s^2 3d^2$; or [Ar] $4s^2 3d^2$

4.20) K: $1s^2 2s^2 2p^6 3s^2 3p^6 4s^1$; or [Ar] $4s^1$. Mn: $1s^2 2s^2 2p^6 3s^2 3p^6 4s^2 3d^5$; or [Ar] $4s^2 3d^5$

4.21) Ionization potential, or ionization energy, is the energy required to remove an electron from an atom. When an electron is taken from an atom, the ion that is left has a positive charge. The attractive force between the ion and the negatively charged electron must be overcome in removing the electron, and this requires energy.

4.22) The ionization potential of strontium is less than that of calcium because the electrons that must be removed from the neutral atom are farther from the nucleus than they are in calcium.

4.23) Many chemical properties are established by the number of electrons in the highest energy level. Both magnesium ($3s^2$) and calcium ($4s^2$) have two electrons at their highest level.

4.24) (a) Q M; (b) D E        4.25) J L W        4.26) E D G

# CHAPTER 5

5.1) Any element.

5.2) An atom is the fundamental structural unit of an element. A molecule is the smallest stable particle of any pure substance, element or compound.

5.3) Molecular compounds have individual and distinct molecules as their structural unit. Ionic compounds have no structural unit, but are assemblies of positively and negatively charged ions.

5.4) Check with alphabetical list of elements inside back cover.

5.5) Nitrogen, oxygen, hydrogen, fluorine, chlorine, bromine, iodine.

5.6) (a) 2 nitrogen atoms, 1 oxygen atom; (b) 1 phosphorus atom, 3 chlorine atoms; (c) 2 carbon atoms, 4 hydrogen atoms, 1 oxygen atom.

5.7)   Water; carbon monoxide.

5.8)   Atoms gain or lose electrons to form monatomic ions.

5.9)

| ATOMIC NUMBER | ION NAME | ION SYMBOL | ANION OR CATION |
|---|---|---|---|
| 11 | Sodium | $Na^+$ | Cation |
| 53 | Iodide | $I^-$ | Anion |
| 3 | Lithium | $Li^+$ | Cation |
| 20 | Calcium | $Ca^{2+}$ | Cation |
| 16 | Sulfide | $S^{2-}$ | Anion |

5.10)   HCl      5.11)   $HNO_3$—$NO_3^-$—nitrate ion

5.12)   Carbonic acid—$CO_3^{2-}$—carbonate ion      5.13)   Hydroxide ion

5.14)   LiCl; calcium sulfide      5.15)   $NH_4NO_3$; barium carbonate

5.16)   $MgBr_2$; potassium phosphate      5.17)   $Ba_3(PO_4)_2$; ammonium sulfate

5.18)   Two

5.19)   $Ba(OH)_2 \cdot 8\ H_2O$, barium hydroxide octahydrate, barium hydroxide 8-water, or barium hydroxide 8-hydrate.

# ANSWERS TO FORMULA WRITING AND NOMENCLATURE EXERCISE NUMBER 1

1) LiBr, lithium bromide
2) $MgBr_2$, magnesium bromide
3) $NH_4Br$, ammonium bromide
4) $AlBr_3$, aluminum bromide
5) NaBr, sodium bromide
6) $BaBr_2$, barium bromide
7) KBr, potassium bromide
8) $CaBr_2$, calcium bromide
9) $Li_2SO_4$, lithium sulfate
10) $MgSO_4$, magnesium sulfate
11) $(NH_4)_2SO_4$, ammonium sulfate
12) $Al_2(SO_4)_3$, aluminum sulfate
13) $Na_2SO_4$, sodium sulfate
14) $BaSO_4$, barium sulfate
15) $K_2SO_4$, potassium sulfate
16) $CaSO_4$, calcium sulfate
17) LiOH, lithium hydroxide
18) $Mg(OH)_2$, magnesium hydroxide
19) $NH_4OH$, ammonium hydroxide
20) $Al(OH)_3$, aluminum hydroxide
21) NaOH, sodium hydroxide
22) $Ba(OH)_2$, barium hydroxide
23) KOH, potassium hydroxide
24) $Ca(OH)_2$, calcium hydroxide
25) LiF, lithium fluoride
26) $MgF_2$, magnesium fluoride
27) $NH_4F$, ammonium fluoride
28) $AlF_3$, aluminum fluoride
29) NaF, sodium fluoride
30) $BaF_2$, barium fluoride
31) KF, potassium fluoride
32) $CaF_2$, calcium fluoride
33) $Li_2O$, lithium oxide
34) MgO, magnesium oxide
35) $(NH_4)_2O$, ammonium oxide
36) $Al_2O_3$, aluminum oxide
37) $Na_2O$, sodium oxide
38) BaO, barium oxide
39) $K_2O$, potassium oxide
40) CaO, calcium oxide
41) $LiNO_3$, lithium nitrate
42) $Mg(NO_3)_2$, magnesium nitrate
43) $NH_4NO_3$, ammonium nitrate
44) $Al(NO_3)_3$, aluminum nitrate
45) $NaNO_3$, sodium nitrate
46) $Ba(NO_3)_2$, barium nitrate
47) $KNO_3$, potassium nitrate
48) $Ca(NO_3)_2$, calcium nitrate
49) $Li_3PO_4$, lithium phosphate
50) $Mg_3(PO_4)_2$, magnesium phosphate
51) $(NH_4)_3PO_4$, ammonium phosphate
52) $AlPO_4$, aluminum phosphate
53) $Na_3PO_4$, sodium phosphate
54) $Ba_3(PO_4)_2$, barium phosphate
55) $K_3PO_4$, potassium phosphate
56) $Ca_3(PO_4)_2$, calcium phosphate
57) LiCl, lithium chloride
58) $MgCl_2$, magnesium chloride
59) $NH_4Cl$, ammonium chloride
60) $AlCl_3$, aluminum chloride
61) NaCl, sodium chloride
62) $BaCl_2$, barium chloride
63) KCl, potassium chloride
64) $CaCl_2$, calcium chloride
65) $Li_2S$, lithium sulfide
66) MgS, magnesium sulfide
67) $(NH_4)_2S$, ammonium sulfide
68) $Al_2S_3$, aluminum sulfide
69) $Na_2S$, sodium sulfide
70) BaS, barium sulfide

71) $K_2S$, potassium sulfide
72) $CaS$, calcium sulfide
73) $LiI$, lithium iodide
74) $MgI_2$, magnesium iodide
75) $NH_4I$, ammonium iodide
76) $AlI_3$, aluminum iodide
77) $NaI$, sodium iodide
78) $BaI_2$, barium iodide
79) $KI$, potassium iodide

80) $CaI_2$, calcium iodide
81) $Li_2CO_3$, lithium carbonate
82) $MgCO_3$, magnesium carbonate
83) $(NH_4)_2CO_3$, ammonium carbonate
84) $Al_2(CO_3)_3$, aluminum carbonate
85) $Na_2CO_3$, sodium carbonate
86) $BaCO_3$, barium carbonate
87) $K_2CO_3$, potassium carbonate
88) $CaCO_3$, calcium carbonate

## ANSWERS TO FORMULA WRITING EXERCISE NUMBER 2

1) $Ca(OH)_2$
2) $KOH$
3) $Mg(OH)_2$
4) $NH_4OH$
5) $LiOH$
6) $Al(OH)_3$
7) $Ba(OH)_2$
8) $NaOH$
9) $CaBr_2$
10) $KBr$
11) $MgBr_2$
12) $NH_4Br$
13) $LiBr$
14) $AlBr_3$
15) $BaBr_2$
16) $NaBr$
17) $CaSO_4$
18) $K_2SO_4$
19) $MgSO_4$
20) $(NH_4)_2SO_4$
21) $Li_2SO_4$
22) $Al_2(SO_4)_3$

23) $BaSO_4$
24) $Na_2SO_4$
25) $CaF_2$
26) $KF$
27) $MgF_2$
28) $NH_4F$
29) $LiF$
30) $AlF_3$
31) $BaF_2$
32) $NaF$
33) $CaCO_3$
34) $K_2CO_3$
35) $MgCO_3$
36) $(NH_4)_2CO_3$
37) $Li_2CO_3$
38) $Al_2(CO_3)_3$
39) $BaCO_3$
40) $Na_2CO_3$
41) $CaO$
42) $K_2O$
43) $MgO$
44) $(NH_4)_2O$

45) $Li_2O$
46) $Al_2O_3$
47) $BaO$
48) $Na_2O$
49) $Ca(NO_3)_2$
50) $KNO_3$
51) $Mg(NO_3)_2$
52) $NH_4NO_3$
53) $LiNO_3$
54) $Al(NO_3)_3$
55) $Ba(NO_3)_2$
56) $NaNO_3$
57) $Ca_3(PO_4)_2$
58) $K_3PO_4$
59) $Mg_3(PO_4)_2$
60) $(NH_4)_3PO_4$
61) $Li_3PO_4$
62) $AlPO_4$
63) $Ba_3(PO_4)_2$
64) $Na_3PO_4$
65) $CaI_2$
66) $KI$

67) $MgI_2$
68) $NH_4I$
69) $LiI$
70) $AlI_3$
71) $BaI_2$
72) $NaI$
73) $CaS$
74) $K_2S$
75) $MgS$
76) $(NH_4)_2S$
77) $Li_2S$
78) $Al_2S_3$
79) $BaS$
80) $Na_2S$
81) $CaCl_2$
82) $KCl$
83) $MgCl_2$
84) $NH_4Cl$
85) $LiCl$
86) $AlCl_3$
87) $BaCl_2$
88) $NaCl$

## CHAPTER 6

6.1) $3; 6; 3; 3$     6.2) Chromium-52, $^{52}Cr$ or $^{52}_{24}Cr$

6.3) $30; 38; 30;$ $^{68}Zn$ or $^{68}_{30}Zn$; zinc-68

6.4) 1 amu is exactly $^1/_{12}$ of the mass of one atom of carbon-12. It is better than a gram as a unit of mass for atomic particles because the number of grams is so small, ranging from $10^{-24}$ to $10^{-22}$ gram.

6.5) $0.7553 \times 35 = 26.4355$
$0.2447 \times 37 = \underline{\phantom{0}9.0539}$
$\phantom{0.2447 \times 37 = }35.4894 = 35.49$

6.6) $0.5182 \times 107 = \phantom{0}55.4474$
$0.4818 \times 109 = \underline{52.5162}$    Silver, atomic number 47, atomic weight
$\phantom{0.4818 \times 109 = }107.9636 = 107.96$                    107.868

6.7) $39.102; 32.064; 51.996; 118.69$

6.8) $NaCl$ is a compound, not an element.

6.9) Atomic weights refer to elements, molecular weights to molecular compounds and formula weights to ionic compounds.

6.10) That quantity of any species that contains the same number of units as the number of atoms in exactly 12 grams of carbon-12.

6.11) The number of atoms in exactly 12 grams of carbon-12; or the number of atoms in one gram atomic weight of an element.

6.12) The weight in grams of one mole of any species.

6.13) $6.9 + 79.9 = 86.8$ g/mole

6.14) $2(35.5) = 71.0$ g/mole

6.15) $40.1 + 32.1 + 4(16.0) = 136.2$ g/mole

6.16) $14.0 + 4(1.0) + 14.0 + 3(16.0) = 80.0$ g/mole

6.17) $6(12.0) + 6(1.0) = 78.0$ g/mole

6.18) $2(23.0) + 12.0 + 3(16.0) + 10(18.0) = 286$ g/mole

6.19) $(6.9/86.8)100 = 7.95\%$ Li; $(79.9/86.8)100 = 92.0\%$ Br

6.20) $(40.1/136)100 = 29.5\%$ Ca; $(32.1/136)100 = 23.6\%$ S; $(4 \times 16.0/136)100 = 47.1\%$ O

6.21) $(6 \times 1.0/78.0)100 = 7.7\%$ H; $(6 \times 12.0/78.0)100 = 92.3\%$ C

6.22) $(180/286)100 = 62.9\%$ $H_2O$

6.23) $\dfrac{68.4 \text{ g LiBr}}{} \left| \dfrac{1 \text{ mole LiBr}}{86.8 \text{ g LiBr}} \right. = 0.788$ mole LiBr

6.24) $\dfrac{17.2 \text{ g Cl}_2}{} \left| \dfrac{1 \text{ mole Cl}_2}{71.0 \text{ g Cl}_2} \right. = 0.242$ mole $Cl_2$

6.25) $\dfrac{34.1 \text{ g CaSO}_4}{} \left| \dfrac{1 \text{ mole CaSO}_4}{136 \text{ g CaSO}_4} \right. = 0.251$ mole $CaSO_4$

6.26) $\dfrac{0.345 \text{ mole NH}_4\text{NO}_3}{} \left| \dfrac{80.0 \text{ g NH}_4\text{NO}_3}{1 \text{ mole NH}_4\text{NO}_3} \right. = 27.6$ g $NH_4NO_3$

6.27) $\dfrac{1.82 \text{ moles C}_6\text{H}_6}{} \left| \dfrac{78.0 \text{ g C}_6\text{H}_6}{1 \text{ mole C}_6\text{H}_6} \right. = 142$ g $C_6H_6$

6.28) $\dfrac{0.791 \text{ mole Na}_2\text{CO}_3 \cdot 10 \text{ H}_2\text{O}}{} \left| \dfrac{286 \text{ g Na}_2\text{CO}_3 \cdot 10 \text{ H}_2\text{O}}{1 \text{ mole Na}_2\text{CO}_3 \cdot 10 \text{ H}_2\text{O}} \right.$
$= 226$ g $Na_2CO_3 \cdot 10 \, H_2O$

6.29) $\dfrac{1.24 \text{ moles Mg}}{} \left| \dfrac{6.02 \times 10^{23} \text{ atoms Mg}}{1 \text{ mole Mg}} \right. = 7.46 \times 10^{23}$ atoms Mg

6.30) $\dfrac{0.713 \text{ mole Br}_2}{} \left| \dfrac{2 \times 6.02 \times 10^{23} \text{ atoms Br}}{1 \text{ mole Br}_2} \right. = 8.58 \times 10^{23}$ atoms Br

6.31) $\dfrac{29.6 \text{ g Na}}{} \left| \dfrac{6.02 \times 10^{23} \text{ atoms Na}}{23.0 \text{ g Na}} \right. = 7.75 \times 10^{23}$ atoms Na

6.32) $\dfrac{3.40 \text{ g Ca}}{} \left| \dfrac{6.02 \times 10^{23} \text{ atoms Ca}}{40.1 \text{ g Ca}} \right. = 5.10 \times 10^{22}$ atoms Ca

6.33) $\dfrac{38.1 \text{ g N}_2}{} \left| \dfrac{2 \times 6.02 \times 10^{23} \text{ atoms N}}{28.0 \text{ g N}_2} \right. = 1.64 \times 10^{24}$ atoms N

6.34) $\dfrac{0.521 \text{ mole H}_2\text{S}}{} \left| \dfrac{6.02 \times 10^{23} \text{ molecules H}_2\text{S}}{1 \text{ mole H}_2\text{S}} \right.$
$= 3.14 \times 10^{23}$ molecules $H_2S$

6.35) $\dfrac{0.0626 \text{ mole Br}_2}{} \left| \dfrac{6.02 \times 10^{23} \text{ molecules Br}_2}{1 \text{ mole Br}_2} \right.$
$= 3.77 \times 10^{22}$ molecules $Br_2$

6.36) $\dfrac{12.4 \text{ g } N_2}{} \bigg| \dfrac{6.02 \times 10^{23} \text{ molecules } N_2}{28.0 \text{ g } N_2} = 2.67 \times 10^{23}$ molecules $N_2$

6.37) $\dfrac{6.45 \text{ g } CO}{} \bigg| \dfrac{6.02 \times 10^{23} \text{ molecules } CO}{28.0 \text{ g } CO} = 1.39 \times 10^{23}$ molecules $CO$

6.38) $\dfrac{13.6 \text{ g } LiBr}{} \bigg| \dfrac{6.02 \times 10^{23} \text{ units } LiBr}{86.8 \text{ g } LiBr} = 9.43 \times 10^{22}$ units $LiBr$

6.39) $\dfrac{2.35 \times 10^{21} \text{ Ba atoms}}{} \bigg| \dfrac{1 \text{ mole Ba}}{6.02 \times 10^{23} \text{ Ba atoms}} = 3.90 \times 10^{-3}$ mole Ba

6.40) $\dfrac{1.09 \times 10^{23} \text{ Br}_2 \text{ molecules}}{} \bigg| \dfrac{1 \text{ mole Br}_2}{6.02 \times 10^{23} \text{ Br}_2 \text{ molecules}} = 0.181$ mole $Br_2$

6.41) $\dfrac{7.06 \times 10^{23} \text{ atoms He}}{} \bigg| \dfrac{4.00 \text{ g He}}{6.02 \times 10^{23} \text{ atoms He}} = 4.69$ g He

6.42) $\dfrac{4.06 \times 10^{22} \text{ molecules } O_2}{} \bigg| \dfrac{32.0 \text{ g } O_2}{6.02 \times 10^{23} \text{ molecules } O_2} = 2.16$ g $O_2$

6.43) $\dfrac{1.19 \times 10^{23} \text{ units KI}}{} \bigg| \dfrac{166 \text{ g KI}}{6.02 \times 10^{23} \text{ units KI}} = 32.8$ g KI

6.44)  6 and 10 are both divisible by 2, so the empirical formula of $C_6H_{10}$ is $C_3H_5$.
$C_7H_{10}$ cannot be "reduced," and is therefore an empirical formula. It may
be—and, in fact, is—a molecular formula.

6.45–6.49)

|  | ELEMENT | GRAMS | MOLES | MOLE RATIO | FORMULA RATIO | EMPIRICAL FORMULA |
|---|---|---|---|---|---|---|
| 6.45) | C | 40.0 | 3.33 | 1 | 1 | |
| | H | 6.7 | 6.7 | 2 | 2 | $CH_2O$ |
| | O | 53.3 | 3.33 | 1 | 1 | |
| 6.46) | C | 25.3 | 2.10 | 1 | 3 | |
| | H | 3.5 | 3.5 | 1.67 | 5 | $C_3H_5$ |
| 6.47) | Na | 32.4 | 1.41 | 2 | 2 | |
| | S | 22.6 | 0.704 | 1 | 1 | $Na_2SO_4$ |
| | O | 45.0 | 2.81 | 4 | 4 | |
| 6.48) | Mg | 13.2 | 0.543 | 1 | 1 | |
| | Ca | 21.8 | 0.544 | 1 | 1 | |
| | C | 13.0 | 1.08 | 2 | 2 | $MgCaC_2O_6$ |
| | O | 52.1 | 3.26 | 6 | 6 | |
| 6.49) | H | 5.00 | 5.00 | 1 | 1 | HF—Molar weight = |
| | F | 95.00 | 5.00 | 1 | 1 | 20.0 g/mole formula units |

$$\dfrac{40.0 \text{ g compound}}{1 \text{ mole compound}} \bigg| \dfrac{1 \text{ mole formula units}}{20.0 \text{ g compound}} = \dfrac{2 \text{ moles formula units}}{1 \text{ mole compound}}$$

Molecular formula is therefore $H_2F_2$.

# CHAPTER 7

# ANSWERS TO EQUATION BALANCING EXERCISE

1) $4 \text{ Na} + O_2 \rightarrow 2 \text{ Na}_2O$
2) $H_2 + Cl_2 \rightarrow 2 \text{ HCl}$
3) $4 \text{ P} + 3 O_2 \rightarrow 2 P_2O_3$
4) $KClO_4 \rightarrow KCl + 2 O_2$

5) $Sb_2S_3 + 6 HCl \rightarrow 2 SbCl_3 + 3 H_2S$
6) $2 NH_3 + H_2SO_4 \rightarrow (NH_4)_2SO_4$
7) $CuO + 2 HCl \rightarrow CuCl_2 + H_2O$
8) $Zn + Pb(NO_3)_2 \rightarrow Zn(NO_3)_2 + Pb$
9) $2 AgNO_3 + H_2S \rightarrow Ag_2S + 2 HNO_3$
10) $2 Cu + S \rightarrow Cu_2S$
11) $2 Al + 2 H_3PO_4 \rightarrow 3 H_2 + 2 AlPO_4$
12) $2 NaNO_3 \rightarrow 2 NaNO_2 + O_2$
13) $Mg(ClO_3)_2 \rightarrow MgCl_2 + 3 O_2$
14) $2 H_2O_2 \rightarrow 2 H_2O + O_2$
15) $2 BaO_2 \rightarrow 2 BaO + O_2$
16) $H_2CO_3 \rightarrow H_2O + CO_2$
17) $Pb(NO_3)_2 + 2 KCl \rightarrow PbCl_2 + 2 KNO_3$
18) $2 Al + 3 Cl_2 \rightarrow 2 AlCl_3$
19) $4 P + 5 O_2 \rightarrow 2 P_2O_5$
20) $NH_4NO_2 \rightarrow N_2 + 2 H_2O$
21) $3 H_2 + N_2 \rightarrow 2 NH_3$
22) $Cl_2 + 2 KBr \rightarrow Br_2 + 2 KCl$
23) $BaCl_2 + (NH_4)_2CO_3 \rightarrow BaCO_3 + 2 NH_4Cl$
24) $MgCO_3 + 2 HCl \rightarrow MgCl_2 + CO_2 + H_2O$
25) $2 P + 3 I_2 \rightarrow 2 PI_3$
26) $2 PbO_2 \rightarrow 2 PbO + O_2$
27) $2 Al + 6 HCl \rightarrow 2 AlCl_3 + 3 H_2$
28) $Fe_2(SO_4)_3 + 3 Ba(OH)_2 \rightarrow 3 BaSO_4 + 2 Fe(OH)_3$
29) $2 Al + 3 CuSO_4 \rightarrow Al_2(SO_4)_3 + 3 Cu$
30) $2 KClO_3 \rightarrow 2 KCl + 3 O_2$
31) $3 Mg + N_2 \rightarrow Mg_3N_2$
32) $2 C_6H_{14} + 19 O_2 \rightarrow 12 CO_2 + 14 H_2O$
33) $3 FeCl_2 + 2 Na_3PO_4 \rightarrow Fe_3(PO_4)_2 + 6 NaCl$
34) $Li_2O + HOH \rightarrow 2 LiOH$
35) $2 HgO \rightarrow 2 Hg + O_2$
36) $CaSO_4 \cdot 2 H_2O \rightarrow CaSO_4 + 2 H_2O$
37) $2 C_3H_7CHO + 11 O_2 \rightarrow 8 CO_2 + 8 H_2O$
38) $NaHCO_3 + HCl \rightarrow NaCl + H_2O + CO_2$
39) $Bi(NO_3)_3 + 3 NaOH \rightarrow Bi(OH)_3 + 3 NaNO_3$
40) $FeS + 2 HBr \rightarrow FeBr_2 + H_2S$
41) $Zn(OH)_2 + H_2SO_4 \rightarrow ZnSO_4 + 2 HOH$
42) $P_4O_{10} + 6 H_2O \rightarrow 4 H_3PO_4$
43) $C_4H_9OH + 6 O_2 \rightarrow 4 CO_2 + 5 H_2O$
44) $CaC_2 + 2 H_2O \rightarrow C_2H_2 + Ca(OH)_2$
45) $3 CaCO_3 + 2 H_3PO_4 \rightarrow Ca_3(PO_4)_2 + 3 CO_2 + 3 H_2O$
46) $PCl_5 + 4 H_2O \rightarrow H_3PO_4 + 5 HCl$
47) $CaI_2 + H_2SO_4 \rightarrow 2 HI + CaSO_4$
48) $C_3H_7COOH + 5 O_2 \rightarrow 4 CO_2 + 4 H_2O$
49) $Mg(CN)_2 + 2 HCl \rightarrow 2 HCN + MgCl_2$
50) $(NH_4)_2S + HgBr_2 \rightarrow 2 NH_4Br + HgS$

## QUESTIONS

7.1) Two molecules of benzene react with fifteen molecules of oxygen to produce twelve molecules of carbon dioxide and six molecules of water.

7.2) $2 Ca (s) + O_2 (g) \rightarrow 2 CaO (s)$

7.3) $4 P (s) + 5 O_2 (g) \rightarrow P_4O_{10} (s)$; also $P_4 (s) + 5 O_2 (g) \rightarrow P_4O_{10} (s)$

7.4) $2 K (s) + F_2 (g) \rightarrow 2 KF (s)$

7.5) $Si (s) + 2 Cl_2 (g) \rightarrow SiCl_4 (s)$

7.6) $3 Mg (s) + N_2 (g) \rightarrow Mg_3N_2 (s)$

7.7) $2 NaCl (s) \rightarrow 2 Na (s) + Cl_2 (g)$

7.8) $2 HgO (s) \rightarrow 2 Hg (l) + O_2 (g)$

7.9) $H_2CO_3 (aq) \rightarrow H_2O (l) + CO_2 (aq)$ (Note: "Carbonic acid" exists only in water solution. $CO_2$ is a partly soluble gas.)

7.10) $2 H_2O_2 (l) \rightarrow 2 H_2O (l) + O_2 (g)$

7.11) $C_3H_8 (g) + 5 O_2 (g) \rightarrow 3 CO_2 (g) + 4 H_2O (l)$

7.12) $2 C_2H_2 (g) + 5 O_2 (g) \rightarrow 4 CO_2 (g) + 2 H_2O (l)$

7.13) $2 CH_3CHO (l) + 5 O_2 (g) \rightarrow 4 CO_2 (g) + 4 H_2O (l)$

7.14) $C_{12}H_{22}O_{11} (s) + 12 O_2 (g) \rightarrow 12 CO_2 (g) + 11 H_2O (l)$

7.15) $Mg (s) + H_2SO_4 (aq) \rightarrow MgSO_4 (aq) + H_2 (g)$

7.16) $Ba (s) + 2 HOH (l) \rightarrow Ba(OH)_2 (aq) + H_2 (g)$

7.17) $Zn (s) + 2 AgNO_3 (aq) \rightarrow 2 Ag (s) + Zn(NO_3)_2 (aq)$

7.18) $Mg (s) + NiCl_2 (aq) \rightarrow Ni (s) + MgCl_2 (aq)$

7.19) $Br_2 (l) + 2 NaI (aq) \rightarrow I_2 (aq) + 2 NaBr (aq)$

7.20) $AgNO_3 (aq) + KBr (aq) \rightarrow AgBr (s) + KNO_3 (aq)$

7.21) $Pb(NO_3)_2 (aq) + CuSO_4 (aq) \rightarrow PbSO_4 (s) + Cu(NO_3)_2 (aq)$

7.22) $MgCl_2 (aq) + 2 NaF (aq) \rightarrow 2 NaCl (aq) + MgF_2 (s)$

7.23) $Na_2S (aq) + 2 AgNO_3 (aq) \rightarrow 2 NaNO_3 (aq) + Ag_2S (s)$

7.24) $2 NaOH (aq) + MgBr_2 (aq) \rightarrow 2 NaBr (aq) + Mg(OH)_2 (s)$

7.25) $KOH (aq) + HNO_3 (aq) \rightarrow KNO_3 (aq) + HOH (l)$

7.26) $Mg(OH)_2 (s) + 2 HCl (aq) \rightarrow MgCl_2 (aq) + 2 HOH (l)$

7.27) $Zn (s) + H_2O (g) \rightarrow ZnO (s) + H_2 (g)$

7.28) $BaO (s) + H_2O (l) \rightarrow Ba(OH)_2 (aq)$

7.29) $3 FeO (s) + 2 Al (s) \rightarrow 3 Fe (s) + Al_2O_3 (s)$

7.30) $Fe_2O_3 (s) + 3 CO (g) \rightarrow 2 Fe (s) + 3 CO_2 (g)$

## CHAPTER 8

8.1) $\dfrac{3.91 \text{ moles } CO_2}{} \left| \dfrac{1 \text{ mole } CaCO_3}{1 \text{ mole } CO_2} \right. = 3.91 \text{ moles } CaCO_3$

8.2) $\dfrac{0.284 \text{ mole } CaCO_3}{} \left| \dfrac{1 \text{ mole } CO_2}{1 \text{ mole } CaCO_3} \right. = 0.284 \text{ mole } CO_2$

8.3) $\dfrac{0.462 \text{ moles } CaCO_3}{} \left| \dfrac{1 \text{ mole } CO_2}{1 \text{ mole } CaCO_3} \right| \dfrac{44 \text{ g } CO_2}{1 \text{ mole } CO_2} = 20.3 \text{ g } CO_2$

8.4) $\dfrac{54.1 \text{ g } CaCO_3}{} \left| \dfrac{1 \text{ mole } CaCO_3}{100.1 \text{ g } CaCO_3} \right| \dfrac{1 \text{ mole } CaCl_2}{1 \text{ mole } CaCO_3} = 0.540 \text{ mole } CaCl_2$

8.5) $\dfrac{95.2 \text{ g } NaCl}{} \left| \dfrac{1 \text{ mole } NaCl}{58.5 \text{ g } NaCl} \right| \dfrac{1 \text{ mole } Na_2SO_4}{2 \text{ moles } NaCl} \left| \dfrac{142 \text{ g } Na_2SO_4}{1 \text{ mole } Na_2SO_4} \right.$

$$= 116 \text{ g } Na_2SO_4$$

8.6) $\dfrac{1.09 \text{ g } AgBr}{} \left| \dfrac{1 \text{ mole } AgBr}{188 \text{ g } AgBr} \right| \dfrac{2 \text{ moles } Na_2S_2O_3}{1 \text{ mole } AgBr} \left| \dfrac{158 \text{ g } Na_2S_2O_3}{1 \text{ mole } Na_2S_2O_3} \right.$

$$= 1.83 \text{ g } Na_2S_2O_3$$

8.7) $\dfrac{12 \text{ g KHC}_4\text{H}_4\text{O}_6}{} \left| \dfrac{1 \text{ mole KHC}_4\text{H}_4\text{O}_6}{188 \text{ g KHC}_4\text{H}_4\text{O}_6} \right| \dfrac{1 \text{ mole NaHCO}_3}{1 \text{ mole KHC}_4\text{H}_4\text{O}_6} \Big|$

$$\dfrac{84.0 \text{ g NaHCO}_3}{1 \text{ mole NaHCO}_3} = 5.4 \text{ g NaHCO}_3$$

8.8) $\dfrac{2.69 \text{ g MnO}_2}{} \left| \dfrac{1 \text{ mole MnO}_2}{86.9 \text{ g MnO}_2} \right| \dfrac{1 \text{ mole Cl}_2}{1 \text{ mole MnO}_2} \left| \dfrac{71.0 \text{ g Cl}_2}{1 \text{ mole Cl}_2} \right| = 2.20 \text{ g Cl}_2$

8.9) $\dfrac{45.0 \text{ g CHCl}_3}{} \left| \dfrac{1 \text{ mole CHCl}_3}{119 \text{ g CHCl}_3} \right| \dfrac{3 \text{ moles Cl}_2}{1 \text{ mole CHCl}_3} \left| \dfrac{71.0 \text{ g Cl}_2}{1 \text{ mole Cl}_2} \right|$

$$= 80.5 \text{ g Cl}_2$$

8.10) $\dfrac{175 \text{ g C}_{17}\text{H}_{35}\text{COONa}}{} \left| \dfrac{1 \text{ mole C}_{17}\text{H}_{35}\text{COONa}}{306 \text{ g C}_{17}\text{H}_{35}\text{COONa}} \right|$

$$\dfrac{3 \text{ moles NaOH}}{3 \text{ moles C}_{17}\text{H}_{35}\text{COONa}} \left| \dfrac{40.0 \text{ g NaOH}}{1 \text{ mole NaOH}} \right| = 22.9 \text{ g NaOH}$$

8.11) $\dfrac{255 \text{ g Cl}_2}{} \left| \dfrac{1 \text{ mole Cl}_2}{71.0 \text{ g Cl}_2} \right| \dfrac{1 \text{ mole O}_2}{2 \text{ moles Cl}_2} \left| \dfrac{32.0 \text{ g O}_2}{1 \text{ mole O}_2} \right| = 57.5 \text{ g O}_2$

8.12) $\dfrac{6.90 \text{ g CO}_2}{} \left| \dfrac{1 \text{ mole CO}_2}{44.0 \text{ g CO}_2} \right| \dfrac{1 \text{ mole C}_6\text{H}_{12}\text{O}_6}{6 \text{ moles CO}_2} \left| \dfrac{180 \text{ g C}_6\text{H}_{12}\text{O}_6}{1 \text{ mole C}_6\text{H}_{12}\text{O}_6} \right|$

$$= 4.70 \text{ g C}_6\text{H}_{12}\text{O}_6$$

8.13) $\dfrac{48.3 \text{ g (CaSO}_4)_2 \cdot \text{H}_2\text{O}}{} \left| \dfrac{1 \text{ mole (CaSO}_4)_2 \cdot \text{H}_2\text{O}}{290 \text{ g (CaSO}_4)_2 \cdot \text{H}_2\text{O}} \right|$

$$\dfrac{2 \text{ moles CaSO}_4 \cdot 2\text{H}_2\text{O}}{1 \text{ mole (CaSO}_4)_2 \cdot \text{H}_2\text{O}} \left| \dfrac{172 \text{ g CaSO}_4 \cdot 2\text{H}_2\text{O}}{1 \text{ mole CaSO}_4 \cdot 2\text{H}_2\text{O}} \right| = 57.3 \text{ g CaSO}_4 \cdot 2\text{H}_2\text{O}$$

8.14) $\dfrac{12{,}800 \text{ g N}_2\text{O}}{} \left| \dfrac{1 \text{ mole N}_2\text{O}}{44.0 \text{ g N}_2\text{O}} \right| \dfrac{1 \text{ mole NH}_4\text{NO}_3}{1 \text{ mole N}_2\text{O}} \left| \dfrac{80.0 \text{ g NH}_4\text{NO}_3}{1 \text{ mole NH}_4\text{NO}_3} \right|$

$$= 23{,}300 \text{ g NH}_4\text{NO}_3$$

8.15) $\dfrac{5.00 \text{ tons SO}_2}{} \left| \dfrac{1 \text{ ton-mole SO}_2}{64.1 \text{ tons SO}_2} \right| \dfrac{4 \text{ ton moles FeS}_2}{8 \text{ ton-moles SO}_2} \Big|$

$$\dfrac{120 \text{ tons FeS}_2}{1 \text{ ton-mole FeS}_2} \left| \dfrac{100 \text{ tons ore}}{12.1 \text{ tons FeS}_2} \right| = 38.7 \text{ tons ore}$$

8.16) $\dfrac{3.94 \text{ g Zn}}{} \left| \dfrac{1 \text{ mole Zn}}{65.4 \text{ g Zn}} \right| \dfrac{1 \text{ mole H}_2}{1 \text{ mole Zn}} \left| \dfrac{24.5 \text{ } \ell \text{ H}_2}{1 \text{ mole H}_2} \right| = 1.48 \text{ } \ell \text{ H}_2$

8.17) $\dfrac{5.19 \times 10^4 \text{ g CH}_4}{} \left| \dfrac{1 \text{ mole CH}_4}{16.0 \text{ g CH}_4} \right| \dfrac{1 \text{ mole C}_2\text{H}_2}{2 \text{ moles CH}_4} \left| \dfrac{26.0 \text{ g C}_2\text{H}_2}{1 \text{ mole C}_2\text{H}_2} \right|$

$$= 4.22 \times 10^4 \text{ g C}_2\text{H}_2 \text{ theoretical}$$

$$\dfrac{3.81 \times 10^4 \text{ g}}{4.22 \times 10^4 \text{ g}} \left| \dfrac{100}{} \right| = 90.3\% \text{ yield}$$

8.18) $\dfrac{7.00 \times 10^6 \text{ g Cu}_2\text{S}}{} \left| \dfrac{1 \text{ mole Cu}_2\text{S}}{159 \text{ g Cu}_2\text{S}} \right| \dfrac{2 \text{ moles Cu}}{1 \text{ mole Cu}_2\text{S}} \Big|$

$$\dfrac{63.5 \text{ g Cu theoretical}}{1 \text{ mole Cu}} \left| \dfrac{61.2 \text{ g Cu actual}}{100 \text{ g Cu theoretical}} \right| = 3.42 \times 10^6 \text{ g Cu actual}$$

8.19) $\dfrac{3.85 \text{ g CH}_4}{} \left| \dfrac{1 \text{ mole CH}_4}{16.0 \text{ g CH}_4} \right| \dfrac{1 \text{ mole CHCl}_3}{1 \text{ mole CH}_4} \left| \dfrac{120 \text{ g CHCl}_3}{1 \text{ mole CHCl}_3} \right|$

$$= 28.9 \text{ g CHCl}_3 \text{ theoretical}$$

$$\dfrac{23.0 \text{ g CHCl}_3 \text{ actual}}{28.9 \text{ g CHCl}_3 \text{ theoretical}} \left| \dfrac{100}{} \right| = 79.6\% \text{ yield}$$

8.20) $\dfrac{325 \text{ kg NaOH actual}}{} \ \bigg| \ \dfrac{100 \text{ kg NaOH theoretical}}{92.0 \text{ kg NaOH actual}} \ \bigg|$

$\dfrac{1 \text{ mole NaOH}}{0.0400 \text{ kg NaOH}} \ \bigg| \ \dfrac{1 \text{ mole Na}_2\text{CO}_3}{2 \text{ moles NaOH}} \ \bigg| \ \dfrac{0.106 \text{ kg Na}_2\text{CO}_3}{1 \text{ mole Na}_2\text{CO}_3}$

$$= 468 \text{ kg Na}_2\text{CO}_3$$

8.21) $\dfrac{40.0 \text{ g NH}_3}{} \ \bigg| \ \dfrac{1 \text{ mole NH}_3}{17.0 \text{ g NH}_3} = 2.35 \text{ moles NH}_3;$
(available)

$\dfrac{2.35 \text{ moles NH}_3}{} \ \bigg| \ \dfrac{3 \text{ moles O}_2}{4 \text{ moles NH}_3} = 1.76 \text{ moles O}_2$
(required)

$\dfrac{40.0 \text{ g O}_2}{} \ \bigg| \ \dfrac{1 \text{ mole O}_2}{32.0 \text{ g O}_2} = 1.25 \text{ moles O}_2;$
(available)

$\dfrac{1.25 \text{ moles O}_2}{} \ \bigg| \ \dfrac{4 \text{ moles NH}_3}{3 \text{ moles O}_2} = 1.67 \text{ moles NH}_3$
(required)

Oxygen is the limiting reagent.

$\dfrac{1.25 \text{ moles O}_2}{} \ \bigg| \ \dfrac{6 \text{ moles H}_2\text{O}}{3 \text{ moles O}_2} \ \bigg| \ \dfrac{18.0 \text{ g H}_2\text{O}}{1 \text{ mole H}_2\text{O}} = 45.0 \text{ g H}_2\text{O released}$

2.35 moles NH$_3$ available
$-$ 1.67 moles NH$_3$ used
0.68 mole  NH$_3$ left

$\dfrac{0.68 \text{ mole NH}_3}{} \ \bigg| \ \dfrac{17.0 \text{ g NH}_3}{1 \text{ mole NH}_3} = 12 \text{ g NH}_3 \text{ left}$

8.22) $\dfrac{19.6 \text{ g C}}{} \ \bigg| \ \dfrac{1 \text{ mole C}}{12.0 \text{ g C}} = 1.63 \text{ moles C available}$

$$\simeq 1.63 \text{ moles ZnO required}$$

$\dfrac{135 \text{ g ZnO}}{} \ \bigg| \ \dfrac{1 \text{ mole ZnO}}{81.4 \text{ g ZnO}} = 1.66 \text{ moles ZnO available}$

$$\simeq 1.66 \text{ moles C required}$$

Carbon is the limiting reagent.

$\dfrac{1.63 \text{ moles C}}{} \ \bigg| \ \dfrac{1 \text{ mole Zn}}{1 \text{ mole C}} \ \bigg| \ \dfrac{65.4 \text{ g Zn}}{1 \text{ mole Zn}} = 107 \text{ g Zn produced}$

1.66 moles ZnO available
$-$ 1.63 moles ZnO used
0.03 mole  ZnO left

$\dfrac{0.03 \text{ mole ZnO}}{} \ \bigg| \ \dfrac{81.4 \text{ g ZnO}}{1 \text{ mole ZnO}} = 2 \text{ g ZnO left, to 1 significant figure}$

Alternative calculation:

$\dfrac{19.6 \text{ g C}}{} \ \bigg| \ \dfrac{1 \text{ mole C}}{12.0 \text{ g C}} \ \bigg| \ \dfrac{1 \text{ mole ZnO}}{1 \text{ mole C}} \ \bigg| \ \dfrac{81.4 \text{ g ZnO}}{1 \text{ mole ZnO}} = 133 \text{ g ZnO used}$

135 g ZnO available $-$ 133 g ZnO used = 2 g ZnO left.

8.23) All calculations in ton-moles.

$\dfrac{3.00 \text{ tons S}_2\text{Cl}_2}{} \ \bigg| \ \dfrac{1 \text{ mole S}_2\text{Cl}_2}{135 \text{ tons S}_2\text{Cl}_2} = 0.0222 \text{ mole S}_2\text{Cl}_2;$

$\dfrac{0.0222 \text{ mole S}_2\text{Cl}_2}{} \ \bigg| \ \dfrac{1 \text{ mole CS}_2}{2 \text{ moles S}_2\text{Cl}_2} = 0.0111 \text{ mole CS}_2 \text{ required}$

$$\frac{1.00 \text{ ton CS}_2 \quad | \quad 1 \text{ mole CS}_2}{| \quad 76.2 \text{ tons CS}_2} = 0.0131 \text{ mole CS}_2;$$

$$\frac{0.0131 \text{ mole CS}_2 \quad | \quad 2 \text{ moles S}_2\text{Cl}_2}{| \quad 1 \text{ mole CS}_2} = 0.0262 \text{ mole S}_2\text{Cl}_2 \text{ required}$$

$S_2Cl_2$ is the limiting reagent.

$$\frac{0.0222 \text{ mole S}_2\text{Cl}_2 \quad | \quad 1 \text{ mole CCl}_4 \quad | \quad 154 \text{ tons CCl}_4}{| \quad 2 \text{ moles S}_2\text{Cl}_2 \quad | \quad 1 \text{ mole CCl}_4} = 1.71 \text{ tons CCl}_4$$

$$\begin{array}{r} 0.0131 \text{ mole CS}_2 \text{ available} \\ - 0.0111 \text{ mole CS}_2 \text{ used} \\ \hline 0.0020 \text{ mole CS}_2 \text{ left} \end{array}$$

$$\frac{0.0020 \text{ mole CS}_2 \quad | \quad 76.2 \text{ tons CS}_2}{| \quad 1 \text{ mole CS}_2} = 0.15 \text{ ton CS}_2 \text{ left}$$

8.24)  1 mole Mg $\simeq$ 1 mole $MgCl_2 \cdot 6\,H_2O \simeq$ 2 moles $Cl^- \simeq$ 2 moles AgCl

a) $$\frac{4.01 \text{ g AgCl} \quad | \quad 1 \text{ mole AgCl} \quad | \quad 1 \text{ mole Mg} \quad | \quad 24.3 \text{ g Mg}}{| \quad 144 \text{ g AgCl} \quad | \quad 2 \text{ moles AgCl} \quad | \quad 1 \text{ mole Mg}}$$

$$= 0.338 \text{ g Mg}$$

b) $$\frac{4.01 \text{ g AgCl} \quad | \quad 1 \text{ mole AgCl} \quad | \quad 1 \text{ mole MgCl}_2 \cdot 6\,H_2O \quad |}{| \quad 144 \text{ g AgCl} \quad | \quad 2 \text{ moles AgCl} \quad |}$$

$$\frac{204 \text{ g MgCl}_2 \cdot 6\,H_2O}{1 \text{ mole MgCl}_2 \cdot 6\,H_2O} = 2.84 \text{ g MgCl}_2 \cdot 6\,H_2O$$

$$\frac{2.84\,g}{5.25\,g} \times 100 = 54.1\% \text{ MgCl}_2 \cdot 6\,H_2O$$

# CHAPTER 9

9.1)  K·    ·P̈:    :B̈r:

9.2)  Ga:    ·Pb:

9.3)  F, Cl, Br, I, At

9.4)  $Na^+$, $Mg^{2+}$ and $Al^{3+}$ are isoelectronic with Ne. $S^{2-}$ and $Cl^-$ are isoelectronic with Ar; less common, $P^{3-}$.

9.5)  $O^{2-}$ and $F^-$. Less common, $N^{3-}$.

9.6)  $S^{2-}$, $K^+$, $Ca^{2+}$. Less common, $P^{3-}$ and $Sc^{3+}$.

9.7)  
$$K· \searrow \atop K· \nearrow \,+\, \ddot{S}: \;\to\; {K^+ \atop K^+} \left[ :\ddot{S}: \right]^{2-} \quad K_2S$$

9.8)  $S^{2-}$

9.9)  $Ca^{2+}$, $K^+$, $Cl^-$, $Br^-$

9.10)  For noble gas atoms, and the monatomic ions that are isoelectronic with them, size decreases with increasing atomic number. Larger number of protons in nucleus gives it greater attracting force for electrons and pulls them in more closely.

9.11)  Ions are formed when neutral atoms lose or gain electrons. The electron(s) that is(are) lost by one atom is(are) gained by another, effectively a transfer of electrons from one atom to another. The attraction between the ions produced makes up the "ionic bond." Covalent bonds are formed when a pair of electrons is shared by the two bonded atoms. Effectively the electrons belong to both atoms, spending some time near each nucleus.

9.12) K—Cl bond is ionic, formed by "transferring" an electron from a potassium atom to a chlorine atom. Cl—Cl bond is covalent, formed by two chlorine atoms sharing a pair of electrons.

9.13) :C̈l· + ·Ï: → :C̈l: Ï:

9.14) Nonmetal atoms are usually 1, 2 or possibly 3 or 4 short of an "octet" of electrons, and achieve that octet most easily by gaining the missing electrons. When two nonmetal atoms combine, the easiest way for both atoms to reach the octet is to gain each other's electrons, or share them, forming a covalent bond. If the second atom is a metal, however, it has 1, 2 or possibly 3 electrons more than an octet. It reaches the octet by giving its electrons to the nonmetal, becoming a positive ion itself, and making the nonmetal atom a negative ion, and forming an ionic bond.

9.15) The energy of the system decreases. The change is exothermic, as the energy lost by the system passes to the surroundings.

9.16) A bond that has a symmetrical distribution of electrical charge is nonpolar; if the charge distribution is unsymmetrical the bond is polar. Bonds between identical atoms are completely nonpolar.

9.17) Cl—Cl (nonpolar); Br—Cl (essentially nonpolar); I—Cl (polar); F—Cl (polar)

9.18) In Br—Cl and I—Cl, chlorine is the more electronegative; in F—Cl, fluorine is the negative pole.

9.19) "Electronegativity is a measure of the relative ability of two atoms to attract the pair of electrons forming a single covalent bond between them." Because noble gases do not normally form bonds, electronegativity numbers are not assigned to them.

9.20) A multiple bond exists when two atoms are bonded by two or three pairs of electrons, forming two or three covalent bonds.

9.21)

H—B̈r:

HBr

H—S̈:
|
H

H₂S

H—P̈—H
|
H

PH₃

9.22)

:F̈—Ö:
|
:F̈:

OF₂

:C≡O:

CO

$$\left[\begin{array}{c} :\overset{..}{O}: \\ | \\ :\overset{..}{O}-S-\overset{..}{O}: \\ | \\ :\overset{..}{O}: \end{array}\right]^{2-}$$

SO₄²⁻

9.23)

$$\left[ :\overset{..}{C}l-\overset{..}{O}: \right]^{-}$$

ClO⁻

$$\left[\begin{array}{c} :\overset{..}{O}: \\ | \\ :\overset{..}{O}-Br-\overset{..}{O}: \\ | \\ :\overset{..}{O}: \end{array}\right]$$

BrO₄⁻

:Ö:
|
H—Ö—S—Ö—H
|
:Ö:

H₂SO₄

9.24)

C₄H₁₀:

H H H H
| | | |
H—C—C—C—C—H
| | | |
H H H H

$C_4H_8$:  H—C=C—C—C—H  or  H—C—C=C—C—H

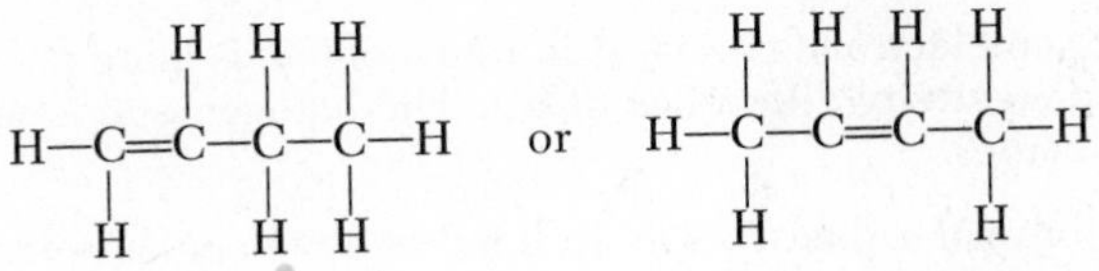

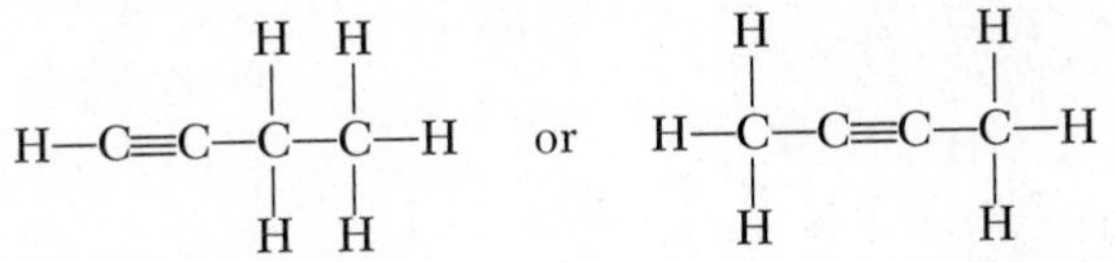

$C_4H_6$  H—C≡C—C—C—H  or  H—C—C≡C—C—H  or

C=C—C=C  or  C=C—C—C—H

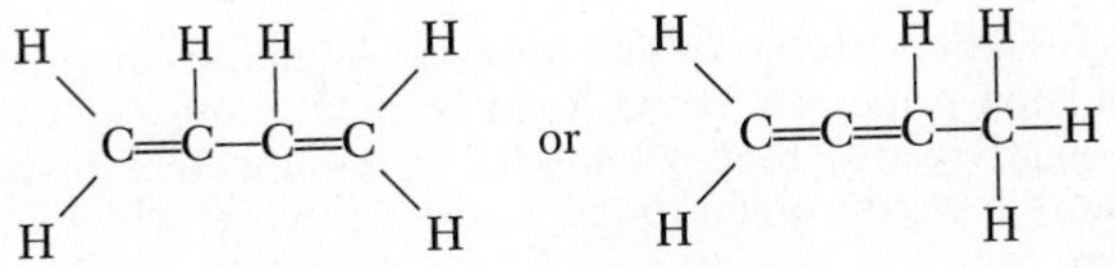

9.25)

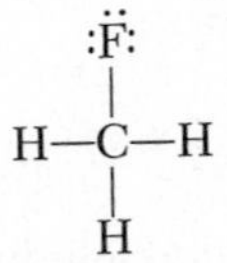

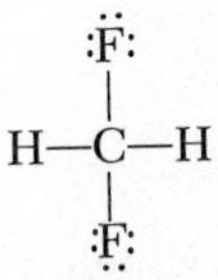

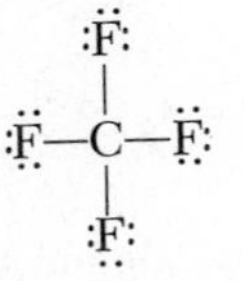

:F:
|
H—C—H
|
H

CH₃F

:F:
|
H—C—H
|
:F:

CH₂F₂

:F:
|
:F—C—F:
|
:F:

CF₄

9.26)  See page 466.

9.27)  $C_5H_{10}$:

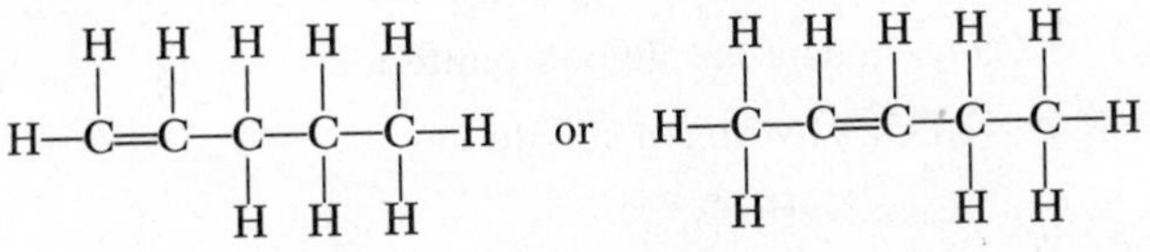

H—C=C—C—C—C—H  or  H—C—C=C—C—C—H

9.28)  $C_3H_6O$:

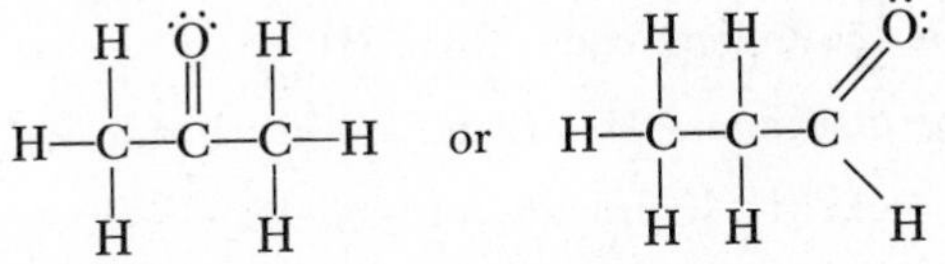

H—C—C—C—H  or  H—C—C—C=O:

9.29)  HCOOH:

H—C
  ‖O:
  \
  O—H

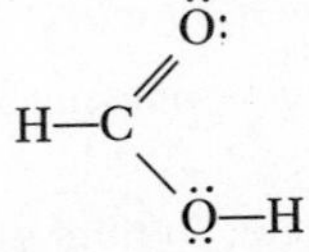

9.30)  If the total number of electrons is *odd*, they cannot arrange themselves so that each atom is surrounded by eight electrons, an *even* number.

| COMPOUND | ELECTRON PAIR GEOMETRY | MOLECULAR GEOMETRY |
|---|---|---|
| 9.31) BeH₂ | Linear | Linear |
| CF₄ | Tetrahedral | Tetrahedral |
| OF₂ | Tetrahedral | Bent |
| 9.32) IO₄⁻ | Tetrahedral | Tetrahedral |
| ClO₂⁻ | Tetrahedral | Bent |
| CO₃²⁻ | Planar triangular | Planar triangular |

9.33)  Bonds are distributed symmetrically around carbon atom, so total molecular charge distribution is uniform.

9.34) HCl, with an electronegativity difference of 0.9, is more polar than HI, with an electronegativity difference of 0.4. The halogen end is more negative in both molecules.

9.35) $H_2O$ is more polar than $H_2S$. Both bond and molecular polarity decrease for the hydrides of column VIA from oxygen to tellurium. $H_2Te$ has zero electronegativity difference in its bonds, and the molecule is essentially nonpolar.

9.36)

Water                                    Methanol

Bond angles around oxygen atoms are approximately equal, according to electron repulsion principle. H—O bond is much more polar than C—O bond (electronegativity differences 1.4 vs 0.4). Bonding electrons are therefore displaced more toward oxygen in the water molecule, which is the more polar of the two.

# CHAPTER 10

10.1) Calcium ion; chromium(III) ion; zinc ion; phosphide ion; bromide ion

10.2) $Li^+$; $NH_4^+$; $N^{3-}$; $F^-$; $Hg^{2+}$

10.3) Nitric acid; sulfurous acid; $HClO_4$; $H_2SeO_4$

10.4) $SO_4^{2-}$; $ClO_2^-$; iodate ion; hypobromite ion

10.5) $HCO_3^-$; $H_2PO_4^-$; hydrogen sulfate ion

10.6) $K_2S$; $Cu(NO_3)_2$; $NaHCO_3$

10.7) Magnesium sulfite; aluminum fluoride; lead carbonate [or lead(II) carbonate]

10.8) Sulfur dioxide; dinitrogen oxide; $PBr_3$; HI

10.9) $HSO_3^-$; $KNO_3$; manganese(II) sulfate; sulfur trioxide

10.10) Bromate ion; nickel(II) hydroxide (or nickel hydroxide); AgCl; $SiF_6$

10.11) $TeO_4^{2-}$; $FePO_4$; sodium acetate; hydrogen sulfide (or dihydrogen sulfide)

10.12) Monohydrogen phosphate ion (or hydrogen phosphate ion); copper(II) oxide; $Na_2C_2O_4$; $NH_3$

10.13) HClO; $CrBr_2$; potassium hydrogen carbonate; sodium dichromate

10.14) Cobalt(III) oxide; sodium sulfite; $HgI_2$; $Al(OH)_3$

10.15) $Ca(H_2PO_4)_2$; $KMnO_4$; ammonium iodate; selenic acid

10.16) Mercury(I) chloride; periodic acid; $CoSO_4$; $Pb(NO_3)_2$

10.17) $UF_3$; $BaO_2$; manganese(II) chloride; sodium chlorite

10.18) Potassium tellurate; zinc carbonate; $CrCl_2$; $HC_2H_3O_2$

10.19) $BaCrO_4$; $CaSO_3$; copper(I) chloride; silver nitrate

10.20) Sodium peroxide; nickel carbonate [or nickel(II) carbonate]; FeO; $H_2S$ (aq)

10.21) $Zn_3P_2$; $CsNO_3$; ammonium cyanide; disulfur decafluoride

10.22) Dinitrogen trioxide; lithium permanganate; $In_2Se_3$; $Hg_2(SCN)_2$

# CHAPTER 11

11.1) An ideal gas consists of widely spaced independent particles moving in straight lines until colliding with each other or walls without loss of energy.

11.2) Evidence that gas particles are moving.

11.3) Suggests no loss of energy in collisions.

11.4) Gas particles are widely separated compared to same number of particles close together in liquid state.

11.5) When gas particles are pushed close to each other, the intermolecular attractions become significant. The molecules are no longer independent. The ideal gas model is violated, so the gas does not behave ideally.

11.6) Particles move in straight lines until they hit something—eventually the walls of the container, thereby filling it.

11.7) Pressure, volume, temperature, quantity.

11.8) 1 atmosphere is the name for the pressure that will support a column of mercury 760 millimeters high. This is the pressure exerted at sea level on a "normal" day. The *torr* is a name for one millimeter of mercury as a pressure measurement.

11.9) $P_g = 747$ mm Hg $+ (729 - 263)$ mm Hg $= 1213$ mm Hg

11.10) $\dfrac{3.15\ \ell \quad\left|\quad 0.940\ \text{atm}\right.}{6.26\ \text{atm}} = 0.473\ \ell$

11.11) Absolute zero is the temperature at which the average particle translational kinetic energy is zero. This presumably means no particle movement.

11.12) $328° + 273° = 601°K$

11.13) $90° - 273° = -183°C$

11.14) $\dfrac{85{,}600\ \ell \quad\left|\quad (25 + 273)°K\right.}{(15 + 273)°K} = 88{,}600\ \ell$

11.15) $\dfrac{912\ \text{torr} \quad\left|\quad (27 + 273)°K\right.}{(68 + 273)°K} = 802\ \text{torr}$

11.16) $\dfrac{73.4\ \text{ml} \quad\left|\quad 824\ \text{torr} \quad\right|\quad (23 + 273)°K}{749\ \text{torr} \quad\left|\quad (43 + 273)°K\right.} = 75.6\ \text{ml}$

11.17) $\dfrac{1.62\ \ell \quad\left|\quad 1.02\ \text{atm} \quad\right|\quad (3 + 273)°K}{4.86\ \text{atm} \quad\left|\quad (23 + 273)°K\right.} = 0.317\ \ell$

11.18) Standard temperature and pressure. These are arbitrary standards at which gases may be compared. Standard temperature is 0°C; standard pressure is 760 mm Hg, or 1 atmosphere.

11.19) $\dfrac{47.9\ \text{ml} \quad\left|\quad 718\ \text{torr} \quad\right|\quad 273°K}{760\ \text{torr} \quad\left|\quad (26 + 273)°K\right.} = 41.3\ \text{ml}$

11.20) $\dfrac{46.9\ \text{ml} \quad\left|\quad 760\ \text{torr} \quad\right|\quad (273 + 24)°K}{738\ \text{torr} \quad\left|\quad 273°K\right.} = 52.5\ \text{ml}$

11.21) $V = \dfrac{nRT}{P} = \dfrac{0.16\ \text{mole} \quad\left|\quad (22 + 273)°K \quad\right|\quad \dfrac{62.4\ (\ell)\ (\text{torr})}{(\text{mole})\ (°K)}}{751\ \text{torr}} = 3.92\ \ell$

11.22) $n = \dfrac{PV}{RT} = \dfrac{9.40\ \text{atm} \quad\left|\quad 2.55\ \ell\right.}{0.0821\ \dfrac{(\ell)\ (\text{atm})}{(\text{mole})\ (°K)} \quad\left|\quad (273 + 20)°K\right.}$

$= \dfrac{9.40\ \text{atm} \quad\left|\quad (\text{mole})\ (°K) \quad\right|\quad 2.55\ \ell}{0.0821\ (\ell)\ (\text{atm}) \quad\left|\quad (273 + 20)°K\right.} = 0.996\ \text{mole}$

11.23) $\quad MW = \dfrac{gRT}{PV} = \dfrac{5.89 \text{ g}}{745 \text{ torr}} \left| \dfrac{(22 + 273)°K}{2.68 \ell} \right| \dfrac{62.4 \ (\ell) \ (torr)}{(mole) \ (°K)} = 54.3 \text{ g/mole}$

11.24) $\quad P = \dfrac{gRT}{V(MW)} = \dfrac{25.0 \text{ g } SO_2}{2.15 \ell} \left| \dfrac{0.0821 \ (\ell) \ (atm)}{(mole) (°K)} \right| \dfrac{1 \text{ mole } SO_2}{64.1 \text{ g } SO_2} \left| \dfrac{293°K}{} \right.$

$$= 4.36 \text{ atm}$$

11.25) $\quad T = \dfrac{PV(MW)}{gR} = \dfrac{24.0 \text{ atm}}{624 \text{ g } O_2} \left| \dfrac{21.2 \ \ell}{} \right| \dfrac{32.0 \text{ g } O_2}{1 \text{ mole } O_2} \left| \dfrac{(mole) \ (°K)}{0.0821 \ (\ell) \ (atm)} \right.$

$$= 318°K = 45°C$$

11.26) $\quad g = \dfrac{PV(MW)}{RT} = \dfrac{798 \text{ torr}}{282°K} \left| \dfrac{1.75 \ m^3}{} \right| \dfrac{1000 \ell}{1 \ m^3} \left| \dfrac{0.00400 \text{ kg He}}{1 \text{ mole He}} \right|$

$$\dfrac{(mole) \ (°K)}{62.4 \ (\ell) \ (torr)} = 0.317 \text{ kg He}$$

11.27)

| ELEMENT | GRAMS | MOLES | MOLE RATIO |
|---|---|---|---|
| C | 85.6 | 7.13 | 1 |
| H | 14.4 | 14.4 | 2 |

Empirical formula: $CH_2$
Molar weight of empirical formula: 14.0 g/mole

$$MW = \dfrac{gRT}{VP} = \dfrac{1.69 \text{ g}}{1 \ell} \left| \dfrac{62.4 \ (\ell) \ (torr)}{(mole) \ (°K)} \right| \dfrac{298°K}{750 \text{ torr}} = 41.9 \text{ g/mole}$$

$$\dfrac{41.9 \text{ g}}{1 \text{ mole}} \left| \dfrac{1 \text{ mole formula units}}{14.0 \text{ g}} \right. = 3 \text{ moles formula units/mole}$$

Molecular formula, $C_3H_6$

11.28) The volume occupied by one mole.

11.29) $\quad \dfrac{16.9 \ \ell}{} \left| \dfrac{1 \text{ mole}}{22.4 \ \ell} \right. = 0.754 \text{ mole}$

11.30) $\quad \dfrac{1.83 \text{ g}}{1 \ell} \left| \dfrac{22.4 \ \ell}{1 \text{ mole}} \right. = 41.0 \text{ g/mole}$

11.31) $\quad \dfrac{0.937 \text{ g}}{0.744 \ \ell} \left| \dfrac{22.4 \ \ell}{1 \text{ mole}} \right. = 28.2 \text{ g/mole}$

11.32) $\quad \dfrac{2.65 \text{ g } HgO}{} \left| \dfrac{1 \text{ mole } HgO}{217 \text{ g } HgO} \right| \dfrac{1 \text{ mole } O_2}{2 \text{ moles } HgO} \left| \dfrac{22.4 \ \ell \ O_2}{1 \text{ mole } O_2} \right.$

$$= 0.137 \ \ell \ O_2$$

11.33) $\quad \dfrac{85.0 \text{ ml } H_2}{} \left| \dfrac{1 \text{ mole } H_2}{22,400 \text{ ml } H_2} \right| \dfrac{1 \text{ mole Mg}}{1 \text{ mole } H_2} \left| \dfrac{24.3 \text{ g Mg}}{1 \text{ mole Mg}} \right.$

$$= 0.0922 \text{ g Mg}$$

11.34) $\quad \dfrac{3.26 \text{ g Mg}}{} \left| \dfrac{1 \text{ mole Mg}}{24.3 \text{ g Mg}} \right| \dfrac{1 \text{ mole } O_2}{2 \text{ moles Mg}} = 0.0671 \text{ mole } O_2$

$$V = \dfrac{0.0671 \text{ mole } O_2}{} \left| \dfrac{(25 + 273)°K}{752 \text{ torr}} \right| \dfrac{62.4 \ (\ell) \ (torr)}{(°K) \ (mole)} = 1.66 \ \ell \ O_2$$

11.35) $\quad n = \dfrac{4.16 \ \ell \ H_2}{} \left| \dfrac{1.65 \text{ atm}}{(243 + 273)°K} \right| \dfrac{1 \ (°K) \ (mole)}{0.0821 \ (\ell) \ (atm)} = 0.162 \text{ mole } H_2$

$$\dfrac{0.162 \text{ mole } H_2}{} \left| \dfrac{1 \text{ mole } CuO}{1 \text{ mole } H_2} \right| \dfrac{79.5 \text{ g CuO}}{1 \text{ mole } CuO} = 12.9 \text{ g CuO}$$

11.36) Equal volumes of gases at the same temperature and pressure contain equal numbers of molecules. Such volumes are therefore in the same ratio as the ratio of their numbers of molecules. Equation coefficients, which represent the ratio of numbers of molecules, must also correspond to the equivalent ratio of volumes.

11.37) $\dfrac{1.28 \,\cancel{\ell \, O_2}}{} \left|\, \dfrac{2 \,\ell \, H_2}{1 \,\cancel{\ell \, O_2}} \right. = 2.56 \,\ell \, H_2$

11.38) $\dfrac{28.3 \,\cancel{\ell \, H_2O}}{} \left|\, \dfrac{1.46 \,\cancel{atm}}{2.19 \,\cancel{atm}} \,\right|\, \dfrac{(460 + 273)\cancel{{}^\circ K}}{(540 + 273)\cancel{{}^\circ K}} \,\right|\, \dfrac{1 \,\ell \, CO}{1 \,\cancel{\ell \, H_2O}} = 17.0 \,\ell \, CO$

11.39)   The total pressure exerted by a gaseous mixture is the sum of the partial pressures of the components of the mixture: $P = p_1 + p_2 + p_3 + \ldots$. See Equation 11.26 and related text (page 264) for explanation.

11.40)   $P = 0.364 \text{ atm} + 0.108 \text{ atm} + 0.529 \text{ atm} = 1.001 \text{ atm}$

11.41)   $P_{H_2} = 751 \text{ torr} - 40.0 \text{ torr} = 711 \text{ torr}$

# CHAPTER 12

12.1)   Because of the distances between molecules in the gas phase, a given quantity occupies a much larger volume as a gas than as a liquid. Density is mass/volume. Therefore the larger volume of the gas yields a lower density.

12.2)   Air can be compressed by pushing the widely spaced molecules closer together. Liquid molecules are already close, and therefore cannot be compressed.

12.3)   Substances with strong intermolecular attractions tend to have low vapor pressures. The strong attractions make the evaporation rate slow at a given temperature. A relatively low vapor concentration—and thus a low equilibrium vapor pressure—is therefore sufficient to make the condensation rate equal to the evaporation rate.

12.4)   Molar heat of vaporization is the amount of heat required to vaporize one mole of a liquid at its boiling point.

12.5)   Motor oil is more viscous than water. From this it may be predicted that intermolecular attractions are stronger in motor oil, as strong attractions lead to internal resistance to liquid flow, which is the property called viscosity.

12.6)   Surface tension is greater in mercury than in water, indicating stronger intermolecular attractions.

12.7)   Soap reduces intermolecular attractions so the soapy water is able to penetrate fabrics and cleanse them throughout.

12.8)   Dipole forces are electrostatic attractions between polar molecules, the negatively charged region of one molecule being attracted to the positively charged region of another molecule. Dispersion, or London, forces are weaker electrostatic attractions between molecules that are generally nonpolar, but which become "temporary dipoles" through instantaneously shifting electron clouds. Hydrogen bonds are exceptionally strong intermolecular attractions between the hydrogen that is bonded to a strongly electronegative element in a polar molecule and the negative region, often the same highly electronegative element, of an adjacent molecule. The hydrogen atom becomes a small highly concentrated region of positive charge that can approach quite closely the negatively charged region of the nearby molecule.

12.9)   HBr and $NF_3$: dipole; $C_2H_2$: dispersion; $C_2H_5OH$: hydrogen bonds

12.10)  The high melting points of ionic compounds suggest correctly that ionic bonds are stronger than dipole forces. Both are electrical in character, ionic forces arising from a nearly complete transfer of electrons, and dipole forces from nonsymmetrical distribution of electrical charge within molecules.

12.11)  $CCl_4$, because it is larger, as suggested by its higher molecular weight.

12.12)  $H_2S$, because it is slightly more polar.

12.13) Fluorine, oxygen and nitrogen. The electronegativity difference between hydrogen and these elements is large enough to shift the bonding electron pair away from the hydrogen atom. If the molecule is polar, a hydrogen bond develops between the hydrogen atom of one molecule and the fluorine, oxygen, or nitrogen atom of another.

12.14) (a) Dispersion forces; (b) dipole forces

12.15) a, c, d

12.16) $C_6H_{14}$, a larger molecule with higher molecular weight than $C_3H_8$, will have stronger intermolecular attractions and therefore higher melting and boiling points.

12.17) Opposing changes are occurring at equal rates.

12.18) Evaporation at constant rate begins immediately when liquid is introduced. At that time condensation rate is zero. Net rate of increase in vapor concentration is a maximum, so rate of vapor pressure increase is a maximum at start. Later condensation rate is more than zero, but less than evaporation rate. Net rate of increase in vapor concentration is less than initially, so rate of vapor pressure increase is less than initially. At equilibrium, evaporation and condensation rates are equal. Vapor concentration and therefore vapor pressure remain constant.

12.19) All of the liquid evaporated before the vapor concentration was high enough to yield a condensation rate equal to the evaporation rate. At lower than equilibrium vapor concentration, the vapor pressure is lower than the equilibrium vapor pressure.

12.20) Concentration is measured in moles/liter, n/V. Solving ideal gas equation for concentration, $\frac{n}{V} = \frac{p}{RT} = \left(\frac{1}{RT}\right)$ (p). At constant temperature 1/RT is a constant, so pressure and concentration are proportional.

12.21) (a) Second and third boxes have greatest pressure—the equilibrium pressure. (b) First box probably has least, having all evaporated before reaching equilibrium. Only possible exception is if vapor pressure just reached equilibrium pressure as last molecule evaporated in first box.

12.22) Boiling point is temperature at which vapor bubbles form spontaneously anywhere in a liquid. This is the temperature at which the vapor pressure equals the pressure above the liquid. See Figure 12.10, page 285.

12.23) Gas, because vapor pressure is greater than surrounding pressure.

12.24) High boiling liquids have strong intermolecular attractions, and therefore require high energy to escape from the liquid to form a gas. Evaporation rate is therefore slow, quickly equaled by condensation rate at low vapor concentration, or vapor pressure.

12.25) More energy is required to vaporize X, so it would have the higher boiling point and lower vapor pressure.

12.26) Crystalline solids have rigid and precise geometric structures because of highly ordered and recurring patterns, whereas amorphous solids have some freedom to move because of irregular structure.

12.27) A, molecular; B, metallic

## CHAPTER 13

*Values in brackets, [], are answers expressed in SI energy units, joules (J), or kilojoules (kJ).*

13.1) Vertically, temperature; horizontally, energy

13.2) D, E and F     13.3) C

13.4) The solid sample melts entirely to a liquid, all at constant temperature K.

13.5) O – N

13.6) $\dfrac{325\text{ g}}{}\ \bigg|\ \dfrac{0.84\text{ cal}}{\text{(g) (}^\circ\text{C)}}\ \bigg|\ \dfrac{(-19-22)^\circ\text{C}}{}\ \bigg|\ \dfrac{1\text{ kcal}}{1000\text{ cal}} = -11\text{ kcal}\ \ [-46\text{kJ}]$

11 kcal must be removed.

13.7) $\dfrac{22.63\text{ cal}}{(5.624\text{ g}) (32.4-18.6)^\circ\text{C}} = 0.292\text{ cal/(g)(}^\circ\text{C)}\ \ [1.22\text{ J/(g)(}^\circ\text{C)}]$

13.8) $\dfrac{58.6\text{ cal}}{14.9^\circ\text{C}}\ \bigg|\ \dfrac{1\text{ (g) (}^\circ\text{C)}}{0.092\text{ cal}} = 43\text{ g}$

13.9) The total heat flow in the calorimeter, made up of the energy absorbed by the water and the energy lost by the metal, equals zero.

$\dfrac{100\text{ g H}_2\text{O}}{}\ \bigg|\ \dfrac{1.00\text{ cal}}{\text{(g) (}^\circ\text{C)}}\ \bigg|\ \dfrac{(29.6-23.1)^\circ\text{C}}{}\ +$

$\dfrac{39.0\text{ g metal}}{}\ \bigg|\ \dfrac{c\text{ cal}}{\text{(g) (}^\circ\text{C)}}\ \bigg|\ \dfrac{(29.6-98.2)^\circ\text{C}}{} = 0$

$6.5 \times 10^2 - 2.68 \times 10^3\ c = 0$

$c = 0.24\text{ cal/(g)(}^\circ\text{C)}\ \ [1.0\text{ J/(g)(}^\circ\text{C)}]$

13.10) $\dfrac{3785\text{ g}}{}\ \bigg|\ \dfrac{540\text{ cal}}{1\text{ g}}\ \bigg|\ \dfrac{1\text{ kcal}}{1000\text{ cal}} = 2.04 \times 10^3\text{ kcal}\ \ [8.54\text{ kJ}]$

13.11) $\dfrac{25.0\text{ g}}{}\ \bigg|\ \dfrac{74\text{ cal}}{1\text{ g}} = 1.9 \times 10^3\text{ cal} = 1.9\text{ kcal}\ \ [7.9\text{ kJ}]$

13.12) $\dfrac{2.32 \times 10^3\text{ cal}}{50.0\text{ g}} = 46.4\text{ cal/g}\ \ [194\text{ J/g}]$

13.13) $\dfrac{5.72\text{ kcal}}{65.8\text{ g C}_7\text{H}_8}\ \bigg|\ \dfrac{92.0\text{ g C}_7\text{H}_8}{1\text{ mole C}_7\text{H}_8} = 8.00\text{ kcal/mole}\ \ [33.5\text{ kJ/mole}]$

13.14) $\dfrac{12.5\text{ kcal}}{}\ \bigg|\ \dfrac{1\text{ mole NH}_3}{5.22\text{ kcal}}\ \bigg|\ \dfrac{17.0\text{ g NH}_3}{1\text{ mole NH}_3} = 40.7\text{ g NH}_3$

13.15) $\dfrac{75.0\text{ g}}{}\ \bigg|\ \dfrac{32.7\text{ cal}}{1\text{ g}}\ \bigg|\ \dfrac{1\text{ kcal}}{1000\text{ cal}} = 2.45\text{ kcal}\ \ [10.3\text{ kJ}]$

13.16) $\dfrac{437\text{ cal}}{31.2\text{ g Sn}}\ \bigg|\ \dfrac{1\text{ kcal}}{1000\text{ cal}}\ \bigg|\ \dfrac{119\text{ g Sn}}{1\text{ mole Sn}} = 1.67\text{ kcal/mole}\ \ [6.99\text{ kJ/mole}]$

13.17) $\dfrac{1.50\text{ kcal}}{}\ \bigg|\ \dfrac{1000\text{ cal}}{1\text{ kcal}}\ \bigg|\ \dfrac{1\text{ g Ag}}{21.0\text{ cal}} = 71.4\text{ g Ag}$

13.18) $\dfrac{225\text{ g}}{}\ \bigg|\ \dfrac{1.00\text{ cal}}{\text{(g) (}^\circ\text{C)}}\ \bigg|\ \dfrac{16^\circ\text{C}}{}\ +\ \dfrac{225\text{ g}}{}\ \bigg|\ \dfrac{80\text{ cal}}{1\text{ g}}\ +$

$\dfrac{225\text{ g}}{}\ \bigg|\ \dfrac{0.49\text{ cal}}{\text{(g) (}^\circ\text{C)}}\ \bigg|\ \dfrac{12^\circ\text{C}}{} = 2.29 \times 10^4\text{ cal removed}\ \ [9.58 \times 10^4\text{kJ}]$

13.19) $C_3H_8\text{ (g)} + 5\ O_2\text{ (g)} \rightarrow 3\ CO_2\text{ (g)} + 4\ H_2O\text{ (l)} + 531\text{ kcal}$

$C_3H_8\text{ (g)} + 5\ O_2\text{ (g)} \rightarrow 3\ CO_2\text{ (g)} + 4\ H_2O\text{ (l)}\qquad \Delta H = -531\text{ kcal}$

13.20) $CaO\text{ (s)} + H_2O\text{ (l)} \rightarrow Ca(OH)_2\text{ (s)} + 15.6\text{ kcal}$

$CaO\text{ (s)} + H_2O\text{ (l)} \rightarrow Ca(OH)_2\text{ (s)}\qquad \Delta H = -15.6\text{ kcal}$

13.21) $2\ Al_2O_3\text{ (s)} + 3\ C\text{ (s)} + 516\text{ kcal} \rightarrow 4\ Al\text{ (s)} + 3\ CO_2\text{ (g)}$

$2\ Al_2O_3\text{ (s)} + 3\ C\text{ (s)} \rightarrow 4\ Al\text{ (s)} + 3\ CO_2\text{ (g)}\qquad \Delta H = +516\text{ kcal}$

13.22) $$\frac{4.00 \times 10^3 \text{ g } C_4H_{10}}{} \left| \frac{1 \text{ mole } C_4H_{10}}{58.0 \text{ g } C_4H_{10}} \right| \frac{1380 \text{ kcal}}{2 \text{ moles } C_4H_{10}}$$

$$= 4.76 \times 10^4 \text{ kcal} \quad [1.99 \times 10^5 \text{ kJ}]$$

13.23) $$\frac{5.5 \times 10^4 \text{ kcal}}{} \left| \frac{1 \text{ mole } C_6H_{14}}{990 \text{ kcal}} \right| \frac{86.0 \text{ g } C_6H_{14}}{1 \text{ mole } C_6H_{14}} = 4.8 \times 10^3 \text{ g } C_6H_{14}$$

# CHAPTER 14

14.1) Particle size. Dispersed particles in a solution are molecular or ionic in size, whereas particles in colloids or suspensions contain many molecules or ions, or perhaps large macromolecules.

14.2) Solute is customarily dispersed in solvent. When two substances are dissolved in each other in nearly equal quantities, either may be considered the solvent and the other the solute.

14.3) If solute A is very soluble, its 10 gram per 100 gram of solvent concentration may be quite *dilute* compared to the possible concentration. If solute B is only slightly soluble, its 5 grams per 100 grams of solvent may be close to its maximum solubility, and therefore *concentrated*. Dilute and concentrated are relative terms that may be used to compare different concentrations of a given solute-solvent mixture, but not different mixtures.

14.4) Drop a small amount of solute into the solution. If the solution is unsaturated, the solute will dissolve; if saturated, it will simply settle to the bottom; if supersaturated, it will probably cause additional precipitation of solute from the solution.

14.5) Miscibility, miscible and immiscible generally refer to a solution of one liquid in another. Their meaning corresponds with solubility, soluble and insoluble as applied to solutions in general.

14.6) Attractions between solute particles and solvent molecules are the primary forces promoting solution. If they are stronger than interparticle attractions within the solute and within the solvent, solution will occur.

14.7) If more solute than that required to reach solubility is placed into a given amount of solvent, the solute will dissolve until the solution is saturated. Dissolving and precipitation continue to occur between the saturated solution and the yet undissolved solute, the system being in a state of equilibrium. By pouring off the solution you isolate a saturated solution from the excess solute.

14.8) Stirring increases the net dissolving rate.

14.9) Carbon tetrachloride because both solute and solvent are nonpolar.

14.10) Cyclopentane. Cyclopentane is nonpolar and methanol is polar. If the spot failed to dissolve in a polar solvent (water) there is a better chance it will dissolve in a nonpolar solvent than a second polar solvent.

14.11) The statement is true only to the extent that an increase in air pressure increases the partial pressure of carbon dioxide. Solubility of a gas depends upon the partial pressure of the solute gas, not the total gas pressure above the solution.

14.12) $$\frac{2.78 \text{ g NaCl}}{32.4 \text{ g solution}} \left| \frac{100}{} \right. = 8.58\% \text{ NaCl}$$

14.13) $$\frac{75.0 \text{ g solution}}{} \left| \frac{4.00 \text{ g } H_3BO_3}{100 \text{ g solution}} \right. = 3.00 \text{ g } H_3BO_3$$

14.14) $$\frac{16.9 \text{ g } C_6H_{12}O_6}{0.0875 \text{ kg } H_2O} \left| \frac{1 \text{ mole } C_6H_{12}O_6}{180 \text{ g } C_6H_{12}O_6} \right. = 1.07 \text{ m } C_6H_{12}O_6$$

14.15) $\dfrac{10.0 \text{ g NaCl}}{0.0900 \text{ kg H}_2\text{O}} \left| \dfrac{1 \text{ mole NaCl}}{58.5 \text{ g NaCl}} \right. = 1.90 \text{ m NaCl}$

14.16) $\dfrac{}{0.0500 \text{ kg H}_2\text{O}} \left| \dfrac{1.50 \text{ moles HCOOH}}{1 \text{ kg H}_2\text{O}} \right| \dfrac{46.0 \text{ g HCOOH}}{1 \text{ mole HCOOH}}$

$= 3.45 \text{ g HCOOH}$

14.17) $\dfrac{0.150 \text{ } \ell}{} \left| \dfrac{0.125 \text{ mole AgNO}_3}{1 \text{ } \ell} \right| \dfrac{170 \text{ g AgNO}_3}{1 \text{ mole AgNO}_3} = 3.19 \text{ g AgNO}_3$

14.18) $\dfrac{0.750 \text{ } \ell}{} \left| \dfrac{0.480 \text{ mole H}_2\text{C}_2\text{O}_4 \cdot 2\text{H}_2\text{O}}{1 \text{ } \ell} \right| \dfrac{126 \text{ g H}_2\text{C}_2\text{O}_4 \cdot 2\text{H}_2\text{O}}{1 \text{ mole H}_2\text{C}_2\text{O}_4 \cdot 2\text{H}_2\text{O}}$

$= 45.4 \text{ g H}_2\text{C}_2\text{O}_4 \cdot 2\text{H}_2\text{O}$

14.19) $\dfrac{16.2 \text{ g (NH}_4)_2\text{SO}_4}{0.3000 \text{ } \ell} \left| \dfrac{1 \text{ mole (NH}_4)_2\text{SO}_4}{132 \text{ g (NH}_4)_2\text{SO}_4} \right. = 0.409 \text{ M}$

14.20) $\dfrac{56.2 \text{ g CuSO}_4 \cdot 5 \text{ H}_2\text{O}}{0.5000 \text{ } \ell} \left| \dfrac{1 \text{ mole CuSO}_4 \cdot 5 \text{ H}_2\text{O}}{249.5 \text{ g CuSO}_4 \cdot 5 \text{ H}_2\text{O}} \right. = 0.451 \text{ M}$

14.21) $\dfrac{0.0453 \text{ } \ell}{} \left| \dfrac{0.378 \text{ mole KNO}_3}{1 \text{ } \ell} \right. = 0.0171 \text{ mole KNO}_3$

14.22) $\dfrac{1.24 \text{ moles H}_2\text{SO}_4}{} \left| \dfrac{1000 \text{ ml}}{18 \text{ moles H}_2\text{SO}_4} \right. = 69 \text{ ml}$

14.23) $\dfrac{65.6 \text{ g KOH}}{1.50 \text{ } \ell} \left| \dfrac{1 \text{ mole KOH}}{56.1 \text{ g KOH}} \right| \dfrac{1 \text{ eq KOH}}{1 \text{ mole KOH}} = 0.780 \text{ N KOH}$

14.24) $\dfrac{0.284 \text{ mole Na}_3\text{PO}_4}{1 \text{ } \ell} \left| \dfrac{2 \text{ eq Na}_3\text{PO}_4}{1 \text{ mole Na}_3\text{PO}_4} \right. = 0.568 \text{ N Na}_3\text{PO}_4$

14.25) $\dfrac{11.9 \text{ g H}_3\text{PO}_4}{0.100 \text{ } \ell} \left| \dfrac{1 \text{ mole H}_3\text{PO}_4}{98.0 \text{ g H}_3\text{PO}_4} \right| \dfrac{2 \text{ eq H}_3\text{PO}_4}{1 \text{ mole H}_3\text{PO}_4} = 2.43 \text{ N H}_3\text{PO}_4$

14.26) $\dfrac{0.0500 \text{ } \ell}{} \left| \dfrac{0.114 \text{ eq}}{1 \text{ } \ell} \right. = 0.00570 \text{ eq HCl}$

14.27) $\dfrac{0.00500 \text{ eq H}_3\text{PO}_4}{} \left| \dfrac{1 \text{ } \ell}{0.200 \text{ eq H}_3\text{PO}_4} \right. = 0.0250 \text{ } \ell = 25.0 \text{ ml}$

14.28) $\dfrac{90.0 \text{ g H}_2\text{C}_2\text{O}_4}{1 \text{ mole H}_2\text{C}_2\text{O}_4} \left| \dfrac{1 \text{ mole H}_2\text{C}_2\text{O}_4}{2 \text{ eq H}_2\text{C}_2\text{O}_4} \right. = 45.0 \text{ g H}_2\text{C}_2\text{O}_4/\text{eq}$

14.29) $\dfrac{0.0500 \text{ } \ell}{} \left| \dfrac{0.424 \text{ mole BaCl}_2}{1 \text{ } \ell} \right| \dfrac{1 \text{ mole BaCrO}_4}{1 \text{ mole BaCl}_2} \left| \dfrac{253 \text{ g BaCrO}_4}{1 \text{ mole BaCrO}_4} \right.$

$= 5.36 \text{ g BaCrO}_4$

14.30) $\dfrac{2.10 \text{ g Na}_3\text{PO}_4}{} \left| \dfrac{1 \text{ mole Na}_3\text{PO}_4}{164 \text{ g Na}_3\text{PO}_4} \right| \dfrac{3 \text{ moles AgNO}_3}{1 \text{ mole Na}_3\text{PO}_4} \left|\right.$

$\dfrac{1000 \text{ ml}}{0.246 \text{ mole AgNO}_3} = 156 \text{ ml}$

14.31) $\dfrac{2.4 \text{ } \ell \text{ H}_2}{} \left| \dfrac{1 \text{ mole H}_2}{22.4 \text{ } \ell \text{ H}_2} \right| \dfrac{6 \text{ moles NaOH}}{3 \text{ moles H}_2} \left| \dfrac{1000 \text{ ml}}{6.2 \text{ moles NaOH}} \right. = 35 \text{ ml}$

14.32) $\text{HCl (aq)} + \text{NaOH (aq)} \rightarrow \text{HOH (l)} + \text{NaCl (aq)}$

$\dfrac{0.0100 \text{ } \ell \text{ NaOH}}{} \left| \dfrac{0.862 \text{ mole NaOH}}{1 \text{ } \ell \text{ NaOH}} \right| \dfrac{1 \text{ mole HCl}}{1 \text{ mole NaOH}} \left|\right.$

$\dfrac{1}{0.0168 \text{ } \ell \text{ HCl}} = 0.513 \text{ M HCl}$

14.33) $\dfrac{0.0149 \text{ } \ell}{} \left| \dfrac{0.518 \text{ mole AgNO}_3}{1 \text{ } \ell} \right| \dfrac{1 \text{ mole NiCl}_2}{2 \text{ moles AgNO}_3} \left| \dfrac{1}{0.0100 \text{ } \ell} \right.$

$= 0.386 \text{ M NiCl}_2$

14.34) $MgCl_2$ (aq) + 2 NaOH (aq) → $Mg(OH)_2$ (s) + 2 $MgCl_2$ (aq)

$$\frac{0.0500\,\ell}{} \quad \Big| \quad \frac{0.240 \text{ mole } MgCl_2}{1\,\ell} = 0.0120 \text{ mole } MgCl_2 \text{ available;}$$

$$\frac{0.0120 \text{ mole } MgCl_2}{} \quad \Big| \quad \frac{2 \text{ moles NaOH}}{1 \text{ mole } MgCl_2} = 0.0240 \text{ mole NaOH required}$$

$$\frac{0.0500\,\ell}{} \quad \Big| \quad \frac{0.420 \text{ mole NaOH}}{1\,\ell} = 0.0210 \text{ mole NaOH available;}$$

$$\frac{0.0210 \text{ mole NaOH}}{} \quad \Big| \quad \frac{1 \text{ mole } MgCl_2}{2 \text{ moles NaOH}} = 0.0105 \text{ mole } MgCl_2 \text{ required}$$

NaOH is the limiting reagent.

$$\frac{0.0210 \text{ mole NaOH}}{} \quad \Big| \quad \frac{1 \text{ mole } Mg(OH)_2}{2 \text{ moles NaOH}} \quad \Big| \quad \frac{58.3 \text{ g } Mg(OH)_2}{1 \text{ mole } Mg(OH)_2}$$

$$= 0.612 \text{ g } Mg(OH)_2$$

14.35) $$\frac{140 \text{ gal}}{} \quad \Big| \quad \frac{3.785\,\ell}{1 \text{ gal}} \quad \Big| \quad \frac{1.13 \text{ moles } HNO_3}{1\,\ell} = 599 \text{ moles } HNO_3 \text{ available}$$

$$\simeq 599 \text{ moles } NaHCO_3 \text{ required}$$

$$\frac{61,200 \text{ g } NaHCO_3}{} \quad \Big| \quad \frac{1 \text{ mole } NaHCO_3}{84.0 \text{ g } NaHCO_3} = 729 \text{ moles } NaHCO_3 \text{ available}$$

$$\simeq 729 \text{ moles } HNO_3 \text{ required}$$

$HNO_3$ is the limiting reagent.

$$\frac{599 \text{ moles } HNO_3}{} \quad \Big| \quad \frac{1 \text{ mole } CO_2}{1 \text{ mole } HNO_3} \quad \Big| \quad \frac{22.4\,\ell\,CO_2}{1 \text{ mole } CO_2} = 1.34 \times 10^4\,\ell\,CO_2$$

14.36) $$\frac{25.0 \times 0.324}{21.9} = 0.370 \text{ N NaOH}$$

14.37) $$\frac{0.512 \text{ g } H_2C_2O_4 \cdot 2H_2O}{0.0162\,\ell \text{ NaOH}} \quad \Big| \quad \frac{1 \text{ mole } H_2C_2O_4}{126 \text{ g } H_2C_2O_4 \cdot 2H_2O} \quad \Big| \quad \frac{2 \text{ eq NaOH}}{1 \text{ mole } H_2C_2O_4}$$

$$= 0.502 \text{ N NaOH}$$

14.38) $$\frac{0.452 \text{ g base}}{0.0318\,\ell \text{ HCl}} \quad \Big| \quad \frac{1\,\ell \text{ HCl}}{0.169 \text{ eq}} = 84.1 \text{ g/eq}$$

14.39) A colligative property of a solution is one that is independent of the identity of the solute. Specific gravity, with opposite effects from different solutes, is not a colligative property.

14.40) $$\frac{96.1 \text{ g } C_2H_6O_2}{0.100 \text{ kg } H_2O} \quad \Big| \quad \frac{1 \text{ mole } C_2H_6O_2}{62.0 \text{ g } C_2H_6O_2} = 15.5 \text{ m } C_2H_6O_2$$

$\Delta T_b = 0.52 \times 15.5 = 8.1°C \qquad T_b = 100.0 + 8.1 = 108.1°C$

14.41) $$\frac{4.34 \text{ g } C_6H_4Cl_2}{0.0700 \text{ kg naphthalene}} \quad \Big| \quad \frac{1 \text{ mole } C_6H_4Cl_2}{147 \text{ g } C_6H_4Cl_2} = 0.422 \text{ m } C_6H_4Cl_2$$

$\Delta T_f = (-6.9)(0.423) = -2.9°C; \qquad T_f = 80.2 - 2.9 = 77.3°C$

14.42) $$m = \frac{\Delta T}{K_b} = \frac{(100.89 - 100.00)}{0.52} = 1.71 \text{ m}$$

14.43) A solution of A will be a conductor of electricity, whereas a solution of B will be a nonconductor. A yields ions in solution; B yields molecules.

14.44) A conducting solution must contain ions. "Electricity" is carried through the solution by positive ions moving toward the negative electrode and negative ions toward the positive electrode.

14.45)  $2 Ag^+ (aq) + CO_3^{2-} (aq) \rightarrow Ag_2CO_3 (s)$

14.46)  $Al^{3+} (aq) + PO_4^{3-} (aq) \rightarrow AlPO_4 (s)$

14.47)  $2 H^+ (aq) + SO_3^{2-} (aq) \rightarrow H_2O (l) + SO_2 (g)$

14.48)  $H^+ (aq) + OH^- (aq) \rightarrow HOH (l)$

14.49)  $NH_4^+ (aq) + OH^- (aq) \rightarrow NH_3 (aq) + HOH (l)$

14.50)  $H^+ (aq) + C_7H_5O_2^- (aq) \rightarrow HC_7H_5O_2 (aq)$

14.51)  $Ca^{2+} (aq) + 2 F^- (aq) \rightarrow CaF_2 (s)$

14.52)  $2 H^+ (aq) + Ba(OH)_2 (s) \rightarrow Ba^{2+} (aq) + 2 HOH (l)$

14.53)  $2 Al (s) + 6 OH^- (aq) \rightarrow 3 H_2 (g) + 2 AlO_3^{3-} (aq)$

14.54)  $H_3PO_4 (aq) + OH^- (aq) \rightarrow H_2PO_4^- (aq) + HOH (l)$     (14.8)
$H_3PO_4 (aq) + 2 OH^- (aq) \rightarrow HPO_4^{2-} (aq) + 2 HOH (l)$     (14.9)
$H_3PO_4$ is marginal as a strong or weak acid. It has been regarded as a weak acid in this answer.

# CHAPTER 15

15.1)  Sufficient energy and proper orientation.

15.2)  Rate increases if concentration of A is increased, and decreases if concentration of B is decreased. Both reasons: reaction rates vary directly with reactant concentrations.

15.3)  Kinetic energies of the particles making up any sample of matter vary greatly, although the *average* kinetic energy, expressed by temperature, remains constant. Only those particles with highest kinetic energies are capable of reaction-producing collisions. This will be but a fraction of the total.

15.4)  Negative. See Figure 13.5, page 363.

15.5)  See Figure 15.2, page 364. The activated complex position is at the highest point of the curve.

15.6)  Activation energy is the minimum energy particles must have, usually in the form of kinetic energy, to participate in a reaction-producing collision. It also represents the increase in potential energy required to reach the activated complex point and pass the potential energy barrier. At a given temperature, the reaction with the lower activation energy would proceed faster because a larger portion of the reactant molecules would have sufficient energy to satisfy the activation energy requirement.

15.7)  A catalyst is a substance that increases the rate of a chemical reaction by lowering the activation energy.

15.8)  At the beginning when both reactants are at their highest concentration.

15.9)  Nitrogen and hydrogen concentrations decrease, ammonia concentration increases.

15.10)  $\dfrac{[SO_3]^2}{[SO_2]^2[O_2]}$

15.11)  $\dfrac{[CH_4][H_2S]^2}{[H_2]^4[CS_2]}$

15.12)  $[Cd^{2+}][OH^-]^2$

15.13)  $\dfrac{[H^+][NO_2^-]}{[HNO_2]}$

15.14) $\dfrac{[Ag^+][CN^-]^2}{[Ag(CN)_2^-]}$

15.15) An equilibrium equation may be written in two directions that yield equilibrium constants that are reciprocals of each other.

15.16) Only a small portion of the fluoride ion present will be converted to HF; or most HF is converted to $F^-$. [HF] and/or [OH$^-$] must be very small compared to [F$^-$].

15.17) Because the equilibrium constant is very small, the reaction is favored in the reverse direction at equilibrium.

15.18) $K = \dfrac{[SO_3]^2}{[SO_2]^2[O_2]} = \dfrac{0.60^2}{(0.90)^2(0.45)} = 0.99$

15.19)

| HF (aq) | ⇌ | H$^+$ (aq) | + | F$^-$ (aq) | |
|---|---|---|---|---|---|
| I | 1.0 | | 0 | 0 | $K = \dfrac{[H^+][F^-]}{[HF]} = \dfrac{(0.0265)^2}{1.0} = 7.0 \times 10^{-4}$ |
| R | −0.0265 | | +0.0265 | +0.0265 | |
| E | 1.0 | | 0.0265 | 0.0265 | |

15.20)

| H$_2$ (g) | + | I$_2$ (g) | ⇌ | 2 HI (g) | |
|---|---|---|---|---|---|
| I | 0 | | 0 | 3.00 | $K = \dfrac{[HI]^2}{[H_2][I_2]} = 51.5$ |
| R | +y | | +y | −2y | $= \dfrac{(3.00 - 2y)^2}{y^2} = 51.5$ |
| E | y | | y | 3.00 − 2y | $\dfrac{3.00 - 2y}{y} = 7.18$ |

$[H_2] = [I_2] = 0.327$

$[HI] = 3.00 - 2(0.327) = 2.35$ $\qquad\qquad$ y = 0.327

15.21) $CaCO_3 \text{ (s)} \rightleftharpoons Ca^{2+} \text{ (aq)} + CO_3^{2-} \text{ (aq)}$ $\qquad$ $K_{sp} = [Ca^{2+}][CO_3^{2-}] = 8.7 \times 10^{-9}$

Let s = solubility of $CaCO_3 = [Ca^{2+}] = [CO_3^{2-}]$: $\qquad$ $s^2 = 8.7 \times 10^{-9}$

$s = 9.3 \times 10^{-5}$

15.22) $MgF_2 \text{ (s)} \rightleftharpoons Mg^{2+} \text{ (aq)} + 2\,F^- \text{ (aq)}$ $\qquad$ $K_{sp} = [Mg^{2+}][F^-]^2 = 6.5 \times 10^{-9}$

Let s = solubility of $MgF_2 = [Mg^{2+}]$; $[F^-] = 2s$: $\qquad$ $(s)(2s)^2 = 6.5 \times 10^{-9}$

$s = 1.2 \times 10^{-3}$

15.23) Le Chatelier's Principle says the shift will be in the direction that will counteract partially the initial change. The shift must therefore be in the direction that would reduce [NO], which is in the reverse direction.

15.24) If $Cl_2$ is removed, the shift is in the direction in which $Cl_2$ will be produced. That direction is forward.

15.25) An increase in $[C_2H_3O_2^-]$ would be counteracted by a shift in the direction in which the ion is consumed—the reverse direction.

15.26) If volume of a gaseous equilibrium is reduced, the equilibrium will shift in the direction of fewer molecules. In the reverse direction two molecules combine to form one.

15.27) An increase in volume causes the equilibrium to shift in the direction of more molecules. A shift in the forward direction has one gaseous molecule producing two.

15.28) If heat is added, the equilibrium will shift in the direction in which heat is absorbed—in the reverse direction.

15.29)  Heating the system would cause a shift in the direction in which heat is absorbed—in the forward direction, thereby favoring the formation of HI.

15.30)  a and d, forward; b and c, reverse

# CHAPTER 16

16.1)  Acids, $H^+$; bases, $OH^-$

16.2)  Arrhenius acid is source of $H^+$, or a proton; Brönsted-Lowry acid is a proton donor. Concepts are in agreement. Arrhenius base is a source of $OH^-$ ion; Brönsted-Lowry base is a proton acceptor. $OH^-$ is one of many examples of Brönsted-Lowry bases, so Arrhenius is more limited concept.

16.3)  HOH and $H_2CO_3$; $H_2O$ and $CO_3^{2-}$

16.4)  Forward: acid, $HNO_2$; base, $CN^-$. Reverse: acid, HCN; base, $NO_2^-$.

16.5)  $HSO_4^-$ and $SO_4^{2-}$; $C_2O_4^{2-}$ and $HC_2O_4^-$

16.6)  $H_2PO_4^-$ and $HPO_4^{2-}$; $HCO_3^-$ and $H_2CO_3$

16.7)  A Lewis acid must have an empty valence orbital, capable of forming a covalent bond by accepting an electron pair from a Lewis base. This unshared electron pair characterizes a Lewis base.

16.8)  The hydrogen ion is capable of accepting an electron pair, and the hydroxide ion is capable of donating an electron pair to form a chemical bond. These Arrhenius acid and base ions therefore satisfy the Lewis concept of acids and bases.

16.9)

Aluminum in $AlCl_3$ is surrounded by three electron pairs. It is therefore capable of accepting a fourth pair, which qualifies it as a Lewis acid. The chloride ion has four unshared pairs of valence electrons. When one of these (see color electron pair) becomes a bonding pair with an aluminum chloride molecule (see color bond in $AlCl_4^-$), the chloride ion has served as a Lewis base.

16.10)  The sulfur in $SO_3$ acts as a Lewis acid by accepting an electron pair from the oxide ion, a Lewis base.

16.11)  A strong acid donates protons readily, a weak acid reluctantly. Strong acids are at the top of Table 16.1, weak acids at the bottom.

16.12)  $CO_3^{2-}$; $H_2PO_4^-$; $SO_4^{2-}$; $Br^-$

16.13)  $HC_7H_5O_2$ (aq) + $SO_4^{2-}$ (aq) $\rightleftarrows$ $C_7H_5O_2^-$ (aq) + $HSO_4^-$ (aq)  Reverse

16.14)  $H_2C_2O_4$ (aq) + $NH_3$ (aq) $\rightleftarrows$ $HC_2O_4^-$ (aq) + $NH_4^+$ (aq)      Forward

16.15)  $H_3PO_4$ (aq) + $CN^-$ (aq) $\rightleftarrows$ $H_2PO_4^-$ (aq) + HCN (aq)      Forward

16.16)  Water ionizes very slightly and does not produce enough ions to light an ordinary conductivity device. With a sufficiently sensitive detector, water displays a very weak conductivity.

16.17)  Strongly acidic, pH less than 4; weakly acidic, 4–6; neutral, or close to neutral, 6–8; weakly basic, 8–10; strongly basic, above 10. Ranges are arbitrary. If pH of a solution is x, then hydrogen ion concentration is $10^{-x}$ mole/$\ell$.

16.18)   $[OH^-] = \dfrac{10^{-14}}{10^{-6}} = 10^{-8}$; pOH = 8;

pH = 14 − 8 = 6; $[H^+] = 10^{-6}$

16.19)   pH = 14 − 3 = 11; $[H^+] = 10^{-11}$;

$[OH^-] = \dfrac{10^{-14}}{10^{-11}} = 10^{-3}$; pOH = 3

16.20)   $[H^+] = 10^{-1}$; $[OH^-] = \dfrac{10^{-14}}{10^{-1}} = 10^{-13}$;

pOH = 13; pH = 14 − 13 = 1

16.21)   pOH = 9; pH = 14 − 9 = 5;

$[H^+] = 10^{-5}$; $[OH^-] = \dfrac{10^{-14}}{10^{-5}} = 10^{-9}$

16.22)   $[OH^-] = \dfrac{1.0 \times 10^{-14}}{3.4 \times 10^{-9}} = 2.9 \times 10^{-6}$

pOH = 6 − log 2.9 = 5.54

pH = 14.00 − 5.54 = 8.46

$[H^+] = 10^{-8.46} = 10^{0.54-9} = 3.5 \times 10^{-9}$

16.23)   pH = 14.00 − 11.82 = 2.18

$[H^+] = 10^{-2.18} = 10^{0.82-3} = 6.6 \times 10^{-3}$

$[OH^-] = \dfrac{1.0 \times 10^{-14}}{6.6 \times 10^{-3}} = 1.5 \times 10^{-12}$

pOH = 12 − log 1.5 = 11.82

16.24)   pOH = 1 − log 2.6 = 0.59

pH = 14.00 − 0.59 = 13.41

$[H^+] = 10^{-13.41} = 10^{0.59-14} = 3.9 \times 10^{-14}$

$[OH^-] = \dfrac{1.0 \times 10^{-14}}{3.9 \times 10^{-14}} = 0.26$

16.25)   $[H^+] = 10^{-12.05} = 10^{0.95-13} = 8.9 \times 10^{-13}$

$[OH^-] = \dfrac{1.0 \times 10^{-14}}{8.9 \times 10^{-13}} = 1.1 \times 10^{-2}$

pOH = 2 − log 1.1 = 1.96

pH = 14.00 − 1.96 = 12.04

## CHAPTER 17

17.1)   Oxidation is loss of electrons, or increase in oxidation number; reduction is gain of electrons or decrease in oxidation number. From the standpoint of the first definition, electrons lost by one species must be gained by the other. Oxidation and reduction therefore must be simultaneous processes.

17.2)   a, b, and c, oxidation; d, reduction

17.3)   $Ni^{2+} + 2\,e^- \rightarrow Ni$

$\dfrac{Mg \rightarrow Mg^{2+} + 2\,e^-}{Ni^{2+} + Mg \rightarrow Ni + Mg^{2+}}$

17.4)  $+2, -1, +1, +5$

17.5)  $+5, -3, +7, +3$

17.6)  (a) Copper reduced from $+2$ to $0$; (b) cobalt reduced from $+3$ to $+2$

17.7)  (a) Sulfur oxidized from $+4$ to $+6$; (b) P oxidized from $-3$ to $0$

17.8)  (a) F oxidized from $-1$ to $0$; (b) Mn reduced from $+6$ to $+4$

17.9)  Hydrogen is the reducing agent, and copper oxide is the oxidizing agent.

17.10)  $MnO_2$ is the oxidizing agent, $Cl^-$ the reducing agent.

17.11)  Reactions are similar in that each transfers a subatomic particle, a proton transfer for acid-base and an electron transfer for redox. Further similarities are in use of terms "strong" and "weak," and the fact that some species can function in either role, i.e., acid or base, or oxidizer or reducer. A significant difference is that all acid-base reactions are single proton transfer reactions, though there may be successive proton transfer reactions, whereas many redox reactions involve transfer of two or more electrons.

17.12)  $Ag^+$ is a stronger oxidizer than $H^+$, based on their relative positions in Table 17.1. This means that $Ag^+$ has a stronger attraction for electrons than does $H^+$.

17.13)  $Al, H_2, Fe^{2+}, Cl^-$

17.14)  $Ni + Zn^{2+} \rightleftarrows Ni^{2+} + Zn$      Reverse

17.15)  $2\,Fe^{3+} + Co \rightleftarrows 2\,Fe^{2+} + Co^{2+}$      Forward

17.16)  $\frac{1}{2}\,O_2 + 2\,H^+ + Ca \rightleftarrows H_2O + Ca^{2+}$      Forward

17.17)  
$$\frac{\begin{array}{l} SO_4^{2-} + 4\,H^+ + 2\,e^- \rightarrow SO_2 + 2\,H_2O \\ (Ag \rightarrow Ag^+ + e^-)\,2 \end{array}}{SO_4^{2-} + 4\,H^+ + 2\,Ag \rightarrow SO_2 + 2\,H_2O + 2\,Ag^+}$$

17.18)  
$$\frac{\begin{array}{l} NO_3^- + 10\,H^+ + 8\,e^- \rightarrow NH_4^+ + 3\,H_2O \\ (Zn \rightarrow Zn^{2+} + 2\,e^-)\,4 \end{array}}{NO_3^- + 10\,H^+ + 4\,Zn \rightarrow NH_4^+ + 3\,H_2O + 4\,Zn^{2+}}$$

17.19)  
$$\frac{\begin{array}{l} Cr_2O_7^{2-} + 14\,H^+ + 6\,e^- \rightarrow 2\,Cr^{3+} + 7\,H_2O \\ (Fe^{2+} \rightarrow Fe^{3+} + e^-)\,6 \end{array}}{Cr_2O_7^{2-} + 14\,H^+ + 6\,Fe^{2+} \rightarrow 2\,Cr^{3+} + 7\,H_2O + 6\,Fe^{3+}}$$

17.20)  
$$\frac{\begin{array}{l} (MnO_4^- + 4\,H^+ + 3\,e^- \rightarrow MnO_2 + 2\,H_2O)\,2 \\ (2\,I^- \rightarrow I_2 + 2\,e^-)\,3 \end{array}}{2\,MnO_4^- + 8\,H^+ + 6\,I^- \rightarrow 2\,MnO_2 + 4\,H_2O + 3\,I_2}$$

17.21)  
$$\frac{\begin{array}{l} 2\,BrO_3^- + 12\,H^+ + 10\,e^- \rightarrow Br_2 + 6\,H_2O \\ (2\,Br^- \rightarrow Br_2 + 2\,e^-)\,5 \end{array}}{\begin{array}{l} 2\,BrO_3^- + 12\,H^+ + 10\,Br^- \rightarrow 6\,Br_2 + 6\,H_2O \\ BrO_3^- + 6\,H^+ + 5\,Br^- \rightarrow 3\,Br_2 + 3\,H_2O \end{array}}$$

## CHAPTER 18

18.1)  Radioactivity is the spontaneous emission of rays, or particles, from the nucleus of an atom.

18.2)  Alpha particle, or $\alpha$ particle, a helium nucleus, $_2^4He$; beta particle, or $\beta$ particle, an electron, $_{-1}^{0}e$; and gamma ray, or $\gamma$-ray, a high-energy electromagnetic ray, similar to x-rays.

18.3)  The gas is ionized.

18.4) A cloud chamber is a vessel containing air that is supersaturated in a vapor. As radioactive emissions pass through the chamber, they ionize the air. The vapor then condenses on the ions, leaving a visible "track" behind the emitted particle.

18.5) Transmutation of an element is the change of an atom from one element to another, resulting from a change in the number of protons in the nucleus of the atom.

18.6) Emission of an alpha particle reduces the atomic number by 2 and reduces the mass number by 4.

18.7) $^{212}_{82}Pb \rightarrow {}^{212}_{83}Bi + {}^{0}_{-1}e$; $\qquad$ $^{231}_{90}Th \rightarrow {}^{231}_{91}Pa + {}^{0}_{-1}e$

18.8) $^{228}_{90}Th \rightarrow {}^{224}_{88}Ra + {}^{4}_{2}He$; $\qquad$ $^{222}_{86}Rn \rightarrow {}^{218}_{84}Po + {}^{4}_{2}He$

18.9) The half-life of a radioactive substance is the time required for half of the sample to decay. The fraction of a sample remaining after the passage of six half-lives is $(\frac{1}{2})^6$, or $\frac{1}{64}$.

18.10) With a half-life of 22 years, 66 years is 3 half-lives. Hence 100 g $\times (\frac{1}{2})^3 =$ 12.5 grams.

18.11) From 1950 to 2006 is two half-lives: 56 years at 28 years per half-life. 600 grams $\times (\frac{1}{2})^2 = 150$ grams remaining.

18.12) Working backwards, ten minutes before reaching 0.120 gram (one half-life earlier) the mass was 0.240; 20 minutes, 0.480; 30 minutes, 0.960.

18.13) $\frac{1}{8}$ is $(\frac{1}{2})^3$, indicating an age of 3 half-lives, or $3 \times 5720 = 17,160$ years.

18.14) Chemical properties depend on the electrons outside the nucleus, which are the same for all isotopes of an element. Nuclear change depends on a particular isotope of an element. In radioactivity, for example, one isotope of an element may emit alpha particles, another isotope may emit beta particles, and a third isotope may be stable.

18.15) $UCl_4$. Because of the lower molar weight of $UCl_4$, there are more moles of uranium in 100 grams of $UCl_4$ than in 100 grams of $UBr_4$. Only the radioactive element, uranium, contributes to radioactivity. The radioactivity of 0.10 mole of $UCl_4$ will be the same as that of 0.10 mole of $UBr_4$, inasmuch as both samples contain the same number of moles of uranium.

18.16) Lead is the stable end product of natural radioactive decay series. It is constantly being produced in natural radioactivity.

18.17) Nuclear bombardment involves directing a nuclear particle to strike another nucleus, producing a nuclear reaction.

18.18) Cyclotron and linear accelerator, both mentioned in text. Others are the betatron and synchro-cyclotron.

18.19) The transuranium elements are the elements having atomic numbers greater than 92. They appear as the actinium series of elements at the bottom of the periodic tables in this text.

18.20) $^{99}_{43}Tc$ $\qquad$ 18.21) $^{239}_{94}Pu$ $\qquad$ 18.22) $^{59}_{27}Co$

18.23) A fission reaction is one in which a large nucleus splits into two nuclei of intermediate mass.

18.24) The conversion of mass to energy in a nuclear reaction results in the enormous amount of energy resulting from a nuclear change. This energy is potential in character, stored as "binding energy" that holds the positively charged protons in the nucleus together.

18.25) The major advantage of nuclear energy as a source of electricity is the abundant supply of fuel. Furthermore, the fuel cost is, at the time of this writing, competitive with fossil fuels and more dependable than fossil fuels.

18.26) A fusion reaction is one in which two light nuclei combine to form a nucleus of greater mass, in contrast with the fission reaction in which a heavy nucleus splits into two lighter nuclei.

# CHAPTER 19

19.1) Excluding carbonates, cyanides and the oxides of carbon, plus a few other substances, organic chemistry is the chemistry of carbon. Inorganic chemistry is the chemistry of the other elements plus the above exclusions.

19.2) A compound consisting of hydrogen and carbon.

19.3) A hydrocarbon containing only single bonds, with each carbon bonded to four other atoms, is an alkane. A homologous series is one in which each member differs from those next to it by the same structural unit. Alkanes differ by a —CH$_2$— unit, having the general formula C$_n$H$_{2n+2}$.

19.4) A condensed, or line, formula lists on a single line the structural components of an organic compound. It is better than molecular formulas in conveying some idea of structure, class of organic compound, and distinction between possible isomers of compounds with identical molecular formulas.

19.5) 

$$CH_3CH_2CH_2CH_2CH_2CH_2CH_2CH_2CH_3 \text{ or } CH_3(CH_2)_7CH_3$$

19.6) Isomerism is the existence of two or more compounds of the same molecular formula, but different structural formulas and different physical and chemical properties. Of the compounds listed (a) and (e) are isomers, both having the molecular formula C$_9$H$_{20}$.

19.7) Alkane hydrocarbon minus one hydrogen. Named by alkane prefix designating number of carbon atoms, plus -*yl* suffix.

19.8) Octane. It is possible to count out an 8-carbon chain.

19.9) 

19.10)

19.11)  3-ethylpentane

19.12)  3-chloropentane

19.13)  1-bromo-4,5-dichlorohexane

19.14)

$$
\begin{array}{c}
\text{Br} \\
| \\
-\text{C}-\text{C}- \\
| \\
\text{Br}
\end{array}
$$

19.15)

$$
\begin{array}{c}
\text{Br} \qquad\quad \text{Br}\ \text{Br} \\
|\quad\ |\quad\ |\quad\ |\quad\ | \\
-\text{C}-\text{C}-\text{C}-\text{C}-\text{C}- \\
|\quad\ |\quad\ |\quad\ |\quad\ | \\
\text{Br}
\end{array}
$$

19.16)

$$
\begin{array}{c}
\text{Br}\ \text{Br} \\
|\quad\ |\quad\ |\quad\ | \\
-\text{C}-\text{C}-\text{C}-\text{C}- \\
|\quad\ |\quad\ |\quad\ | \\
\text{Cl}\ \text{Br}
\end{array}
$$

19.17)  A double bond between two carbons. $C_nH_{2n}$.

19.18)  H—C≡C—H   Acetylene, or ethyne

| | | |
|---|---|---|
| H—C≡C—C—H | H—C≡C—C—C—H | H—C—C≡C—C—H |
| Propyne | 1-butyne | 2-butyne |

(with H atoms shown on the carbons)

19.19)  The double bond is between the first and second carbons in 1-hexene, the second and third in 2-hexene, and the third and fourth in 3-hexene.

19.20)  Different structural arrangements around a double bond.

19.21)  *cis*-2-pentene

19.22)

$$
\begin{array}{c}
\text{H}\ \text{H}\ \text{H}\ \text{H}\ \text{H}\ \text{H} \\
|\ \ |\ \ |\ \ |\ \ |\ \ | \\
\text{H}-\text{C}-\text{C}-\text{C}-\text{C}-\text{C}=\text{C} \\
|\ \ |\ \ |\ \ |\ \quad\ | \\
\text{H}\ \text{H}\ \text{H}\ \text{H}\quad\text{H}
\end{array}
$$
  1-hexene

19.23)

$$
\begin{array}{c}
\text{H}\ \text{H}\ \text{H}\ \text{H}\ \text{H} \\
|\ \ |\ \ |\ \ |\ \ | \\
\text{H}-\text{C}-\text{C}-\text{C}-\text{C}=\text{C} \\
|\ \ |\ \ |\ \quad\ | \\
\text{H}\ \text{H}\ \text{H}\quad\text{H}
\end{array}
$$
  1-pentene

19.24)  In an addition reaction one bond of a double or triple bond opens and new atoms are added by bonding to adjacent carbons without displacing other atoms. In a substitution reaction in a saturated compound, atoms already bonded must be displaced to make room for new atoms.

19.25)  $C_3H_6 + H_2 \rightarrow C_3H_8$

19.26)  The chemical combination of two or more monomers. The number so combined may be many thousands.

19.27)

$$
\begin{array}{c}
\text{H}\ \text{Cl} \quad \text{H}\ \text{Cl} \quad \text{H}\ \text{Cl} \qquad \text{H}\ \text{Cl}\ \text{H}\ \text{Cl}\ \text{H}\ \text{Cl} \\
|\ \ | \qquad |\ \ | \qquad |\ \ | \qquad\quad |\ \ |\ \ |\ \ |\ \ |\ \ | \\
\text{C}=\text{C}\ +\ \text{C}=\text{C}\ +\ \text{C}=\text{C}\ \rightarrow\ -\text{C}-\text{C}-\text{C}-\text{C}-\text{C}-\text{C}- \\
|\ \ | \qquad |\ \ | \qquad |\ \ | \qquad\quad |\ \ |\ \ |\ \ |\ \ |\ \ | \\
\text{H}\ \text{H} \quad\ \text{H}\ \text{H} \quad\ \text{H}\ \text{H} \qquad\ \text{H}\ \text{H}\ \text{H}\ \text{H}\ \text{H}\ \text{H}
\end{array}
$$

19.28) Aromatic compounds have ring structures; aliphatic compounds have open chain structures.

19.29) (a) and (c), *m*-dichlorobenzene or 1,3-dichlorobenzene; (b) *p*-dichlorobenzene, or 1,4-dichlorobenzene

19.30) Hydroxyl group, —OH. R—OH, or ROH.

19.31)

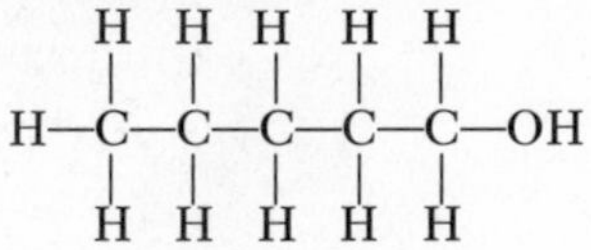

$$\underset{\displaystyle \begin{array}{ccccc} H & H & H & H & H \end{array}}{\overset{\displaystyle \begin{array}{ccccc} H & H & H & H & H \end{array}}{H-C-C-C-C-C-OH}}\quad \text{Primary}$$

19.32)

| O | O | O |
|---|---|---|
| H   H | R   H | R   R′ |
| Water | Alcohol | Ether |

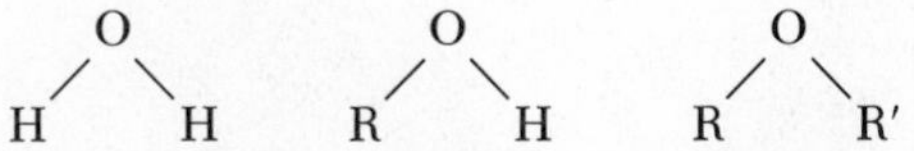

19.33)

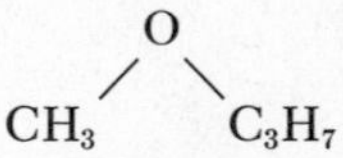

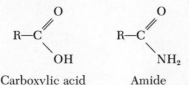

19.34) An aldehyde has at least one hydrogen bonded to a carbonyl group, whereas a ketone has two alkyl groups bonded to the carbonyl group. Aldehydes and ketones both have a carbonyl group.

19.35) RCOOH

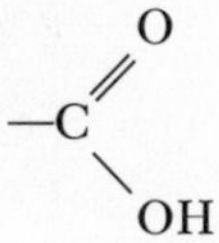

$$-C\overset{\displaystyle O}{\underset{\displaystyle OH}{}}$$

19.36) Carboxyl groups are polar, and hydrogen bonding is present. This leads to relatively strong intermolecular attractions and therefore high boiling points.

19.37) $RCOOH \rightarrow RCOO^- + H^+$

19.38) An amine is a substituted ammonia in which one or more hydrogens of ammonia are replaced by an alkyl group.

19.39) Dimethylpropylamine

19.40) In an amide the —OH part of the carboxyl group is replaced by a —NH₂ group:

$$R-C\overset{\displaystyle O}{\underset{\displaystyle OH}{}}\qquad R-C\overset{\displaystyle O}{\underset{\displaystyle NH_2}{}}$$

Carboxylic acid        Amide

# INDEX

Page numbers in *italics* refer to illustrations; those followed by t refer to tables.

**557**